DROWNING BY ACCIDENT

Why so many people drown

Dr E. A. Meinhard MD FRCPath

Matador
Unit E2 Airfield Business Park,
Harrison Road, Market Harborough,
Leicestershire. LE16 7UL
Tel: 0116 2792299
Email: books@troubador.co.uk
Web: www.troubador.co.uk/matador
Twitter: @matadorbooks

ISBN 978 1800464 988

British Library Cataloguing in Publication Data.
A catalogue record for this book is available from the British Library.

Printed and bound by CPI Group (UK) Ltd, Croydon, CR0 4YY
Typeset in 11pt Adobe Jenson Pro by Troubador Publishing Ltd, Leicester, UK

Matador is an imprint of Troubador Publishing Ltd

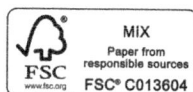

ACKNOWLEDGMENTS

My father was superintendent of our local swimming pool. He taught me and my sister – and hundreds of other children – to swim, to dive and to keep ourselves safe in the water. I still treasure the Royal Life Saving Society's Award of Merit that I won as a teenager.

More recently, paramedics of the Oxfordshire Ambulance Service supervised me and my practice staff as we gave cardiopulmonary resuscitation to a Resusci Anne manikin lying on the waiting room floor.

I thank the lifeguards, lifeboat volunteers, ambulance crews, police officers and firefighters who help drowning victims reach hospital alive, and the doctors, nurses and technicians who battle to repair the damage caused by submersion.

I am indebted to the marine biologists, physiologists, clinicians, researchers and epidemiologists who shared their expertise in books and academic papers. I also thank the reporters whose newspaper articles recorded so many tragic deaths.

I thank Sir Brian Smith and Edwin Palmer, who helped me to explain the dangers of scuba diving, and Matthew Parris, who gave me permission to quote his description of cold-related swimming failure in his memoir 'Chance Witness'.

I am very grateful to the librarians at the British Medical Association, the Wellcome Collection and the British Library – especially the staff of the Science Reading Room, who retrieved a number of journals preserved in deep-frozen underground caches far from London.

I value the encouragement given by David Walker at the Royal Society for the Prevention of Accidents, Tim Ash at the Royal National Lifeboat Institution and Stuart Haynes at the Royal Life Saving Society.

I give my thanks to my husband, Tom, and my family and friends who spurred me on with their continued interest over a number of years, to Sophie Bristow Symonds for her skilled editing, and to Troubador Publishing for turning my computer file into a book.

Elizabeth Meinhard MD FRCPath

CONTENTS

INTRODUCTION

WHY DO SO MANY PEOPLE DROWN?

In Britain more than 600 people die by drowning every year.[1] In the USA more than 4,000 people suffer fatal drowning accidents every year.[2] Those who drown are mostly young and fit, cut off within moments in the midst of vigorous living. Brief reports of their untimely deaths appear in local newspapers serving crowded cities, seaside resorts, country villages, old towns with canals and new towns near gravel pits. A paragraph on an inside page gives the name, the age and the place. Sometimes there is a photograph: a face, usually smiling. The report of a drowning accident. Another drowning accident. They happen all the time.

The scale of the loss of life goes largely unrecognised. Local newspapers restrict their reports to accidents that happen in their area of circulation. National newspapers limit their coverage to stories with graphic witness accounts, multiple fatalities or celebrity victims. Television newscasts air only the most eye-catching film clips: the feverish search for a missing schoolgirl swept down a flooded river, or the desperate rescue of survivors winched from a foundering yacht. This means that people only become aware of a very small sample of all drowning accidents, and each sudden tragedy is taken as an exceptional, isolated and unforeseeable trick of fate.

This lack of awareness has an unfortunate result. When we go swimming, we overestimate how safe we are. We do not realise how frequently drowning accidents occur, how easily they happen, and how quickly they can kill. Our mistaken expectations are compounded by the reputation of swimming as a particularly safe sport. Dr Malcolm Read, medical adviser to the British Olympic Association, writes, 'Swimming is regarded as the ideal form of exercise, because it is so injury-free.'[3]

Swimming merits its reputation for safety within the protective limits of supervised swimming pools. Swimming boosts confidence in children, improves health in pregnant mothers, fosters relaxation in stressed businessmen, and preserves function in retirement. Swimming is kind to asthmatics and gentle to arthritics. Footballers swim as they recover from knee surgery. Smokers swim as they recover from heart attacks. The young build backbone and the old retain muscle mass if they persist with their regular twenty lengths of the pool.

Swimmers can exert themselves, without overexertion. The water cushions them, supporting their bodies and massaging their limbs. Their muscles pull and relax rhythmically, in balance with the activity of their lungs and hearts. When they need to rest, they can stand in the shallow end, or hold onto the side of the pool, recovering their breath and their energy before returning to action. Few swimmers are injured at the highest levels of the sport, whether they are racing in club galas or international championships. They avoid the blisters, bruises, sprains and fractures so common on the playing field.

There have been a number of attempts to classify sports by their comparative risk of injury. In 2001 the American Academy of Pediatrics considered the amount of physical contact between participants and divided sports into three groups. Boxing and rugby were considered 'high risk' contact/collision sports. Skiing and riding were classified as 'medium risk' limited-contact sports. Swimming was listed together with bowling and golf as a 'low risk' non-contact sport.[4]

In a study of fatal sports injuries in Britain, Professor Nigel Warburton noted that boxing accounted for 3 deaths in the years 1986 to 1992, compared with 28 deaths while horse riding, 40 during ball games, 54 while mountaineering, 69 from air sports, and 77 from motor sports.[5] He did not include swimming in his study – although the official statistics of the years 1986 to 1992 record 3,065 fatal drowning accidents.[6] A recent Australian study found that more people drowned while taking part in recreational swimming and water sports than the total number of deaths in all other sports combined.[7] In America in 2006–2008, 14 competitors died during nationally sanctioned triathlons. All but one of the deaths occurred during the swimming phase of the race.[8]

In most sports, the majority of injuries are trivial, serious injuries are unusual, and fatal accidents are the exception. Swimmers are conspicuous by their absence in sports injury clinics crowded by runners with stress fractures, rowers with prolapsed intervertebral discs, gymnasts with dislocating shoulders, and tennis players with overuse injuries. Textbooks on sports medicine may include a few paragraphs on the shoulder pain common in competitive swimmers who practise their power strokes for several hours a day,[9] but few even mention drowning. Textbooks of emergency medicine and intensive care tell another story. Most deal with drowning, along with traffic accidents, burns, falls, stabbings and overdoses.

Sports injuries are usually caused by blows to the body or by excessive wear and tear. Drowning is a different type of injury altogether. It is a form of asphyxiation. After submersion, air cannot get into the victim's lungs, the lungs cannot extract oxygen, oxygen cannot get to the brain. Unconsciousness soon follows. Water enters the lungs. Breathing stops. For a time the heart will go on beating, but not for long. We cannot live without oxygen for more than a few minutes. Drowning happens fast.

In addition to the appalling number of deaths from drowning, an even greater number of victims suffer non-fatal drowning (formerly known as 'near-drowning').[10] They are pulled from the water more dead than alive. Although many make a full recovery, some suffer crippling lung damage from the effects of inhaled water, and others are blighted by irreversible brain damage.

Moreover, every year long-term hospital beds in Spinal Injury Units are filled by young men – almost all these victims are young men – who dive into shallow water, dash their heads against the bottom and break their necks.[11] The resulting damage to the spinal cord causes permanent paralysis (paraplegia), sometimes affecting all four limbs (tetraplegia). An expert in sports injuries highlighted the danger: 'Diving head first into water is the commonest cause of para- and tetraplegia.'[12]

It may seem odd that swimming is everywhere praised as low risk and injury-free, whereas swimming actually has the worst record for death and serious injury of any sport. How can we explain this paradox? To do so we need only realise that there are two conflicting sets of statistics. The claims of safety relate to supervised indoor swimming: the evidence of danger comes from verdicts after coroners' inquests on deaths from drowning in 'open water' and the sea. The safety claims are widely publicised while the evidence of danger is filed away once the inquest verdicts are pronounced, so that teenage boys expect the level of safety usual in the local swimming pool when they swim in a flooded quarry.

This reputation of safety makes it possible for swimmers to forget, play down, or ignore the dangers of frigid lakes, swollen rivers, incoming tides or outgoing rip currents. In a recent light-hearted article featuring swimming as 'the safest way to all-round fitness' the *Financial Times* quoted Sharon Davies, Olympic silver medallist: 'There really are no health downsides. Apart from drowning, of course.'[13]

Swimming manuals do not flag up the dangers. They describe the basic strokes and dives, or give coaching advice and training schedules. Their emphasis is on organised lessons and practice sessions in staffed swimming pools. The particular dangers of swimming in lakes, rivers, or the sea are outside their scope – mentioned briefly, if at all. Even skilled swimmers are lulled into a false sense of security.

Books on scuba-diving cover the considerable dangers as well as the great delights of the sport, spell out the penalties of risk-taking, and emphasise the compelling importance of safety drills and checklists. In contrast, books on water sports focus on technical aspects, with only a paragraph or two on self-preservation. Yet anyone who sets out over water is at risk of falling into it. Canoeists capsize into rivers. Windsurfers are blown out to sea. Yachtsmen fall overboard. Survival may depend upon whether they chose to put on a life jacket, or on whether they practised their self-rescue routines beforehand in calm inshore waters. In an emergency there is no time to read the instructions.

Books on teaching babies and toddlers to swim usually limit comments on safety to warnings against leaving small children unattended. A delightful exception is Virginia Hunt Newman's book *Teaching an Infant to Swim*,[14] which emphasises 'safety first, last, and always'. In her native California many drowning accidents involve young children who live in houses with private swimming pools. She writes, 'The allure of water is a very dangerous thing to a small child who cannot protect himself.' Her emphasis is not on the formal swimming strokes, but on teaching toddlers to dog paddle to safety should they fall into the water. The Australian Council for the Teaching of Swimming and Water Safety also publishes excellent books which blend relevant safety messages with enthusiastic advice covering children's progress from their first experience of water[15] through to efficient stroke development.[16]

Let us be clear. We all need to be able to swim. Water surrounds us. The British Isles have over 11,000 miles of shoreline, with countless bays, headlands, harbours and jetties. Twice each day the tides create their own specific dangers, racing over sand flats and churning below cliffs. Our rivers are kept full and flowing by a complex system of reservoirs and sluices, which store and discharge water to order. Hundreds of miles of canals form a network that dates back to the Industrial Revolution. Over the years long stretches fell into stagnant neglect. Now enthusiastic latter-day navigators have restored many canals, with their locks and towpaths. Increasing numbers of narrowboats cruise along waterways and crowd permanent canal-side moorings. Imaginative regeneration of underused docks in major cities and coastal towns has transformed derelict wastelands into pleasant neighbourhoods. Young families make their homes in quayside flats converted from redundant warehouses. The mirrored surface of the water reflects listed facades, willow trees and rippling light. But beneath the surface, the water is murky and deep.

Parents often hold the false expectation that a child in difficulties would splash and call for help. The phrase 'going down for the third time' suggests a few minutes of grace, at least, before rescue becomes impossible. Fifty years ago, an observant lifeguard on a New York beach realised how fast children drowned if they could not swim.[17] Once they stepped out of their depth, they were able to keep their heads above water for less than a minute. Moreover, their desperate attempts to breathe meant that they rarely called out for help, so that nearby family, friends and onlookers remained completely unaware that they were drowning. It is still too little understood that children often disappear under the surface in a moment, with scarcely a sound. And once they are submerged, the water silences their cries.

The growth in foreign travel increases the risk of drowning. Children have ready access to unguarded swimming pools at their hotels, campsites and holiday villas. Parents take their families to vast, crowded water parks with artificial wave machines and giant water slides. They spend days sunbathing on beaches where rip currents form in the inviting surf. A parent coping with several children may easily lose sight of the daredevil who wants to explore, or the slowcoach who gets left behind. Yet shared supervision often means inadequate supervision, with each carer assuming that someone else is also monitoring the children, so that nobody reacts in time.

Teenagers, released from supervision, let their guard down and take risks that they would not take at home. Many realise too late that drinking on holiday decreases competence, lulls protective common sense and fuels feelings of invulnerability. On adventure holidays, misguided thrill seekers climb into rubber inflatables and launch themselves down white-water rapids, unaware that the locals have a strong economic incentive to say nothing about disasters earlier in the season. Few anglers realise that drowning is the commonest cause of death while angling.[18]

Year after year, statisticians record a similar pattern of drowning accidents, and the shocking death toll is repeated. Babies drown in their baths. Toddlers drown in garden ponds. Schoolboys fall off home-made rafts. Teenagers strike out too far from the shore. Pensioners wade into rivers to save their dogs. Forty per cent of drowning victims lose their lives within two metres of safety.[19] A quarter drown in water less than one metre deep.[20] Three quarters of those who drown are male.[21]

Among children who are still too young to go to school, twice as many boys as girls die from drowning. This fatal bias becomes more marked as children grow to maturity. For boys, the drowning rate reaches a peak in their late teens and continues into their early twenties, even though the majority of those who drown know how to swim. For girls, in contrast, the preschool years are the most dangerous, and as girls get older, they are less likely to drown. The resulting statistics are sobering – ten times as many young men drown as young women.

People court trouble through lack of knowledge or excess of daring. Some play thoughtless pranks, ignore warning signs or take needless risks. Others turn an emergency into a catastrophe by taking the wrong action, or taking no action at all. On rare occasions swimmers may be overwhelmed by entirely unpredictable forces of nature, but most suffer the consequences of their own miscalculations or the misjudgements of others.

There is a pressing need to learn from these recurring drowning accidents. By exploring the errors of judgement, by sharing the insights gained, it may be possible to avoid yet more lethal repetitions. An expert in sports medicine commented on 'the stupidity and lack of common sense which figure substantially in many leisure drownings.'[22] But most of us want to learn how to protect ourselves and our children. The better we understand how easy it is to drown, the fewer will drown.

For safety, we must all learn to swim, and make sure that our children learn to swim. But swimming lessons come with no guarantee against drowning. Swimmers should know how much they are capable of, and stay within their limits. They need to understand how dangerous water can be, and not trust to luck. It is important to remember that more swimmers than non-swimmers die by drowning.[23] In fact, many drowning victims were able to swim very well indeed.

Why do so many victims make similar errors of judgement?

Why do so many play out identical fatal scenarios?

Why do people drown when they know how to swim?

People drown because they do not understand the risks, so they are not able to avoid them. They overestimate the strength and endurance of their bodies. They underestimate the hidden power of moving water and the chilling effect of cold water.

'He swims like a fish!' they say of a strong swimmer, but it is only a figure of speech. A fish is perfectly designed to glide through the water, while we are designed to walk upright on land. A fish is able to extract oxygen from the water passing over its gills, but we need constant access to the air. However well we swim, we cannot swim like a fish.

So this is where we will begin. We will describe some important differences between the body of a fish and the body of a human swimmer, differences that make swimming so simple for a fish and so much more difficult and dangerous for us, and for our children.

PART ONE

WATER IS A HOSTILE ENVIRONMENT

Our bodies are designed for life on land. Once in the water, we are all vulnerable to some degree. We have to learn to cope with the unfamiliar problems of buoyancy, balance and drag.

Our bodies must adjust to the pressure of the water, whether we remain at the surface or dive underwater. Our eyes and ears are designed to respond to light and sound travelling through the air. Water blurs our sight and dulls our hearing. We develop motion sickness when tossed by waves.

The resistance of the water makes swimming hard work – much harder work than swimmers expect. Water extracts heat from our bodies, leading to swimming failure and hypothermia, while a sudden entry into water below 15°C causes cold shock.

Our lives depend on a constant supply of oxygen from the air. When we sink beneath the surface of the water, we cannot reach the air. Once our supply of oxygen is cut off, only minutes remain before we die.

CHAPTER ONE

WEIGHED DOWN AND BUOYED UP

Fish float with a minimum of effort because their bodies have almost the same density as the water they swim in.[1] They are supported by the water, which presses against them from all directions. Narrow struts of cartilage are enough to stiffen their fins. Their bones are light: they have no limb bones, shoulder blades, collarbones or pelvic skeleton. Only a twentieth of their weight is due to bone.

Archimedes discovered more than two thousand years ago that an object immersed in water is buoyed up by a force equal to the weight of the water it displaces. Despite the lightness of its bones, a fish weighs slightly more than the weight of the water it displaces. This means that the fish would sink without a buoyancy device tucked under its backbone – a long gas-filled balloon called the swim bladder. By pumping gas into its swim bladder, the fish increases its size until its body displaces its own weight of water. In this way, the downward pull of the force of gravity is balanced by the upward thrust of the force of buoyancy.

The swim bladder allows the fish to keep its station in the water without constant effort. By adding 7% to the volume of its body, a trout is able to hang as though weightless in a freshwater stream. Since seawater is slightly more dense than freshwater, a herring reaches 'neutral buoyancy' by adding only 5% to its volume. The juvenile free-swimming fry of flatfish such as plaice or turbot have functioning swim bladders. When they settle on the seabed as adults, they have no need for buoyancy devices, and their swim bladders shrink to small remnants.

We live on land, and thin air provides no support. We need a scaffolding of strong, rigid bones as props against the unopposed pull of gravity, as levers that our muscles pull on, and as armour-plating around the delicate tissues of the brain and spinal cord. Our sturdy ribs must cope with the ceaseless work of breathing. When we walk, each leg in turn bears our entire weight. When we run or jump, our bones must withstand even greater forces. Almost a fifth of our body weight is due to bone[8], so that a ten-stone swimmer is weighed down by two stones of bone.

When we enter the water, we are bound to sink unless we displace our own weight of water. Yet we face two problems. First, our bones are so heavy that we must displace a volume of water considerably larger than the volume of the body. Second, we have to breathe, and whenever we lift our mouths above the surface to reach the air, we displace even less water than the volume of the body. Luckily, we can use our lungs as buoyancy devices. The lungs normally contain about three litres of air. If we take a strong inward breath – widening the ribcage, pulling down the diaphragm and pushing out the belly – we can inflate the lungs to their full six-litre capacity. In this way, we can displace an extra three litres of water, sufficient for the upthrust of buoyancy to carry us to the surface and to keep us floating there.

During their early swimming lessons, children have to adjust to the unaccustomed pressure of the water on the chest and abdomen. Their breathing is restricted at first because it takes extra effort to expand the chest. They need time and practice before they learn to float by increasing the volume of the air in their lungs.

Even with fully inflated lungs, a child's mouth lies very close to the surface of the water. There is a narrow margin between floating high enough in the water to reach the air and floating low enough to displace sufficient water to retain buoyancy.

A beginner often puts too much effort into trying to propel his head and shoulders above the surface of the water. But the higher he lifts himself, the less of his body is immersed, and if he displaces too little water, he becomes less buoyant. Then, inevitably, he sinks down below the surface and has to make vigorous efforts to get his head above the water once more. This alternate bobbing up and dipping down is too demanding to continue for long. He soon becomes exhausted.

Any loss of air from the lungs results in loss of buoyancy. A boy who enters the water with a misjudged jump or dive may be winded on impact with the water. With less air in his lungs, his buoyancy is reduced, yet he cannot re-expand his lungs until he pulls himself up to the surface. A swimmer in difficulties cannot shout for help without expelling a stream of air across his vocal cords. If he panics and screams, he empties his lungs, putting himself in immediate jeopardy.

◆ ◆ ◆

A fish keeps its swim bladder inflated, but we cannot keep our lungs inflated all the time. We have to renew the air by breathing in – and out. On land, we usually breathe twelve times per minute, taking about half a litre of air into the lungs and breathing half a litre out again a few seconds later. When we are swimming, we have to breathe faster and deeper to provide our muscles with the extra oxygen they need. Yet, vigorous exhalation leaves as little as one litre of air in the lungs[14] – a volume of air totally inadequate as a buoyancy aid.

A swimmer has to breathe in at precisely the right moment in the stroke. And he must expel air quickly, just before the next in-breath. Even elite swimmers occasionally inhale some water during training and have to be hauled out to recover on the poolside. (Butterfly is the stroke that causes the most problems: as swimmers tire, their arms fail to lift their mouths clear of the water throughout each gasp for breath.) Whatever stroke a swimmer uses in open water, it is all too easy to take in a little water with a lungful of air, especially when caught by a river current or tossed in choppy seas. But when water enters the throat, it causes coughing. Then the swimmer loses buoyancy, and a struggle to stay at the surface quickly becomes a struggle to survive.

◆ ◆ ◆

When children first learn to swim, external buoyancy aids can give them a helpful lift. Buoyancy aids are not recent inventions. In ancient Rome, boys lay on cork floats or wrapped their arms round inflated pigskins. Nowadays, children slip on inflatable armbands or wear flotation vests with inlaid buoyancy strips. With their chests higher in the water and their faces protected from splashes, they feel less apprehensive, remain more relaxed, breathe in more deeply, and learn to float more easily. As soon as confidence and proficiency allow, their armbands can be deflated little by little, their buoyancy strips slipped out one by one. But beginners wearing buoyancy aids belong in the shallow end, always under close supervision.

Armbands give only a slight lift. Life jackets (or 'personal flotation devices': PFDs) provide much more. Cork vests for lifeboat crews were first used in 1854, under the auspices of the Royal National Lifeboat Institution. Cork made way for kapok (fibres from the seed pods of the Java Cotton Tree), then synthetic foam. In the 1930s, the American Navy introduced the inflatable life jacket affectionately nicknamed 'the Mae West' after the well-endowed actress. During the Second World War, many sailors and airmen were saved by their 'Mae West' life jackets, including two future American presidents – John Kennedy and George Bush Senior.

Modern life jackets are varied in design, to serve different needs. Each design has its merits and drawbacks. What is appropriate to wear on a skiff in a landlocked bay would be inadequate on a racing yacht in colder, rougher water further from the shore. Deep-sea fishermen need extra lift to counteract the weight of their sea boots, to keep them at the surface when they are exhausted, injured or unconscious, and to hold them face up until rescue arrives.

A life jacket weighs far less than the water it displaces, making it extremely buoyant. This means that the life jacket tends to float higher in the water than the person who is wearing it. A loose-fitting life jacket can endanger the wearer by allowing him to slip down inside it. In his book *River Safety*, Stan Bradshaw quotes a canoeist who almost drowned when he capsized into a fast-flowing river: 'I quickly realised that my vest was not keeping my head completely out of the water. The water line was right at my lips, and each time I drew a breath, I took in water.' It took him a moment to realise what was wrong: 'My vest was too big and rode up on my body. I went out next week and bought a new vest.' Life jackets must fit snugly. In particular, children need child-sized life jackets, with a crotch strap to keep the jacket in place and a flotation collar to ensure that they float face up.

Most importantly, life jackets have to be worn to do their work. Even the best-chosen life jacket will save nobody if it is stowed in a locker. Most leisure boating accidents are within sight of land, and the people on board too often assume that, if they are thrown into the sea, they will be able to swim to the shore or tread water until they are rescued. Unfortunately, without the added buoyancy of a life jacket, their chances of survival are greatly reduced. The coldness of the water saps their strength. Within a few minutes they sink below the surface. Coastguards report that 80% of the victims who drown in boating accidents are not wearing a life jacket.

When we swim we have to be buoyant. When we swim for our lives we have to remain buoyant. Whether we use the air in our lungs or the gas in our life jackets to stay afloat, we have to ensure that our faces rise above the surface. We need to breathe, and we cannot breathe water.

CHAPTER TWO

BALANCE OF FORCES

A baby's first unaided steps are rightly celebrated as an important milestone. In the next few years, this balancing act is perfected, allowing the child to run and jump, self-correcting without a moment's conscious thought. Later, the youngster graduates from tricycle to bicycle, learning to keep the centre of gravity poised over the narrow strip where the tyres press against the ground.

This land-learned sense of balance is inadequate when a young child enters the water. Non-swimmers need constant close supervision, even when they are merely standing in the shallows. The sturdiest toddler may be toppled by a sudden, unexpected change in water pressure caused by a breaking wave or a retreating undertow. When paddling, the child's upper body passes unimpeded through the air, whereas leg movements are hampered by the resistance of the water. Children tend to lean forward, and the more they hurry, the more unbalanced they become. They have to adapt their stance and their stride, otherwise their centre of gravity will move out in front of their toes, and they will tumble into the water.

Falls which would be trivial on land can have dire consequences in the water. Young children can drown in moments if they fall while paddling. Their arms are not strong enough to lift them more than a few inches from the bottom. The water hinders their movements, but gives them no steadying support or handhold as they try to haul themselves upright. They may fall repeatedly, flailing about underwater, unable to find their feet, unable to right themselves, even when well within their depth. Every year drowned toddlers are discovered face down in puddles, shallow ponds and rock pools.

◆ ◆ ◆

Whether they are walking about on land or wading through the water, children counter the pull of gravity by pressing their feet against the ground. When they paddle waist-deep, however, they displace enough water to bring the force of buoyancy into play.[1] The upthrust of the water gives non-swimmers a sensation of partial weightlessness, novel and exciting at first, but also destabilising and disorienting. They are neither anchored to the bottom nor properly afloat. Any buffet or stumble can pitch them headlong.

During their early swimming lessons, children are liable to capsize whenever they lift their feet from the tiles on the bottom of the swimming pool. Inflated armbands help beginners to keep their balance in the water[2] (just as bicycle stabilisers prop up wobbling young cyclists when they are learning to ride). Using their natural tendency to lean forward when walking through the water, children are taught to glide a little way before catching hold of their teacher or the side of the pool. Then they learn how to return to a standing position by tucking up their knees and lifting their heads before planting their feet securely on the bottom.[3] After mastering this simple pivoting action, they no longer panic if they begin to lose their balance. Instead of grabbing at the water when they feel unstable, they drop their feet to the tiled floor of the pool. Skilled swimmers quickly forget their initial instability in the water, but children are particularly vulnerable during this early, tentative stage before they master the art of swimming.

Once children know how to float, how to glide, and how to regain their footing, they are ready to start learning the swimming strokes that will propel them through the water. But their problems with balance continue. When they are immersed in water, two opposing forces act on their bodies. The force of gravity pulls them downwards, while the force of buoyancy pushes them upwards.[4] They have to make constant adjustments to their body alignment because the position of the centre of gravity changes as they move their arms and legs in sequence, and the position of the centre of buoyancy changes as they breathe in and out.[5]

There is an added complication. The legs are heavy columns of bone and muscle, whereas the chest is largely filled with air. As a consequence, the force of gravity and the force of buoyancy act at different points in the body. The centre of gravity is located at hip level, while the centre of buoyancy (which is the centre of gravity of the water displaced by the swimmer) is located at waist level. The pull of gravity and the lift of buoyancy create a turning force or torque, a constant tip-tilting that pulls their legs downwards and pushes their shoulders upwards.[6]

This tendency to rotate affects all swimmers, but it is most marked in strongly built, athletic men.[7] The heavier the leg bones, and the more muscled the thighs and calves, the stronger the torque, and the more readily the legs will sink. And if a swimmer allows his legs to trail downwards, they act as brakes, increasing the effort needed to move through the water.

Beginners often increase the action of the torque. They want to keep their faces out of the water, so they arch their necks backwards and tense their shoulders. They try to increase their speed, so they thump their hands downwards onto the surface of the water. Unfortunately, these actions disturb the balance of their bodies. Their legs sink, and with each awkward lunge of the arms, self-taught youngsters soon exhaust themselves in their early attempts to swim crawl. Their inefficient zigzag progress usually peters out after a few strokes.

◆ ◆ ◆

Such problems with balance would make life impossible for a fish. Sir James Gray, the eminent zoologist, observed: 'One of the most striking facts about a fish in movement is its ability to move on an even keel, and to change its direction of movement rapidly without losing balance.'[8]

A fish's gas-filled swim bladder is tucked beneath its spine, protected between paired columns of muscle. This neat arrangement ensures that the fish's centre of gravity lies directly above its centre of buoyancy. Relatively little energy is needed to maintain the fish's balance.

The fish's body is streamlined to help it pierce the water, but its fins make use of water resistance to stabilise its position, to control its movements in three dimensions, and to propel its body through the water. Fins do not have the neat, narrow ball-and-socket junctions found in human limbs. They have wide attachments along the fish's body, jutting outwards, firm and flexible. The leading edges of its fins slice through the water, while their broad surfaces hold on to the water or push against it. The vertical, dorsal and anal fins prevent the fish from rolling like a log or yawing like a weathervane. The paired pectoral and pelvic fins counter any tendency to pitch up and down like a see-saw. Extending backwards from the spinal column, the caudal fin pushes the fish through the water and acts as a rudder.

The fins of different species vary remarkably in size and shape to suit their styles of swimming. A speeding barracuda's narrow, backward-pointed fins steady its track like the fletches of an arrow. An eel's extensive frill of dorsal and anal fins doubles its pushing power as it snakes through the water on its long-distance travels. A manta ray hang-glides in the ocean suspended between huge triangular pectoral fins. A shark's stiff, angled pectoral fins create lift like the wings of an aeroplane. A lumpsucker uses its fused ventral fins as a suction disc to fasten itself to a rock in the swirling water of the intertidal zone.[9]

◆ ◆ ◆

To improve their skill, young swimmers need hours of practice in the protected, placid environment of the public swimming pool. Once they have mastered the strokes and progressed to somersaults and racing turns, they feel well-merited pride in their swimming ability. But they expose themselves to danger when they venture into the hidden currents of a river, or enter the surf at a holiday beach, where waves roll towards the shore and rip currents rush out to sea.

An adventurous teenager may find turbulent water enticing and exhilarating, but its power can too easily overpower him. His arms are poor stabilisers compared to the fins of a fish. If he is out of his depth, he cannot regain his balance by dropping his feet to solid ground. He may grab at the water, but it slips through his fingers. The strokes he perfected in the swimming pool count for nothing if the current is too strong for him and he cannot reach the safety of the land.[10]

CHAPTER THREE

SHAPES IN THE WATER

On land, the air that surrounds us scarcely hinders our movements. We notice air resistance only if a sudden gust of wind catches us off balance, or if we travel at speed. We meet with much greater resistance when we move through water. Swimming is far more taxing than jogging. Even wading knee-deep calls for a surprising amount of muscular effort.

It is hardly surprising that water holds us back, since water is 800 times denser than air.[1] Moreover, water molecules tend to stick to each other. A pull of electrical attraction links the hydrogen atoms in each water molecule with the oxygen atoms of neighbouring water molecules in fleeting, ever-changing 'hydrogen bonds'.[2] This attraction between molecules makes water fifty times more viscous than air.[3] Fish are built to cope with the powerful drag forces caused by the density and viscosity of the water, but for human swimmers, drag is a major problem.

Both fish and people have to contend with three types of drag. Form drag relates to the shape and alignment of the swimmer. Viscous drag is related to the attraction between water molecules. Wave drag is related to the surface tension at the boundary between water and air.

FORM DRAG. Fish are streamlined. Whatever their size and shape, their bodies are designed to help them to slip through the water.[4] Minnows and marlins pierce the water with their sharp snouts, their heads gradually widen as far as the gill slits, then their bodies taper gently to narrow, vertical tail fins. Stingrays are flattened plates, their bodies seamlessly fused to their wide, fleshy pectoral fins. Angel fish are narrowed discs, their bodies extended above and below by dorsal and anal fins with sharp leading edges that cut into the water like knife blades.

Our bodies are not streamlined. We cannot slip through the water like a fish. To make progress, we have to push the water out of the way, and water is heavy. Form drag acts most strongly where the blunt contours of a swimmer's head and shoulders push against the water.[5] And as pressure rises in front of the swimmer, his forward movement lowers the pressure behind him, tugging him backwards and sucking in eddies of water to backfill the void. A swimmer minimises form drag when his body lies horizontally in the water. Water resistance increases whenever his legs trail downwards.

VISCOUS DRAG. As a fish swims along, water sticks to the surface of its body and travels with it, creating viscous drag.[6] For the fish, this water displacement is confined to a narrow boundary layer surrounding its body, where water molecules slide over each other in orderly 'laminar flow'.

To reduce friction, fish are slippery. Eels have tight glassy skin, herring are covered by a smooth tiling of overlapping scales, and bream have a thick coat of mucus. The rough skin of a shark, so abrasive that it was formerly used as sandpaper, might appear to be an anomaly. Yet sharkskin, with its covering of tiny teeth, or denticles, each with a bony core and backward-pointing spine, appears to cut drag by trapping and stabilising a very narrow boundary layer around the shark's body.[7]

As a human swimmer moves through the water, a wide boundary layer envelopes his irregular, inconstant shape. Whichever stroke he uses, water swirls round his neck and cascades over his buttocks onto the backs of his thighs. His outstretched arms and legs propel the water into twisting eddies. This turbulent boundary layer is several centimetres thick in places, and it travels with him, pulled along at his expense, as if he were dressed in a transparent, shape-shifting 'Michelin Man' suit. Thus, to make headway in the water, the swimmer must supply the energy needed to move a considerable volume of water as well as himself.[8] And the more he thrashes the water, the greater the turbulence around him, the wider the boundary layer, the greater the viscous drag, and the harder his task.

WAVE DRAG. At the surface of the water, wave drag is greater than form drag and viscous drag combined.[9] This is due to the force of surface tension, which acts as if the water were covered by a taut elastic skin. Flying fish take advantage of surface tension by soaring out of the sea and skimming fifty feet or more like the flattened pebbles in a game of 'ducks and drakes', outracing their water-bound predators impeded by wave drag.[10]

Most fish swim at a chosen depth below the surface. A Canadian study established that Chinook salmon fight their way upstream in Alaskan rivers by swimming well below the surface in fast-flowing water far from the bank.[11] Selecting this route means that they have to swim against a stronger current, but they avoid the wave drag they would encounter in the shallows.

We need to breathe air, so we must swim high in the water, and at the surface we meet with wave drag. As a swimmer moves through the water, his head pushes up a bow wave in front of him, while a wake of crests and ripples spread out behind him, all lifted against the downward pull of gravity and surface tension. If the swimmer doubles his speed when swimming at the surface, viscous drag doubles, form drag increases fourfold, wave drag increases eightfold.[12]

When racing, an elite swimmer avoids the energy cost of wave drag by swimming below the surface for the first part of every length.[13] He uses the momentum from his racing dive, and from each racing turn, to extend his underwater glide, trimming seconds from his time.[14] When a fast-paced freestyler returns to the surface, such a marked bow wave builds up in front of his head that he is able to turn his head to the side and breathe in the following trough.[15]

◆ ◆ ◆

As swimmers increase their skill, they travel further for a given amount of effort.[16] Part of their success comes from reducing the effects of drag. They lie more level in the water. Their legs and feet add impetus to the stroke, instead of acting as brakes. In breaststroke, they pierce the water in every glide. In crawl, they roll with each stroke, boring through the water rather than ramming into it, leading with each arm in turn, not with the broad surface of the front of the chest.

Elite swimmers reduce viscous drag by shaving off their body hair before a race.[17] They wear tight, shiny swimming caps. The fabric of their swimsuits has been recognised as an important factor in competition ever since nylon and Lycra replaced wool. In the 2000 Olympic Games in Sydney, swimmers were first allowed to compete in 'drag-reducing' bodysuits that covered them from neck to ankle.[18] Within a few days, 15 world records were broken, 13 of them by swimmers wearing the newly authorised bodysuits.[19] Ian Thorpe won three Olympic Gold Medals in a bodysuit that encased his arms as well as his legs.[20]

At first, the designers of bodysuits focused on reducing viscous drag. Speedo developed Fastskin, claiming that the new fabric had a drag-reducing effect similar to the denticles in the skin of a shark.[21] By 2003, interest had moved to reducing form drag, using the elasticity of the fabric to smooth the curves of the swimmer's body and to provide compression to his labouring muscles. Ian Thorpe took part in wind tunnel tests of an Adidas design that aimed to minimise turbulence by channelling water more smoothly over the swimmer's back and buttocks.[22] In the 2004 Olympic Games in Athens, Thorpe wore an Adidas swimsuit and won two more Gold Medals.[23]

At the 2008 Olympic Games in Beijing, 25 Olympic records were broken, 23 by swimmers wearing the Speedo LZR Racer bodysuit,[24] a complex garment with welded seams, corset-like areas of high compression supporting the swimmer's torso and thighs, and panels of water-repelling polyurethane over the curves of the body to minimise drag. Michael Phelps won 8 Gold Medals wearing his custom-fitted LZR Racer.[25] Within a year, at the 2009 World Swimming Championships in Rome, rival firms brought out swimsuits made entirely of polyurethane.[26] The result was an even greater fall in drag,[27] and swimmers broke 43 world records.[28]

Swimmers reported that they floated higher in the water when wearing a bodysuit, and coaches noticed that swimmers' feet lay closer to the surface.[29] Small air bubbles were trapped inside the bodysuit and within the weave of the fabric. Although Australian researchers claimed that Fastskin bodysuits significantly reduced drag without increasing buoyancy,[30] journalists suggested that bodysuits made swimmers more buoyant, but only long enough to complete a race.[31] The new 'plastic' bodysuits definitely increased buoyancy.[32] Swimmers were encased in a thin, flexible sheet of polyurethane foam, a compact layer of tiny bubbles filled with carbon dioxide gas. They lay slightly higher in the water, which reduced form drag against the head and shoulders. Ankle-length bodysuits lifted swimmers' legs towards the surface, a particular benefit for muscular, heavy-boned men. By 2010, FINA, the international swimming authority, feared that races had become competitions between swimsuit manufacturers rather than between swimmers.[33] Full-length neck-to-foot bodysuits, zip closures and non-woven fabrics were banned.[34]

◆ ◆ ◆

If drag is a problem for elite swimmers, it is an even greater problem for swimmers with less skill, who soon tire as they battle the density and viscosity of water. They push broad-fronted against the water while their slapping arms and legs churn the water into a froth of useless bubbles.

Many drowning accidents begin when people fall into the water fully clothed.[35] Pockets of air trapped between layers of fabric provide a degree of buoyancy. Indeed, bulky, multilayered winter clothing may briefly give a lift as great as a personal flotation device, allowing victims to stay at the surface for a couple of minutes at least.[36] However, the air is soon dislodged as victims struggle in the water. Then as they try to swim to safety, their sodden garments spread out and catch the water, hindering their progress by increasing drag.

It is sobering to remember that swimmers must supply all the energy needed to push the water out of the way, to set water swirling around them, to carry along the sizeable boundary layer of water surrounding them, to lift a bow wave, to create a wake – and to make progress through the water. In still water, even weak swimmers using awkward, ineffectual strokes will be able to propel themselves for a time, although they will soon tire. When the water itself is in motion, in a river or in the sea, swimmers face hugely increased demands on their strength and skill. Fish can go with the flow, but swimmers have to reach the safety of the riverbank or the beach.

Only in a dive does a swimmer approach the streamlined shape of a fish. He is able to push off against solid ground, propelled through the air by the forceful contractions of strong muscles in his legs and back, arms outstretched in front of him, fingers spearing the water. As he enters the water, his speed is faster than at any other time during his swim. But as soon the swimmer begins his stroke, his body returns to its complex, drag-inducing shape in the water.

CHAPTER FOUR

SAFE ENTRY

From their earliest visits to the local swimming pool, children need to know how to enter the water safely.[1] Toddlers must be trained to stay back from the edge of the swimming pool unless they are holding an adult's hand. During their first few visits, they are gently lifted into the water. Later, they are helped as they walk down the steps into the shallow end of the pool. Then, still under close supervision by parents and swimming teachers, they are taught how to jump into shallow water, on cue.

Children assume that water is soft and yielding. At bath time, they watch it trickling between their fingers. In the swimming pool, they feel its gentle pressure on their legs. But surface tension acts like a skin on top of the water, resisting penetration. Children are surprised by the sudden smack of pain if they bellyflop into the water from the edge of the pool. The impact of a pancake landing from a higher take-off point can easily wind them, and they may be stunned if they strike the surface face first.

Once children learn to pierce the water surface instead of colliding with it, they must be taught how to dive safely. To avoid the possibility of a collision with the tiles at the bottom of the pool, they begin with a plunge dive into water 5 feet deep (deeper than a child's total length with arms extended above the head.) They are trained to protect themselves by locking their hands together and keeping their outstretched arms braced in front of their heads throughout the dive.[2] They are shown how to arch their backs, lift their heads and angle their arms upwards to ensure that their underwater trajectory quickly brings them back to the surface.

There are few diving-related injuries in children younger than ten years old.[3] The danger comes as they grow older, heavier, and stronger. Water deep enough for the safe entry of a ten-year-old will be too shallow for a teenager, who weighs more and whose legs propel him more forcefully into the water. A swimmer who takes a running dive plunges even faster and further. In 2008, a Canadian study of diving-related spinal cord injuries described accidents to young men who broke their necks when they dived in at the deep end of the pool and slammed into the up-slope between deep water and the shallower water in the rest of the swimming pool.[4]

In competitive swimming, swimmers stand on starting blocks, enter the water with a flat racing dive, and glide some way under water to minimise wave drag. The piked racing dive became popular in the 1970s. Pushing off from the starting blocks with all their strength, swimmers bent their bodies at the waist just before they entered the water. The aim was to dive in at a steeper angle and glide further under water before coming to the surface.[5] Unfortunately, the water was usually less than 4 feet deep at the shallow end of the swimming pool. Between 1976 and 1984, at least 25 young Americans suffered serious spinal cord injuries when they hit the tiles at the bottom of the pool.[6] Once the problem was recognised, it was solved by moving the starting blocks from the shallow end to the deep end of the pool – or by building deeper racing pools. The water in Beijing's Olympic racing pool is 3 metres deep throughout its length.[7]

In 1965, a report by the Amateur Swimming Association (ASA) suggested that diving boards placed at the deep end of British public swimming pools contributed to many serious accidents.[8] Swimmers who took off from a diving board penetrated deeper below the surface, and their flight often took them too far up the pool. There was an additional risk of injury when divers and swimmers shared the same pool: divers sometimes plunged down onto the swimmers below.[9]

After publication of the ASA report, many diving boards were removed from the deep ends of swimming pools, while others were re-sited in separate, deeper diving pools. The results were very positive. In 1980, research at the National Spinal Injuries Centre at Stoke Mandeville Hospital concluded, 'Injuries due to diving into the deep end of swimming pools appear to have been eliminated in Great Britain.'[10] The study warned, however, that swimmers continued to break their necks after diving into the shallow end of swimming pools.

◆ ◆ ◆

Children are not taught vertical dives until their standard of swimming is good enough to take them into an official diving pool with a qualified diving instructor.[11] The diving pool is at least 5 metres (16 feet) deep. Yet even with this depth of water, the divers' momentum will carry them to the bottom unless they perform a 'save',[12] which continues the rotation of the dive under water before they steer up towards the surface. Adequate water depth and meticulous safety-conscious techniques are needed to avoid a collision with the floor of the pool.

The higher they climb, the faster they are moving when they enter the water and the deeper they penetrate the water. Competitive divers refine their entry during hours of practice. In former years, a diver sliced through the water surface with stiff pointed fingers. Nowadays, the diver bends one hand backwards at the wrist, grasps the fingers with the other hand, and turns the palms towards the water.[13] The resulting 'flat hand' punches a hole in the surface of the water through which the rest of the body follows.[14] The diver then separates his hands and pulls hard on the water, creating a partial vacuum behind him that sucks down any entry splash.

In a successful dive, only a ring of bursting bubbles marks the point of entry, surrounded by widening circles of ripples. This 'rip entry' (so-called because of the noise it makes) earns high marks, and more importantly, it minimises the impact of the water on the diver's body, an important consideration when diving from a 10-metre platform, because the diver enters the water at a speed of 35 miles per hour.[15]

Competitive divers know that an awkward entry hurts. Vigilant coaching and their own skill and discipline ensure that few competitive divers suffer catastrophic accidents, but minor injuries are frequent, especially during training. Divers pull muscles at take-off and as they twist and somersault during their flight. They sprain their wrists and dislocate their shoulders when they hit the surface of the water at speed. Imperfect alignment on entry can burst an eardrum.

The eyes of divers are subjected to a sudden rise in pressure on every entry. In countless practice dives over the course of years, these repeated impacts may result in detachment of the retina, the fragile light-sensitive layer of cells at the back of the eyeball. Jua Jingjing, described as the queen of Chinese diving, won two silver medals in the 2000 Sydney Olympics. In 2001, she had surgery to repair a detached retina in her right eye. Risking blindness, she won two Gold Medals at the 2004 Athens Olympics and two Gold Medals at the 2008 Beijing Olympics. By 2008, she needed further surgery for retinal detachments in both eyes.[16]

After winning two British Elite Junior diving titles, seventeen-year-old Chris Mears travelled to Australia to compete in the 2009 Youth Olympic Festival. He caught glandular fever while he was training in Sydney. The impact of a dive ruptured his enlarged spleen and he almost died.[17] Despite an abdominal scar twelve inches long, he reached the finals of two springboard events at the 2012 London Olympics. Then, at the 2016 Rio Olympics, Chris Mears and his diving partner, Jack Laugher, won Gold Medals in the 3-metre synchronised springboard competition.[18]

CHAPTER FIVE

DIVING UNDER PRESSURE

Air presses down on the surface of the sea at 'atmospheric pressure': 1 kilogram per square centimetre (more than 14 pounds per square inch). Beneath the surface, pressure rises rapidly with increasing depth. Only 10 metres down, the pressure has doubled to 2 atmospheres, and each further 10 metres of depth increases the pressure by another atmosphere.

Skin-divers and scuba-divers are subject to dramatic changes in water pressure during every dive. Body tissues are virtually incompressible, being largely composed of water.[1] But the lungs are full of air. As divers descend, the water pressure rises, compressing the air in their lungs. Then, as they ascend to the surface, the water pressure falls, and the compressed air in their lungs expands.

When skin-divers take a breath at the surface before they dive, they fill their lungs with air at atmospheric pressure. If they dive to a depth of 10 metres, water pressure doubles, compressing the air in their lungs to half of the volume it had at the surface.[2] As they descend, the buoyancy provided by their air-filled lungs is gradually lost. The deeper they dive, the denser they become and the heavier they feel. Although the air in their lungs gradually re-expands to its former volume as they ascend, getting back to the surface is hard work. Japanese abalone fishermen, who dive to a depth of 20 metres many times a day, need companions on the surface to haul them up by rope.[3]

In the extreme sport of free diving, skin-divers hurtle down to depths of a hundred metres or more on weighted sleds. Their lungs shrink down 'to the size of lemons'. In 2003 Tanya Streeter, the supremely fit and photogenic champion of 'No Limit' free diving, told *The Times*, 'Muscle strength allows me to expand my ribcage to pack in more oxygen and kick back from great depths.'[4] She works out in the gym for three months before each dive and wears metre-long flippers to maximise the thrust of her powerful legs. Yet she is not strong enough to return to the surface unaided. She has to rely on the lifting power of a rescue airbag fixed to the sled. Once the airbag is inflated and released, it rushes upwards, expanding as its goes, pulling her up to the surface.

◆ ◆ ◆

Skin-divers cannot stay underwater for long. They have to return to the surface before their breath runs out.[5] The snorkel allows swimmers to breathe air while their faces are submerged, but the snorkel tube must be short. The longer the tube, the more stale exhaled air lies between the swimmer and fresh air at the surface. And only a short distance below the surface, the pressure of the water surrounding the swimmer's chest makes breathing through the snorkel impossible. To draw breath into the lungs, the swimmer must expand the chest by contracting muscles between the ribs, but the intercostal muscles are designed to function at atmospheric pressure. They are not strong enough to swing the ribcage outwards against the constricting grip of underwater pressure. Dale Sheckter, editor of *California Diving News*, comments that trying to use a snorkel 'even just two feet down would be like trying to breathe with a 300 pound gorilla on your chest.'[6]

Jacques Cousteau, who introduced diving to a wider public, wrote of his boyhood attempt to breathe underwater;[7] 'I read a wonderful story of a hero who hid from villains by breathing through a hollow reed from the river bottom.' He decided to try and do the same: 'I put a length of garden hose through a block of cork, took the breathing end in my mouth, clutched a stone, and jumped into the swimming-pool. I couldn't suck a breath. I abandoned my hose and stone and made off, frantically, for the surface.'

Twenty years later, Cousteau launched the scuba – the Self-Contained Underwater Breathing Apparatus – and solved the problem of breathing underwater.[8] Scuba-divers carry their air supply with them. Cylinders contain air compressed to 300 atmospheres, while the scuba regulator delivers the air to the mouthpiece at a pressure that matches the pressure of the surrounding water,[9] allowing divers to inhale the air despite the high water pressure around them.

Cousteau wrote vividly of the danger of inhaling air under high pressure. Sixty metres below the surface, he developed nitrogen narcosis. He found himself hallucinating in a half-drunken, half-anaesthetised state which he called *'ivresse des grandes profondeurs'* or 'rapture of the deep'.[10] His description of his ascent suggests a continuing wild exaltation: 'I rose through the twilight zone at high speed, and saw the surface pattern in a blaze of platinum bubbles and dancing prisms. It was impossible not to think of flying to heaven.' His companions on the dive were also affected by nitrogen narcosis. In a similar dive later in the year, a valued colleague lost his life.

The pioneers of scuba-diving breathed pure oxygen. They did not realise that they risked the sudden onset of convulsions and unconsciousness only 6 metres below the surface. (Six metres down, at a pressure of 1.6 atmospheres, oxygen is toxic to the brain.)[11] Cousteau himself was almost a victim of oxygen toxicity.[12] He described chasing after a bream: 'He was hanging forty-five feet down. I descended and the fish backed away, keeping a good distance. Then my lips began to tremble uncontrollably. My eyelids fluttered. My spine bent backwards like a bow. With a violent gesture I tore off the weight belt and lost consciousness.' His body was spotted floating on the surface. He was pulled aboard the dive dinghy and resuscitated.

With the introduction of modern scuba equipment, divers no longer breathe pure oxygen. They usually breathe air. Atmospheric air contains 21% oxygen, so the partial pressure of oxygen at the surface is .21 atmospheres. At a depth of 50 metres, however, the water pressure rises to 6 atmospheres, and if the scuba regulator delivers air, the partial pressure of oxygen is 1.26 atmospheres, uncomfortably close to toxicity levels.

Instead of breathing air, scuba-divers can choose a gas mixture suitable for the dive they are planning. Each formulation offers its own special benefits and each carries its own particular risks. Nitrox 40, for instance, contains 40% oxygen and 60% nitrogen, allowing a longer dive time, but with a risk of oxygen toxicity if divers go deeper than 30 metres. In contrast, deep sea divers working at a depth of 600 metres may use a gas mixture containing only 2% oxygen,[13] which would cause asphyxiation if breathed at the surface.

◆ ◆ ◆

Scuba-divers are subjected to 'squeezes' as they descend. Diving masks cover the nose as well as the eyes, allowing divers to top up the air beneath the mask as they descend. Otherwise, the compression of the air beneath the mask may cause bloodshot eyes and bruised eyelids.[14] If the entrance to a nasal sinus is blocked, a partial vacuum may lead to pain, bleeding and headache.[15] A diver may even suffer sudden toothache due to compression of the gas in a decayed tooth.[16]

Most 'squeezes' are no more than a nuisance, but 'middle ear squeeze' is potentially dangerous.[17] At the surface, both sides of the taut, delicate eardrum are equally exposed to atmospheric pressure. During descent, the increasing water pressure in the ear canal pushes against the outer side of the eardrum, while the falling volume of air in the middle ear pulls the eardrum inwards. To equalise the pressure on the two sides of the eardrum – to clear the ears – a diver must pump air repeatedly from the throat into the middle ear through the connecting Eustachian tube.[18] Its airtight entrance in the throat has to be opened actively, by swallowing, by jaw movements or by blowing out against closed nostrils (a brief inflation of the tubes called the Valsalva manoeuvre).

The Eustachian tubes are nearly four centimetres long, and they are easily blocked by mucus or inflammation.[19] A diver with a cold risks injury to the eardrums – or worse – during the descent. If he cannot clear his ears, the eardrums will bulge inwards. If he ignores the pain and continues his descent, he may rupture an eardrum. This is a life-threatening injury under water. Cold water rushes through the hole in the eardrum, flooding the middle ear. The sudden fall in temperature disturbs the balance sensors in the inner ear, triggering vertigo. The diver quickly becomes disorientated.[20]

◆ ◆ ◆

Fish are very sensitive to the pressure of the water.[21] As well as having a defined geographical spread, most species tend to swim at a preferred depth. As long as they 'maintain their station', the water pressure remains constant, and the amount of gas in the swim bladder needs little adjustment. To maintain neutral buoyancy, fish only need to top up the gas from time to time.

Fish that live in shallow water or swim high in the water table, like trout or herring, stay buoyant by coming to the surface, swallowing air, then forcing the air along a narrow duct that connects the gullet to the swim bladder.[22] To lower their buoyancy, they deflate the swim bladder by burping up unwanted air.

Fish that swim lower in the water table, like pike or cod, have no easy access to air at the surface. The duct between gullet and swim bladder shrivels to a thread. Instead of swallowing air at the surface, they renew the gas needed for buoyancy by extracting oxygen from their own blood as it flows through a 'gas gland', a patch of specialised blood vessels in the wall of the swim bladder.[23] When a cod swims downwards, the water pressure increases, and the gas in its swim bladder is compressed. To avoid any loss of buoyancy, the blood vessels in the gas gland release enough extra oxygen to return the swim bladder to its former volume. Conversely, when the cod swims upwards, the water pressure decreases, and the gas in the swim bladder expands. To avoid excessive buoyancy, an appropriate amount of oxygen is reabsorbed from the swim bladder.

The swim bladder provides a clear survival advantage for many species of fish – neutral buoyancy saves energy. A number of mid-ocean fish are able to adjust the amount of gas in the swim bladder in order to benefit from 'nocturnal ascent'. [24] They spend the daylight hours in relative safety hundreds of metres down in cool, dark waters, under the enormous pressure of 40 atmospheres or more. Each night they rise to feed on flora and fauna near the surface at pressures as low as 2 atmospheres. At dawn, they sink again into the depths. Despite being subjected to huge changes in pressure during each nightly ascent and each subsequent descent, the fish hold the volume of the swim bladder constant. As the fish rise, some of the expanding gas is extracted from the swim bladder. As they sink, the gas gland releases extra oxygen into the swim bladder. These buoyancy adjustments during extreme changes of pressure take several hours.

Mid-ocean predators, such as tuna or shark, have no swim bladder.[25] They have traded neutral buoyancy for speed of manoeuvre. They have no time to adjust to changes in water pressure as they chase their prey into the depths and up again. In contrast, fish with a swim bladder suffer serious consequences if they ascend too rapidly. Fishermen have a name for the unhappy outcome. They call it 'gassing up'.[26] At a depth of 20 metres, a seabass is under a pressure of 3 atmospheres. If it is hooked and reeled to the surface, its swim bladder swells to 3 times its normal volume, cramming the fish's body cavity and forcing its stomach and intestines out of its mouth.

The consequences can be just as dire for a scuba-diver who forgets to breathe out, actively and continuously, as he returns to the surface at the end of a dive.[27] At a depth of 20 metres, his scuba delivers air at a pressure of 3 atmospheres, the same pressure as the water around him. During his ascent, the water pressure lessens, and his safety depends on how well he copes with the inevitable threefold expansion of the air in his lungs.

If he fails to exhale enough of the expanding air, he risks the desperate emergency of 'burst lung'. The expanding air distends, distorts and tears the delicate tissues of his lungs.[28] Air under pressure puffs up his neck, compresses his heart, or explodes into the pleural space between lung and chest wall, causing his lungs to collapse.[29] Bubbles of air force their way into torn blood vessels in the lungs and are carried to the left side of his heart, where they are pumped round his body like so many micro-calibre pellets of lead shot fired at random. Wherever these 'air emboli' come to rest, they block blood vessels, and trigger the formation of blood clots.[30] If they enter a coronary artery, they may bring about a heart attack. If they arrive in the brain, small areas of brain tissue are destroyed, leading to unconsciousness, convulsions, paralysis or death.

The volume of gas doubles as a scuba-diver ascends through the last 10 metres to the surface,[31] so that even a comparatively shallow dive may end in serious injury. It is even possible to suffer 'burst lung' when learning to scuba-dive in the deep end of a swimming pool.[32] (A rise of less than two metres can balloon the lungs to bursting point if a novice scuba-diver submerges, fills his lungs with pressurised air, then holds his breath while returning to the surface.)

◆ ◆ ◆

As well as breathing out all the way to the surface, scuba-divers must come up slowly. Those who ascend too fast risk the onset of decompression sickness, which is not caused by the expansion of air in their lungs, but by the formation of gas bubbles in their tissues.

On land at sea level, our bodies contain about a litre of dissolved nitrogen[33] as well as small amounts of oxygen, carbon dioxide and argon. During a dive, under increased pressure, a greater volume of gas dissolves in the scuba-diver's body. A fifth of the air delivered by the scuba mouthpiece is oxygen, which is captured in the lungs and used by his tissues. However, four-fifths of the air is nitrogen, which collects in the diver's body, particularly in fat.[34] In tissues with a generous blood supply, such as the brain or the kidneys, nitrogen levels rise relatively quickly. In tissues with fewer blood vessels, such as cartilage or tendon, the build-up of nitrogen takes longer. The deeper the dive and the longer it lasts, the more nitrogen accumulates in the diver's tissues.

If he descends to a depth of 50 metres, the currently accepted safe limit when using compressed air,[35] he breathes air delivered at 6 atmospheres of pressure, the same pressure as the surrounding water. By the time he has completed his planned time at depth, his body will hold six times more nitrogen than it held at the surface.[36]

The pressure on the diver's body 50 metres below the surface is the same as the pressure inside an unopened bottle of champagne.[37] The weight of the water above him keeps the nitrogen in solution in his body tissues, just as the cork in the champagne bottle keeps the carbon dioxide in the wine. As yet, there are no bubbles.

If the diver takes his time during his ascent to the surface, the nitrogen will cause no harm. Rising slowly, meeting with lower and lower pressure as he rises, he clears his body of excess nitrogen by breathing it away.[38] If he rises too quickly, however, he gives himself too little time to rid his body of the nitrogen lurking in his tissues. As soon as pressure falls too low to hold the nitrogen in solution, it is released as a myriad of tiny bubbles, just as uncorking a bottle of champagne releases its fizz. The bubbles grow larger as the diver ascends, crushing tissue and blocking blood vessels.[39] Expanding bubbles combine with their neighbours, prising tissues apart. Veins draining affected tissues become loaded with bubbles. In severe cases, so many bubbles enter the diver's veins that the right side of his heart is filled with froth.[40]

Decompression sickness takes many guises, depending on how many bubbles are released, and where they form.[41] The commonest manifestations have been honoured with nicknames: bubbles in bones and muscles cause 'the bends'; bubbles in the lung capillaries cause 'the chokes';[42] bubbles in the inner ear cause sudden deafness and 'the staggers'; bubbles under the skin cause 'fleas', 'diver's itch' and 'marbling';[43] bubbles in the spinal cord cause a range of neurological disorders, with severity ranging from fleeting pins and needles, through 'diver's palsy', up to permanent paralysis; bubbles in the diver's brain cause sudden confusion, slurred speech and loss of vision.[44] With the onset of convulsions, his condition is likely to deteriorate rapidly.

Divers who carry out hard physical work under water seem to be particularly susceptible to severe episodes of decompression sickness.[45] Their muscular activity consumes a large amount of oxygen and produces a large amount of carbon dioxide, which tends to build up in the diver's body in addition to the accumulating nitrogen (and carbon dioxide comes out of solution even more readily than nitrogen).[46] The mechanical to-and-fro of muscle action may also intensify bubble formation, just as champagne froths if the bottle is shaken before the cork is popped.

◆ ◆ ◆

Decompression sickness was first recognised in the nineteenth century in men building bridges and tunnels. Engineers kept water out of submerged or subterranean work sites by pumping air at high pressure into huge chambers called caissons (French: 'big box'). Construction workers spent their days labouring inside these caissons, entering and exiting via air locks. In 1866, work began on the Eads Bridge across the Mississippi at St Louis. Engineers planned to excavate down to bedrock more than 30 metres below the riverbed. Early in the construction process, air pressure in the caissons reached two atmospheres, and some of the workers were struck by pains in their bones. The pains appeared after the end of the shift and eased on returning to work. Workers were said to have 'the bends' because their contorted posture resembled the 'Grecian bend' silhouette of the fashionable bustle-wearing women of the time.[47] As the workers dug deeper, it became more difficult to hold back the water. The air pressure in the caissons was increased to four atmospheres or more. A sixth of the workers became ill, and fourteen died of the newly recognised Caisson Disease.

In 1878, the French physiologist Paul Bert explained the cause of Caisson Disease.[48] He put dogs into pressure chambers. When he took them out again he found bubbles in their blood. He realised that caisson workers were not injured by working under increased pressure, but by returning too abruptly to atmospheric pressure. In 1890, decompression chambers were introduced during the excavation of London's Blackwall Tunnel and New York's East River Tunnel, initially to treat workers suffering from 'the bends',[49] and later to protect workers at the end of their working day by slowing their transition from high pressure inside the caisson to atmospheric pressure outside. With the growth of commercial and military deep-sea diving, symptoms of 'the bends' were noticed in many divers, and Caisson Disease was renamed decompression sickness.

All scuba-divers are at risk of developing decompression sickness if they ascend too fast. Yet diagnosis is difficult because the symptoms are so diverse. Holidaymakers who fly home within 24 hours of scuba-diving are exposed to reduced cabin pressure and they sometimes develop symptoms during the flight.[50] Indeed, minor episodes of decompression sickness probably happen more often than divers like to admit, even to themselves.

◆ ◆ ◆

In 1997, an editorial in the *British Medical Journal* noted that 'silent' bubbles can be detected in the venous blood of many divers, and warned that divers may suffer brain damage even without a recognised episode of decompression sickness.[51] Soon afterwards, studies from Germany[52] and Belgium[53] reported an increased the risk of brain damage in scuba-divers with a common heart defect called 'patent foramen ovale'.

A baby still in the womb is supplied with oxygen via the umbilical cord. The baby's circulating blood bypasses its developing airless lungs by flowing through the foramen ovale[54] (Latin: 'oval opening'), a gap in the thin wall that separates the two upper chambers of the heart (the right atrium and the left atrium). Normally, soon after the baby's first breath, a flap closes over the foramen ovale, separating the right side from the left side of the heart. The right side of the heart now pumps blood to the lungs for oxygenation, whereas the left side of the heart pumps oxygenated blood round the body. In about one person in four, the flap leaks, allowing some blood to flow directly from the right side of the heart to left side.[55] This 'right-to-left shunt' is often very slight or intermittent. Someone with a patent foramen ovale may have no symptoms and remain unaware of any problem – unless they take up scuba-diving.

When a diver ascends too fast, the blood in his veins returns to the right side of the heart carrying a dangerous load of gas bubbles. Normally, all this blood is pumped to the lungs, where most of the bubbles are caught in the fine mesh of the lung capillaries and go no further,[56] although the diver may suffer from 'the chokes'. However, if the diver has a patent foramen ovale, some of the returning venous blood, carrying its load of gas bubbles, will flow through the heart defect and reach the left side of the heart. Then each heartbeat will pump bubbles of nitrogen to all parts of the diver's body, bubbles capable of blocking off blood vessels and damaging tissue just like the air emboli in 'burst lung'. If the diver coughs, the pressure rises inside his chest, impeding the circulation through his lungs, so that an even greater volume of bubble-laden blood flows from the right side directly to the left side of the heart.[57] The diver ascends to the surface head up, so that rising nitrogen bubbles are swept up into the carotid and vertebral arteries and carried to his brain.

In 1994, Dr Peter Glanvill gave a vivid description of the onset of decompression sickness in his own body.[58] His dry suit ballooned as he returned to the surface at the end of a dive. As the dive boat took him to the shore, he developed a tight band of pain around his chest, and noticed a numb, tingling sensation from mid-chest down. He spent several hours in a decompression chamber before he recovered. A year later, after a short, shallow dive, patches of skin on his thigh and foot became numb, and he needed a second session in a decompression chamber. He agreed to have a contrast echocardiogram as an experimental procedure. Frothy liquid was injected into the back of his hand. Looking at the echo screen, he could see the progress of the bubbles as they entered his veins and 'watched the bubbles take a deadly short cut' as they passed through his heart, revealing his previously unsuspected patent foramen ovale. He was shaken to find that a minor flaw in his heart, which caused him no problems on land, had greatly increased his risk of serious injury when diving. A quarter of scuba-divers are in the same high risk group – and unaware of any danger.

◆ ◆ ◆

Urgent treatment is needed if a surfacing scuba-diver develops symptoms due to over-expansion of gas in the lungs[59] or to over-rapid ascent.[60] The diver is given 100% oxygen to breathe and transported to the nearest decompression chamber. If he is still alive when he enters the pressure chamber, his chances of survival are good. The pressure inside the chamber is slowly increased, compressing the offending collections of gas and limiting further tissue damage. As he continues to breathe oxygen (rather than air, which contains 79% nitrogen), the excess nitrogen in his body is breathed away and oxygen replaces nitrogen in the shrunken bubbles. The oxygen is captured by the haemoglobin in his bloodstream and used by his tissues. Then the pressure in the decompression chamber is slowly returned to atmospheric pressure. With luck, the injured diver will find himself symptom-free, although his body will be peppered with internal scars.

Slow ascent is the key to avoiding decompression sickness. But how slow must the diver's ascent be? Or rather, how fast can the scuba-diver safely make the journey back to the surface? Guessing and hoping has resulted in many dead and injured divers. A century ago the Royal Navy drafted in a team of scientists to study the problem, led by Professor J. S. Haldane.[61] They put goats into pressure chambers, rather than dogs.[62] The first dive tables were published in 1908, specifying safe ascent times related to the depth of the dive and length of time spent under water. Haldane gave clear recommendations that divers should speed up their descent to shorten the time that they are exposed to high pressure, that they should spend as little time at depth as possible, and that they should ascend in stages, quicker at first, but taking longer as they near the surface. He noted that ascents at uniform speed were 'needlessly slow at the beginning and usually dangerously quick at the end.'[63] His dive tables have been modified over the years, but his advice is still valid today.

Divers can now equip themselves with dive computers no bigger than a wrist watch, but technological gadgets can help only up to a point. Scuba-diving remains a dangerous sport, no matter how much instruction and training divers receive. Deep underwater, an unexpected malfunction or maladjustment of diving gear, a minor defect in lung or heart, a moment's inattention, an elementary mistake, or the onset of panic can all too quickly end in a diving fatality.[64] In five years (2009–2013), the Royal Society for Prevention of Accidents recorded the accidental deaths of 84 divers in British lakes, quarries and coastal waters.[65]

CHAPTER SIX

LOOKING OUT FOR TROUBLE

We are land animals, with senses that help us survive on land. In the water, fish rely on sense organs that differ from ours in many ways. Indeed, fish possess senses which we lack entirely.

When we leave the land and enter the water, our land-adapted senses of sight, touch and hearing present us with impaired information, although we remain largely unaware of the extent of the loss. While we are swimming, we need to be on high alert, particularly in open water or in the sea. Our safety may depend on taking rapid life-and-death decisions, and we have to interpret incomplete and confusing sensations as best we can.

◆ ◆ ◆

We depend on our eyes to inform us about our surroundings and to warn us of danger. But appearances are often deceptive. To children, a mat of pondweed may appear as solid as a lawn – until they try to walk across it. To survivors of a boating accident, the shore may look reassuringly close – until they begin to swim towards it.

At the local swimming pool, there is little ambiguity. The edge is clearly defined, the depth is marked, the water surface is brightly lit, the filtered water is clear, and the tiles on the bottom are sharply visible. However, there are no such certainties when we stand beside a stretch of open water. We encounter visual confusions, and we often guess wrong.

Even before we enter the water, tricks of the light affect what we see. The surface of the water reflects much of the light which falls on it. When the water is calm, its smooth surface acts as a mirror. The reflection of lakeside trees may create a compelling illusion that the lake bed lies as deep below the surface as the trees stand above it. Every year, young men take illusion for reality and suffer paralysis when they plunge into shallow water.

When the surface is broken by ripples and waves, the reflections are fragmented and distorted, disguising what lies underneath and luring the unwary into danger. The summer sunshine glinting from the surface of a northern lake tempts unsuspecting holidaymakers to dive into frigid water. After a few beers on a summer evening, friends walk out of a riverside pub, see the gently undulating pattern of lights shining on the water and decide to cool off with a quick dip, unprepared for the fierce currents that swirl below the surface.

The eye is tricked by refraction as well as reflection. Light travels slower through water than air,[1] and light rays are bent (refracted) at the water surface.[2] Since we assume that light rays travel in straight lines, the water 'loses' a quarter of its depth.[3] To a toddler, a rock pool may look shallow enough for paddling, even when the water is deep enough to drown him.

◆ ◆ ◆

Our sense of sight is so important, and our eyes are so delicate and sensitive, that a protective reflex prompts us to blink in response to a loud noise, a puff of wind, an object approaching the face or the lightest touch on the cornea.[4] We blink every ten seconds or so, sweeping the eyelids over the cornea like windscreen wipers, spreading a protective film of tears and removing particles of dust. The surface of the cornea must stay moist, otherwise it loses its transparency. (Fish have no need to blink since their eyes are naturally bathed in water.)

In their first swimming lessons, children blink constantly as they cope with the unfamiliar sensations of water continually splashing their faces and dripping from their hair. Swimming teachers encourage children to get their faces wet, to put their faces into the water, then to open their eyes under water, coaxing them, in easy stages, to keep their eyes open while swimming.[5]

Children make slow progress if they screw their eyes shut during their early swimming lessons,[6] yet blind children learn to swim very well – as long as they keep their heads above water.[7] To compensate for their lack of sight, they rely on their ability to detect the subtle clues from echoes and reverberations which sighted people ignore.[8] If blind children get water in their ears, their hearing becomes muffled, and they lose much of their sense of position and direction. Within a moment, they may become so severely disoriented that panic sets in.[9] Should this happen, their swimming teacher must be ready to jump into the pool to rescue and reassure them.

With time and practice, blind swimmers get used to water in their ears. Indeed, blind swimmers have competed in every Paralympic Games since they were first established in Rome in 1960.[10] During a race, each blind competitor is helped by a 'tapper',[11] who stands at the end of the pool and touches the swimmer's head or back with a padded rod when the moment arrives for a turn or for the finishing stroke.

◆ ◆ ◆

The eyes of fish and humans have the same general design. Light shines through the transparent cornea, passes through the lens, and falls, in sharp focus, on the light-sensitive cells in the retina at the back of the eyeball. Light reaches a fish's eye after travelling through water, and since its speed scarcely changes as it enters the cornea, there is no appreciable bending of the light at the water/cornea boundary.[12] The cornea is flat, but the lens is spherical, with a dense central zone, capable of the strong refraction needed to focus the light rays on the retina.[13]

The human eye also contains a convex, converging lens, so it might be assumed that the lens plays the major role in focusing the light rays on the retina. In our eyes, however, light reaches our eyes after travelling through the air, and most of the work of refraction and focusing of light rays is done when light reaches the domed air/cornea boundary.[14] We use the lens for fine-tuning. When we read a book, a ring of muscles squeezes the lens, increasing its curvature and its converging power.[15] When we look into the distance, the lens returns to a wider, flatter shape.

As soon as the swimmer's face dips into the water, the accustomed air/cornea boundary becomes a water/cornea boundary. Light is no longer refracted and focused as it enters the eye, and the lens is not powerful enough to compensate.[16] Swimmers with normal 20:20 vision on land are reduced to 20:200 vision when they open their eyes underwater.[17] (During a sight test, people with 20:200 vision can read only the largest letter at the top of the familiar Snellen chart: they are considered 'legally blind'.)[18] Surprisingly, despite this dramatic loss of image quality, swimmers seem unaware of the serious blurring of their vision when they look through water instead of air.[19]

◆ ◆ ◆

Swimmers restore the air/cornea boundary when they put on close-fitting swimming goggles, trapping air in front of their eyes.[20] Even the simplest pair of swimming goggles, with flat plastic eyepieces, improves their vision markedly, whether or not they usually wear spectacles. Corrective eyepieces for near-sightedness, long-sightedness, and astigmatism are optional extras.

Goggles work well for swimming at or near the surface. They protect the eyes from chlorinated water in swimming pools and from salty water in the sea. But goggles are not suitable for scuba-diving. To avoid eye damage from 'squeeze' as they descend, scuba-divers need diving masks that enclose the nose. Although diving masks restore the air/corneal boundary,[21] the light is also bent (refracted) at the boundary between the inner surface of the diving mask and the air trapped inside the mask. Diving masks help scuba-divers to see more clearly, but the flat window of a standard diving mask acts as a magnifying glass,[22] making objects under water appear considerably bigger and closer.

Related to these magnification effects, and probably more important for the diver, the mask transmits only the central third of the normal field of view, edged by optical distortions.[23] On land, the field of view stretches in an arc of 180 degrees from side to side, and the scuba-diver is likely to assume that the wide window of the diving mask affords a similar panorama. Due to refraction, however, the mask allows the diver to see an arc of only 70 degrees, a degree of tunnel vision, with an even greater restriction when looking downwards.

Inspired by the dome-ports of underwater cameras used by film crews, the 'double-dome' diving mask allows a diver a wide field of view in every direction of gaze.[24] However, solving one problem of refraction creates a new one. The curved water/mask boundary becomes a concave, diverging lens, so that a diver with 20:20 vision cannot bring the light rays to a focus on the retina. To benefit from the improved performance of a 'double-dome' diving mask, the diver needs the help of corrective contact lenses.

◆ ◆ ◆

On land, we can see for miles, but visibility through the clearest water is less than 200 feet at best. Water is translucent, rather than transparent. Light is scattered by the water itself, and scattered again as it passes through the bodies of the microscopic life forms that throng the water.[25] Visibility is further impaired by salt, silt, and the bubbles stirred into the upper layers by breaking waves. In the turbid water of a river estuary, visibility is measured in inches.[26]

Much of the light that falls on the surface is reflected back into the air.[27] When the sun is low on the horizon in the early morning and late afternoon, only 20% of the light enters the water.[28] On a sunny day, if the water is calm during a dive, scuba-divers may look upwards and see a circle of brightness on the water surface and a descending cone of light. This is Snell's window,[29] a phenomenon caused by the refraction of light as it strikes the surface of the water.[30] Divers may see fish backlit above them, swimming across the cone of light. (This illuminated contrast is used by underwater photographers and by lurking predators.)[31] Beyond the edge of the circle of brightness, the water surface acts as a downward-facing mirror, reflecting only the small amount of light travelling upwards from the underwater world.[32] Beyond the cone of light, fish loom and vanish through a dim bluish haze.[33]

Light is gradually absorbed as it travels downwards through the water.[34] Only one metre below the surface, half of the light that entered the water has already been absorbed. Ten metres down, only 15% remains, and at a depth of 100 metres, just 1% of light penetrates the darkness. Water absorbs the longer wavelengths of the visible spectrum more rapidly than the shorter wavelengths. Red light fades out entirely within 10 metres, so that red objects appear black. At a depth of 25 metres, only blue light is left.[35]

◆ ◆ ◆

In the retina, two types of light-sensitive cells respond to light,[36] as if the eye were a camera with two speeds of film, the cones for daytime, and the rods for dusk. The cones contain photopsins, visual pigments that are bleached by light, then rapidly renewed. Rods contain rhodopsin or visual purple, which is a thousand times more sensitive to light than photopsin, but its renewal is much slower.[37] If the light is bright enough to bleach photopsin in the cones, it will also bleach rhodopsin in the rods. In near-darkness, however, the rods make vision possible by synthesising rhodopsin faster than light can bleach it.

During the day we see sharp, colourful images. In bright sunlight, we lower our eyelids and shade our eyes to cut out glare. The iris constricts, narrowing the pupil and focusing the light onto a small central area of the retina, the fovea (Latin: pit), where cones are closely packed together.[38] In shadow or at twilight, the iris widens to let in more light, but the colour fades if the light becomes too dim to stimulate the cones. Our night vision depends on rods, and we see only grainy shapes, without detail, in shades of grey.[39]

Fish look out on a half-lit world, their eyes permanently wide open, pupils set at maximum aperture. Many species respond to changes in light intensity by using a process called retino-motor motion.[40] During the day, the eyes becomes light-adapted (the cones move up to the surface of the retina and the rods are pulled down, hidden from the light among cells full of black granules of melanin). At dusk, the eyes becomes dark-adapted (the rods, with their stored rhodopsin, move up towards the surface of the retina, the cones are pulled down, and the melanin-containing cells shrink back out of the way). Sharks have close-packed rods, but no cones. A silvery layer of guanine crystals lies beneath the retina. This tapetum lucidum (Latin: carpet of light) catches the light and reflects it back towards the rods, giving them a double chance of stimulation.[41] Swordfish hunt for squid in the cold ocean depths. They improve their sight and increase their hunting success by warming their eyes with 'heater organs,'[42] eye muscles modified to produce heat instead of movement.[43]

The human eye has no retino-motor movement of cones, rods or melanin granules, and the retina has no underlying tapetum lucidum. For us, dark-adaptation begins only when the light fades, and the build-up of rhodopsin takes at least half an hour.[44] Car drivers have slower reaction times at twilight, with longer stopping distances and a disproportionate number of fatal car crashes.[45]

When we step from a brightly lit building after dark, our eyes need time to adjust to the sudden fall in illumination. Similarly, if scuba-divers descend rapidly from surface sunlight into low-light conditions, they will see almost nothing at first.[46] They lose their sharp, coloured central vision because there is insufficient light to stimulate the cones, while the rods need time to become dark-adapted. Rods are insensitive to the red part of the spectrum, so scuba-divers can start the process of dark-adaptation by putting on red goggles for half an hour before the dive.[47]

◆ ◆ ◆

Photographers use wide-angle lenses, exposure control, colour filters and artificial lighting to capture the spectacular images of fish which we admire in books and magazines.[48] A scuba-diver viewing the scene underwater has the dimmer switch turned down. If the water is murky, or if flippers stir up sediment, the only pointer to the surface may be the upward track of bubbles exhaled from the scuba mouthpiece.

Deep underwater, the diver peers through the misty blue water. His diving mask limits his field of view. A wetsuit encases and insulates his body. The buoyancy of the water dulls his gravity-related position sense. The rasp of air escaping from his mouthpiece is the only sound he hears in Cousteau's 'silent world'. It is hardly surprising that some scuba-divers react to this sensory deprivation with feelings of isolation, confusion, and fright.

CHAPTER SEVEN

FEELING THE VIBRATIONS

We feel something touching our skin only if the contact causes a distortion of microscopic touch receptors shaped like corkscrews and onion bulbs.[1] When we enter the water we are aware of little more than a generalised feeling of slight pressure upon every submerged surface of the body, often pleasurable but rarely very informative.

Fish, in contrast, have the benefit of specialised sensors which detect and map every momentary change in the flow of the surrounding water.[2] Scattered over a fish's head and body are small pits containing neuromasts (Greek: 'sense-hillocks'). Each neuromast contains a tuft of hair cells embedded in a tiny dome of jelly, a cupola, that bulges into the water and sways in response to drag forces in the water. Any movement of the cupola tugs on the hair cells and triggers an impulse in sensory nerves.

A lateral line runs along each side of a fish's body from snout to tail.[3] Its position is marked by a row of scales, each with a central pore. Beneath these perforated scales, a series of neuromasts sit in a shallow water-filled tunnel, protected from the full force of drag, but still in close contact with the surrounding water.

The neuromasts on the fish's skin and in its lateral lines monitor how fast the water is moving, in which direction.[4] They provide the fish with much of the sensory information it needs to modify its speed and direction from moment to moment. They alert the fish to patches of turbulence. They register alterations in water pressure when the fish nears an obstruction, allowing the fish to take evasive action. They assist the fish in its search for food by detecting the movement of nearby prey. And they warn of the approach of predators.

In the distant past, an infolding of the lateral line developed into the vestibular organ, the fish's inner ear.[5] We have no neuromasts in our skin, and when we are swimming, our awareness of the water is rudimentary compared to the sensitivity of a fish. Yet our hearing and balance depend on hair cells which evolved from the vestibular organs of ancestral fish.

◆ ◆ ◆

Our ears are designed to capture sound vibrations travelling through the air.[6] We hear sounds ranging from low-pitched rumblings vibrating at 16 Hertz (16 cycles per second) to high-pitched squeaks vibrating at 20,000 Hertz, but our ears are most sensitive to sounds vibrating between 1,000 and 3,000 Hertz, the familiar frequencies of human speech.[7]

Each ear canal acts like a hearing trumpet, funnelling airborne sound vibrations down to the taut membrane of the eardrum, which vibrates in response. A series of three tiny bones, the ossicles, convey the vibrations from the eardrum across the air-filled middle ear to a second taut membrane, the oval window, which vibrates in turn.[8] The ossicles act as levers, amplifying the airborne vibrations sixtyfold.[9] Without this intensification, the vibrations would be rapidly damped out after entering the fluid-filled cochlea (Greek: snail), the spiral cavity of the inner ear, where the vibrations are decoded. Hair cells lie side by side in orderly sequence in contact with a membrane that oscillates in response to the vibrations[10] (high-frequency vibrations close to the oval window and low-frequency vibrations near the tip of the cochlea.)[11] The movement of the membrane tugs at the appropriate hair cells, stimulates the auditory nerve and allows us to hear the distinctive sounds.

Little airborne sound enters the water.[12] The energy of high-frequency vibrations is soon dissipated due to the density and viscosity of the water.[13] Our hearing suffers as soon as we get water in our ears because we lose the efficient capture and amplification of airborne sound. Water presses on the eardrums, preventing vibration. The ossicles cannot deliver vibrations to the oval window. To reach the inner ear, sound has to penetrate the dense bone of the skull, and most of the vibrations fade away. Despite the sensitivity of the cochlea, only degraded responses are triggered in the auditory nerve. Like children with middle-ear infection or glue ear, swimmers hear only muffled, low-pitched sounds.

On land, we can usually tell where sounds are coming from.[14] Sound travels in air at about 350 metres per second,[15] arriving slightly earlier in the ear turned towards the sound. Moreover, the head absorbs high-frequency vibrations, casting a 'sound shadow' on the opposite ear.[16] Sound travels four times faster in water, at over 1,400 metres per second,[17] arriving virtually simultaneously in both ears. When we are swimming, our hearing is so severely compromised that we cannot tell whether the muffled sounds are coming from the left or right, from ahead or behind us.[18] High-pitched sounds are lost to both ears, so there is no 'sound shadow'.

Surprisingly, swimmers rarely realise the extent of their sensory deprivation – although they are virtually deaf. In the calm waters of a public swimming pool, the unwonted quietness probably contributes to the feeling of relaxation prized by lane swimmers. In open water or the sea, however, such a marked loss of hearing may have serious consequences. Deafness reduces swimmers' awareness of their surroundings, cuts them off from their companions, and distances them from the warning shouts of onlookers or rescuers.

◆ ◆ ◆

For many fish, the sense of hearing is even more important than the sense of sight. Low-frequency sounds can travel hundreds of miles in water,[19] whereas light is largely scattered and absorbed within a couple of hundred feet. Sound vibrations travel through the thin bones of a fish's skull directly into the fluid-filled cavities of the vestibular organs, the fish's inner ears. On the walls of the cavities, patches of close-packed hair cells sit beneath thin plates of chalk called otoliths (Greek: ear stones).[20] Like the domes of jelly in the neuromasts of the fish's skin and lateral lines, the otoliths pick up the vibrations of waterborne sound and tug at the hair cells, triggering responses in the auditory nerve.

Most bony fish can hear low-pitched sounds with vibrations between 20 Hertz and 1,000 Hertz. Some species are sensitive to higher-pitched sounds, amplifying vibrations up to 5000 Hertz with the help of their gas-filled swim bladders: herring even have narrow extensions of the swim bladder ending in air-filled bubbles close to each inner ear.[21] Catfish are sensitive to vibrations up to 13,000 Hertz. They locate their prey at the bottom of muddy ponds thanks to a chain of small bones, the Weberian ossicles, between the swim bladder and the inner ear.[22] The survival value of acute hearing is clearly illustrated in shad, which are sensitive to ultrasound vibrations up to 180,000 Hertz. Their main predators, the dolphins, emit ultrasonic echo-locating clicks as they hunt. Shad detect the clicks and take avoiding action.[23]

Fish can determine whether waterborne sounds come from left or right, from above or below.[24] Sound vibrations push more forcibly upon the neuromasts in the lateral line nearer to the origin of the sound, while the hair cells on the walls of the vestibular organs give differing patterns of response depending on their alignment.

As well as sensing underwater sound, the vestibular organs register the position of the fish's head, turning movements in three dimensions, and changes in the fish's speed.[25] The fish maintains its equilibrium by linking these sensory inputs with the signals from neuromasts on its skin and lateral lines.[26]

◆ ◆ ◆

Our inner ears are embedded in the temporal bones at the base of the skull, with the cochlea beside the vestibular apparatus. Early anatomists, puzzled by the twisting canals and interlinked cavities of the inner ear, grouped them together as 'the labyrinth'[27] (after the subterranean tunnels where the Cretan Minotaur had his lair). Hair cells on the walls of small cavities act as gravity sensors, signalling every tilt of the head, every forward, backward, upward or downward movement, and every acceleration and deceleration. We sense rotary motion with the help of hair cells in the semicircular canals,[28] fluid-filled tunnels at right angles to each other in three planes, horizontal, transverse and sagittal (Latin: like an arrow, front to back). We rely on the wealth of sensory information provided by the vestibular apparatus to orient ourselves in space and to control our posture and balance.[29] The brain links vestibular responses with sensations from muscles, joints and the soles of the feet. The resulting vestibular postural reflexes help us to stabilise the position of the head and body and maintain our equilibrium when we walk, run, jump and turn.[30] By linking vestibular responses with visual sensations, we are able to fix our gaze on an object of interest while turning the head or travelling at speed.[31]

We normally spend our waking hours with the head erect. Our gravity and motion sensors work best when we are upright.[32] Indeed, our upright stance is so important that we can tell if we are standing even one degree off the vertical, and we constantly readjust our balance to return to a precisely vertical equilibrium. When we lie flat, it is usually in order to rest or to sleep.

The fish's body lies horizontal in the water, and its vestibular responses are fine-tuned for swimming. When we are swimming, however, our vestibular system is tip-tilted, so that our gravity and motion sensors are less accurate, and the interpretation of sensory input is less reliable. As a result, it is harder for swimmers to compensate when plunged into turbulent water or dashed by breaking waves. Their self-righting ability is at its weakest just when their lives depends on it.

On land, we are able to assume that the ground stays still beneath our feet, but when we are swimming in moving water, our motion sensors may be taxed to the limit, and beyond. Some people are more susceptible to motion sickness than others, but everyone develops motion sickness if the stimulation is severe and prolonged. Ballet dancers use the technique of 'spotting' to prevent disorientation and dizziness when performing pirouettes.[33] As they spin, they fix their gaze for as long as possible on one spot at eye level at the back of the auditorium, then they turn the head quickly and refix their eyes on the same spot. In rough water, swimmers cannot fix their gaze on a steady object. Their bodies are in constant motion, and everything is in motion around them. Even the horizon appears to move as swimmers are tossed up and down by the waves. The onset of motion sickness usually takes swimmers by surprise, adding to their apprehension and compounding their danger.[34]

Contestants in open-water swimming races often take anti-nausea medication to try to control the symptoms of motion sickness.[35] Even with medication, motion sickness is a major problem for the intrepid band of adventurers who attempt to swim the English Channel[36] and for the crews of the support boats that accompany them. Scuba-divers have to stay alert, so they usually avoid anti-nausea medication, with its sedative side effects. Yet for susceptible divers, nausea remains a nuisance – and a handicap. Some scuba-divers become seasick while still in the boat taking them to the dive site. Others suffer from motion sickness during the dive. Buoyancy vests, weight belts and full gas canisters strapped on their backs alter the distribution of their body weight and disturb their familiar sense of balance. Their eyes search in vain for orienting points of reference underwater. If nausea becomes severe enough to trigger vomiting, the diver must choose whether to remove the scuba mouthpiece or to vomit into it.[37] Neither choice is a good one.

Motion sickness, nausea and vomiting are unpleasant enough, but the sensation of vertigo is truly incapacitating. Vertigo may develop during descent if the diver's eardrum ruptures,[38] or it may affect a diver ascending too fast at the end of a dive,[39] a symptom of decompression sickness caused by a bubble of nitrogen in the labyrinth ('the staggers').[40] Divers have a vivid sensation of spinning and of the water whirling round them. Their eyes jerk back and forth uncontrollably. They cannot coordinate the movements of their limbs. And as they lose their equilibrium, fear turns to panic. An attack of vertigo is disorienting and frightening even when the victim is in bed at home. If vertigo develops underwater, the sudden loss of spatial awareness puts the diver's life in jeopardy.

◆ ◆ ◆

As well as neuromasts in their skin and lateral lines that respond to drag forces in the water, many species of fish have neuromasts that respond to changes in the electrical charge of the jelly that surrounds the hair cells. Scattered over the snouts of sharks and rays are deep, flask-shaped, mucus-filled pits – the ampullae of Lorenzini.[41] The pits act as tiny super-sensitive voltmeters, lined by cells resistant to current flow except for the modified hair cells in their depths. Stefano Lorenzini described his ampullae in 1678, but their purpose remained a mystery until 1971, when Adrianus Kalmijn, the Dutch physiologist, demonstrated that sharks and rays use their sensitivity to electric fields to locate small plaice hidden under a layer of sand.[42]

Sharks can detect both direct and alternating current. Electrically charged acid and salt particles constantly leak through the skin into the surrounding seawater, creating a direct current which flows out in all directions. A flow of blood and body fluids from an injured fish (or a human swimmer) provide a trail of bioelectrical signals in the water. The movement of the gill arches during respiration, the contraction of the muscles and the beating of the heart create pulses of alternating current, helping a shark to home in on its prey. (We use electrocardiograms to record and display electrical signals from the heart.)

Catfish, sturgeon and lungfish also have electroreceptors in their skin,[43] with shorter pits and fewer receptive hair cells than the ampullae of Lorenzini of sharks. Some species of catfish appear to be sensitive to the electrical stresses that build up in rocks before an earthquake. In China and Japan, since ancient times, catfish have been used as living seismographs, their sudden bursts of activity taken as prior warning of an imminent catastrophe.[44]

Several species of fish have electric organs, modified muscle or nerve cells which act like batteries, capable of storing electric charge.[45] The South American knifefish hunts at night in weed-choked tropical rivers. They fire high-frequency low-voltage pulses into the water and use electroreceptors in their skin to find their way, to locate their prey by sensing distortions in the electric field around them, and to detect pulses of electricity produced by potential mates, rivals and predators.

A few species are armed with large electric organs capable of producing powerful discharges that stun or kill their prey.[46] In African rivers, the electric catfish can deliver a shock of 250 volts or more, stronger than the mains current in a domestic plug socket.[47] In South American waterways, electric eels can produce repeated shocks of 600 volts, lethal to anyone who gets too close.[48] Torpedo rays were known in the classical world as 'numbfish' due to their effect on the fishermen who touched them.[49] (Torpedo was their Roman name, from the Latin torpere: to be numb or paralysed. The Ancient Greeks called them narke, from which we derive our word narcosis.)

Torpedo rays are aggressive.[50] In 2006, a commercial diver was inspecting an oil pipeline off the Texas coast when he was stunned by the repeated electric discharges of a torpedo ray.[51] Luckily, he was linked to the surface by his air supply, by sound and by video. After he called for help, his sound contact went silent and his video showed a motionless seabed. A rescue diver brought him to the surface, still unconscious, with water seeping into his diving helmet. He recovered after treatment in a decompression chamber, but he was left with an irregular heartbeat. When his colleagues reviewed the video, they saw a large torpedo ray approaching him, followed by the crackle of four electrical discharges.

◆ ◆ ◆

Air is several billion times more resistant than water to the passage of an electric current, which probably explains why our bodies lack sense organs capable of detecting electrical signals.[52] In the water, our land-adapted senses of sight, touch, hearing, smell, taste, and temperature awareness are all compromised to a greater or lesser degree. Our vestibular sensations of position and balance are rapidly overwhelmed when we are pulled about by strong currents or tossed about by waves. Lifted by the buoyancy of the water, we lose the gravity-related sensations in our joints and the soles of our feet that make us aware of the position and movement of our limbs. In a heated indoor swimming pool, few swimmers notice the extent of their sensory loss, even though their sight is blurred and their ears are stopped with water. In open water or in the sea, swimmers need their senses on full alert. Unfortunately, they often have to rely on confused, inaccurate or degraded information from the sense organs that protected them perfectly adequately on land.

CHAPTER EIGHT

FLEXING THE MUSCLES

Muscles can pull, but they cannot push. They work by contracting, and once contracted they must be stretched out to their former length before they can work again. For us, this alternation of active pulling and passive stretching means that many of our muscles come in opposing pairs. We contract the biceps to bend the elbow; we contract the triceps to straighten it. We contract the hamstrings to bend the knee; we contract the quadriceps to straighten it. The contracting fibres usually run the whole length of the muscle. In the thigh, they may be 30cm long.

We maintain our upright posture, support the weight of the head, pull air into the lungs, operate tools, and move from place to place thanks to the coordinated interplay of hundreds of different muscles, each with particular tasks to fulfil. Walking involves more than fifty muscles in the back, buttocks, thighs, calves and feet.[1] Swimming involves very many more.

◆ ◆ ◆

The muscles of fish are arranged in a different way. The greater part of a fish's body consists of four blocks of muscle that run the entire length from head to tail, two on each side in balanced symmetry.[2] Under the lateral line on either side, a slender column of darker muscle, the lateral band, also runs the length of the fish.[3] All these muscles have a single function. They move the fish through water by bending its body first to one side, then to the other.

These blocks of muscle are divided into thin slices, called myomeres.[4] (We see the myomeres when a fish is cooked and the flesh separates into flakes.) Short, close-packed muscle fibres run between thin fibrous diaphragms that stretch across the fish's body from the vertebrae to the skin. These diaphragms are not flat, but shaped as backward-pointing cones,[5] so that each slice of muscle nestles into the next, stacked in tight formation. This arrangement allows powerful arrays of muscle fibres to coordinate their contractions.

When the fish is swimming, the backbone permits the fish's body to bend only to the right or to the left. On each side in turn, an orderly wave of muscle contraction begins behind the head and travels backwards to the tail, resulting in a forceful lateral displacement of the fish's body. Each myomere contracts for less than 50 milliseconds and then relaxes until its turn comes round again,[6] yet propulsion is continuous. A mackerel can bend its body from one side to the other as often as 170 times in a minute,[7] and every beat of its tail, to the right or to the left, pushes against the resisting water and propels the fish forwards.

In this efficient arrangement, when the muscles on the right side contract, the muscles on the left side, lying on the longer, outer curve of the fish's body, are passively stretched, readying them for contraction. Then, when the muscles on the left side contract, the muscles on the right side are stretched in their turn. This sequence can be repeated for as long as the fish needs to progress through the water.

At first sight, something similar happens when we swim. In crawl, while the right arm pulls backwards through the water, propelling the swimmer forward, the left arm is recovering its forward position. Then, as the left arm pulls in its turn, the right arm recovers. However, the recovery phase for each arm involves lifting the shoulder and arm clear of the water and swinging the arm forward through the air. These recovery movements play no part in propelling the swimmer through the water, but they greatly increase the swimmer's workload.

In breaststroke, the recovery phase is even more problematic. Granted, the arms pull strongly backwards just as the legs are getting into position for their kick, and as the legs are kicking, the arms are getting into position for their pull. Yet strong muscular action is needed to reposition the arms and legs after every pull and before every kick. Moreover, the swimmer's body is reasonably streamlined only during the glide phase of the stroke, whereas the repositioning movements are carried out under water, so that drag and turbulence raise the energy cost of moving through the water.

◆ ◆ ◆

For most of the time, fish swim at a leisurely pace, but when they need to escape from danger or capture prey, their muscles respond at once. A stationary trout can push its speed to 140 feet per second (95 mph) within a twentieth of a second.[8] Its 'fast-track acceleration' produces a thrust equal to four times its body weight as it darts away from an angler's shadow on the water. The trout tires after a few beats of its tail,[9] but its instant, brief reaction is enough to carry it to safety.

Game fish like marlin cover thousands of miles of open ocean while cruising at 5 miles per hour. In short bursts of energetic swimming, they can speed through the water at 50 miles per hour.[10] To withstand the forceful pull of their muscles, their vertebrae are reinforced with sturdy plates and spurs of bone.[11] For better streamlining, the dorsal fin folds down into a groove.[12] Near the tail, the body narrows into a strong fibrous shaft, the caudal peduncle,[13] which has a distinct keel on each side to stabilise water flow, like a spoiler on a racing car.[14] The tail fin is a wide crescent with a span of three feet or more, designed to push more water backwards (increasing thrust) while pushing less water to the side (minimising turbulence).[15]

A charging game fish has enormous kinetic energy. Numerous reports describe swordfish ramming the point of their swords through the planking of a boat.[16] In 2006, during an international tournament off Bermuda, a blue marlin leapt from the water and transfixed an unfortunate fisherman.[17] The marlin got away, and the fisherman recovered after emergency surgery.

◆ ◆ ◆

When we are walking, we rely on friction to hold each foot in place while we swing the body forward step by step, unimpeded by the air. But when we are swimming, we lose our firm footing. We are no longer helped forward by the friction between the feet and the ground. Instead, we are held back by the friction between the jutting contours of the body and the water.

We cannot move through the water until we learn the strokes. Beginners direct most of their efforts to keeping their heads above the surface. Self-taught youngsters arch their backs as they attempt front crawl, exhausting themselves by rapid, feeble flailing with their arms and ineffectual thrashing with their feet. They soon have to stop to recover their strength and catch their breath.

With coaching, swimmers learn to harness the powerful muscles of their upper chest and back to increase the pulling power of their arms.[18] They lie flatter in the water, improving their balance and decreasing drag. When swimming crawl, they present a narrower profile by rolling from side to side, instead of pushing full-fronted against the water. With increasing skill, they learn to stretch forward to catch hold of the water, haul on it, then press it towards their feet.[19] Their stroke rate slows, but each stroke takes them further.

As their technique improves, their speed increases. But the faster they swim, the greater the water resistance, so that swimmers have to put in a surprising amount of effort to travel rather slowly through the water. At the 2016 Olympic Games in Rio, Kyle Chalmers won the Gold Medal for Men's 100-metre Freestyle with a time of 47.58 seconds.[20] Thus, after years of single-minded preparation, rigorous drill and intense competition, in a thrilling, close-fought race that lasted less than a minute, he swam as fast as he possibly could – and briefly reached the speed of 4.7 miles per hour, the speed of a brisk walk.

Meanwhile, on the running track, Usain Bolt won the Gold Medal for the Men's 100-metre Sprint in 9.81 seconds,[21] travelling almost five times as fast as the Gold Medal swimmer. Olympic champions are at the peak of their strength and skill, but while Olympic runners cope only with air resistance, Olympic swimmers have to battle against the much greater resistance of the water.

◆ ◆ ◆

Whatever their level of skill, swimmers can practise their strokes in the warmed water of their local swimming pool. However, swimming in cold open water may trigger a reflex that protects muscles from overexertion. Swimmers suddenly find themselves in the grip of cramp.[22]

Athletes in a number of sporting disciplines suffer from cramp, which seizes the most active muscles during strenuous exercise. Cramp strikes when muscles are contracted, by pulling them even shorter. For runners, the onset of cramp is a disappointment and an embarrassment, rather than a danger. They only have to hobble to the side of the track and work through a set of stretching exercises. Once the offending muscle regains its full length, the pain settles.

Cramp is more frequent in swimmers, and more threatening. The agonising, rigid spasm makes swimming impossible. And by incapacitating swimmers, it threatens their safety. The first sign of cramp may be limited to a few fibres in a powerful muscle in the calf, the gastrocnemius. When we walk, the gastrocnemius contracts in a brief, forceful action that lifts the heel from the ground.[23] Then, as the leg swings forward, the muscle relaxes. In the water, we contract the gastrocnemius to point the toes. But when swimming crawl, we hold the toes pointed throughout the stroke, making the gastrocnemius particularly prone to cramp.

A change to breaststroke is sometimes enough to relax the muscle and prevent the cramp from spreading, but it is usually wiser to get out of the water. If the cramp persists and the swimmer cannot reach the land, the affected muscle must be rested and stretched until the spasm subsides. (The swimmer must float on the back, stick the affected leg into the air, grab the toes and pull them downwards, while massaging the knotted calf muscle with the other hand.) As soon as the cramp subsides, the swimmer should skull towards the shore.

If the swimmer decides to swim on regardless, the pain will get worse. As the cramp spreads and more muscles go into spasm, the swimmer is likely to panic. Yet panic is even more dangerous than cramp. Montague Holbein, a leading open-water swimmer in the early years of the last century, commented about the problem of cramp: 'Losing presence of mind has undoubtedly been responsible for more deaths than the actual seizure.'[24]

◆ ◆ ◆

Fish press the extensive surfaces of their bodies and their fins against the water, making use of water resistance to move through the water. Human swimmers can only press with the palms of their hands, the soles of their feet, and the narrow leading edges of their arms and legs. Frogs, ducks and otters have webbed feet to extent the working surface. The best that human swimmers can do is to cup their fingers together.

Leonardo da Vinci left us a drawing of a rather impractical webbed glove,[25] and in the eighteenth century, Benjamin Franklin, the American statesman, inventor and swimming enthusiast, made an attempt to increase the working area of his hands by using wooden paddles ten inches long:[26] 'I pushed the edges of these forward and I struck the water with the flat surfaces as I drew them back.' He met with only partial success: 'I remember I swam faster with the use of these palettes but they fatigued my wrists.'

Franklin also tried out unwieldy wooden flippers fixed to the soles of his shoes, but he was swimming breaststroke,[27] and he found them unsatisfactory: 'I observed that the stroke is partly given by the inside of the feet and ankles and not entirely with the soles of the feet.' Flippers proved their worth only after the introduction of front crawl. Competitive swimmers use short swim fins during training to increase ankle flexibility and build their kick strength. Snorkellers and scuba-divers use flippers to push themselves through the water, while free divers use extra-long fin blades or tuck both feet into a wider 'monofin.'[28] (Since flippers hold their feet extended, scuba-divers risk developing cramp if they overexert themselves in the chilly water deep below the surface.)[29]

◆ ◆ ◆

In a swimming pool, the swimmer pushes against the water. In a river, the water pushes against the swimmer. The river may look placid enough, but faster currents will be flowing beneath the surface. Apart from the brief intervals of slack water at the turning of the tide, the River Thames flows through London faster than 5 miles per hour.[30] Many other rivers in Britain flow as fast or faster, especially after rain. Swimmers who are caught in a current lose much of their power to control their progress through the water. At best, they have to swim harder and longer than they planned. At worst, they are swept downriver, unable to propel themselves towards the bank. All swimmers are at risk, however expert their technique.

At the seaside, too many people forget the power of moving water, which changes from moment to moment depending on the state of the tide, the phase of the moon, the contour of the beach, the force of the wind, or the presence of a storm over the horizon. Surging waves are stronger than the strongest swimmers. And on every surfing beach, there are rip currents that can catch unwary swimmers in their grip and carry them out to sea.

CHAPTER NINE

GOING THE DISTANCE

Whatever stroke a swimmer chooses, moving through the water involves the coordinated actions of many muscles. Swimming has been called 'the most nearly perfect exercise' partly because it is such hard work.[1] Yet too many swimmers fail to realise just how easy it is to become exhausted, especially when swimming in open water.

Our muscles are composed of two types of muscle fibre that work together in tandem – red muscle fibres and white muscle fibres.[2] Both red and white fibres are packed with glycogen, an insoluble form of glucose.[3] To provide the energy needed for muscle contraction, the stored glycogen is quickly converted to glucose.

Red muscle fibres are aerobic. They consume oxygen in a series of chemical reactions which break down all six high-energy carbon:hydrogen bonds in glucose.[4] The energy released is used for aerobic muscle contraction. Carbon dioxide and water are produced as by-products.

White muscle fibres are anaerobic. Enzymes allow the partial breakdown of glucose, releasing a fraction of its energy for anaerobic muscle contraction.[5] Oxygen is not required at the moment of contraction, but a short-term 'oxygen debt' is created, which soon has to be repaid, with interest.[6] Anaerobic contraction uses up large amounts of glucose, so that the stores of glycogen are soon depleted. Lactic acid and other potentially toxic substances are produced as by-products.

Red muscle fibres contain large amounts of myoglobin[7] (an oxygen-binding pigment) and many mitochondria[8] (the microscopic powerhouses where most of the energy is released). A dense network of capillaries supplies red muscle fibres with oxygenated blood. Red muscle fibres are capable of prolonged, steady activity. At moderate work rates, with adequate supplies of oxygen and glucose, red muscle fibres are capable of repeated muscle contractions, without tiring – for hours if need be.

White muscle fibres are thicker and stronger than aerobic red muscle fibres. They have fewer, smaller mitochondria and no myoglobin.[9] White muscle fibres are capable of immediate strenuous activity, but only for a limited time. Before long, the build-up of acidity slows anaerobic energy production, and muscle contractions become weaker.[10] Pain soon brings muscle activity to a halt. Anaerobic muscle fibres cannot regain their former power until lactic acid is removed and glycogen stores are replenished – the 'oxygen debt' has to be repaid.

◆ ◆ ◆

Some of our muscles contain more red muscle fibres than white, while others contain more white muscle fibres than red. The muscles supporting the spinal column have a greater proportion of red muscle fibres (making it easier to maintain low-intensity contractions throughout the day without fatigue), whereas the muscles of the arms have a greater proportion of white muscle fibres (favouring the brief forceful movements needed to lift heavy weights).[11] At the back of the calf, the soleus muscle lies deep to the gastrocnemius. These two muscles fuse to form the Achilles tendon, but their functions are subtly different. The soleus contains more red muscle fibres than white (suitable for exerting a constant pull on the heel when standing). In contrast, the gastrocnemius contains more white muscle fibres than red[12] (capable of the rapid forceful contractions needed for running and jumping).

Most of us have rather fewer red muscle fibres than white muscle fibres (a red:white ratio of 45:55 is usual).[13] Some people are born with a higher or lower proportion of red fibres than normal. Successful marathon runners are likely to be endowed with a greater proportion of aerobic red muscle fibres (a red:white ratio of 82:18),[14] allowing them to maintain their pace for 26 miles and 385 yards, until they breast the tape, tired and somewhat out of breath. Successful sprinters usually have a preponderance of anaerobic white muscle fibres (a red:white ratio of 37:63),[15] allowing them to dash across the finishing line, exhausted and gasping for air.

Swimming is an endurance sport. Elite swimmers, even those who compete in the shorter 50-metres or 100-metres races, have muscle fibre proportions more like marathon runners than sprinters. (A red:white ratio of 74:26 is common in competitive swimmers.)[16]

◆ ◆ ◆

In fish, red and white muscle fibres are segregated in separate blocks of muscle.[17] We can see the two types of muscle clearly demarcated whenever we eat a kipper. Two slender columns of red aerobic muscle, the lateral bands, adhere to the skin, tucked under four prominent fillets of white anaerobic muscle.

Fish live in an environment where the supply of oxygen is very restricted: a litre of water contains less than 10 millilitres of oxygen.[18] (In comparison, a litre of air contains 210 millilitres of oxygen.) Since aerobic red muscles cannot work without oxygen, the anaerobic white muscles play a crucial role. They allow fish to swim, albeit briefly, when the supply of oxygen is insufficient for their immediate needs.

Fish use aerobic red muscle for unhurried swimming.[19] Relatively little oxygen is needed by a pike lurking in the reeds or a trout holding its station beside a riverbank. The large blocks of anaerobic white muscle are held in reserve,[20] only used for the rapid, forceful swimming needed when the pike drives forward to capture its prey or when the trout senses danger and darts away. Once the fish catches its prey or escapes from the predator, it returns to swimming at a gentler pace, using only the two narrow lateral bands of aerobic red muscle. The ability to unleash powerful anaerobic muscular contractions at a moment's notice is essential for a fish's survival. Otherwise, the fish would starve, or it would be eaten.

Some marine fish have found ways to deal with the high energy cost of swimming long distances at high speed. Tuna have more red muscle than other fish, and during sustained rapid cruising, they satisfy their high demand for oxygen by swimming with their mouths open, ramming the water over their gills.[21] Atlantic cod switch between brief bursts of anaerobic white muscle contractions and longer recuperative glides (burst-and-coast swimming).[22]

Very few species swim at top speed for long. When a fish outstrips its oxygen supply and has to rely on anaerobic muscle contraction, the glycogen stored in its white muscle fibres lasts a couple of minutes at most.[23] Yet a short spell of anaerobic swimming is often enough. A pike captures its prey within seconds, a trout darts to safety with a few swishes of its tail, and a dogfish can travel 600 metres fuelled by two minutes' worth of glycogen.[24]

Nevertheless, glycogen stores are rapidly depleted during anaerobic swimming, and the fish tires rapidly as its supply of glucose runs out, making it extremely vulnerable to predator attack. Several hours may be needed for the removal of lactic acid from the fish's muscles,[25] for the renewal of its depleted glycogen stores,[26] and for the capture of enough extra oxygen to repay its 'oxygen debt'.[27] Some fish fail to recover if they swim to exhaustion. In a much-cited study, Canadian scientists recorded a 40% death rate in rainbow trout after only six minutes of intensive exercise.[28]

◆ ◆ ◆

Compared to fish, we have easier access to oxygen, but the harder we work our muscles, the more oxygen the red fibres consume, and the more lactic acid the white fibres produce. Elite swimmers are particularly good at supplying oxygen to their muscles. In vigorous training sessions, they prepare themselves for the intense muscular activity of racing. They suck in more air than untrained swimmers and empty their lungs faster, ready for the next intake of air.[29] Their hearts pump more blood to their lungs and muscles with every beat.[30] Their muscles become stronger, and the mitochondria grow in size and number. As they push their speed to the limit in repeated short sets, lactic acid becomes a potential source of energy rather than a dead-end by-product of anaerobic contraction.[31] In a so-called 'lactate shuttle', some of the lactic acid produced by white muscle fibres enters the enlarged mitochondria of red muscle fibres. Then, with the help of oxygen, the remaining high-energy carbon:hydrogen bonds in lactic acid are broken down, releasing further energy for muscle contraction.[32]

The aerobic and anaerobic fitness of athletes can be assessed during training by measuring the volume and composition of the air they inhale and exhale, and by recording the level of lactic acid (or lactate) in repeated samples of blood. As the pace of the exercise gets faster, an athlete's oxygen intake per minute increases until it reaches a plateau – the maximal oxygen uptake (commonly called VO2 max).[33] The level of lactate in the athlete's blood rises slowly at first, but before VO2 max is reached, larger amounts of lactate accumulate in the blood. The point where the slow rise of blood lactate changes to a rapid rise is taken as the athlete's lactate threshold.[34] Elite swimmers have both a high VO2 max and a high lactate threshold.[35] But they know that once they dive from the starting blocks, their oxygen intake must keep pace with the demand by their muscles. Incurring a hefty oxygen debt is only worthwhile in the last few strokes of the race.

An average swimmer has a much lower VO2 max. In open water, he is likely to choose breaststroke rather than crawl. Yet sports scientists have shown that a swimmer uses more oxygen when swimming breaststroke than crawl,[36] and much more when holding his head above the surface throughout the stroke instead of placing his face in the water during the glide phase.[37] At rest, the body contains around 2 litres of oxygen ready for use – 500ml in the lungs, almost 1,000ml loosely bound to the haemoglobin in the blood, 300ml held by myoglobin in the muscles, and 250ml dissolved in the tissues. During strenuous swimming, all of this available oxygen is consumed by aerobic muscle contraction within a couple of minutes.[38]

Unlike the red:white ratio of 74:26 in the muscles of an elite swimmer, an average swimmer is likely to have a red:white ratio of 45:55. Whether he uses breaststroke or crawl, the muscles in his arms provide most of the power for his stroke,[39] and they contain a high proportion of white muscle fibres. Practice in his local swimming pool may have improved his strength and endurance, but he has a low lactate threshold and his muscles are not capable of using lactic acid as a source of energy. Since lactic acid is released during anaerobic contraction,[40] he will feel a painful burning sensation in the muscles of his arms as soon as their efforts outstrip their oxygen supply.[41] The gastrocnemius muscles in his calves also contain a high proportion of white muscle fibres, and they soon tire if they are starved of oxygen. As his legs sink lower in the water, drag holds him back, hampering his flagging efforts to reach safety.[42]

In open water, a swimmer is at risk as soon as his demand for energy outpaces his supply of oxygen. Once he begins to feel short of breath, he has already built up a sizeable 'oxygen debt'. Panting, then gasping, straining to stay at the surface, he must repay at least some of his 'oxygen debt' before he can continue to swim. To recover, he must repay his 'oxygen debt' in full, which means that, over the next hour or so, he must capture around 11.5 litres of extra oxygen over and above his continuing moment-to-moment oxygen usage.[43]

Anaerobic muscle contraction soon empties his glycogen stores, and the swimmer tires quickly as the level of glucose in his bloodstream falls.[44] With his physical exhaustion comes mental exhaustion. He cannot think clearly when his brain is starved of glucose.[45] But if he panics, his frantic struggling uses more of the oxygen and glucose that remain in his bloodstream, increasing his exhaustion and his plight.

Children have small, immature lungs, limiting their ability to supply their aerobic muscle fibres with oxygen. Moreover, their muscles do not contain enough glycogen for sustained anaerobic muscle contraction.[46] If they become breathless while swimming, exhaustion soon follows. Unable to swim to safety, unable even to keep themselves at the surface, children sink and drown with startling suddenness.

◆ ◆ ◆

Someone who feels tired on a country walk can sit on a wall, admire the view and eat a bar of chocolate. Someone who feels tired in a swimming pool can hold on to the gulley or climb out of the water and head for the changing room. But someone who tires when swimming in a lake has to go on swimming until he reaches the shore. Few people realise how taxing such a swim can be. Every year, swimmers lose their strength and disappear beneath the surface of idyllic stretches of open water.

CHAPTER TEN

FRESH WATER AND SALT WATER

We are mostly made of water. More exactly, we are mostly made of salty water. The complex chemical processes of life take place in cells which are bathed in blood and tissue fluid containing nine grams of salt per litre. The kidneys control the amount of water and salt lost in the urine from moment to moment, helping to maintain the homeostasis (Greek: 'staying the same') that allows our body tissues to function as they should.

We need to drink a minimum of 1,500 millilitres of water every day.[1] (Approximately 300ml of water evaporates from the skin, 400ml is carried out of the lungs on our breath, 100ml is emptied from the bowel, and the kidneys must produce at least 700ml of urine to wash out the toxic products of the body's chemical processes.)[2] If we drink more water than we need, the kidneys simply produce more urine. If we don't drink enough, the kidneys compensate at first, by producing a smaller volume of urine[3] (under the influence of the anti-diuretic hormone released by the pituitary gland). But we soon become dehydrated.

In hot weather, and during exercise, we lose salty water as sweat, and the lost fluid must be quickly replaced. Few swimmers realise that they sweat during an energetic swim. Professor Louise Burke of the Australian Institute of Sport gives the simple explanation: 'Because they are already wet, swimmers are generally unaware of their sweat losses in the pool.'[4] Inevitably, swimmers swallow small amounts of water as they swim, but not enough to replace the fluid lost in sweat. Researchers noted that members of the Australian National Swimming Team sweated out more liquid during their pre-competition practice sessions than they took in,[5] even though they were 'highly motivated to drink during training, it was compulsory to keep a drink bottle on the pool deck during sessions, and a supply of cool sports drinks was provided'. Yet even mild dehydration impairs sporting performance.[6] A mere 2% shortfall of body water lowers blood pressure and weakens muscle power. Without fluid replacement, every cell in the body is affected, including the cells of the brain.[7]

In triathlons, dehydration is often viewed as a problem that develops during the cycling and running phases of the race, when the competitor's overheated body is cooled by evaporation of the sweat that bathes the skin. In fact, dehydration begins much earlier, during the swimming phase of the race.[8] Sweating is triggered by heat sensors sited both in the skin and in the brain.[9] Although water extracts heat from the skin, the repeated contractions of the swimmer's muscles produce such large amounts of heat that the temperature of the blood rises, alerting the heat sensors in the brain, and stimulating an outpouring of sweat.

◆ ◆ ◆

Fish have the same concentration of salt in their blood as we have – nine grams of salt per litre.[10] They spend their entire lives immersed in water, but fresh water contains very little salt, whereas seawater is very salty. This means that freshwater fish and marine fish must maintain homeostasis in different ways.

Although fish have scaly waterproof skin, the extensive surface of their gills allows water to enter or leave the fish's body, bringing the processes of diffusion and osmosis into play. (Diffusion is a form of molecular mixing that tends to equalise the distribution of any substance dissolved in water. Osmosis is a particular form of diffusion that pulls water molecules from a dilute salt solution through the barrier of a cell membrane into a more concentrated salt solution.)[11]

The water of lakes and rivers has a salt concentration of less than one gram per litre. Because the concentration of salt is greater inside the body of a freshwater fish, some salt diffuses from the fish into the surrounding water, while the process of osmosis pulls a constant stream of water through the gills into the fish's body. To avoid bloating, freshwater fish offload excess water. They drink virtually nothing, while their well-developed kidneys produce large quantities of dilute urine.[12] They maintain the salt concentration of their blood by absorbing salt from their food, and by conserving salt during the formation of urine. In addition, the gills of freshwater fish are studded with specialised 'chloride cells', which capture or recapture salt from the water around the fish.[13]

In contrast to freshwater fish, marine fish are bathed in seawater with a salt concentration of thirty-five grams per litre.[14] Marine fish have no risk of becoming waterlogged. Quite the reverse. The process of osmosis constantly pulls water out of their bodies into the saltier sea.[15] To avoid dehydration, marine fish drink seawater (up to 40% of their body volume daily), absorbing a large amount of salt as they do so.[16] Their small kidneys produce very little urine. Some salt diffuses into their bodies, but the specialised 'chloride cells' on the gills of marine fish pump much larger amounts of salt back into the sea.[17]

Sharks have a different way of coping with the high salt content of seawater. Most fish convert the waste products of protein digestion into ammonia, which diffuses from their gills into the surrounding water.[18] Instead of ammonia, sharks produce urea, which they retain in their bodies. The high concentration of urea (a hundred times higher than the level found in human blood) counteracts the outward osmotic pull of seawater.[19] Sharks have no 'chloride cells' on their gills. Instead, a rectal gland gets rid of excess salt.[20]

Most fish are restricted to life either in fresh water or in the sea.[21] A pike would die in the sea, shrivelled by dehydration and unable to rid itself of excessive salt. A herring would die in a lake, its body bloated by the water absorbed through its gills. The bull shark is one of the few species able to swim from the sea into a river and back again.[22] While it remains in the sea, the bull shark has a high level of urea in its blood. When a bull shark moves into a river, its kidneys respond to the changed osmotic balance by expelling much of the retained urea.

◆ ◆ ◆

On dry land, we avoid the problems of osmotic bloating and dehydration that fish must overcome. When we enter the water for a short swim, the difference between the salt concentrations inside and outside our bodies scarcely matters. On longer swims, however, the difference becomes important, because our skin is not entirely waterproof, particularly after prolonged immersion.

When swimming in fresh water, the pull of osmosis draws water through the skin into the swimmer's body. The kidneys then produce more urine, ridding the body of excess water by a process called diuresis (Greek: 'flowing through'). Diuresis causes a swimmer no problems during brief sessions in a swimming pool, apart from the need to leave the pool to empty the bladder.

Some of the benefits of 'taking the waters' in Bath's Georgian heyday were probably due to diuresis.[23] Fashionable visitors sat for hours immersed up to the neck in warm water. Spa treatment was credited with curing many patients paralysed by lead poisoning (once common, due to contact with lead paint, lead piping, lead-glazed crockery and pewter tankards). The scientific basis of the claims was confirmed in 1986 when lead workers were studied as they sat in a spa bath for three hours, immersed chin-deep in water at 35ºC. Lead loss increased fourfold during the resulting diuresis.[24]

When swimming in the sea, the osmotic pull of the seawater draws water out of the swimmer's body. The longer the swim, the more water is lost through the skin and the more marked the resulting dehydration. On seaside holidays, children often become thirsty – and dehydrated – when they play in rock pools in the intertidal zone, where sun and wind speed the evaporation of seawater, leaving behind an extra salty brine.

Many shipwreck survivors complain of intense thirst,[25] but the age-old taboo against drinking seawater is well founded.[26] If a castaway drinks seawater in a misguided attempt to slake his thirst, he takes in thirty-five grams of salt with every litre. Since his kidneys cannot load his urine with more than twenty grams of salt per litre, his predicament becomes even more desperate.

◆ ◆ ◆

Whether a swimmer is immersed in fresh water or seawater, his body is affected by the pressure of the surrounding water: the hydrostatic pressure. Even while the swimmer is floating at the surface, water pressure acts as an 'anti-gravity suit', squeezing as much as 700ml of blood from his arms and legs into the major blood vessels in his chest.[27] Pressure receptors near the heart detect an apparent increase of blood volume, with the potential to overload the swimmer's circulation. To rid his body this 'excess' fluid, the pituitary gland stops releasing anti-diuretic hormone, and as a result, his kidneys produce an increased volume of dilute urine. This process is called immersion diuresis.[28] When immersed in cold water, the swimmer's body reacts with a protective constriction of the blood vessels, first in the skin, then in the arms and legs. This cold-induced narrowing of blood vessels compounds the squeezing effect due to hydrostatic pressure. The swimmer has cold diuresis in addition to immersion diuresis.[29]

Swimmers who take part in lengthy open-water races lose water due to sweating, immersion diuresis and cold diuresis. If they swim long distances in the sea, their bodies also lose water due to the osmotic pull of seawater. They usually protect themselves from dehydration by taking frequent drinks: experts advise an intake of about half a pint of liquid every quarter of an hour.[30]

Channel swimmers do much of their training in outdoor swimming pools or freshwater lakes, which leaves them unprepared for the full impact of the dehydration that develops over many hours of immersion in salt water. Even with a dedicated escort boat and their own backup team encouraging ad lib consumption of their favourite drinks, many Channel swimmers find it impossible to take in enough liquid to avoid a degree of dehydration.[31] Whatever their choice of refreshment for their arduous undertaking, they are unlikely to follow the example of Captain Webb, the first person to swim the Channel, who kept himself going throughout his twenty-two hours in the sea with 'cod-liver oil, beef tea, brandy, coffee and strong old ale'.[32]

◆ ◆ ◆

Immersion diuresis is a recognised nuisance for scuba-divers encased in neoprene wetsuits, who notice that they produce increased volumes of urine during submersion. The water becomes colder with depth, so that scuba-divers are affected by cold diuresis as well as immersion diuresis. A light-hearted comment in *Scuba Diving Magazine* divided scuba-divers into two schools: those who pee in their suits and those who lie about it.[33]

Immersion diuresis in scuba-divers is not so much a joke as a problem. Divers cannot top up their fluid intake once they are submerged. They often become dehydrated from the combined effect of sweating, immersion diuresis, cold diuresis, the osmotic pull of seawater, and the constant loss of water vapour from their lungs as they breathe dry air from the scuba mouthpiece.

Unfortunately, scuba-divers face a double risk if they develop dehydration. First, dehydration causes muscle weakness, dizziness and confusion, all especially dangerous at depth.[34] Second, dehydration thickens their blood, hampering the circulation and increasing the risk of decompression sickness while returning to the surface.[35] Divers need more time to breathe away the nitrogen that dissolved in their tissues at depth. Bubbles are more likely to form, and once formed, they are more likely to grow, blocking capillaries and causing blood clots.

Divers who take no liquid before a dive are certain to become dehydrated. They still become dehydrated if they drink a pre-dive cup of tea or coffee or a glass of cola or beer. These drinks are all diuretics, stimulating the production of urine. A better choice would be one of the sports drinks formulated to replace the water and salt lost in sweat. French researchers confirmed the protective effect of pre-dive saline-glucose drinks by using a Doppler apparatus to count decompression bubbles in the venous blood of military divers at the end of their dives.[36]

Some divers use a different approach. In 1962, scientists studying the effects of zero gravity during early manned space flight recognised that astronauts developed a marked diuresis similar to the immersion diuresis of swimmers.[37] Since the 1990s, astronauts have taken desmopressin, a synthetic version of anti-diuretic hormone, to control diuresis. In 2005, the American Naval Medical Research Institute confirmed that desmopressin prevents immersion diuresis in scuba-divers.[38]

If the survivor of a boating accident is held at the surface by his life jacket, immersion diuresis soon reduces his blood volume by half a litre or more, and as he waits for rescue, he becomes increasingly exhausted, chilled and dehydrated.[39] If rescuers haul him upright from the water – or more dramatically, if a helicopter winchman hoists him into the air – the pressure of the water surrounding his body is suddenly removed, and the unopposed force of gravity pulls blood from the major vessels in his chest into the veins of his chilled, dangling legs.[40] His reduced blood volume is suddenly inadequate. His blood pressure plummets. Too little blood reaches his brain. Too little blood flows through his coronary arteries. He may lose consciousness or suffer a cardiac arrest. Whenever possible, survivors of prolonged immersion are best lifted lying flat.

◆ ◆ ◆

To stay healthy and active, both fish and men must achieve homeostasis. Freshwater fish maintain their fluid balance while drinking very little water, whereas we need at least 1,500 ml of water every day. Marine fish replace water lost from their bodies by drinking seawater, but we cannot drink seawater because it contains too much salt. Prolonged, energetic swimming makes homeostasis difficult for us to achieve. We become dehydrated even while our bodies are immersed in water.

CHAPTER ELEVEN

CHILLED TO THE BONE

We are warm-blooded. Our vital organs are bathed in blood at a constant temperature of 37°C. Our bodies work at peak efficiency at this 'normal' temperature. The speed of the complex chemical processes of our cells is controlled by enzymes so temperature-sensitive that a change of only 2°C seriously threatens our health. If our temperature rises, we become feverish, exhausted, then delirious. If our temperature falls, we become chilled, sluggish, then unconscious.

We produce heat centrally, mainly in the liver, heart, brain, and muscle. We lose heat from the surface of the body. To keep the temperature steady at 37°C, heat production must be balanced by heat loss. A thermostat in the brain, the thermoregulatory centre, responds to two different alerting signals.[1] Heat-sensitive and cold-sensitive nerve endings in the skin register changes in the temperature of the world around us, while central receptors monitor changes in the temperature of the blood flowing through the brain.

We tailor the rate of heat loss by controlling the amount of blood flowing through the skin.[2] In hot weather or after energetic exercise, the blood vessels in the skin dilate, transporting heat from the body's core to the surface. Blood flow through the skin may rise to 4 litres per minute or more. In cold weather, the blood vessels in the skin constrict, conserving heat. Blood flow through the skin may fall to as little as 20 millilitres per minute.

On land, we can cope with all but the most extreme climates. In summer, we strip down to thin cottons. If we are still too hot, we sweat, losing heat as the sweat evaporates from the skin. In winter, we bundle up in heat-retaining clothes. If we are still too cold, we shiver, producing extra heat from muscle contraction. When the air around our bodies has a temperature between 26°C and 30°C, it is said to be 'thermoneutral' – we neither sweat nor shiver.[3]

Air is a poor conductor of heat, especially when it is trapped between layers of clothing. Even in winter, it is relatively easy for us to maintain a normal core temperature. When we enter the water, however, our land-based temperature controls are inadequate. Water extracts heat from the body very much faster than air.[4] For water to be 'thermoneutral', its temperature must lie between 35°C and 35.5°C[5] – very close to blood heat. Water below 35°C drains heat from the body, and the colder the water, the faster the heat is drained away. Moreover, the rate of heat loss is increased by movement, whether the swimmer moves through the water or water flows past the swimmer.

The water in public swimming pools is usually kept between 25°C and 29°C.[6] Swimmers notice the cold when they first enter the water, but the sensation soon fades. Nevertheless, within a quarter of an hour or so, unless they generate heat by vigorous swimming, they will lose so much body heat to the water that they start to shiver.

In 1973, junior members of a swimming club helped physiologists to investigate the cooling effect of water.[7] The test swim lasted forty minutes in water at 20.3°C. All 28 children had a normal core temperature before they entered the swimming pool, but by the end of their swim, 11 children had a core temperature below 36°C, including 5 with a core temperature below 35°C (indicating the onset of hypothermia). Several swimmers looked so cold that they were told to come out of the water early, including a shivering boy with a core temperature of 34.4°C (3°C lower than his temperature at the start). The researchers commented, 'Some of the falls in temperature were remarkably large, considering that the water temperature, 20.3°C, was well above the usual summer sea temperature around Britain, Canada, and the northern coasts of the United States.' They suspected that many children escape dangerous cooling only because they become uncomfortable enough to come ashore voluntarily or on instruction. They pointed out: 'Enforced prolongation of immersion after this point, as might happen if a child clinging to a float drifted away from shore, would clearly threaten life quickly.'

The temperature of open water is very much lower than the familiar water of public swimming pools. The further north, the colder the water. During late spring and early summer, when mountain streams are swollen with the run-off of melted snow and ice, the water may be near freezing even on the sunniest day. If ponds and lakes are fed by underground springs, they remain frigid all year round. Even in summer, the temperature of the sea round the British Isles rarely rises above 15°C. In winter it falls to 5°C. And in warmer climates, cold currents chill the sea beside the sun-baked beaches of popular holiday resorts.

When we swim in open water, our bodies lose heat to the water much faster than we can replace it, even if the blood supply to the skin is cut down to the minimum, and no matter how hard we exercise our muscles or how much we shiver. Swimmers usually assume that they will be able to swim as strongly in chilly open water as in a public swimming pool. This overconfidence may be fatal. They are likely to discover, too late, that they can swim only a fraction of the distance they found so easy in warmer water.[8] Few swimmers realise how quickly their bodies chill when they are immersed in water. Few recognise how dangerous cold water can be.

In an important study in 1969, researchers had an unexpected opportunity to witness the danger for themselves. The subjects were four fit young men, all claiming to be good swimmers, who set out to swim for twelve minutes in an outdoor swimming pool with water at 5°C.[9] They all failed to complete the swim. One man managed to swim for only one and a half minutes before he abruptly floundered and sank. He was unable to grab the side of the pool only a metre away and had to be hauled out of the water by the physiologists conducting the experiment. Within a minute of his rescue he was fully recovered, alert and cheerful. Yet a similar loss of power in open water would very likely have ended in his death by drowning.

◆ ◆ ◆

Fish are cold-blooded. Their body temperature depends on the temperature of the water around them. The water flowing past their skin and over their gills extracts the heat generated by their metabolic processes. Fish are adapted to cope with this constant heat loss.

Fish have cold receptors in the skin, the lateral line and the brain.[10] Many species respond to a temperature gradient by moving up and down the water table,[11] conserving energy in deeper, cooler water and swimming up to warmer water to catch their prey or to speed the digestion of their food. Some adapt their metabolism to suit changes in water temperature by switching between different versions of temperature-sensitive enzymes.[12] Others offset a slowed metabolic rate during winter by increasing the oxygen uptake in their muscles.[13]

A few species are able to capture and recycle some of the heat generated by muscle contraction.[14] In the great white shark[15] and the bluefin tuna,[16] arteries that carry cold, oxygenated blood towards the muscles are interlaced between veins that carry warm, de-oxygenated blood away from the muscles. This close grouping of blood vessels, called a rete mirabile (Latin: 'wonderful net'), acts as a compact countercurrent heat exchanger, transferring heat from the veins to the arteries, and maintaining the contracting muscles at a temperature 10°C higher than the surrounding ocean[17] – an important competitive advantage since warmer muscles are capable of faster, stronger contractions.

Great white sharks[18] and bluefin tunas[19] also have heat exchangers in the belly, trapping some of the heat produced by chemical processes in the liver, stomach and intestine. The extra warmth speeds their digestion, releasing large amounts of energy from the fat and flesh of their carnivorous diet. The bluefin tuna also has heat exchangers sited close to the skull, warming the brain and eyes, increasing the hunting success of these predatory fish.[20]

◆ ◆ ◆

For humans, sudden entry into water colder than 15°C triggers an extreme reaction, with drastic effects on circulation, breathing and the ability to swim – a response so disabling that it is known as 'cold shock'.[21] The sympathetic nervous system goes into overdrive, preparing for the emergency actions of fight or flight. Stress hormones, including adrenaline, are released into the bloodstream. The pulse rate and blood pressure increase dramatically. Even fit young men develop abnormal heart rhythms, especially if cold water jets into the nose.[22] In older swimmers with coronary artery disease, 'cold shock' may bring on a heart attack. Physiologists have suggested that these dangerous irregularities are due to the simultaneous but strongly conflicting stimulation of two sets of nerves that control the rhythm of the heart.[23] (Impulses from sympathetic nerves increase the heart rate, whereas impulses from parasympathetic nerves slow the heart rate). In a few unlucky individuals, the parasympathetic vagus nerve reacts so strongly that the heart stops beating altogether.

Cold shock triggers an involuntary 'gasp reflex' which takes the swimmer by surprise and which he is helpless to prevent.[24] If he is at the surface, the 'gasp reflex' forces him to suck two or three litres of air into his lungs. (If he falls, jumps, or dives into cold water, the momentum of his entry often carries him beneath the surface, so that the gasp reflex compels him to inhale water instead of air.) The sudden 'gasp reflex' is followed by several minutes of deep, rapid, uncontrollable panting. As much as 60 litres of air per minute may enter and leave the lungs (compared with the normal 6 litres per minute), yet the swimmer often feels short of air. Moreover, this energetic over-breathing may bring on dizziness and muscle spasms (tetany) due to the increased loss of carbon dioxide in the exhaled breath. Not surprisingly, some swimmers panic.

On land, most people can hold their breath for more than a minute. A swimmer suffering from cold shock can hold his breath for ten seconds at most, greatly increasing the risk of inhaling water, especially in the first few minutes, and especially when struggling in rough water. Such a loss of breath control endangers the swimmer. He cannot time his breathing to coincide with the correct moment in the stroke.[25] He loses the ease and efficiency of a practised semi-automatic skill. He strains to reach the air, upsetting his balance. His legs see-saw downwards, slowing his progress. Cold shock is so incapacitating that even skilled swimmers reach breaking point with little warning.

Recent research at the University of Portsmouth suggests that people who fall into cold water should 'Float First' for two or three minutes, spreadeagled on their backs, before attempting to swim to safety.[26] By allowing themselves time to regain control of their breathing and to recover from the worst effects of cold shock, they are less likely to inhale water.

After five minutes or so, cold shock grows less intense. Swimming becomes easier for a time, but the swimmer remains in danger because the water continues to extract heat from his body. The blood vessels in his skin shut down, and within ten minutes, his skin is as cold as the water. He becomes, literally, numb with cold.[27] Blood is routed away from his hands, and his fingers curl and splay apart. (Among Channel swimmers, this sign of chilling is known as The Claw.)[28]

Beneath his skin, a layer of fat acts as insulation, slowing the loss of heat from deeper tissues. Beneath the fat, his muscles act as a further layer of insulation, as long as he remains immobile.[29] If he tries to swim, however, the increased blood flow through his muscles carries heat from his body core closer to his skin. Although the muscle contractions produce some heat, the cold water extracts heat even faster than before.

The blood vessels that supply his arms and legs constrict, a reaction that conserves warmth in the vital organs of his body's core but speeds the cooling of his limbs. His muscles are starved of oxygen because the narrowed blood vessels deliver less blood to his muscles. Lactic acid accumulates in the muscles of his arms and legs, causing pain, fatigue and cramp. As his muscles chill, stiffen and weaken, he develops swimming failure. Then shivering disrupts his strokes.[30]

As water cools, there is an increase in its density,[31] viscosity,[32] and surface tension.[33] This means that the forces of drag are noticeably greater in cold water.[34] At the very time when cold is sapping his strength, the swimmer has to work harder to make progress through the water. If he becomes too tired to swim further, he will soon sink beneath the surface. In open water, more than half of those who drown are capable swimmers within three metres of a safe refuge.[35]

In his memoir *Chance Witness*, Matthew Parris gives a vivid description of his own almost-fatal experience of swimming failure:[36]

I walked alongside the Thames in front of County Hall. The tide was high. Whipped by a winter wind the water was rough. I saw a little boy and girl weeping and staring into the river. I asked them what was wrong. 'Our dog is drowning,' they said. It had scrambled over the parapet and tumbled in. The little dog's head was visible in the swell, swimming in circles, unable to find the steps. It had been in for twenty minutes. I could see it was floundering. I did a stupid thing. I threw off my suit (I only had two) and jumped into the Thames.

Do not ever do this. I lasted a couple of minutes – just long enough to reach the dog, (now swimming towards Westminster Bridge), start swimming towards some submerged steps – and collapse just short of them. Hands reached out and grabbed me. It was as if I were a puppet and someone had cut my strings.

Be warned, should you ever mix with very cold water: you do not fade gradually, allowing time to get out. First the tremendous shock of the cold hits you. Then for a while you are fine and swimming strongly, and you think, 'Amazing! I can do this.' Then – snap – your muscles go. There is no warning.

The River Thames may be dangerously cold, but the water in a lake is usually far colder. In autumn, as the water at the surface cools, it becomes more dense and sinks to the bottom. (The lake is said to 'turn over'.)[37] If the weather is cold enough, this cooling continues until the temperature of the entire lake falls to 4°C, when water reaches its maximum density. In many lakes, the temperature at the bottom remains close to 4°C throughout the year.

During the summer, only the surface layer of water is warmed by the sun. Warmer water is less dense, so it stays at the top. Beneath the surface layer, the water remains dense and cold. There is very little mixing between the upper warmer layer and the colder water below, except in a narrow transition zone called a thermocline, where the temperature of the water suddenly changes.[38]

In a summer heatwave, the surface layer of a lake may warm to 27°C, but beneath the surface layer, the water may be twenty degrees colder. The warmed surface layer lures many unwary swimmers into dangerously cold open water. In an official website, under the heading 'Mediterranean heat but Baltic water!'[39], a Ranger in the Cumbrian National Park warns tourists, 'You may think that it's warm enough to take a plunge, just by dangling in your hands and feet. What people don't realise is that there is a "thermocline barrier" which basically means temperature a foot below the surface is still extremely cold.'

Holidaymakers who decide to dive from their boat in the middle of a lake risk an encounter with water cold enough to render them powerless.[40] Even if swimmers keep to the shallow water at the edge of a lake, they create turbulence that brings cold water swirling up to enfold them. A gentle incline at the shore often angles down to a much colder drop-off. On an outing to Ullswater in September 2006, a fifteen-year-old boy was paddling in knee-deep water close to the shore when he stepped off a rocky ledge into frigid water eighteen feet deep. Two companions tried to save him, but in moments, all three sank into the depths and drowned.[41]

In late summer, the temperature gradients present in deep water are of great interest to anglers because many sport fish feed at the thermocline or just above it.[42] Yet scuba-divers are often taken by surprise by the rapid fall in water temperature as they descend through a thermocline.[43] The air supplied from their scuba becomes suddenly colder and harder to breathe.[44] The air in their buoyancy control vests cools and shrinks. Nitrogen bubbles in neoprene also shrink, so that their wetsuits lose buoyancy and much of their insulating power.[45] Their muscles may seize up in the cold. And the abrupt fall of water temperature in an ear canal can trigger vertigo.

The rapid change of water temperature at the thermocline also causes problems as divers ascend. As they move upwards into warmer water at the end of the dive, their abrupt increase of buoyancy can take them to the surface dangerously fast.[46]

◆ ◆ ◆

Channel swimmers are not permitted to wear wetsuits or head coverings, yet they tolerate immersion in cold seawater for fifteen hours or more as they battle across from Dover to Cap Gris-Nez.[47] Most Channel swimmers have a thick layer of fat beneath their skin, which acts as an efficient insulator and increases their buoyancy. And importantly, Channel swimmers become 'cold-hardened' by a lengthy training regime that involves regular swims in open-air swimming pools throughout the year.[48] (Tooting Bec Lido is a favourite venue for long-distance swimmers.)[49]

To protect Channel swimmers during their arduous swim, the crews of their escort boats watch out for early signs of exhaustion and hypothermia, colloquially known as 'the umbles'.[50] First, the swimmer *grumbles* as he complains of feeling cold. Next, he *fumbles* as his fingers curl and spread into The Claw. Then he *mumbles* as his thinking and speech slow down, and he *stumbles* as his swimming stroke becomes uncoordinated.

Many Channel swimmers have to abandon their attempted crossing, even as they near the French coast. On occasions, exhausted swimmers are so determined to swim on and complete the crossing that they must be overruled by their support team and lifted, protesting, onto the escort boat.[51] Despite rigorous training before their cross-Channel swim and close monitoring during their attempt, nine swimmers have died while trying to swim the English Channel.[52]

◆ ◆ ◆

Survivors of boating accidents who fall into the sea develop swimming failure within a few minutes. An American Coastguard instructor pointed out the vital importance of life jackets clearly and strongly: 'It is impossible to die from hypothermia in cold water unless you are wearing flotation, because without flotation – you won't live long enough to become hypothermic.'[53]

The survivor's life jacket helps to keep him afloat and gives some insulation. He may be able to slow the rate of heat loss slightly by adopting the Heat Escape Lessening Posture (HELP), with his legs bent at hip and knee, and his arms wrapped round his chest, protecting the pulse points where large arteries run close beneath the skin in his neck, armpits and groin.[54] But he will lose heat constantly if his head is uncovered, since the rich network of blood vessels in his scalp do not constrict on cooling. And even if he is wearing thick protective clothing, water constantly extracts heat from his body.

His land-adapted protective responses to the cold are gradually overwhelmed.[55] Floating in the sea for half an hour or so, he loses so much body heat that his core temperature starts to fall. Even a mild degree of hypothermia, with a core temperature of 35°C (only 2°C below normal), greatly lowers his chances of survival.[56] As his blood cools, it becomes more viscous, so that his heart has to work harder to pump blood around his body.[57] His vital organs grow colder, and metabolism slows in every tissue, including his brain. His thoughts become muddled and his memory impaired.

Shivering increases heat production fivefold,[58] but the heat generated by shivering is not enough to replace the heat extracted by the water. As his shivering intensifies, his energy reserves are depleted. He becomes too exhausted to keep his back turned toward the waves. Unless his life jacket is fitted with a splash guard, he will inhale seawater when a wave strikes his face.

If rescue is delayed and the survivor's core temperature falls below 32°C, his shivering stops and even the inadequate amount of heat produced by muscle contraction is lost.[59] He scarcely moves, he hardly breathes, his pupils dilate, and he sinks into a coma. By the time his core temperature has fallen to 28°C, he has become immobile and unrousable. The beating of his heart becomes slow, weak and erratic, easily tipped into the powerless writhing of ventricular fibrillation,[60] when each individual muscle fibre beats at its own rate. Now his heart no longer functions as a pump, and his blood ceases to circulate. At 24°C, stone cold, inert and pulseless, he is at the point of death.[61]

♦ ♦ ♦

Fresh water is most dense at 4°C. Since its density decreases as it cools from 4°C to 0°C, near-freezing water forms a layer at the water surface. When the water reaches freezing point at 0°C, it changes into ice.[62] Below the ice, water remains at 4°C. Freshwater fish are able to overwinter in the chilly but unfrozen water beneath the ice.[63] As the water cools around them, their metabolic activity gradually slows, and they enter into a state of torpor. They use little energy and survive without food for weeks on end.

In the polar regions, the high salt content of seawater lowers its freezing point from 0°C to minus 1.9°C, and as seawater cools, its density increases right up to its freezing point.[64] This means that the temperature of seawater beneath the sea ice in the Arctic and Antarctic is almost 6°C lower than the temperature of fresh water beneath the ice of a frozen lake.

Fish that live in this super-cooled water cannot rely on metabolic slowdown for survival. More dynamic adaptations are needed. Many species of fish living in the polar regions have developed natural antifreezes.[65] Specialised proteins and glycoproteins circulate in their blood and attach themselves to minute ice crystals as soon as they begin to form, preventing the ice seeds from growing. Without antifreeze to protect them, the fish would rapidly freeze solid.

The Antarctic icefish has antifreeze in its blood, but it has no red blood cells.[66] It was once thought that the absence of red cells was a helpful adaptation, allowing the fish's thinned blood to circulate more easily in an environment where cold increases viscosity. More recently, scientists have suggested that the icefish gains no advantage from the loss of its red cells, since its colourless blood can transport only one-tenth as much oxygen as other species of fish. To compensate for the lack of haemoglobin, the icefish has a large heart, wide blood vessels and a fast circulation. Further, the frigid polar seas contain more dissolved oxygen than warmer oceans, so that the icefish is able to capture enough oxygen for its metabolic needs from the chilly seawater flowing through its gills.

✦ ✦ ✦

There is a high risk of 'cold shock' when people enter icy water, whether by accident or on a sudden impulse. On New Year's Eve 2000, a young man was playing ice hockey with his friends on a Canadian lake. At 1.30am, he decided to swim between two holes cut in the ice only two metres apart. He jumped in, but he failed to resurface.[67] Firefighters recovered his body from waist-deep water close to the spot where he disappeared.

Ice-covered ponds have a compelling allure for unwary young children.[68] In our changeable British climate, the ice is rarely thick enough to support their weight. Children who fall through the ice are in the gravest peril unless they are rescued within minutes – and rescue is difficult and dangerous. Dogs also run onto frozen ponds, and if they fall through the ice, their owners often try to rescue them, ignoring the risk of falling through the ice themselves. To discourage dog owners from putting themselves in danger, the fire service accepts call-outs to rescue dogs from frozen ponds, treating such canine emergencies as useful training for the rescue of people.[69] Half of the people who lose their lives in frozen ponds were trying to rescue another person or a dog.[70]

When children fall into icy water, they cool far more rapidly than adults.[71] Their body surface is larger in proportion to their size, with a thinner layer of insulating fat beneath their skin, so that cold water extracts heat at a faster rate. The scalp and skull are thin, allowing rapid cooling of the surface of the brain. They often swallow large volumes of cold water. Life-threatening hypothermia may develop within minutes.

Paradoxically, this rapid cooling sometimes protects children from death,[72] just as pre-operative cooling protects patients during heart surgery.[73] A few children survive prolonged submersion in water colder than 15°C: the rapid slowing of metabolic activity in the brain allows them to live through what would otherwise have been fatal oxygen deprivation.[74] In contrast, children submerged for ten minutes in water above 15°C usually suffer irreversible brain damage.

In 1988, the *Journal of the American Medicine Association* carried the report of a two-year-old girl who survived after being submerged in frigid water for sixty-six minutes.[75] Her core temperature fell to 19°C. Her small body quickly entered a state of suspended animation. She received prolonged, skilful resuscitation at the scene and in hospital, and made a remarkable recovery, without serious brain damage.

Such a positive outcome is very much the exception after a long submersion. Nevertheless, attempts at resuscitation are always worthwhile, even when a drowning victim appears to be dead – icy cold, unresponsive, not breathing, pulseless, with fixed, dilated pupils. Rescuers should begin active cardiopulmonary resuscitation (CPR), and transfer the victim to a hospital intensive care unit as speedily as possible. Once warmth is returned to the tissues of the body, the victim's metabolic processes may gradually resume their normal functions. Indeed, so many apparently lifeless corpses have made a full recovery, that rescuers must remember the adage: 'No one should be pronounced dead until they are warm and dead.'[76]

CHAPTER TWELVE

CAPTURING OXYGEN

Robert Boyle was one of the founders of the Royal Society. In 1660, he described a series of experiments 'to satisfie ourselves in some measure, about the account upon which Respiration is so necessary to the Animals, that Nature hath furnish'd with Lungs.'[1] He put snails, insects, eels, birds or mice into a glass vessel connected to a vacuum pump, extracted the air, and recorded their few remaining minutes of life – to the last gasp.

A hundred years later, when Joseph Wright of Derby painted his arresting 'An Experiment on a Bird in the Air Pump',[2] scientists were still groping towards an understanding of respiration.[3] Robert Priestley is credited with discovering oxygen in 1774 by using a convex lens (a burning glass) to focus sunlight onto mercuric oxide in a closed retort,[4] but he did not realise that the novel gas he produced made up a fifth part of the air we breathe.

It was the French chemist Antoine-Laurent Lavoisier who gave oxygen its name,[5] recognised its presence in the air, and showed that when an equal amount of oxygen was consumed by the burning of a candle or the breathing of a guinea pig, an equal amount of heat was released. Lavoisier concluded that respiration must be a controlled version of combustion.[6] Without oxygen, the fire of life would be snuffed out.

Between 1789 and 1792, in the midst of the French Revolution, Lavoisier succeeded in measuring how much oxygen a man used while sitting, working and eating. Yet his renown as a chemist failed to save him from denunciation by Marat.[7] In 1794 the guillotine brought his studies to a sudden, savage end.

◆ ◆ ◆

Fish and men are alike in needing a constant supply of oxygen. Gills and lungs are oxygen extractors. Both are triumphs of design, although the designs are utterly different. Gills and lungs both provide a huge surface area within a limited space, making it possible for red cells in the blood to capture oxygen from the water or from the air. Red cells are packed with haemoglobin, a pigment that forms a loose, reversible combination with oxygen – to form oxy-haemoglobin.[8] So effective is the haemoglobin in binding oxygen that the red cells are capable of transporting thirty to a hundred times more oxygen than would be possible if oxygen merely dissolved in the blood.[9]

As the blood circulates, oxyhaemoglobin gives up its oxygen to any tissue that needs it. Oxygen is consumed in the mitochondria of every cell, allowing the release of the energy necessary to continue the chemical processes of life. The more active the tissues, the more oxygen is consumed by its cells. The brain, in particular, cannot function without a constant supply of oxygen.

◆ ◆ ◆

Oxygen makes up 21% of air, but little dissolves in water. Even in white-water rapids, when water and air tumble together down a mountain gorge, fresh water contains only 1% of oxygen.[10] Seawater contains even less. Since warm water holds less oxygen than cold water, the oxygen content of many tropical rivers is lower than 0.5%.[11] Fish could not survive without an efficient means of extracting oxygen from the water.

Water is sucked into a fish's mouth and pumped through narrow gaps between the gill arches at each side of the throat. Projecting from the gill arches are combs of long filaments[12] covered by thousands of delicate folds called lamellae (Latin: 'little leaves'). The filaments act as fine sieves, forcing the water into close contact with the lamellae.[13] A thin layer of cells is all that separates water flowing past the lamellae from blood flowing inside them – a working surface that may be sixty times greater than the surface area of the fish's body.[14]

All the blood pumped by the fish's heart surges straight to the gills.[15] In a classic countercurrent system, water flows outwards over the lamellae, while blood flows in the opposite direction within the lamellae.[16] At every point along the lamellae, oxygen in the water diffuses into the fish's blood, where it is captured and held as oxyhaemoglobin in the red cells. After less than half a second of contact, the blood leaving the fish's gills contains twenty-five times as much oxygen as the water that entered the gills, while the water leaving the gills contains hardly any oxygen at all.[17] The newly oxygenated blood travels on round the fish's body, unloading its cargo of oxygen to any tissue where oxygen has been consumed by metabolic activity. Then the partly de-oxygenated blood returns to the heart, to be pumped again through the gills.

When a fish is resting, its need for oxygen is low, but the more active the fish, the more oxygen it needs. When a pike launches an attack on an unwary perch, it increases its oxygen capture tenfold[18] by pumping more water over its gills and increasing the flow of blood through the gill lamellae.[19] Mackerel have many more lamellae on their gill filaments than the slow-moving toadfish, and twice as much haemoglobin in their blood.[20] And importantly, fish of all species can defer the immediate demand for oxygen by using powerful anaerobic white muscles during short bouts of active swimming, an arrangement that allows the fish extra time – after the brief flurry of activity is over – to capture vital oxygen from the water.[21]

Fish are cold-blooded, which means that their body temperature is the same as the temperature of the water they swim in. In warmer water, their metabolism becomes more active, and they need more oxygen. However, warm water contains less oxygen, so fish have to extract more oxygen from oxygen-depleted water. Oxygen levels are also low in the presence of decaying plants or blooms of algae, even when the water is cold. Fish trapped in warm or stagnant water tend to swim near the surface, extracting whatever oxygen has diffused from the air into the uppermost layer of water.[22] There may be forty times more oxygen in the air above an ornamental pond than in its water, and goldfish sometimes supplement their oxygen supply by gulping bubbles of air.

A number of freshwater species have evolved special adaptations (accessory respiratory organs) to extract supplementary oxygen from the air.[23] Electric eels fill their mouths with air and absorb oxygen through the roughened, vascular lining.[24] Siamese fighting fish pump air over thin bony plates above the first gill arch.[25] Weatherfish swallow air and absorb oxygen in the gut.[26] Bowfins in America and bichirs in Africa absorb oxygen from air trapped inside the swim bladder.[27]

Most fish with accessory respiratory organs remain in the water and continue to rely on their gills to capture the greater part of the oxygen that they need. A few species are able to leave the water briefly and move about on the land. While they are out of the water, they protect their gills from drying out on exposure to the air.[28] Mudskippers absorb oxygen through their skin and the lining of their mouths as they scuttle over roots in mangrove swamps. They keep their gills moist by holding water in enlarged gill chambers.[29] Walking catfish wait for rainy weather before setting off overland.[30] Vascular outgrowths above their gill arches absorb oxygen from the air, and their gill filaments are stiffened to hold their shape when the catfish leaves the water.[31]

Lungfish can breathe air: the swim bladder evolved into simple lungs.[32] Australian lungfish rely on their gills in well-oxygenated water, but in the stagnant water of drying ponds, they swim up to the surface and suck air into their lungs. African and South American lungfish depend on their lungs for oxygen capture, while their gills are reduced to small vestiges.[33]

Gills cannot function in air. They are designed to work in water, where every filament contributes to the enormous surface area for the capture of oxygen. As soon as gills lose the support of the water, their neat arrays of separate filaments collapse under their own weight,[34] like bladderwrack at low tide. Most of their working surface is lost. The filaments twist and stick together, falling into jumbled, sodden layers, glued together by mucus. The lamellae desiccate and congeal, soon becoming impermeable to oxygen. The capillaries shut down, interrupting blood flow. Water contains less than a twentieth of the oxygen in air, but most fish suffocate within minutes if they are taken out of the water.

◆ ◆ ◆

Our lungs are close-packed sponges of delicate air cells called alveoli (Latin: 'little cavities'), more than 300 million alveoli,[35] with a combined surface area of more than 70 square metres.[36] The alveoli lie deep inside the chest, separated from the air by the trachea and twenty or so divisions and subdivisions of branching, narrowing bronchi. To maintain our vital supply of oxygen, we need frequent changes of air. We take around twelve breaths every minute, so people who live to the age of three score years and ten will fill their lungs with air more than 440 million times.

We renew the air inside the lungs by expanding and contracting the chest cavity. To breathe in, we contract muscles in the diaphragm and between the ribs, pulling down the dome of the diaphragm, swinging out the ribs and breastbone, and sucking in about 500 millilitres of air.[37] To breathe out, we simply relax the muscles in the diaphragm and between the ribs, and elastic recoil returns the lungs to their unexpanded volume, effortlessly expelling 500 millilitres of air. This flow of 500 millilitres of air into and out of the chest is known as the tidal volume.

Children rely mainly on the downward movement of the diaphragm to pull air into their lungs. And they breathe faster. At rest, a five-year-old child takes about twenty breaths a minute, while a toddler may take thirty breaths a minute.[38] The younger the child, the smaller the lungs and the lower the tidal volume.

A thin film of moisture covers the oxygen-absorbing surfaces of the lungs. Specialised cells produce a detergent-like substance called surfactant, which lowers the surface tension within the alveoli.[39] Without surfactant, the attraction between water molecules would pull the alveolar walls inwards, causing them to shrink, collapse and stick together. The lungs would become stiff and unyielding, and adequate oxygen capture would become impossible. Surfactant allows the lungs to expand easily when we breathe in, so that relatively little muscular effort is needed to inhale. And, just as important, surfactant allows the alveoli to preserve their size and spherical shape when we breathe out, so that the lungs remain filled with air.[40]

Each breath brings air into close contact with our blood. A close-meshed network of tiny blood vessels, the pulmonary capillaries, wrap the alveoli (an estimated 19,000 miles of capillaries in total).[41] The right ventricle of the heart pumps blood into the pulmonary arteries, then through alveolar capillaries, where the haemoglobin in the red cells captures its full quota of oxygen.[42] The air we breathe contains 21% oxygen. In less than a second, only 13.4% oxygen remains in the alveolar air and the blood leaves the lungs with a haemoglobin oxygen saturation of 97%, its bright crimson colour indicating the high concentration of oxyhaemoglobin.[43]

The oxygenated blood does not continue round the body as it does in the simpler, slower circulation of the fish. Instead, it returns, via the pulmonary veins, to the left side of the heart, where the vigorous pumping of the left ventricle sends blood to every part of the body, most importantly to the brain. The left ventricle is more muscular than the right ventricle,[44] and the blood pressure in the aorta is much higher than the blood pressure in the pulmonary artery

(120/80 mm Hg compared to 25/10 mm Hg).[45] As the arterial blood circulates, the tissues extract the oxygen they need from the oxyhaemoglobin in the red cells. The oxygen saturation falls to 75%, and the colour of the blood gradually changes from crimson to dark purple-red. This normal 'de-oxygenated' venous blood then returns to the right side of the heart to be pumped again through the lungs for renewed oxygenation. Together the pulmonary blood vessels contain only 9% of the blood in our bodies,[46] yet the two sides of the heart pump in lockstep. Minute by minute, equal volumes of blood flow through the two sides of the heart.[47] When we are sitting quietly, the right ventricle pumps 5 litres of blood through the lungs every minute, while the left ventricle pumps 5 litres of blood round the rest of the body.[48] Since the body contains only 5 litres of blood in total, this means that the equivalent of the entire volume of blood in the circulation is pumped through the lungs, and round the body, every minute.[49]

When we take exercise, we need more oxygen. We breathe faster and deeper. Instead of taking twelve breaths per minute, we double or triple our breathing rate. Instead of a tidal volume of 500 millilitres, we pull in and force out 2 or 3 litres of air with every breath.[50] During vigorous exercise, air exchange rises dramatically from 6 litres of air per minute to 120 litres per minute or more.[51] Athletes achieve even higher rates of air exchange.

At the same time, the heart beats harder and faster. The right ventricle pumps more blood through widened alveolar capillaries in the lungs, while the left ventricle pumps more blood through the active muscles.[52] At rest, the heart pumps about 70 millilitres of blood with each beat: this stroke volume rises to 110 millilitres per beat during exercise.[53] At rest, the pulse rate is 70 beats per minute, rising to 120 beats per minute during exercise. Athletes in competition may increase their cardiac output sixfold.[54]

With training, athletes can draw even more air into the lungs, carry more oxygen round the body, and release more oxygen into the mitochondria of their oxygen-depleted muscles. An important measure of sporting fitness is the athlete's maximum oxygen uptake per minute per kilogram of body weight during intense exercise (the VO2 max).[55] Elite swimmers are particularly good at servicing the oxygen supply chain.[56] They have larger lungs and larger hearts than usual.[57] During races, the practised movements of the diaphragm and ribs force much greater volumes of air into and out of their lungs, much faster than normal.[58] Blood circulates faster through the alveolar capillaries, speeding the capture of oxygen.[59] Their muscles respond to years of intensive training by extracting oxygen more efficiently from the blood and by using lactic acid as an additional source of energy.[60]

Skilled swimmers are able to enjoy the relaxed awareness that they are breathing more deeply than normal, in rhythm with the stroke. However, athletes with a running or a cycling background who decide to train for a triathlon often struggle to master the breath control needed for swimming. Runners and cyclists can breathe in and out whenever they choose, but swimmers must time each inhalation for precisely the right moment in their stroke, and they must be quick about it, filling their lungs as soon as they break the surface, breathing out just before the next intake of air.

Most people are limited in their ability to capture enough oxygen for even a moderately vigorous swim. They cannot progress very far through the water before they find themselves out of breath. Their flagging muscles soon need more oxygen than they can capture, while their laboured breathing interferes with their efforts to swim. They have to take a breather. Yet treading water is almost as strenuous as continuing to swim. If they tire in open water, it is all too easy for swimmers to find themselves struggling for air.

◆ ◆ ◆

For oxygen capture, our lungs must draw in air and the pulmonary capillaries must transport blood. We have no store of oxygen in the body, apart from the partially de-oxygenated alveolar air in our lungs and small amounts of oxygen bound to haemoglobin in the blood and to myoglobin in the muscles – enough for a minute or two at most. Without the continued formation of oxyhaemoglobin, no oxygen can be carried to the brain or to the heart. The small amount of oxygen dissolved in water is enough to maintain the slower, cold-blooded metabolism of fish, but is insufficient to maintain human life. Besides, we cannot extract oxygen from water as fish can. Like the experimental animals in Robert Boyle's air pump, we need a constant supply of oxygen from the air. If the supply is cut off, the fire goes out. Only minutes remain before we die.

PART TWO

A CLOSER LOOK AT DROWNING ACCIDENTS

Drowning is not an instant event, but a process that begins when someone has to struggle to stay at the surface of the water. Non-swimmers and swimmers alike have to be able to reach the air. Once they sink below the surface, they lose their vital supply of oxygen. Unconsciousness soon follows. When they try to breathe, they draw water into their lungs instead of air. Inhaled water injures their lungs. Lack of oxygen injures their brain. Without early rescue and resuscitation, these injuries rapidly cause further injuries in an inescapable progression to certain death.

In the cinema, drowning is often portrayed as a drama of struggling, splashing and yelling.[1] Yet more than fifty years ago, an observant lifeguard collected evidence that people often sink and drown so quickly and silently that nearby family, friends and onlookers remain completely unaware of the tragedy taking place right beside them.[2] Dr Frank Pia worked for years as a senior lifeguard on New York's Orchard Beach,[3] a mile-long crescent of imported sand known as 'The Bronx Riviera'. Facing east on Long Island Sound, Orchard Beach is protected from the surf that pounds the Atlantic coast, making it relatively safe for swimmers. The shore slopes gently, but sandbars alternate with troughs on the seabed. Non-swimmers who step from a sandbar into a trough suddenly find themselves 'out of their depth' – and in immediate danger of drowning.

Frank Pia rigged up a cine camera in an observation tower and filmed real-life accident victims 'from the beginning of their difficulty, through its development, to their subsequent rescue.' He noticed marked differences in behaviour between swimmers and non-swimmers who were struggling to survive in the water. Swimmers could float, wave and call for help. They were able to keep their heads above water – for a limited time – but they could not get back to safety unaided. They needed to be rescued quickly, before they exhausted their strength and began to sink.

Non-swimmers did not know how to float. Instead of waving, they tried to stay at the surface by pressing their arms down on the water with frantic, ineffective paddling movements (the 'instinctive drowning response'). They rarely called for help because they had no breath to spare and no time to fill their lungs. Without immediate rescue, non-swimmers usually sank beneath the surface in less than a minute. Toddlers sank within twenty seconds.

The well-worn phrase 'Going down for the third time' implies two or three minutes, at least, to effect a rescue. It is still too little understood that young children may disappear under the surface in a moment. And once submerged, the water will silence their cries.

◆ ◆ ◆

At every age, more males drown than females.[4] Among children who are still too young to go to school, twice as many boys as girls die from drowning. This fatal bias becomes more marked as children grow to maturity. For boys, the drowning rate reaches its peak in their late teens and continues into their early twenties, even though the majority of those that drown know how to swim. For girls, in contrast, the preschool years are the most dangerous, and as girls get older, they are less likely to drown. The resulting statistics are sobering – ten times as many young men drown as young women.[5]

In Britain, a third of fatal drownings involve people over fifty years of age: twice as many men as women are victims of drowning.[6] In Australia, too, a third of fatal drowning accidents involve people over fifty years of age: three-quarters of these older victims are men.[7]

In the chapters that follow, we will consider where drowning accidents occur, who are most at risk, and why the danger is so often underestimated. Then we will follow a typical drowning accident, explaining the physical changes in the body of a young man from the beginning of his difficulty, through its development, to the moment when his heart stops beating.

CHAPTER THIRTEEN

DANGER IN THE GARDEN

Young children are fascinated by water, but when the water is close to their home, their fascination can lead to sudden tragedy.

GARDEN PONDS

Most garden ponds are artificial creations. A simple pond holds water in a moulded liner, edged by gravel or paving stones, enhanced and partly obscured by bordering foliage. More elaborate ponds have pump-aided waterfalls, underwater ledges to hold plant containers, and deeper central areas where fish swim among the water lilies.

Garden ponds attract young children. The dancing reflections intrigue them. They love to dabble their fingers in the water. They lean over the edge, peering through the waterweed, grabbing at bubbles or goldfish, stretching for floating toys. Yet if they lean too far, if they overbalance, no protective barrier separates them from the water. And a pond need not be deep to be deadly.

Children fall into ponds after slipping on wet flagstones or tripping on uneven brickwork. After rain, an eco-friendly edging of recycled railway sleepers becomes as lethal as a pirate's plank. If the water is covered in duckweed, or choked with aquatic plants, a child may mistake the surface of the pond for solid ground, and run straight into the water.

Ponds appear safer when they are surrounded by a low wall, perhaps built for adults to sit on. Yet they remain just as enticing. Even if children cannot blunder straight into the water, they can lie across the parapet, dip their arms into the water, lift their feet from the ground – and see-saw head first into the water. Or they can scramble up onto the wall, determined to walk around the pond, only to lose their balance and tumble in.

Sometimes children enter the water on purpose, to retrieve a ball or to explore a waterfall. They find it easy enough to get into the pond, but much more difficult to get out, even when the water is shallow. At the edge of the pond, overhanging plants obstruct their exit. Their feet slither on the plastic pool liner or sink into the mud and plant debris which collect at the bottom of the pond. Waterweed tangles round their ankles. Children have limited muscular strength and coordination. If they topple over, they struggle to regain their feet.

A carefully tended pond adds to the beauty of a garden. Yet any gain in appearance must be judged against the risk to children. In 2002, the *British Medical Journal* published a study warning that the number of children drowning in garden ponds had almost doubled in the previous decade:[1] 11 children drowned in garden ponds in 1988–1989, while 21 children drowned in garden ponds in 1998–1999. The researchers commented on this disturbing trend: 'The rise in the number of drownings in garden ponds may be due to an increase in the number of water features in gardens, perhaps as a result of popular garden programmes on television.'

Gardeners rarely consider their pond to be a hazard, especially if the water is too shallow to reach a child's knees. A small pond close to the house is usually left unprotected. Sometimes, the owner cordons off a larger, deeper pond in a secluded corner of the garden behind a sturdy palisade, but the installation of a fence does not remove all danger. Gardeners may be careful to lock the gate behind them, only for a visitor or a workman to leave it open. Visible horizontal rails are an invitation to climb over the fence, while ingenious and enterprising children find ways to squeeze between the slats or wriggle underneath, thereby defeating the best efforts to protect them. In Dover in 2009, a three-year-old boy managed to scale a six-foot garden wall by clambering onto a forgotten picnic table, a discarded trampoline and a stack of unused garden furniture – only to drown in the pond in the garden next door.[2]

A few safety-conscious gardeners fit rigid metal grilles across the surface of their ponds,[3] but grilles give no protection unless they are securely fixed on all sides, low enough to prevent a child from crawling underneath, high enough to stay clear of the water surface after heavy rain, and strong enough to bear the weight of a child who decides to jump up and down in the middle of it. Chicken wire is not sufficiently sturdy: it sags in the middle and lifts at the edges.[4] Netting placed across a pond may be enough to protect fish from a predatory heron, but it will not bear the weight of a child.[5] New netting droops like a flimsy hammock, while old netting gives way.

The Royal Society for the Prevention of Accidents (RoSPA) advises parents who have a pond in their garden to fill it in or convert it into a sandpit for a few years, until their children are at least six years old.[6] A well-known gardening expert built a five-foot fence of chestnut paling around the medieval fishpond in his garden. When he moved house, he decided against a pond, explaining, 'The children were worryingly small.'[7]

◆ ◆ ◆

Toddlers are just beginning to understand the world around them. They cannot imagine that water might drown them.[8] They cannot assess risk. Warnings mean little to them. As they grow, they become more active, adventurous and impetuous. They learn by doing, and they rarely sit still for long.[9] One moment they are settled happily, playing with their toys on the kitchen floor, and the next moment they are running the length of the garden. Young children want to explore their surroundings, but if they go off on their own, they meet dangers on the way.[10] Since young children cannot be expected to keep themselves safe, adults must protect them from harm.

When a garden has a pond in it, someone has to keep the child in sight, but unbroken supervision is not easy to achieve. Parents are busy with household tasks. A babysitter may cope well with a placid, obedient child, but fail to keep control of a tough-minded, energetic youngster. Grandparents underestimate a child's gain in mobility since the last visit. Friends and neighbours often have several children competing for their attention. When they see that the toddler is playing with an older playmate, they relax their vigilance. Yet the older child may lose interest in the game and leave the toddler alone. Or the younger child may tire of boisterous play and wander away.

The best and surest prevention of drowning accidents in young children is the constant friendly presence by a responsible adult. Parents must keep track of toddlers, watch what they are doing moment by moment, and stay close enough to grab them. Babysitters need clear instructions to do the same.

Parents need to be particularly vigilant when their children are playing in someone else's garden rather than their own.[11] In 2000, a study by the Department of Trade and Industry reported that 62 preschool children – 49 boys and 13 girls – drowned in garden ponds in the United Kingdom during the years 1992 to 1999: 53 were toddlers less than two years old. While 11 children drowned in ponds in their own gardens, 51 children drowned in garden ponds belonging to neighbours, relatives or friends, including 24 youngsters who 'drowned in a neighbour's pond after wandering away from their own home or the home of the people they were visiting.'[12] In 2001 in Warwickshire, a three-year-old girl attending a village playschool walked out of an open door, through an unbolted gate, across a car park, along a road, and into a nearby garden, where she drowned in an algae-covered pond.[13]

◆ ◆ ◆

If a toddler falls into a pond unobserved, he is likely to drown, even if he is 'within his depth'. On land he can stretch out his hands to break his fall, but the same protective gesture will not hold him at the water surface for a moment. The impetus of the fall will plunge him below the surface, and he will go on falling, a little more slowly, until he reaches the bottom, shocked, disoriented and helpless. The chill of the water often triggers a gasp reflex, forcing the child to take an inward breath – of water.[14] After the initial splash, there is no noise to alert rescuers.

Time passes before the child is missed. Then the search usually starts in the house. The minutes tick by. The search becomes increasingly frantic. By the time someone remembers the pond and finds the small body floating in the water, it is often too late to save the child's life. Since every second of delay makes survival less likely, it is best to check on the pond early in the search.

When a young child drowns in a garden pond, the fatal accident is usually due to a brief lapse of adult supervision. A distraught carer tells the paramedics who respond to the 999 call:

> 'I just went indoors for a minute, and when I came out he had disappeared.'[15]
> 'She must have opened the patio door.'[16]
> 'I thought he had gone upstairs to his bedroom.' [17]
> 'I was trying to settle the baby.'[18]

At an Inquest in Chester in 2002, the Coroner recorded a verdict of Accidental Death on a toddler who drowned in a garden pond. He told the court that he had to deal with similar deaths twice a year: 'What appears to be a very innocent ornament is so often a deadly trap to young children.'[19]

Over the course of four months in 2008, Leicester Royal Infirmary admitted five young children found floating unconscious in garden ponds.[20] Three of the children died. The Matron of the children's intensive care unit said, 'It is heartbreaking. All parents know that once children are crawling or walking they are into everything and naturally curious about everything with no idea of the dangers. This is why it is vitally important that parents take adequate precautions and never leave small children alone near water, even for a minute.'

WATER BUTTS AND BUCKETS

Every water container in the garden presents a potential danger to young children. In 1992–1999, 8 children were discovered drowned in water butts, in dustbins and planters holding a few inches of stagnant rainwater or in a small paddling pool left unemptied on a lawn.[21] In Winchester in 2007, a fifteen-month-old toddler came upon a makeshift water tank sunk into the ground by a previous occupant of the family home. He managed to remove the lid. Then he pitched head first into the water and drowned.[22] In Fife in 2016, a man who bred ornamental koi carp rented his house to a young family. Soon afterwards, the parents found their two-year-old twins drowned in a large commercial fish tank in a lean-to adjoining the house.[23]

A particular danger in America is the sturdy 5-gallon bucket, brought home as containers of food, paint or industrial products, then reused for soaking clothes, mopping floors or washing cars. Usually the victim is an active, inquisitive infant aged nine to twelve months who is learning to walk. He crawls over to the bucket unobserved, and holds onto it as he pulls himself upright. The weight of the water in the bucket makes it stable, so that it does not fall over. The baby is top-heavy, and the rim of the bucket is level with his midriff. As he leans into the bucket to investigate the soapy water, he tips head first into the bucket. He cannot lift himself out, no matter how hard he struggles. If he is not rescued immediately, he will drown, even if the bucket is almost empty.

A study by the Cook County Institute of Forensic Medicine, Chicago, reviewed all 49 drowning deaths between 1985 and 1989 in children younger than three years of age, and found that 12 toddlers, a quarter of the total, had drowned in 5-gallon buckets.[24] Prompted by this study, the US Consumer Product Safety Commission (CPSC) identified 'at least 67 drowning deaths in buckets during the years 1985 to 1987'[25] and issued a press release warning: 'Large Buckets Are Drowning Hazards For Young Children.'

Few 5-gallon buckets are used in British homes, but in 1996, a government report listed 15 deaths over twelve years involving children who drowned in buckets used to soak nappies or as fermentation containers for winemaking.[26]

A SWIMMING POOL IN THE GARDEN

For children, a swimming pool in the garden is the biggest draw of all. In America, especially in California, 'backyard pools' are common. And in California, the most frequent cause of accidental death in preschool children is drowning in a backyard pool.[27] In litigation after drowning accidents in America, the swimming pool is sometimes cited as an 'attractive nuisance', at once enticing and dangerous to a child.[28] Indeed, in 2001, an American economist calculated that a swimming pool in the garden poses one hundred times more risk to a child than a gun in the house.[29]

Relatively few families in Britain have a swimming pool in their garden, yet as early as 1982, a study in the *British Medical Journal* warned that during 1975–1979, 34 children drowned in domestic swimming pools before their seventh birthday, including 5 children who drowned while trespassing and 5 who drowned in the rainwater that had collected in the bottom of swimming pools emptied for the winter.[30]

The risk of drowning and near-drowning accidents is at its highest during the first six months after a family installs their own swimming pool or moves into a house with a swimming pool in the garden.[31] It is much safer if everyone in the family knows how to swim, including the youngest members, who are most at risk of drowning.[32] Early swimming lessons teach toddlers how to cope with a sudden immersion, hold their breath underwater, pull themselves up to the surface, roll over and float on their backs, and how to propel themselves to the side of the pool using a simplified dog-paddle.[33]

Inflated armbands and buoyancy vests help toddlers to keep their balance in shallow water, but children who need 'flotation' also need watchful parents close by. Blow-up rafts easily drift out of the shallows. If non-swimmers fall in, jump in, slip through the centre of an inflatable ring, or pedal a tricycle over the edge of the paved pool surround, they plummet to the bottom of the pool at once, without a sound and without a struggle.[34] Their chances of survival are slim unless someone nearby reacts immediately and pulls them out of the water. In America, three-quarters of drowned children are found in the water less than five minutes after they were seen alive and well by an adult responsible for their safety.[35]

Even when parents are in the water with their children, they need to stay within grabbing distance. It is all too easy for non-swimmers to slip beneath the surface while their parents' attention is momentarily focused elsewhere. In 2000, a couple were briefly distracted as they chatted together while standing close to their three-year-old son in the shallow end of a swimming pool. The father described his shock when he turned to find the boy flailing helplessly underwater: 'Another half minute and he could have died – within reach of my leg.'[36]

An American doctor described how his wife's quick reaction saved their three-year-old grandson from drowning.[37] The boy was starting to learn to swim, closely supervised in the shallow end of a pool, and wearing armbands to increase his buoyancy. After swimming, the family settled down to a picnic beside the swimming pool. Then, without his armbands, the boy ran to the pool. 'What happened next was both dramatic and enlightening,' wrote the doctor. 'He jumped feet first into five feet of water and literally sank like a lead weight straight to the bottom. No cries for help, no flailing arms, no nothing.' His wife understood the danger, dived in immediately, and pulled the boy up to the surface, gasping but unhurt.

A young woman recalled falling into the family's Oxfordshire swimming pool when she was five years old, before she learned to swim.[38] She had a vivid memory of gliding rapidly towards the bottom of the pool, head first, with a stream of bubbles emerging from her mouth and surging upwards past her cheeks. She was rescued by her father, who was standing beside her when she fell into the water, but she had a narrow escape because her next memory was of vomiting pints of water as she lay gasping for breath beside the pool.

◆ ◆ ◆

Whenever there are children in the water or near the pool, at least one adult should be on pool duty. If no adult is available to act as a lifeguard, the pool must be out of bounds. This is a difficult rule for children to follow, so great is the lure of the swimming pool. Australian researchers warned that 20% of drowning and near-drowning accidents take place in winter: 'On a wintry day, however uninviting a cold pool may look in adult eyes, to a toddler an exposed water hazard may be fatally seductive.'[39] Toddlers have even squeezed themselves through cat flaps in their eagerness to reach the swimming pool in the garden.[40]

The poolside is a relaxing place for adults, with their sunloungers and sundowners, and accidents happen when adults get absorbed in their books, their laptops, and their conversations on mobile phones,[41] or when they become so relaxed that they fall asleep beside the pool. It is important that an adult takes moment-to-moment responsibility for the safety of any children in the water. Merely 'keeping an eye on them' is far too casual. Children's lives may be lost if their appointed lifeguard fails to guard them.

Even after they have learned to swim, children must have constant adult supervision. Without spoiling their fun or stopping their play, safety rules must be explained to children, then strictly enforced. (The same rules must apply, with even closer supervision, when children bring friends home to swim with them.) No running along the edge. No pushing others into the water. No ducking. No play fights. No sham calls for help. No holding on to another child in the water because each child can easily pull the other beneath the surface.[42] No competitions in underwater swimming or breath-holding, because shallow water blackout strikes without warning.[43] No diving in, because domestic swimming pools are usually far too shallow for safe diving.[44] Children should never be allowed to swim alone or with only another child for company.[45] And as their strokes improve and they become more confident in the water, every child needs to understand that anyone can drown, however well they swim.

When children are invited to a party with a swimming pool in the garden, an attentive adult is needed for pool duty from first arrival to last departure.[46] The party-givers cannot act as efficient lifeguards when they are busy greeting the guests and organising food and games. So parents should consider going to the party too. Instead of merely delivering their children to the party and returning at home-time to pick them up, parents can take their turn at lifeguard duty beside the pool, helping their hosts to hold a successful party, and ensuring the safety of every child. In Dorset in 2004, a mother arrived to collect her five-year-old son from a birthday party to find him lying drowned at the bottom of the swimming pool.[47]

◆ ◆ ◆

Just as a safety gate on the landing stops a toddler from falling downstairs, a safety fence round a swimming pool blocks children from unsupervised access to the water. The protective benefits of swimming pool fences were recognised in Australia more than fifty years ago. Ian Scott, director of research at Kidsafe (formerly the Child Accident Prevention Foundation of Australia) pointed out that Australia was one of the first countries to construct significant numbers of domestic pools, and 'among the first to show what was to be a virtual epidemic of toddler drowning.'[48] In 1960 in northern Queensland, the local government of Mulgrave Shire introduced a stringent regulation that all swimming pools had to be surrounded by fences.[49] Over the next ten years, no children drowned in fenced pools. During the early 1970s, Canberra and the surrounding Australian Capital Territory also brought in swimming pool fencing regulations, which proved just as successful.[50]

Despite the evidence that fencing round swimming pools saved lives, attempts to introduce similar regulations elsewhere in Australia were contested and delayed for decades.[51] People who advocated laws compelling pool owners to fence their pools had heated arguments with those who considered that such laws infringed civil liberties, and insisted that parental supervision was enough. At a public meeting in Sydney in 1992, a bereaved mother who championed pool fencing was taunted with the cruel question, 'Where were you when your child drowned?'[52]

By 1998, there was clear evidence that a child who lives in a house with an unfenced swimming pool in the garden is four times more likely to drown than a child barred from access to the family swimming pool by a surrounding fence.[53] Further research confirmed that pools fenced on all four sides were twice as safe as pools which allowed even limited access from the house via a lockable door.[54] Fencing round domestic swimming pools is now mandatory in Australia, New Zealand and some American states. Nevertheless, many pool owners fail to comply with the law, and many local authorities fail to enforce it.[55] And in Britain, there is still no law requiring childproof fencing round family swimming pools.[56]

To be effective, a fence round a swimming pool should be at least four feet high, with a self-closing, self-latching gate that opens outwards so that a child cannot push the gate open.[57] Adults have a responsibility to check that the gate is securely closed whenever they leave the pool area. If the gate is left open, the fence is worse than useless. At an inquest in Hampshire in 2009, the foster father of a seventeen-month-old girl admitted that he failed to shut the gate after carrying out repairs on the swimming pool, although eight children were playing nearby. He had intended to return to the pool, then he became involved with tasks in the house and forgot about the gate – until he heard a commotion outside and saw his son running towards the swimming pool where the little girl was floating unconscious in the water. He tried to resuscitate her, but it was too late.[58]

Temporary above-ground pools are rarely surrounded by an adequate safety fence, and parents frequently decide to leave access steps in place, ready for a later swim. Children like to explore, and they don't wait for permission before they start to climb up the steps.[59] Even if parents remember to remove the steps, children are surprisingly adept at climbing up and over the side of the pool.[60] While visiting friends in Norwich in 2008, a two-year-old boy played in the garden beside an above-ground pool containing water three feet deep. His father removed the access steps, then he sat chatting over a cup of coffee, out of sight of the pool. A short time later, the boy was discovered lying unconscious in the water. He died in hospital after five days in intensive care.[61] A brick found near the pool may have acted as a stepping stone. In Southampton in 2011, a two-year-old boy drowned in a portable pool that his parents bought only the day before to replace a smaller, more accessible paddling pool.[62] In America between 2001 and 2009, 209 people drowned in portable swimming pools: 94% of the victims were less than five years old.[63]

◆ ◆ ◆

A rigid, securely fixed swimming pool cover acts as a protective barrier between a child and the water,[64] as long as the cover is strong enough to support the weight of the child and provided that a responsible adult always closes the cover before leaving the pool area. However, a number of children have drowned or suffered permanent brain damage when they swam under a partly open pool cover or walked across a flimsy cover that dipped beneath them and pitched them into the water.

Solar blankets are designed to float on the surface of the pool, trapping the warmth of the sun and preventing heat loss due to evaporation. They provide no protective barrier whatsoever.[65] Rather the reverse. Children who attempt to walk across the flimsy layer of bubble-wrap polythene are enfolded in a sinking plastic pit and held fast as the water pours in around them.

Electronic alarms can provide a helpful secondary layer of protection when installed in addition to fencing around the swimming pool. A perimeter alarm sounds when someone passes through an infrared beam beside the pool. An immersion detector sounds when someone jumps in, falls in or pushes off from the side for an unauthorised swim. A personal wristband alarm sounds when immersed in water. But alarms are useless unless they are in good working order – and switched on. Although they may give parents some warning that their child has entered the pool area or disturbed the water, they put no physical barrier between the child and the pool.[66] A child who sets off a perimeter alarm is already dangerously close to the water. A child who sets off an immersion detector or a wristband detector is already in the water. By the time rescuers respond to the alarm, the child may have already disappeared below the surface of the pool. Even in daylight, it can be difficult to spot a child lying motionless on the bottom. At dusk or after nightfall, it may be impossible to see anything, or anyone, underwater.

◆ ◆ ◆

It might be expected that the number of swimming pool drownings in Britain would rise in line with the increasing number of private swimming pools. Instead, the number of fatal drowning accidents has decreased. In 1988–1989, 26 children drowned: in 1998–1999, 9 children drowned.[67] And this welcome fall has continued – 2 children drowned in private swimming pools in 2005.[68]

The owners of domestic swimming pools have stepped up their efforts to keep their children safe by applying several layers of protection.[69] Fences and pool alarms decrease the likelihood of a child entering the water unobserved. Earlier and more focused swimming lessons give children a greater chance to surface and reach the side of the pool, and allow parents more time to react and rescue them. Life rings and 'shepherd's hooks' are kept on the poolside. Many parents have increased their lifesaving skills and learned the techniques of first aid and resuscitation, in case the worst should happen and a child is found submerged. Mobile phones allow faster contact with emergency services, and earlier arrival of paramedics. And importantly, there is increased awareness that constant vigilance is needed to prevent such avoidable and devastating accidents from destroying the lives of children and the happiness of the families involved.

CHAPTER FOURTEEN

OPEN WATER

While their children are young, parents have the task of watching over them constantly to keep them safe. As the years go by, parents gradually prepare their children to keep themselves safe. While encouraging their increasing independence, parents must warn children of possible dangers, including the danger of drowning.

Before children are allowed to venture out unsupervised, parents should explain, clearly and firmly, the particular dangers of swimming in open water. Warnings are useless if they are too vague to put children on their guard. Parents should not mince their words, although few are likely to express themselves as forcefully as the naval commander in Arthur Ransom's classic book *Swallows and Amazons*. In reply to a letter asking him to allow his four children to sail alone on a Cumbrian lake and camp out on an island, the absent father sent a telegram: 'BETTER DROWNED THAN DUFFERS IF NOT DUFFERS WONT DROWN'.[1] Since they were enterprising and disciplined, keen sailors and strong swimmers, his tongue-in-cheek message was meant to convey his confidence in them – he expected them to keep themselves safe and he would be disappointed in them if they didn't. They survived. But accidents happen. And not only to duffers. Over five years, 2009–2013, the Royal Society for the Prevention of Accidents (RoSPA) recorded 422 fatal drowning accidents in open water: 62 in ponds, 162 in lakes, 35 in reservoirs and 163 in canals.[2]

PONDS AND LAKES

A garden pond is a dangerously accessible drowning risk for toddlers. Rural ponds are natural adventure playgrounds for older children. They hide in woodland dens or chase each other along the water's edge, pretending that they have to fight for their lives, or swim for their lives. Inevitably, some children return home scratched or bruised, but fatal accidents are rare – except for drowning.

In 2000, three children were swinging from ropes tied to a tree overhanging a Lincolnshire pond when they lost their grip and plunged into the water. One youngster struggled to the bank and a friend pulled him to safety, but two girls disappeared below the surface.[3] A police helicopter carrying thermal imaging equipment had to be called in to locate their bodies. The pond, once the site of ironstone and gravel extraction, was shallow at its margin, sloping steeply to deep water. Only a week before the children died, a local man saw boys swinging on the ropes and tried to warn them of the danger: 'I said I hoped they could swim because if they fell in it's fifteen feet deep. They told me to mind my own business.'

Eight years later, a twelve-year-old boy drowned in the same pond after toppling from a flimsy raft made of polythene sheeting.[4] His friends ran to summon help, but difficult access delayed the arrival of the emergency rescue team.

Youngsters who swim confidently in the relatively warm water of their local swimming pool almost always overestimate their ability to swim in the ponds and lakes within reach of their homes. Yet open water is often cold enough to chill their muscles to a state of numbed powerlessness within minutes. In June 2010, three boys set off to swim 50 yards to an island in a Yorkshire lake. Halfway across, one of the boys shouted, 'I can't do it!' His friends tried to pull him to the shore, but he slipped from their grasp and sank.[5] He was the third teenager to drown in the lake in recent years.

Many lakes are fed by streams at one end and drained by streams at the other end. When currents run beneath the surface, an easy swim in one direction becomes a desperate struggle on the return journey. In times of drought, the water margins shrink back from their usual limits. Steep drop-offs into deep water lie closer to the water's edge, so that even paddling through the shallows is perilous for weak swimmers. And the water itself is not the only hazard. Children jump into shallow water and impale themselves on discarded bicycles. They dive into shallow water and dash their heads against rocks on the lake bed.

Stretches of open water are often choked with waterweed. If their legs become entangled in underwater fronds, young swimmers panic and thrash about. They lack the mental control and the physical skill to free themselves and scull gently away from danger. In 2008, on a fishing trip to Knott Hill Nature Reserve near Manchester, two teenagers decided to ignore the warning signs and race each other across a disused reservoir. As they neared the opposite shore, they were caught up in a bed of reeds.[6] A friend could only watch, horrified, as they struggled and drowned.[7]

In winter, as the temperature falls towards zero, youngsters must be reminded to keep off the ice on frozen ponds.[8] In Stirlingshire in January 2001, two five-year-old boys were playing together near a duck pond only 400 yards from their homes. One of the boys told the fatal accident enquiry, 'It was a big bit of ice. I couldn't break it with my feet. So we went on.' His friend called out to the ducks, 'We're not going to hurt you.' Moments later, the ice gave way.[9]

◆ ◆ ◆

Young children fail to recognise the dangers of open water, but teenagers, particularly teenage boys, are often fully aware of the dangers and test themselves despite the risks. Well-judged parental advice is easily forgotten when a group of boys go off on their own. Peer pressure often overcomes common sense[10]. In February 2008, on the evening when their fifteen-year-old son drowned, his parents thought that he was playing computer games with four of his friends. Instead, they sat round a campfire drinking cider. Then they set off on two makeshift rafts to paddle on a lake in the dark. Both rafts capsized and only four of the boys managed to swim to the shore[11].

Adventurous, exuberant, physically active boys who enjoy playing outdoors with their friends are more liable to have accidents than their quieter, more timid classmates [12].Risk-taking increases when boys compete for the leadership of their group or when friends play truant from school[13]. Some impulsive, disobedient boys revel in breaking the rules and defying authority. They may have succeeded in getting themselves into and out of a series of scrapes and escapades before a misjudgement in the water ends in tragedy. Shock and regret overwhelm their families and friends when these vivid young lives are cut short.

◆ ◆ ◆

Young men who drown in open water are often strong swimmers who put themselves at risk on a sudden impulse. If alcohol is involved, they are more likely to take a chance, and they are less able to cope as they flounder and sink. In August 2010, at a wedding reception beside a Shropshire lake, the intoxicated best man decided to round off the evening with a solitary late-night swim. He failed to return[14]. After three days of searching, his body was found in the lake.

In London in March 2011, a first-year student at Imperial College told friends that once his exams were over, he planned to celebrate by jumping from a bridge across the nearby Serpentine. He was last seen alive as he left the students' union after an end-of-term drinking game.[15] Two weeks later, a woman riding a hired pedalo came upon his body in the Serpentine.

During every summer heatwave, people try to cool off by swimming in open water. They expect to emerge refreshed, but sudden immersion in cold water may trigger an immediate gasp reflex that leads to fatal water inhalation, and longer immersion in cold water may cause their muscles to seize up, so that they find themselves unable to swim a few yards to safety. In July 2013, a national newspaper reported 13 drowning deaths during a brief spell of hot weather.[16] A fit young sportsman posed for a photograph as he enjoyed the sunshine beside Cantref Reservoir in the Brecon Beacons. A few minutes later, he jumped into the water. He failed to resurface.[17] It was several hours before emergency agencies found his body at the bottom of the reservoir.

The Cumbrian lakes are intensely cold, even in midsummer.[18] In July 2004, a man stood on the Ullswater shore, watching his daughter and her friend as they played in their new inflatable rubber dinghy. He decided to join the fun, swam out to the dinghy, and began to push them along. His wife described him as a powerful swimmer, but within minutes he became so chilled that he lost his strength. Without a struggle, he slipped beneath the surface of the water, leaving the girls drifting in the dinghy. His body was found lying in the silt at the bottom of the lake.[19]

Kayakers and canoeists also face the dangers of cold shock and swimming failure if they fall into the frigid water. After his kayak capsized on Ullswater in June 2010, an eighteen-year-old man resurfaced and began to swim to safety, but his strength faded and he drowned before he reached the shore. He was not wearing a life jacket.[20] At the inquest, the coroner stressed the difficulty of getting back into a capsized canoe. He urged people going out on the lake to wear a life jacket as a safety precaution.[21]

Even a canoeist who succeeds in reaching the shore may suffer serious consequences. In January 2015, a businessman was thrown into the water when his canoe overturned on Ullswater. He appeared unhurt after swimming 50 yards to the shore, although he told his friend that he thought he had swallowed some water. Two hours after his immersion in the lake, he collapsed and died at the wheel of his car. A heart attack was suspected. At post-mortem, however, the pathologist found that inhaled water had severely damaged his lungs.[22] His death was recorded as due to secondary drowning.[23]

RESERVOIRS

Reservoirs look calm, safe and inviting. They are actually very deep, very cold and very dangerous.[24] Beneath the surface, systems of intakes and outflows divert volumes of water through unseen underwater culverts.[25] In 2010, a university student was swimming with friends in a Scottish reservoir when he was caught by a strong undercurrent, swept away and drowned.[26] In September 2014, seven teenagers were walking beside a Lancashire reservoir when they spotted a wooden raft tied to a jetty. As they jumped on board to investigate, the raft drifted away from its mooring. They tried to paddle back, but a current soon took them 50 yards from the water's edge. After a hour marooned in the middle of the reservoir, two boys decided to swim to shore to get help. They were totally unprepared for the coldness of the water. They were soon struggling to stay at the surface. One boy managed to return to the raft, so disabled by the cold that his friends had to haul him out of the water. The other boy sank and drowned.[27]

As water leaves a reservoir, sluice gates and spillways present a double danger. Upstream, the current propels leaves, branches and people towards the gate. Swimmers may be pinned against unyielding wood and metal, then pulled downwards into a 'trash catcher' grille guarding the exit flow. Downstream, the water surges down its spillway with the speed of a millrace, forceful enough to wash away anyone caught in its path. In 2009, while working on a small hydroelectric scheme, a Scottish landowner was trapped and drowned in a sluice draining water from a loch on his estate.[28]

United Utilities owns and operates more than 180 reservoirs in the north-west of England,[29] including Thirlmere in Cumbria, which supplies Manchester with much of its water. While enjoying a barbecue with friends on the banks of Thirlmere in July 2001, a young man decided to go for an evening swim. He failed to rejoin the party. Police and mountain rescue teams searched through the night. A Sea King helicopter was called in to sweep the area with searchlights. After two more days of searching, police divers found his body forty feet down on the bed of the lake.[30] He was only 20 yards from the water's edge. A spokesman for the water company said, 'We run information campaigns every year warning people not to swim in reservoirs, and yet every year people do and there are tragedies.'[31]

In 2011, after eight people had drowned in reservoirs in the north-west within four years, the water company attempted to raise awareness of the danger by running a campaign on Facebook that showed floral tributes left at the scene of a drowning tragedy.[32] In 2012, after the ninth death, the water company made a short dramatic video: *It's Not a Game*.[33] Aimed at teenagers, free-to-view on YouTube, the video begins with a youngster's casual decision to swim in a reservoir and ends with an underwater camera sequence as he drowns – game over and no second chance.[34]

On a visit to Thirlmere in 2013, a young woman and her brother entered the water close to one of the prominent warning notices on the lake shore:[35]

DANGER – KEEP OUT OF THE WATER
NO SWIMMING – BEWARE DEEP WATER –
BEWARE SUDDEN DROP
PEOPLE DIE IN RESERVOIRS

They swam out to an island 40 yards from the shore. Then, as they tried to swim back, they were overcome by the cold. A family picnicking beside the lake spotted the woman struggling in the water and managed to pull her into the shallows. By the time they lifted her from the water, she had stopped breathing. They kept her alive with mouth-to-mouth resuscitation until the air ambulance arrived, and she recovered in hospital. Her brother sank and drowned only a few yards from the water's edge. Recording a verdict of Accidental Death, the Coroner emphasised the danger of swimming in the Cumbrian lakes: 'The one particular message is that people must think very carefully before entering the lakes in the Lake District. They are all known to be extremely cold. Swimming in places like this can be fatal – even if you spend a short time in the water and are only swimming short distances.'[36]

In 2016, a safety campaign by Welsh Water warned about the danger of swimming in reservoirs and reminded people that 'No Swimming' notices are there to protect them. A video – *One Last Breath* – dealt with the deadly effect of jumping into cold water and the grief of the family and friends of two teenagers who drowned.[37] The campaign was supported by the mother of a teenage boy who drowned in Pontsticill Reservoir ten years before.[38] A friend of her son recalled the day he died: 'There were warning signs everywhere but we ignored them. The evening before, we were all in the reservoir. The scariest thing is that there were eleven of us swimming. As a fifteen-year-old, you just think that it wouldn't happen to me or to any of us. But it could have been any of us – or all of us.'

The Land Reform (Scotland) Act 2003 grants the 'Right to Roam', allowing everybody access to most of the land and to inland water throughout Scotland, including reservoirs. According to the Scottish Outdoor Access Code, landowners are expected to comply by 'not purposefully or unreasonably preventing, hindering, or deterring people from exercising access rights on and off paths and tracks,'[39] and they are forbidden from 'erecting a sign or notice worded in a way which intimidates or deters the public'.[40] To comply with the law, Scottish Water removed 'No Swimming' notices at reservoirs.[41] At Glen Mill Reservoir in Ayrshire in August 2014, a young angler decided to go for a swim at the end of a sunny day spent fishing. He was soon in difficulty. His friend jumped into the reservoir to save him. They both drowned.[42]

CANALS

Between 1760 and 1830, thousands of miles of canals were built in Britain, allowing barges to bring raw materials to factories and take finished goods to market. The water in a canal looks reassuringly placid, but these long, narrow stretches of man-made open water have their own special dangers for anyone walking beside a canal, cycling on the towpath, travelling along a canal in a narrowboat, or tying up at a wharf.

Most canals are only 3 to 4 feet deep beside their banks, deepening in the middle to 5 to 6 feet, where the passage of boats disturbs the mud on the bottom. A cyclist who skids on the towpath and falls into the water usually escapes with a ducking,[43] but canals are deep enough to drown children every year,[44] and too many people risk a fatal plunge as they stagger along a towpath in the dark after a visit to a canal-side pub.[45]

The early canals were wide, flat ditches a few miles long, lined with a layer of clay to prevent the water from trickling away into the soil. The introduction of locks allowed canals to climb up and down hills. By 1804, the 92 locks on the Rochdale Canal allowed barges to carry loads across the Pennines.[46] At Caen Hill in Wiltshire, a flight of 16 locks takes the Kennet and Avon Canal up a 1:30 incline, the steepest flight of locks in the world.[47]

A complex system of water management is at work to safeguard the water levels along each stretch of the canal. Most locks have a 'rise' of 6 feet or more. Every time a boat enters a lock, whether the lock is filling or emptying, whether the boat is moving uphill or downhill, water moves along the canal towards the valley – water which must be replaced to keep the canal in action. Huge amounts of water are involved, often pumped from holding reservoirs or from the lower levels of the canal itself. A typical wide lock on the Grand Union Canal holds 50,000 gallons of water,[48] and to supply the needs of summer boating, 4 million gallons of water is pumped into the canal every day as it crosses the Chiltern Hills at the Tring Summit.[49]

Working locks attract passers-by, but getting too close to the edge of the lock is risky, especially for children. Powerful flows of water enter and leave the locks. When a lock is filling, water rushes from the upper stretch of the canal through underground conduits low on the sides of the lock and through underwater sluices, called paddles, in the upper gates.[50] The swirling inflow would buffet the strongest swimmer. When a lock is emptying, the paddles are opened in the lower gates, and the outflowing current is powerful enough to suck people below the surface and hold them there. To add to the danger, most locks have a bypass waterway that flows through a 'trash catcher' grille and over a weir.

When a lock is full, with the water surface at the same level as the uphill stretch of the canal, the water is at least 9 feet deep, and often very much deeper (Tuel Lane Lock on the Rochdale Canal is almost 20 feet deep when full).[51] Non-swimmers who fall into the deep, cold water of a full lock stand little chance of survival unless they are rescued immediately.[52]

When the lock is empty, with the water surface at the same level as the downhill stretch of the canal, the water is little more than 3 feet deep, just enough to float boats clear of the base of the lock. But now there is a drop of at least 7 feet to the surface of the water. Anyone falling into an empty lock is likely to plunge beneath the surface, where they may collide with the brick floor of the lock or find themselves trapped in the water, unable to scale the lock's vertical walls.[53] Lock-keepers use long-handled rakes called kebs to clear debris from the canal – and to rescue people.[54] During the course of a single year, the lock-keeper at Little Wittenham in Oxfordshire pulled 17 people from the water after they fell into the lock.[55]

Canals are working waterways and it is never safe to swim in them, although thrill-seeking children are often willing to take the risk. In Wigan in 2005, a twelve-year-old boy drowned while playing on an inflatable dinghy in the canal.[56] Only a year later, Wigan youngsters were dicing with death in a new craze of 'canal surfing' – treading water at the bottom of a flight of locks while their pals cranked open lock gates to release a torrent of thousands of gallons of water.[57] To protect the boys from their risky games, security guards had to be appointed to patrol the flight during the school holidays. In 2007 the Wigan Fire Service backed a summer project running free canoeing and kayaking sessions to 'teach young people how to stay safe on the borough's canals', and to 'raise awareness of the dangers of open water swimming'.[58]

When railways and motorways took away their traffic, many canals fell into disrepair, but almost 2,000 miles of navigable canals still criss-cross the country. Canals carry fewer industrial payloads nowadays. Instead, narrowboats and cruisers carry people, many of them unused to boats and waterways, many of them careless about safety. Everything seems to be under control – until an accident happens.

Holidaymakers rarely wear life jackets while cruising on a canal, but children, at least, should wear life jackets at all times when aboard, even when the boat is tied up beside a mooring stage.[59] A narrowboat is a confining space for children, an ill-equipped adventure playground surrounded by water deep enough to drown them. When the boat is underway, children should not be allowed to sit on the roof, lean over the side, or inch their way along the gunwale. When the boat is moored, they should have a steadying parental hand to support them as they step from the boat onto the bank. Unsupervised children may fall overboard in a moment, disappearing under the water without a sound, their absence only noticed when it is too late to save them.[60]

◆ ◆ ◆

For much of the length of the canal, their hired narrowboat carries holidaymakers calmly along, slowing the pace of their busy lives. At locks, however, they need to pay attention. An Inland Waterways Association of Ireland website discussion on safe operation of locks included the informed comment, 'Locks are inherently dangerous; not dangerous only if people are careless or inexperienced or stupid or in too much of a hurry, though of course any of these conditions increases the danger.'[61]

When people hire a narrowboat or cruiser for a holiday boating trip, they are given brief instructions on how to cope with the locks.[62] It all sounds simple enough. When travelling upstream, enter an empty lock, close the lower gates, fill the lock with water, open the upper gates and travel onward. When travelling downstream, enter a full lock, close the upper gates, empty the lock, open the lower gates, and travel onward.

To operate the lock, someone on the boat has to climb onto the canal bank, swing the gate shut behind the narrowboat and wind open the paddles to fill or to empty the lock. Then, after the water has reached the correct level, the operator has to walk to the gate in front of the narrowboat and swing it open. These manoeuvres are carried out very close to the water's edge. Most people cope without incident, but some are not so lucky. A blow from a gate's balance beam, a jerk from a snagging rope, a slip on a muddy towpath, or a false step on the narrow walkway above the lock gates can send an unlucky holidaymaker plunging over the edge of the lock into the water.

To the rest of the party waiting on a narrowboat, nothing much appears to be happening as the lock fills or empties, but below the surface, thousands of gallons of water gush in or drain out within a few minutes. As the lock fills, the narrowboat may collide with the side of the lock or the downhill gate. On a family boating holiday in 2009, a woman was standing at the tiller when their narrowboat slammed into the gate. The impact caught her off balance. She fell over the handrail. The swirling water carried her under the boat, where she was fatally injured by the blades of the propeller.[63]

As water drains from the lock, the bow of a narrowboat sometimes 'hangs up' on the cill beneath the gates before nosediving into the water. In 1998, on the Leeds–Liverpool Canal, one of the rope fenders on front of a narrowboat snagged on the top of a gate, holding the bow suspended in mid-air. Then the boat splashed down into the lock and sank. Four people drowned,[64] trapped in the cabin.[65]

In October 2013, an elderly couple survived an accident which might have killed them both. The husband was navigating a narrowboat through a lock on a Birmingham canal while his wife operated the lock gates. He heard her call out, and saw her fall fifteen feet into the lock, where she lay unconscious, face down in the water. He jumped into the lock, swam over to her, and lifted her head above the surface. A passer-by helped him to pull her back onto the boat. His wife said, 'I remember falling and a splash, and the next thing I knew I was waking up in hospital.' Meanwhile, her 72-year-old husband suffered a heart attack brought on by exertion and fright. He, too, was admitted to hospital, where he recovered after successful surgery.[66]

DROWNING ACCIDENTS INVOLVING ANGLERS

During 2009–2013, 30 anglers drowned while fishing in lakes, reservoirs and canals.[67] Anglers frequently slide down the bank, tip off folding stools, slip from moving pontoons, or fall asleep while night-fishing and roll into the water.[68] Their thick layered clothes become waterlogged, impeding both their swimming and their exit from the water.[69] Anglers are at risk whatever their age. In 2007, a fishing expedition ended in tragedy when a nine-year-old boy waded into a lake to retrieve his only float. He was soon out of his depth, and despite rescue attempts by his brother and a friend, he sank and drowned.[70] In 2009, as a special treat, a 64-year-old angler took his grandson camping beside a fishing lake in the Fens. They pitched their tent only five feet from the water's edge. The boy woke in the night to find himself alone. He raised the alarm, but before help arrived, he found his grandfather floating lifeless in the lake.[71]

When fishing from a rowing boat, an angler usually pays more attention to his rod than his oars, especially when he has a fish on his line. Simply by standing up, he makes the boat less stable. He risks a capsize if he leans over the side while reeling in his catch or stretches too far while sweeping the fish into his net.[72] If the boat is unsecured, it may drift away, leaving the angler marooned, and a number of anglers have drowned as they waded into the water, trying to retrieve their escaping boat.[73] A careless jump when boarding or disembarking can push a small boat out from its mooring, so that the angler falls into the water between boat and bank. On a fishing trip with his father near Sandwich in August 2019, a six-year-old boy fell into the River Stour as he tried to jump from a jetty into the boat.[74] His father and two others immediately dived into the river in a rescue attempt, but he was swept away. His body was found four days later, after extensive searches in the river and along its banks.

Despite their risk of falling into the water, few anglers wear life jackets.[75] In 2002, an experienced fisherman took his two-year-old grandson fishing on Loch Awe in the Scottish Highlands. The boat was close to the shore, in calm conditions. They were not wearing life jackets, but when the little boy fell overboard, his grandfather jumped into the water to try and save him, and they both drowned.[76] In March 2011, two teenagers were fishing from an inflatable dinghy in the middle of a lake, without lifejackets. The dinghy sprang a leak and sank beneath them. Only one of youngsters managed to reach the shore.[77]

At the start of a fishing trip, the lake may be calm and the weather clear, but conditions often change with little warning. In Argyll in March 2009, four young men were enveloped by fog as they rowed across Loch Awe to their campsite. A friend on shore heard them calling for help. Then their boat capsized and sank.[78] In stormy weather, the wind on a lake can whip up waves big enough to swamp small boats. In August 2009, a well-known naturalist and his young son drowned during a squall on Loch Maree in the Scottish Highlands. Their canoe was found at the edge of the lake, overturned and empty.[79]

◆ ◆ ◆

The great majority of those who drown in open water are schoolboys and young men. Most of the victims could swim, but they failed to understand that swimming in lakes, reservoirs and canals is far more demanding and dangerous than swimming in their local swimming pool.

CHAPTER FIFTEEN

'WILD SWIMMING' AND OPEN-WATER RACES

'Wild swimming' has become fashionable. Open-water races have become commercial. Both encourage risk-taking, and the risks are downplayed.

The writer Roger Deakin began the vogue for 'wild swimming' with his book *Waterlog: a swimmer's journey through Britain*.[1] His farmhouse in Suffolk had a deep moat, and for many years he enjoyed 'breaststroking up and down the thirty yards of clear, green water'. Then he set off on an extended swimming expedition in idyllic settings from the Scilly Isles to the Hebrides.

Roger Deakin claimed that he was inspired by John Cheever's unsettling short story 'The Swimmer'.[2] Cheever's tale opens with Neddy Merrill sipping gin beside a friend's swimming pool in affluent upstate New York. Merrill decides to swim to his home via a string of swimming pools belonging to friends and acquaintances. But as we follow his progress, the season changes ominously from summer to autumn. The strength goes from his arms and shoulders. In place of his usual flutter kick, his legs manage only a hobbled sidestroke. He feels bewildered, then stupefied with exhaustion. When he reaches his home, he finds it decayed and deserted.

In 'The Swimmer', as the mood of the story darkens, John Cheever asks why Neddy Merrill was determined to complete his journey even if it meant putting his life at risk, but in *Waterlog*, Roger Deakin set limits to his risk-taking. He preferred to feel the water on his skin, but after finding the sea 'scaldingly cold', he ordered a made-to-measure wetsuit.[3] He noticed the current tugging at his legs as he waded in the sea on the North Norfolk coast and decided that swimming would not have been a good idea. He swam out to inspect the wreck of a trawler grounded on a sandbar on the Sussex coast. Two men were building sea defences, and when they saw him swim further out into the open sea, they hurried down the beach towards him. He realised that they thought he intended 'leaving a pile of clothes behind and disappearing for good'. When he apologised and thanked them, they said they were also worried about rip tides.

Early in the book, he was intrigued by a nautical chart showing the Corryvreckan whirlpool, a maelstrom off Jura, the island where George Orwell wrote *1984*. Two hundred pages further on, he arrived in Jura and saw the whirlpool 300 yards offshore – 'a low-pitched, continuous seething of brawling waves'. His swimming trunks stayed in his rucksack, and he told us, 'Only the deer saw me turn away from Corryvreckan and make my way slowly back uphill.'[4]

In 2004, five years after *Waterlog* was published, a group from the South London Swimming Club decided to leave Tooting Bec Lido behind and spend five days in the Scilly Isles. The plan was to swim from island to island, without wetsuits, led by an established Channel swimmer, who paddled beside them in a canoe. It was fortunate that a motorboat went with them on the third outing, because two swimmers had to be hauled out of the water and rushed to hospital suffering from hypothermia.[5] The chilled swimmers were re-warmed, and, undeterred, the group swam on. The leader said that he might suggest wetsuits on future trips.

Since Roger Deakin's *Waterlog*, several books have been published that encourage 'wild swimming' in lakes, rivers, waterfalls and quarries, with listings and photographs of recommended locations.[6] Unfortunately, they give a minimum of advice on safety, some of it inaccurate. Indeed, a number of the suggested locations are places where people have died of drowning, although no mention is made of these fatalities. For instance, in *Wild Swimming*, the listings include:

+ Loughrigg Tarn in Cumbria
  A large wooden cross stands on its bank, a memorial to a boy who drowned in the tarn on a school trip in 1960 when his legs became entangled in the reeds.[7]
+ Peel Island and Brown Howe, Lake Coniston
  In August 2007, a young man drowned 20 metres from the shore near Coniston Sailing Club.[8] In June 2010, a nineteen-year-old student staying at the nearby Coniston Hall campsite drowned soon after he entered the lake.[9] In July 2013, a young artist drowned in North Bay.[10] In all three Inquests, the Coroner emphasised the coldness of the water.
+ The waterfalls and plunge pools of Pontneddfechan in Wales
  In 1998, an adventure instructor was swept to his death when he went to the rescue of a schoolboy who stumbled while crossing the river.[11] In 2001, a seventeen-year-old student was sucked into a whirlpool and drowned when an upstream rainstorm increased the volume of water flowing down the narrow gorge.[12] In 2002, a sixteen-year-old boy jumped into the pool beneath Gwladys Falls, panicked and drowned.[13]
+ Llyn Gwynant, a lake at the foot of Snowdon
  An identical view of the Elephant Rock in Llyn Gwynant appears on a 'wild swimming' website[14] and a BBC report of the death of a twenty-year-old man from Manchester, a strong swimmer, who drowned in the lake in July 2006.[15] At the

Inquest, the Coroner warned of the danger of swimming in cold mountain waters even in high summer.

- Linn of Dee and Linn of Quoich, Braemar
  In Aberdeenshire, the River Dee flows down from the Cairngorms to form a series of rapids and rocky pools near Braemar. A granite memorial at Linn of Dee[16] records the drowning of a young couple in 1927 when a woman fell into the river and her fiancé dived in to try and save her.[17] In 2006, at the nearby Linn of Quoich, a ten-year-old girl slipped into the river and drowned. A young man who tried to save her was carried more than a hundred yards downstream by the fierce current.[18]

- Loch Lomond
  In May 2008, a young man waded out to retrieve a boy's football. He was less than ten feet from the shore when a strong undercurrent dragged him under the water.[19] A month after his death, as the Water Rangers' patrol boat sailed past the pebble cove where the young man drowned, the Senior Water Ranger told a reporter that he could not think of a beach where there had not been a fatality at some point.[20] There were 10 drowning deaths in Loch Lomond between June 2007 and June 2008.[21]

- Gullet Quarry in Herefordshire
  The quarry provided much of the stone for nearby Malvern. After stonecutting ended about fifty years ago, the excavation filled with water. The description in the 2008 edition of *Wild Swimming* reads: 'Deep, green, spring-fed quarry lake in great natural amphitheatre in cleft of Malvern Hills. View millions of years of geology with birdlife circling above. Shelving entrance, sunny in the afternoon. Danger signs abound (the water is deep) so take care and don't swim alone.'[22] The book fails to explain the reason for the danger signs – 6 people died of drowning at Gullett Quarry between 1973 and 2002.[23]

In the summer of 2010, police and the fire and rescue services became so concerned about young people using social network sites to plan swims in rivers and quarry pools, including Gullett Quarry, that they issued formal warnings about the danger.[24] The advice of a police inspector was emphatic, 'We urge everyone, irrespective of how strong a swimmer they are, not to be tempted to enter any of these waters no matter how hot it gets – it could be a matter of life or death.'

Gullet Quarry featured again, even more prominently, in the 2013 edition of *Wild Swimming*. Just as the book appeared, however, there were two more deaths in Gullet Quarry within a week.[25] The victims were both young men, swimming with friends, who suffered sudden swimming failure as they set out to cross the lake. They sank into the depths before anyone could reach them. A young woman dived beneath the surface to try and help the second victim, but found the water was pitch black 'like someone had turned the lights off.' The coroner praised the people who went to their rescue. 'I was very impressed with the number of people who tried to assist Russell and Justas,' he said. 'People tried to help despite knowing the dangers they faced.'[26]

Malvern Hills Conservators closed the access to Gullett Quarry and erected signs bearing the clear warning: 'Two young men have drowned in this lake in recent weeks. Swimming here is dangerous and illegal. Do Not Enter The Water.'[27] However, 'Gullet Quarry, Ledbury' still appeared on a website,[28] and people continued to swim in the quarry.[29]

On 18 July 2013, a few days after the young men became the seventh and eighth victims of drowning at Gullet Quarry, the *Guardian* carried an article with the title: 'Wild swimming isn't dangerous, but our behaviour around water can be.'[30] The writer was Daniel Start, author of *Wild Swimming*. After admiring the beauty of Gullet Quarry, its spring-fed water warmer than the sea, its lack of currents or underwater obstructions, he went on to imply that the danger was due to the behaviour of 'young lads on hot weekends who tempt each other, often after a few beers, to jump from higher and higher ledges.' In fact, neither of the two young men who drowned there in 2013 had jumped into the water, and neither had been drinking.

Wild swimming enthusiasts find it hard to admit that open water may be too cold for safe swimming. When swimmers first immerse themselves in open water, its low temperature triggers a powerful reaction in the cold-receptors in their skin. If they stay in the shallows, they grow accustomed to the cold and the warning signals calm down, even though the water continues to extract heat from their bodies.[31] And when they set off to swim across the quarry, the cold water extracts heat faster than before, so that swimmers may be incapacitated by sudden swimming failure in the deepest part of the quarry.[32]

Paradoxically, as long as they don't suffer immediate cold shock, the young lads who jump into Gullett Quarry are probably at less risk than the swimmers – because the jumpers do not stay in the water long enough to become chilled. Of the eight victims who drowned in the quarry, only one had been jumping in, but he surfaced safely after his jump. At the inquest in 2002, the coroner read a police statement to the court: 'He was seen to jump off the cliff feet first and emerged from the water to a cheer from those watching. He was then observed to swim towards a strip of land known as The Slip, before getting into difficulties.'[33]

In July 2013, after fatal drowning accidents in Coniston Water and Thirlmere, the Cumbrian police warned about the dangers of wild swimming,[34] reminding anyone who planned to jump from bridges or swim in the region's rivers, lakes and reservoirs that 'cold temperatures, strong currents, fatigue and over-estimating swimming ability all contribute to the risk'. Spokesmen for the Lake District Search and Mountain Rescue Association, the Cumbria Fire and Rescue Service, and the Maritime and Coastguard Agency emphasised that 'even strong, experienced swimmers can quickly get into difficulties.' All of these trained professionals urged caution: 'Cooling off by going for an outdoor swim might seem appealing but we've already seen it can have tragic consequences.'

Some wild swimming websites discount any such advice, forgetting that their dismissive attitude may encourage people to ignore the risk and put their lives in jeopardy. An opinion piece called 'Dangers in the water – the lies and the statistics: UK Drowning Statistics 2002' included a photograph of a car being dragged out of a river, with the caption 'Only a very small fraction of UK drownings are related to swimming'.[35] The article suggested that distorted, misleading statistics overstated the number of people who drowned while swimming in open water, and it ended with the comment: 'We believe that RoSPA is largely responsible for propagating this myth.'

A similar website published a series of articles under the label 'Fraud's Corner' which contested the safety advice given to the public by local councils, emergency services, the Environment Agency and the Royal Society for the Prevention of Accidents (RoSPA). One of the articles, published online in July 2010, had the title: 'The truth about reservoir swimming'.[36] The writer complained that swimming was banned in most English and Welsh reservoirs. One water authority (United Utilities) was chosen for particular censure because it had recently 'banned swimming in any water owned by the company. As a result, the 2010 UK Ironman will not take place in Appleton Reservoir, Bolton, as it did last year.' The writer probably did not know that a sixteen-year-old boy drowned in Appleton Reservoir in June 2009.[37] With a group of friends, he had climbed over two locked gates and jumped into water more than 20 feet deep. As he struggled in the water, his friends tried to lift him to the surface. 'I had hold of him. We were both getting pulled under the water. It was hard to keep him up,' one of his friends told the Inquest. 'We all tried to dive under but it was too deep and cold.'[38] Police divers were needed to locate the victim's body. (As for the 2010 Ironman UK Triathlon, the swimming leg was simply moved a few miles to the lake in Pennington Flash Country Park,[39] the chosen venue since then.)

◆ ◆ ◆

The water in swimming pools is warmed to around 28°C, whereas the water in ponds, lakes and reservoirs in Britain is usually well below 15°C.[40] Someone who makes a sudden entry into such cold open water is likely to suffer 'cold shock' – a gasp reflex followed by several minutes of panting for breath, together with a racing pulse and a dramatic increase in blood pressure. The skin on the face has the greatest concentration of cold receptors, so swimmers can lessen the effect of cold shock by splashing water on their faces before they enter the open water,[41] and they should get in slowly, in two stages – first to the waist, then to the neck.[42] 'Cold shock' is intensified if they plunge straight in.

After half a dozen uncomfortable entries into open water, swimmers gradually become 'cold-hardened'.[43] They are less troubled by the cold as they enter the water, the cold shock response becomes less marked, their bodies cope with longer immersions, and the onset of shivering is delayed. However, 'cold-hardened' or not, every moment that swimmers remain immersed in cold water, heat is drained from their bodies. Children can also become 'cold-hardened' to a certain extent,[44] but they lose body heat even more quickly than adults.[45]

As soon as swimmers enter cold water, their bodies react to minimise heat loss. First, the circulation in their skin shuts down.[46] The ulnar nerve lies just under the skin at the elbow and the wrist, and it is soon numbed by the cold. The muscles of the hands lose their strength, so that the fingers spread apart and bend. (This important warning sign is known to Channel swimmers as 'The Claw'.)[47] Then the blood vessels supplying the muscles in their arms and legs constrict. This reflex constriction limits heat loss from the limbs, and preserves the temperature of the body's core at 37°C for as long as possible,[48] but less oxygen arrives to fuel aerobic contraction, lactic acid accumulates due to anaerobic contraction, and the swimmers' strokes lose power.[49]

Hypothermia develops insidiously, sapping strength, and slowing thought processes.[50] Mild hypothermia, with a core body temperature around 35°C, is common in experienced open-water swimmers.[51] As long as swimmers monitor their performance moment by moment, and get out of the water if they become breathless, if their fingers start to spread, or if they notice that they are feeling colder, they will suffer no long-lasting harm. But swimmers are in immediate danger if they begin to shiver.[52] Their quivering muscles produce heat, but not enough to compensate for the heat they are losing to the water. The work of shivering will quickly exhaust them. Hypothermia can kill.[53]

It is important to remember that no matter how cold swimmers feel while they are in the water, their temperature always falls even lower once they climb out onto dry land.[54] This 'afterdrop' is due to the transfer of central body heat to the chilled outer layers of their bodies as their cold-constricted blood vessels relax. A seasoned open-water swimmer described the tingle in his skin and muscles when warm blood from the body's core surges into his cold limbs:[55] 'Warm blood flowing back into your extremities also carries with it a sense of euphoria every time.' (Could this sense of euphoria be a factor in the addictive lure of open-water swimming?)

◆ ◆ ◆

People who take part in open-water racing cannot avoid the chilling effect of the water,[56] and they are wise to pay attention to the danger.[57] Most swimmers wear wetsuits during open-water races, choosing thicker-gauge neoprene for better insulation and extra buoyancy.[58] Wetsuits have to fit, but without limiting the expansion of the chest and abdomen or restricting the movement of arms and legs.[59] Brightly coloured swimming caps increase the visibility of swimmers in the water while cutting down the rate of heat loss from their heads.[60] Tinted, non-misting swimming goggles protect the eyes and help swimmers to check that they are staying on course. Some competitors clip on eye-catching tow floats that bob behind them as they swim.[61]

Open-water swimming races have become increasingly popular. They attract large numbers of entrants, including many who are unprepared for the demands of swimming in cold water. In September 2008, in the first Great North Swim, more than 2,000 people swam a mile in Windermere.[62] In 2009, 6,000 people took part.[63] In 2010, blooms of toxic blue-green algae forced the cancellation of the Great North Swim,[64] but 3,000 people entered the Great East Swim in Afton Water reservoir in Suffolk. A young woman had to be helped out of the water. She was given emergency treatment at the scene, but she died shortly after arriving in hospital.[65]

A year later, in June 2011, 10,000 swimmers entered the Great North Swim in Windermere. Competitors had been asked to confirm that they could swim a mile in an indoor pool, but the majority had no experience of open-water swimming. In groups of 300 or so, they set off at intervals to swim half a mile, a mile or two miles in extremely cold water. An experienced open-water swimmer remarked, 'I went down to have a look and the shock in people's eyes was phenomenal.'[66] Sixty swimmers were unable to complete their chosen distance and two swimmers were airlifted to hospital: a 46-year-old man died after failing to respond to lakeside CPR, while a man in his sixties survived after reaching hospital in a critical condition.[67] Experts urged future competitors to get their bodies acclimatised by practising in the cold water of an outdoor pool before competing in open-water races.[68]

In the Great North Swim in June 2013, a woman was rescued from the lake and airlifted to hospital, seriously ill with a suspected heart attack. She survived.[69] In June 2014, a man began the two-mile swim, only to collapse before he completed the distance. Emergency doctors treated him at once and an air ambulance took him to hospital. He was pronounced dead soon after his arrival.[70]

Even cold-hardened swimmers lose their lives in open-water races. In 2010, the *Journal of the American Medical Association* (*JAMA*) published a brief research paper, 'Sudden Death During the Triathlon', which reported 14 deaths of competitors in officially sanctioned triathlons between January 2006 and November 2008: 13 of the 14 deaths occurred during the swimming phase of the race.[71] The authors calculated that competitors were twice as likely to die in triathlons as in marathons. In August 2011, 2 swimmers drowned during a New York City triathlon,[72] prompting an article in *Scientific American*: 'Why Is Swimming the Most Deadly Leg of a Triathlon?'[73]

As the popularity of triathlons increases, more triathlon-associated deaths are reported. In USA and Canada between October 2008 and November 2015, 10 people died during triathlon training and 48 contestants died during triathlon races – 58 deaths in all, with 42 deaths during the swimming phase of the race.[74]

Open-water races begin with a mass stampede of jostling swimmers.[75] Contestants barge and swim over each other in their efforts to lead the pack or to position themselves on the inside of a turn. They may be forced to hold their breath as they cope with the aggressive action of other competitors. Olympic Silver Medallist Keri-Anne Payne finished a race in the

Hudson River, New York, with a black eye and bloody nose from a kick in the face.[76] Not surprisingly, some swimmers panic.[77]

As contestants plunge into cold open water, the sudden release of adrenaline into the bloodstream may trigger an abnormal heart rhythm.[78] Throughout the race, their hearts must beat faster and pump harder than normal. In choppy water, they usually change to head-up freestyle, a much more demanding stroke than their usual crawl. The strenuous exertion may be too much for swimmers who have minor, undiagnosed heart problems. In all 13 swimming fatalities reported in 'Sudden Death During the Triathlon', the declared cause of death was drowning.[79] However, heart abnormalities were identified in 7 of the 9 victims who had post-mortem examination: one was born with an abnormal coronary artery, and six had muscular thickening of the left ventricle, including a victim with a pre-race diagnosis of Woolf-Parkinson-White syndrome (which can cause palpitations and sudden fainting.)[80]

The swimmers' thrashing legs and flailing arms churn the water surface, splashing faces and drenching the air. As they fill their lungs with air, they cannot avoid inhaling droplets of water. When droplets of water enter the lungs, however, they trigger two reflexes that interfere with lung function: small airways close off (which limits oxygen capture)[81] and small blood vessels in the lungs constrict (which raises pulmonary blood pressure and increases the work of the right side of the heart).[82]

Worse, some competitors in open-water races and triathlons develop sudden severe breathlessness caused by fluid leaking into their lungs from the over-filled pulmonary capillaries surrounding the alveoli, a condition known as swimming-induced pulmonary oedema (SIPE),[83] triggered by the combination of strenuous exercise and immersion in cold water. The exertion of vigorous swimming greatly increases blood flow through the lungs. At the same time, an excessive volume of blood is forced into the chest by the pressure of the water on the swimmers' immersed arms and legs, by the cold-related constriction of blood vessels in their skin and muscles, and by compression by close-fitting wetsuits. As swimmers become increasingly breathless, they cough up frothy mucus, often stained with blood.

This dangerous condition was first described in scuba-divers[84] and in young soldiers taking part in time trials while swimming in the sea.[85] It is now increasingly recognised in mass swimming events. In North Carolina, anaesthetists and physiologists at Duke University studied 36 swimmers who had experienced an episode of SIPE.[86] A number had significant medical conditions affecting the heart and lungs, and the researchers suggested that swimmers who develop SIPE should be offered tests of heart and lung function. The Canadian triathlete Katherine Calder-Becker described her own experience of repeated episodes of SIPE throughout ten years of international racing success,[87] revealing the strong drive that motivates many open-water swimmers to continue their demanding sport.

In 2004, a 36-year-old man with well-controlled insulin-dependent diabetes was training for a local triathlon in Pembrokeshire. Wearing a wetsuit, he had been swimming in the sea for only ten minutes when he began to feel short of breath. He swam on for half a mile, but he became more breathless. When he coughed up blood-stained mucus, he got out of the water.[88] His family doctor heard crackles on the right side of his chest, and abnormal breath sounds were still present when he was admitted to hospital. His chest X-ray, taken twelve hours after the episode, was normal, but he was found to have mild left ventricular hypertrophy and two leaking heart valves. He improved quickly on the ward. After discharge from hospital, he returned to triathlon training.

For people who decide to test themselves by taking part in open-water races and triathlons, there is an extra element of risk. As they battle with each other and themselves, their emotional commitment often hardens into a grim determination to win, to achieve a personal best, or simply to get to the finishing line.[89] They don't want to be seen as quitters, so they continue swimming even when they are flagging, even after they realise that they ought to withdraw from the race.

Moreover, when swimmers signal that they need help, it takes time for lifeguards to reach them and more time to get them out of the water and provide treatment. Lifeguards cannot see contestants' faces, and if they fail to notice distress signals, rescue and resuscitation may be fatally delayed. Lying face down in the water, a swimmer who loses consciousness may sink beneath the surface unobserved. In June 2010, a first-time competitor disappeared while taking part in a triathlon in Philadelphia, unseen by the lifeguards patrolling in kayaks. After two days of searching, his body was recovered from the river.[90]

In October 2010, the swimming world was shocked when a popular young champion, at the peak of his fitness, drowned unnoticed near the finishing line of the Open Water 10-kilometre World Cup in Dubai.[91] A search began only after one of his teammates finished the race and reported him missing. His body was found two hours later. The Task Force appointed by the Fédération Internationale de Natation (FINA) noted that the sea was very warm (30°C) and the air even warmer (38°C). The FINA Report concluded that heat exhaustion, fatigue and dehydration led to loss of consciousness before he died of drowning.[92] Sonar units are now available to alert rescue boats as soon as a competitor stops moving or sinks below the surface.[93]

♦ ♦ ♦

The first Olympic 10km Marathon Swim races were held in Beijing in 2008. The venue was the purpose-built rowing basin.[94] Four years later, the 2012 Olympic Marathon Swim races were held in the Serpentine in London's Hyde Park, where the calm, shallow water was warm enough to allow competitors to dispense with wetsuits.[95] In 2016, the Marathon Swim races were held in the sea beside Rio de Janeiro's Copacabana Beach.

As part of the preparation for the Rio Olympics, the 12 highest-ranked Brazilian open-water swimmers were invited to take part in an official 10km race, offshore. Researchers measured the swimmers' eardrum temperature immediately before and after the race. At 21°C, the seawater was considered relatively warm. Nevertheless, at the end of the race, after two hours in the water, 10 of the swimmers were hypothermic: 3 had mild hypothermia with a temperature of 35–34°C, while 7 had moderate hypothermia with a temperature of 34–30°C.[96] These results confirm that strong, skilled, cold-hardened swimmers are at risk of cold-related swimming failure and hypothermia even when competing in tropical waters.

Two days before the Olympic 10km Marathon Swim races at Copacabana Beach, waves washed away the starting pontoon. The contestants had to wade out from the shore, then swim to the starting line.[97] Newspaper coverage made no mention of hypothermia. Instead, the emphasis was on dramatic disqualifications in both races when swimmers attempted to win medals by shoving racers underwater, by grabbing hold of their legs and by blocking their finishing sprints.[98]

CHAPTER SIXTEEN

HARD LANDINGS

When swimmers dive or jump into the water, the further they fall, the faster their speed of entry. They may be winded or stunned if they strike the surface of the water, and they may be badly injured if they collide with the bottom of the swimming pool, the riverbed or the sea floor.

SPINAL INJURY DUE TO DIVING INTO SHALLOW WATER

Most swimming pools, especially those in hotels and private houses, are too shallow for safe diving.[1] Without the presence of a lifeguard to keep discipline, however, swimmers sometimes find it impossible to resist the temptation to dive in. Tragically, a casual dive may result in severe injury. As the diver plunges downwards, his momentum carries him to the bottom. His outstretched arms buckle, giving him little protection. His head thuds against the tiles. The impact may cause a scalp wound, a degree of concussion, or worse.

If he smashes full-tilt onto the bottom of the swimming pool, his head comes to a sudden stop, while his body continues its descent. His head is forced violently forwards until his chin thumps into his breastbone.[2] The bones in his neck – his cervical vertebrae – crush downwards one upon another. If the vertebrae fail to withstand the concentrated force of compression, the young man will break his neck. (The victim is almost always a young man.)

The possibility of spinal injury should be considered if the diver complains of neck pain, appears stunned or has bruises on the forehead, and any initial spinal damage is easily aggravated by further movements of his neck. Lifeguards are trained to minimise the risk of paralysis by using a back board and protective collar to prevent the diver's head from lolling backwards or twisting to one side as they lift him from the water.[3]

The damage may be limited to a wedge of collapsed bone or a 'teardrop' fracture chipped from the surface of a vertebra. Sometimes two or more vertebrae are affected. On occasion, an entire vertebra is smashed to pieces – a burst fracture – and jagged fragments of bone are propelled backwards towards the spinal cord. If the supporting spinal ligaments rupture at the point of greatest stress, the vertebrae are jerked out of line in a crippling fracture dislocation[4] and the spinal cord may be crushed or pierced by splintered bone.

Normally the spinal cord is very well protected by the stacked barrel-shaped vertebral bodies that lie in front of the cord and by the armoured barrier of vertebral arches that enclose the cord at the back and sides. But any damage to the cord has serious consequences because the spinal cord carries all instructions from the brain to the body and all sensations from the body to the brain. The diver may become the victim of immediate and permanent paralysis from the neck down.

Most diving-related fractures affect the 5th or 6th cervical vertebrae. A few swimmers break the first or second cervical vertebra at the top of the neck, where most of the nodding and turning movements of the head take place.[5] The spinal canal is wider at this level, so the spinal cord may be spared immediate injury. However, early recognition and treatment is crucial because fractures of these upper cervical vertebrae are usually very unstable. The swimmer appears to have escaped with nothing more than a stiff neck, until a shrug or an upward glance precipitates a sudden life-changing injury to the spinal cord.

A child is sometimes paralysed by damage to the spinal cord even though the X-ray report reads 'No bony injury'.[6] The ligaments tying the cervical vertebrae together are still so elastic that they can stretch as much as five centimetres, whereas the child's spinal cord can stretch no more than five millimetres. If his head is forced violently forwards, the delicate nerve fibres which run the length of the spinal cord may be wrenched and torn.

In 1991, an editorial in the *British Medical Journal* warned that at least 60 patients had been admitted to spinal injuries units in the United Kingdom over the previous two years with quadriplegia (paralysis of all four limbs) as a result of diving injuries.[7] According to Howard Cloward, a distinguished American neurosurgeon: 'Probably no accident or illness is more devastating than quadriplegia from a cervical spine injury. A strong, healthy person in the vigour of youth is injured in a diving accident and suddenly finds himself totally paralysed and helpless for the rest of his life, completely dependent on others for all his bodily needs, yet still retaining a keen and alert mind. The psychological and emotional impact of such a catastrophe is tremendous.'[8]

Talks, leaflets, and videos for schools have raised awareness of diving-related spinal injury. Nevertheless, it is one thing to point out to schoolchildren that diving into shallow water is risky. It is quite another thing for teenagers and young adults to remember the risk and restrain their urge to dive in. Especially if they taught themselves to dive. Especially if they are on holiday. And especially if they have been drinking alcohol.

Too often, young men dive into ponds, lakes, and rivers without the slightest idea of the depth of the water. Indeed, they may assume that a shallow pond must be deep enough for diving, simply because its murky water hides the bottom.[9] If they crash down onto tangles of tree roots or discarded supermarket trolleys, they may be trapped beneath the surface, unconscious or paralysed, unable to swim up to the air.[10] Unless their predicament is recognised at once, they will drown.

Swimmers wading out from the shore often plunge forward into oncoming waves, expecting to surface after the wave has rolled past. But if they strike their heads against a ridge in the seabed, they discover that wet sand is as unyielding as paving stones. Surfers risk injury if they ignore signs warning them of 'Dangerous Shore Break'.[11] A surfer who rides the crest of a wave that continues to build when nearing the beach may be carried forwards and downwards (surfers speak of 'going over the falls'), before tumbling head first as tons of water cascade onto the shore. An older surfer with arthritic changes in the neck risks injury even in less testing waves.[12] Bony spurs narrow his spinal canal. If he strikes his head on the sand, or twists his neck in an awkward landing, these rough outgrowths of bone may damage his spinal cord.

JUMPING INTO RIVERS, QUARRIES AND THE SEA

Jumping feet first into water from a height is less dangerous than diving in, and someone who makes a risky jump is less likely to break his neck than someone who makes a risky dive. Nevertheless, injuries are common after foolhardy jumps. The traditional May Morning celebration in Oxford brings crowds into the city centre. In 2005, over one hundred people, mostly university students, jumped 25 feet from Magdalen Bridge into the River Cherwell. They landed in water less than 3 feet deep.[13] Forty were injured. Twelve had to be admitted to hospital with broken legs, shattered ankles and penetrating wounds from jagged debris on the riverbed.

In *Waterlog*, Roger Deakin describes an afternoon spent watching a succession of young bikers jump 40 feet from the parapet of the fourteenth-century Devil's Bridge arching over the River Lune in Cumbria.[14] There was a lot of shouting, and his first thought was that the spectators were trying to persuade a young man not to jump. He wrote, 'Swimming closer, I realised to my horror that the crowd *were egging him on*'. Although he was intrigued by 'a more or less continuous procession of lemming leapers', he was not tempted to jump himself.

During a heatwave in July 2012, a young man jumped from the Devil's Bridge and failed to surface.[15] The Coroner recorded a verdict of Accidental Death. He said, 'People have jumped off that bridge on a regular basis over many years. It's a dangerous thing to do.' He asked people to think twice before jumping: 'You think of water as flexible and easy to jump into, but it could be almost like hitting a solid object from that height, and occasionally, someone gets it wrong and enters the water not quite at the position they intend to. I think that's what happened to Darrell. He went into the cold water, gasped and passed out, and if you pass out under water you're in serious trouble. He was unconscious immediately and would have known nothing.'

Two days after the death at Devil's Bridge, a fifteen-year-old boy jumped 20 feet from a disused railway bridge into the River Neme in Northamptonshire.[16] As he swam towards the riverbank, the current swept him away. Firemen found his body half a mile downstream.

◆ ◆ ◆

Frigid water collects in the workings of abandoned quarries. Beneath the surface, part-cut blocks, discarded machinery and remains of former workshops lie submerged and unseen. These quarry lakes are known to be dangerous, but for many young people, the attraction outweighs the risk, especially in warm weather. Youngsters meet at their chosen site and prove their mettle by taking turns to leap from the crater rim into the water. Unfortunately, some discover that sudden cold shock can, literally, take their breath away.[17] Others set off to swim to safety, but fail to reach the water's edge.[18]

In July 2003, a sixteen-year-old boy and four of his friends climbed over two sets of security fences surrounding the disused Holcombe Quarry in Somerset. He jumped 60 feet from the rim of the quarry and disappeared beneath the water.[19] Police divers recovered his body the next day. Holcombe Quarry was the scene of a second fatal accident in June 2010. A nineteen-year-old man was one of a group of thirty people who ignored warning signs, evaded security guards, and arrived at the edge of the quarry to jump and swim. He surfaced after his jump, but he appeared disoriented and swam in circles. Before his friends could reach him, he sank.[20] Again, police divers were needed to remove his body from the water.

In May 2009, a fifteen-year-old boy drowned in the Blue Lagoon, an abandoned limestone quarry at Harpur Hill near Buxton in Derbyshire.[21] The water is not only cold and deep. It is also intensely alkaline due to residues of calcite (quicklime) in the stone.[22] Official warning signs erected round the quarry show photographs of lime burns, with the message: 'Think! Would you swim in ammonia (pH 11.5) or bleach (pH 12.6)? Well these are similar to the Blue Lagoon (pH 11.3)!'

Undeterred, people continued to swim in the quarry lake, and after pictures of its turquoise water were posted on the internet, they arrived in increasing numbers to picnic, party, jump and swim. In 2013, the district council decided to take drastic action. They dyed the water black. After the water changed colour, fewer people came to the quarry.[23] 'It's not pretty anymore,' a local woman said. 'They don't think they're on holiday in the Bahamas anymore. They know they're in Harpur Hill. It was absolutely beautiful to look at but was horrendously dangerous.'

◆ ◆ ◆

The risk-taking continues at the seaside. Piers project into deeper water, but it is never a good idea to jump from a pier. Beneath the wooden walkway, the water swirls between the cast-iron pillars, its depth changing from moment by moment as the tides come in and out. In July

2007, four men jumped into the sea from the end of Clacton Pier. Nearby were official signs warning people not to jump. The assistant manager tried to persuade two of their friends not to follow them over the rail. He explained that there was debris in the water under the pier and danger from rip tides. He thought he had convinced them not to jump, but the moment he turned away, he heard two splashes.[24] Both men were in their forties and had been drinking heavily. The current pulled them under the pier, then out to sea. They struggled briefly before they lost consciousness. When the coastguards reached them, they were floating face down in the water. An air ambulance took them to hospital, where one man was pronounced dead soon after arrival. His companion was admitted to the intensive care unit, but he died five days later.

On a Sunday afternoon in 2008, after a couple of drinks with his friends, a young man ran the length of Southsea Pier and jumped into the sea. It was low tide. He fell 30 feet into 3 feet of water.[25] He hit the seabed and broke his neck in three places. He was dragged unconscious from the sea and airlifted to hospital. Complex orthopaedic surgery succeeded in stabilising his neck. Although he regained some power in his arms, he was paralysed from the waist down, for life. Three years after his accident, he decided to counter the appeal of the madcap jumps posted on the internet.[26] Confined to a wheelchair, he appeared in a powerful video, speaking without self-pity of his limited life and the grief and disruption suffered by his family. Kent Police uploaded 'The Dangers of Tombstoning' on YouTube,[27] and the Royal National Lifeboat Institute gave him a special award.[28]

TOMBSTONING BECOMES A CRAZE

Jumping into water from bridges and piers is not a new activity, but its popularity soared after it was given a catchy new name – 'tombstoning' – a term first coined by a journalist in 1995 after she watched people jumping into the sea off Tomb Rock at Wembury on the South Devon coast.[29] Young people used their mobile phones to record daredevil jumps and posted their video clips on the internet.[30] Tombstoning soon became a craze.[31] Boys as young as 10 launched themselves from the harbour walls of Arbroath into the path of fishing boats returning to port.[32] A young man doing backflips into Newhaven harbour was carried out to sea by the current. Floating semi-conscious, he was fortunate to be spotted by the keen-eyed skipper of a local dive boat, who hauled him on board.[33]

On the Dorset coast, the huge natural limestone arch of Durdle Door juts from the cliffs into the sea. In 2007, a young man climbed up its crumbling rock face. Then, urged by three companions, he jumped a hundred feet from the top of the arch.[34] His face and chest smacked down on the surface of the water and he disappeared beneath the waves. He lay on the seabed 13 feet down, unconscious and convulsing. His life was saved by a swimmer who brought him to the surface. He was pulled onto an inflatable dinghy, bleeding from the mouth, his face blue from lack of oxygen. An off-duty doctor treated him on the beach, and a rescue helicopter airlifted him to hospital. Watching the rescue, a coastguard deplored the tombstoning craze: 'Young men are dying pointlessly, and getting seriously injured every summer, in a bid to impress their friends.'[35]

On Plymouth Hoe, conspicuous yellow warning signs bear the message: 'No Diving or Tombstoning – Danger of Death'. Teenagers ignore the warnings. Instead, they stand on top of the sea wall and jump 65 feet into a small inlet called Dead Man's Cove. In a variant of skinny-dipping called 'moonstoning', some lads strip off their clothes and jump naked into the sea.[36] Most tombstoners attempt a streamlined feet-first entry. Others increase the impact on their own bodies and on their audience by 'bellystoning', leaping face down from the sea wall and smacking themselves flat against the unyielding surface of the sea.[37]

Cliff-jumping youngsters assume that they will be cushioned by the water. They forget about the tides as they hurtle down into swirling water which may not be deep enough to spare them from collision with the seabed.[38] If they make a clean entry into deep water, they may plummet so far underwater that they cannot get back to the surface before their breath runs out. Even if they manage to regain the surface, the rapid chilling of their muscles hampers their return to safety.

As more people jumped into rivers, quarries and the sea, the number of accidents increased. In five years (2004–2008), 12 people were killed while tombstoning. Then, in the next five years (2009–2013), the number of deaths doubled to 24.[39] Many more survive tombstoning accidents only to spend the rest of their lives in a wheelchair. The Maritime and Coastguard Agency and the Royal National Lifeboat Institution (RNLI) are repeatedly summoned to rescue the victims when jumps go wrong. In 2014, the emergency services joined forces with the Royal Society for the Prevention of Accidents to launch the public awareness campaign: 'Tombstoning. Don't jump into the unknown.'[40]

A Plymouth-based journalist spent his childhood summers in the 1980s jumping 30 feet off a Cornish harbour wall. He had a simple explanation for the attraction of jumping into the sea: 'In those days we didn't call it tombstoning. We called it "having fun".'[41] Yet high spirits can end in tragedy when groups of teenagers mistake bravado for bravery, seemingly oblivious to the dangers. Indeed, the dangers are probably part of the attraction of tombstoning. Apprehension before the jump is transformed, after the jump is completed, to a buzz of euphoria at having cheated death. Unfortunately, not everyone manages to cheat death.[42] A spokesman for the RNLI commented, 'It's called tombstoning for a reason – get it wrong and you end up with a tombstone.'[43]

JUMPING FROM HIGH BRIDGES

Most people who tombstone off harbour walls expect to survive. But a few structures become magnets for people who are determined to die. The Clifton Suspension Bridge crosses the

Avon Gorge 245 feet above the River Avon. A leap from the bridge is almost always fatal, although a young woman survived in 1885 when her crinoline acted as a parachute and slowed her fall.[44]

In 1996, researchers wrote: 'The presence of the suspension bridge in Bristol contributes significantly to the local pattern of suicide. Local residents are twice as likely to commit suicide by jumping than are other residents in England and Wales.'[45] Two years later, safety barriers were fitted along each side of the bridge's central span.[46] During the five years before the barriers were installed, 41 people jumped to their deaths from the bridge: during the five years after their installation, 20 people jumped. Although the number of suicides halved, the number of people who attempted to climb onto the bridge structure remained unchanged. Security staff patrol the bridge day and night. They explained that the safety barriers 'bought time' to intervene, to dissuade would-be suicides from jumping, and to organise their protection and treatment. (It is possible that these interventions contributed to the slight fall in the overall suicide rate in Bristol after 1998.)

The Golden Gate Bridge stretches more than one and a half miles across San Francisco Bay, 220 feet above the surface of the sea,[47] carrying six lanes of traffic, with walkways for bicycles and pedestrians on each side. The original design specified a guard rail five and a half feet high on the outer sides of the walkways, canted inward to increase the difficulty of climbing over the protective barrier, specifically to deter potential suicides. However, for aesthetic reasons, the design was modified. The height of the guard rail was lowered to four feet. Within three months of the bridge's opening in 1937, the first suicide victim climbed over the rail and leaped into the sea.[48] Since then more than 1,600 people have killed themselves by jumping from Golden Gate Bridge.[49]

While the bridge was being built, a steel safety net was suspended beneath it. The net saved the lives of 19 construction workers, but it was removed when the bridge was completed.[50] The provision of safety barriers has been under debate for decades. Campaigners point out that safety barriers have reduced deaths from other iconic sites,[51] such as the Empire State Building and the Eiffel Tower. They contend that the absence of safety barriers allows depressed or lonely people, often on a sudden impulse, to jump to a quick near-certain death.

Opponents claim that a safety barrier would spoil the look of the bridge. They object to the cost. And they argue that there would be little to gain from an anti-suicide barrier: 'If a person is bent on suicide he will find a way and inexorably go someplace else to kill himself.' This assumption was shown to be mistaken more than thirty years ago by Dr Richard Seiden, a clinical psychologist at the University of California at Berkeley.[52] He studied the records of all 515 people who attempted to jump from the bridge between 1937 and 1971 but were prevented from jumping. He was able to confirm that 451 were still alive, while only 64 had died. In all except one case, the cause of death was on record: 31 died of natural causes, 7 died in accidents, 25 committed suicide, including 7 who returned to jump off the bridge. Thus,

94% of those who were prevented from jumping to their deaths were either still alive or had died of natural causes. Only 6% had died a violent death.

Dr Seiden wrote, 'The popular mythology holds that one is gently swallowed by the waves to die by drowning.' The reality is rather different, and much more brutal. After a fall lasting four seconds, victims hit the water at 75 miles per hour.[53] A few drown, but most die from their impact on the water surface, their bodies shattered by multiple blunt-force injuries. Fewer than 30 jumpers have survived the fall. Some survivors acknowledged that they regretted their decision to jump – even as they fell. 'I still see my hands coming off the railing,' one young man recalled. 'I instantly realized that everything in my life that I'd thought was unfixable was totally fixable – except for having just jumped.'[54]

In 2016, there were 39 confirmed suicides at the Golden Gate Bridge, and patrol officers persuaded 184 people not to jump. In 2017, with an increasing number of suicidal people arriving at the bridge and five extra patrol officers appointed, there were 33 confirmed suicides, but patrol officers made 245 successful interventions.[55] Work has now begun on a safety barrier, a stainless steel net suspended twenty feet below the bridge, scheduled for completion in 2021.[56] Meanwhile, every two weeks or so, someone clambers over the low rail and smashes into the water.

CHAPTER SEVENTEEN

SWEPT DOWNSTREAM

Our daily encounters with water suggest that it is easily controlled. Water runs into the bathroom sink at the turn of a tap, trickles between our fingers, and flows gently away. But the tap delivers a column of water less than one centimetre wide, with a flow rate measured in cubic centimetres per second. Rivers, in contrast, deliver huge volumes of water, with a flow rate measured in cubic metres per second. And water is heavy. A cubic metre of water weighs 1,000 kilograms. Water on the move pushes hard against anything in its path, and the faster it flows, the greater its impact. We underestimate the power of moving water at our peril.

We admire the beauty of gliding brooks and quiet backwaters, unaware of the forceful currents swirling beneath the tranquil surface. Along the course of a river there are particular danger points, sites of repeated accidents, where people make serious miscalculations and suffer the consequences. In the five years 2009 to 2013, there were 604 fatal drowning accidents in Britain's streams and rivers.[1]

UPLAND STREAMS AND RIVERS

Even in a shallow mountain stream, the force of the current demands respect. If hikers choose a crossing point where the stream narrows, they have to deal with deeper water and a stronger current. When they wade across, their safety depends on the steadying effect of the friction between the soles of their boots and the rocks they step on. Rocks are polished by the action of the water and coated with a glaze of slippery algae and moss, so the impact of a few inches of tumbling water can be enough to overcome the friction that holds their boots in place.

A mountain stream is particularly dangerous after a rainstorm in its catchment area. Clear water turns murky as the run-off increases, and the sudden rise in water volume can take even seasoned locals by surprise. Hikers in a walking group see safety in numbers. Once their leader has crossed a stream in spate, the others tend to assume that they can follow without risk, forgetting that the water could sweep them away. In June 2007, on a route march in heavy rain near the end of basic training, thirty young soldiers were crossing a ford in the swollen Risedale

Beck near Catterick. The surging water knocked seven soldiers off their feet. A seventeen-year-old recruit was caught by the current, swept downstream, and drowned.[2]

Once a mountain stream swells to become a river, the water appears to flow more slowly. But appearances are deceptive. The speed of the current usually increases from the river's source to its mouth. Although the gradient lessens, the river grows deeper and wider,[3] fed by its many tributaries. Gravity pulls the water in two directions – downwards towards the riverbed and downhill towards the sea. Water pushes into crevices and sucks away particles of soil, scouring the bed of the river and eroding its banks. Its downhill progress is impeded by projecting rocks and waterweeds growing along its course. The flow is faster in the centre of the channel.

When wading knee-deep in an upland river, a fisherman is well advised to walk upstream. The current then presses against the front of his legs, bracing and locking the knees. A weighted wading stick helps him to keep his balance and to probe the riverbed for boulders and potholes.[4] In 2010, an eminent biochemist slipped while salmon-fishing in the River Spey and the fast-flowing water carried him away. His body was found six miles downstream.[5]

In New Zealand, people drown while fording rivers every year, and hikers are warned against 'get-home-itis'.[6] Instead of risking a dangerous crossing as daylight fades, they are advised to search for a safe place to cross or consider making camp for the night – even if they have work commitments the next day.

WATERFALLS

A waterfall forms when the path of a watercourse crosses from hard, resistant bedrock onto softer, crumbling stone. People are drawn to the spectacle of the tumbling, thundering water, but unfenced viewpoints beside the falls are unsuitable places for exploration. In 2009, a young man and his three-year-old daughter fell from a hillside track overlooking Nantcol Waterfall near Harlech. They were swept 50 feet down a series of water chutes, and both were drowned.[7] In 2010, a walker was taking photographs near the brink of High Force waterfall in Teesdale when he slipped into the water. The current carried him over the rim of the falls and he plunged 70 feet to his death at the base of the cascade.[8]

Books on 'wild swimming' list waterfalls as special attractions – despite their undoubted danger for swimmers.[9] At the base of a waterfall, a pool is carved out by the grinding action of fallen boulders and by plummeting water with its burden of grit. The water in the pool is usually very cold, surprisingly deep, and often extends beneath jutting underwater ledges. In May 2015, on a family picnic to celebrate his birthday, a teacher jumped into the pool beneath Low Force, a waterfall on the River Tees in County Durham, but he failed to reappear.[10] After an intense search by fire crews and mountain rescue teams, his body was found the next day at the bottom of the pool. At the inquest, the pathologist gave his opinion that the likely cause of death was 'cold shock'.

Ceunant Mawr Waterfall at Llanberis has been described as one of the most impressive waterfalls in Wales.[11] Water from the southern slopes of Snowdon plunges more than a hundred feet down a ravine into a pool thirty feet deep. The falling water churns the pool into whirlpools and recirculating currents, which can catch hold of unlucky swimmers and pull them beneath the surface. In June 2015, four friends set off on an early morning walk to see the sunrise. Then they decided to swim in the pool beneath the waterfall. Two men were caught in a deadly whirlpool and drowned.[12] Members of the Llanberis Mountain Rescue Team risked their lives as they rescued the two exhausted, traumatised survivors,[13] who were airlifted to hospital suffering from hypothermia.

While on holiday in the Scottish Highlands in July 2017, three young Londoners arrived at Lower Bruar Falls and decided to swim. Locals warned them to stay away from the cave beside the waterfall, but the strong current washed them into the water-filled cave.[14] One of the swimmers managed to rescue a struggling friend, but the other friend slipped out of reach. The emergency services soon arrived, but they decided that it was too dangerous to go into the water. Instead, they called out divers to search the cave the next day.

GORGES AND CANYONS

Flowing water carves gorges through sandstone and limestone rocks, but wet slippery rock formations and cold churning water make a potentially lethal combination. In 2006, a fifteen-year-old girl was killed while 'gorge jumping' at a supervised summer camp for children at an activity centre in Scotland.[15]

French cavers invented the sport of canyoning by following the course of mountain streams plunging into gorges. Canyoners face a succession of challenges as they scramble over boulders, splash through watercourses, abseil down waterfalls and jump into rapids.[16] They usually protect themselves with wetsuits, boots and helmets, but accidents are frequent, both in Britain and on the continent. In 2002, on an adventure holiday in the Alpes-Maritimes in south-eastern France, a young Scottish lawyer was dashed against rocks by the fast-flowing water of the River Loup, then sucked into a whirlpool, where he drowned.[17] A father underestimated the risks of this unregulated extreme sport when he took his children canyoning in the French Alps in 2006. His thirteen-year-old daughter was abseiling down a steep rock face when she fell 20 feet into a rock pool. Her safety rope snagged on an underwater branch, holding her underwater as she struggled to free herself. By the time a guide reached her, she had drowned.[18]

The inherent danger of canyoning is compounded by rainfall on any of the mountain slopes that drain into the gorge. From one moment to the next, the run-off from a rainstorm can transform a gentle stream into a raging torrent. It is vitally important that canyoners listen to the weather forecast before they set out on their descent. So far, no groups in Britain have been hit by flash floods, but there have been horrific accidents abroad. In Switzerland in 1999, a party of 52 young people were abseiling down a series of waterfalls in Saxeten Gorge when a thunderstorm several miles away released a cloudburst that carried trees and rocks down the mountainside.[19] A wall of water 7 metres high swept down the narrow gorge as far as Lake Brienz, carrying everything in its path: 21 of the canyoners were killed.

In New Zealand in 2008, on a canyoning trip from one of the country's best-known outdoor pursuits centres, 10 schoolchildren were trapped with their teacher and an instructor on a ledge above the flooding River Mangatepopo. As the water rose to engulf them, they inflated their life jackets and tried to save themselves by floating down the river. Six children and their teacher were swept to their deaths[20]. A fifteen-year-old boy was lucky to survive after the torrent carried him over a dam and dumped him against a jumble of logs, his boots torn off and his helmet split in two.

In August 2018, flash floods killed 5 canyoners, including their guide, as they abseiled into the Zoicu canyon in Corsica. Two weeks later, 11 canyoners were washed away and killed by a flash flood in the Raganello Gorge in southern Italy.[21]

CAVING

Limestone rocks are slowly dissolved as rainwater seeps through the soil and trickles into rocky crevices. In time, surface streams disappear. Instead, the water travels below ground, hollowing out extensive cave systems before emerging at distant springs. Cavers and potholers explore the mysterious dry watercourses carved by ancient underground rivers. They clamber down sinkholes, wriggle along narrow crevices called 'crawls' and 'squeezes', and explore subterranean caverns thousands of years in the making. Cavers face injury due to falls, rockfalls, hypothermia and entrapment, but the most frequent cause of death is drowning.

One cave – Porth yr Ogof in the Brecon Beacons – has been the site of eleven fatal drowning accidents.[22] The River Mellte, flowing into the largest of the cave's fifteen entrances, has carved out numerous underground pools, sumps and flooded passageways. Ten of the victims drowned in one particular resurgence pool, where the river rises close to its exit from the cave.[23] Beneath its rippling surface, the pool is 7 metres deep and shaped like a deadly bottle dungeon. Spurs of rock jut out from its walls, trapping anyone who falls into the frigid, swirling water.

Any downpour sends rainwater funnelling down into the local cave system. In 1967, a rainstorm in Yorkshire led to Britain's worst caving accident. Six expert young cavers set out to explore Mossdale Caverns in Upper Wharfedale. After they entered the caves, torrential rain began to fall, flooding the narrow underground passages. Desperate work by hundreds of volunteers dammed and diverted Mossdale Beck. A score of fire engines pumped water from the cave throughout the night. It took fellow cavers more than five hours of difficult and hazardous crawling before they came upon five bodies jammed in a passage two feet wide and ten inches high. Hoping against hope that the sixth caver had survived, the rescuers continued their search, but two days later they found his body.[24] Since it was impossible to bring the victims to the surface, the coroner ordered that they should remain in the cave where they died. Mossdale Caverns were sealed. Three years later, their bodies were moved to a higher cavern, later named The Sanctuary.

In December 2007, after several days of rain, rising water trapped two potholers in an underground streamway in the Long Churn cave system in Ribblesdale, only a few miles from Mossdale Caverns. They both drowned as they tried to return to safety.[25] At the inquest, the coroner warned of the danger of going into the cave during rain. Only one month later, in January 2008, a downpour trapped 15 cavers in Long Churn.[26] Two of the leaders swam through floodwater to summon help, leaving 11 students and two supervisors behind. Four hours went by before the water receded and cave rescuers were able to reach the group and lead them out of the cave.

In a series of serious incidents between May and October 2008, more than 50 chilled and frightened people were trapped underground for hours, standing knee-deep in icy water until they were located and guided to safety by volunteer cave rescuers summoned from local villages.[27]

The deaths in 2007 and the entrapments in 2008 sharpened awareness that entering a cave or descending a pothole is much more dangerous during wet weather. Cavers now have mobile phones. They can check the local weather forecast before leaving home. If rain is likely, they can contact the other members of their group and reschedule their expedition, instead of meeting near the cave and deciding, 'Well, we're here now. Let's chance it.' And they can recheck the weather forecast at the last moment, just before they make the fateful decision to enter the cave system. People used to die in caving accidents almost every year, but since December 2007, there have been no caving fatalities in Britain.[28]

WHITE-WATER RAPIDS

Cavers and potholers know that their sport is dangerous, and try to minimise the risks. Where white-water rapids are concerned, however, the all-too-apparent risks appear to be a large part of the attraction. As canoeists and kayakers increase their skill, their risk-taking keeps pace, particularly when club outings involve several young men engaged in friendly rivalry. But a river in spate is not a good place to learn from mistakes.

Rivers become white-water rapids on rocky inclines when extreme turbulence churns air into the water. Where boulders obstruct the flow, the water veers around them in hurtling chutes and swerving eddies, or mounds upstream into pillows and standing waves before tumbling into downstream souse holes.[29] During the Easter holidays in 2010, 5 people were drowned in canoeing and kayaking accidents in Scotland and Ireland.[30] An expert in accident prevention, an experienced kayaker himself, pointed out that canoeists often travel long distances to find the perfect water for paddling and may be tempted to take a risk rather than opt for safety first.[31] He wrote, 'White-water kayakers and canoeists want fast-flowing water. You want that challenge. The difficulty you have is that the dynamics of river levels are so unpredictable that a rise of a foot of water can make a hell of a difference.'

On New Year's Day 2012, a woman was trapped when her kayak became wedged underwater in a Cumbrian mountain torrent. By the time her companions were able to free her, she was deeply unconscious.[32] She died in hospital two days later. In 2016, a kayaker capsized and drowned in a swollen Welsh river, trapped underwater by his foot.[33] Recording a verdict of Accidental Death, the Coroner said that he viewed kayaking in the same way as he viewed mountaineering, 'There is no safe mountain and similarly there is no safe river.' In 2017, a kayaker lost his life wedged under a fallen tree in a 'notorious' stretch of the River Dart[34] where another kayaker drowned in 2009.[35]

◆ ◆ ◆

White-water rafting has grown in popularity from small beginnings in the 1960s to become part of a vast leisure industry. A grading system for white-water rapids gives some guidance on the skill needed to run them. The difficulty is rated from Grade 1: Easy (obstructions obvious and easy to steer... self-rescue easy) to Grade 6: Extreme (definite risk to life... rescue may be impossible).[36] Whatever the rating given to a particular white-water rapid, its degree of difficulty alters with any change in the volume of water sweeping down the incline.

On managed rivers, rafting may be scheduled to take advantage of timed water releases from upstream dams. Elsewhere, water level in the rapids will vary with the seasons. Northern rivers will be swollen by cold meltwater in spring and early summer. Asian rivers increase in force with the onset of the monsoon. Some rapids become hazardous in a drought, when the lower water level leaves more rocks exposed, but most rapids become more turbulent and treacherous in flood.

Once the guide casts off and the raft begins its run, white-water rafters can do little to influence the raft's trajectory, except hold tight, paddle a bit, and hope that their guide knows what he is doing. If their raft hits a rock, dips under a wave, or flips over, rafters are tossed into the torrent, their desperate attempts to swim to safety hampered by the swirling currents and by the reduced buoyancy of water filled with bubbles of air. Some are swept downstream or dashed against the rocks. Others are wedged beneath undercut ledges or pulled underwater by the violent suction of a whirlpool. Fallen trees act as deadly strainers, catching rafters who are pinned against the obstruction by the force of the current.

In 2007, while white-water rafting in the rainforest of northern Queensland, 8 members of a British school party were thrown into the River Tully. A teacher was held fast underwater, his foot jammed in a crevice between rocks.[37] Horrified pupils looked on as the guides struggled to free him and as they tried unsuccessfully to revive him. Over the next two years, on the Tully River alone, 4 further white-water rafters were drowned.[38] Adventure tourists rafting down tropical rivers also make close contact with local wildlife – from man-eating crocodiles to water snails infested with schistosomiasis.[39]

As white-water rafting increases in popularity, more rafters drown. In September 2006, CNN reported that at least 50 people had died in the previous eight months on white-water rapids in the USA.[40] America Outdoors, a national association representing commercial rafting companies, attempted to dismiss this sobering statistic by claiming that only a proportion of the deaths occurred on commercial guided trips.[41] But after a string of white-water rafting fatalities in New Zealand, a report by the New Zealand Maritime Safety Authority placed the blame firmly at the door of the rafting companies,[42] commenting: 'Personnel in the industry are predominantly young men with a desire for excitement and adventure. Many are quite immature and live life "on the edge". This severely affects safety judgements and assessment of client capability.'

With minimum outlay and with little or no regulation, white-water rafting enterprises have been set up in far-flung tourist destinations around the world, a source of desperately needed income – even when the water levels are too high for safety, the equipment is dilapidated and the guides have only rudimentary training. Brochures show full-colour photographs of rubber dinghies packed with eager youngsters in helmets, wetsuits and life jackets, grinning as they run through Grade 4 (Very Difficult) rapids, still grinning as they fall into the water.[43] Slogans proclaim: 'Warning: thrills ahead' or 'Adrenaline rush guaranteed'. Yet when holidaymakers sign their waiver forms and climb into a raft for a trip billed as 'The Ultimate Adventure', they are expecting an unforgettable experience. They are not expecting their trip to be the last thing they will ever do.[44]

THE COMPLEX PATTERN OF CURRENTS IN LOWLAND RIVERS

On straight reaches of the river, the current flows fastest below the surface, mid-stream. But a river rarely runs straight for long.[45] Instead, the water carves out sinuous bends along its course. On the outer side of each bend, the water surges against the bank, undercutting it and scooping the riverbed into deep pools. On the inner side of the bend, where the current is less fierce, the water sheds its suspended load of eroded sand and gravel. As the water swerves around each bend, the surface layers on the outer side plunge downwards towards the riverbed in a spiral secondary circulation which twists across the river towards slacker water opposite. These secondary currents may prevent swimmers from reaching the safety of the riverbank, catching them and tugging them out towards the faster flow in the middle of the river, where they quickly tire.

Sudden transitions from slow-moving to fast-moving water are common. Anglers call them creases and watch out for them from the riverbank, knowing that hungry fish hold their station on the slow side of the crease, ready to snap up floating insects and baited hooks.[46] But swimmers don't notice creases, and they are often taken by surprise by a sudden increase in the force of the current. On a hot summer day in 2013, a young man decided to cool off with a swim in the River Severn. The current caught him and pulled him deep underwater. Divers found his body the next day.[47] In 2017, a teenager drowned in a similar accident in the Thames near Hampton Court Bridge.[48]

People who set off to swim across a river often find that a short direct route is impossible. Instead, the current propels them into a longer diagonal swim. They are likely to find themselves pulled along into deeper, wilder water, with treacherous undercurrents. Within a moment, a brief dip can become a desperate struggle for survival. In 2009, after an outing to celebrate his eighteenth birthday at the Henley Regatta, a student decided to swim across the River Thames with a college friend. His friend found the swim a lot harder than he had expected, but he managed to reach the opposite bank. The student lost his strength in the middle of the river, and although his friend swam back to him and made vigorous attempts to tow him to safety, he sank and drowned.[49]

Even in sunny weather, rivers are usually dangerously cold. In 1999, several school friends set off to swim across the River Thames at Wallingford. Halfway across, one of the boys began to struggle in the water after developing cramp. Another boy managed to grab him briefly before he disappeared beneath the surface.[50] In 2016, a sixteen-year-old boy swam across the River Thames near Abingdon. He struggled to pull himself onto the opposite bank and lay exhausted on the ground. His friends shouted across that they would drive round and pick him up. Instead, he decided to swim back across the river, but he lost power, drifted and sank.[51] Divers found his body two days later.

A river may appear shallow enough for paddling, only to deepen suddenly. In 2010, a group of children were playing in a shallow bay at the edge of the River Kelvin in Glasgow. When a girl stepped off a ledge into deeper water, a boy jumped into the river and saved her. But the current swept him beneath the undercut riverbank, where he drowned.[52] Near Donnington Bridge in Oxford in 2015, a thirteen-year-old boy was playing piggy-in-the-middle with other teenagers in waist-deep water. Then the current carried him out of his depth, and he drifted downstream before disappearing under the water surface.[53]

◆ ◆ ◆

Anyone who falls into a river fully clothed has to contend with the tug of the current on their spreading garments. Non-swimmers drown unless they are rescued immediately. Weak swimmers are often swept away. Even competent swimmers have to struggle to reach the riverbank. And if the bank is steep or hollowed out, it may be difficult to climb out of the water, especially if weighed down by waterlogged clothes.[54] A ten-year survey of fatal accidents in Oxfordshire and West Berkshire recorded that 18 children drowned in the River Thames, 16 after falling into the river.[55]

Soon after arriving for a walking holiday in Grasmere in February 2011, a woman set off for a walk beside the River Rothay. Heavy rain had just cleared, but the path was wet and slippery. Shortly after she left her hotel, a local businessman saw her body floating in the river.[56] At first, he thought he was looking at a cardboard cut-out, then he realised that someone had fallen in and was being swept away. He ran along the riverside path, intending to jump in, but the river was flowing too fast for him, so he called the police. Later in the afternoon, a search and rescue team found the walker's lifeless body beside the island in Grasmere Lake, carried there by the River Rothay. On a warm evening in May 2012, two Oxford teenagers dared each other to jump into the river from Donnington Bridge hand in hand, 'for a joke'. The girl surfaced at once, but her companion could not swim. He struggled briefly before he disappeared underwater.[57]

People are doubly vulnerable after a few beers. In 2011, after a heavy drinking session in a riverside pub, a promising law student tried to swim across the Thames. He was only ten feet from the other bank when he stopped swimming.[58] A rowing instructor threw him a rescue buoy, but instead of grabbing it, he sank and drowned. In 2014, there were three fatal drowning accidents in the Ouse and the Fosse, rivers that flow through York's historic city centre. The young victims had each spent their last evening drinking with their friends. A fine art student and a musician both fell into the water after leaving nightclubs in the early hours. The third victim, a soldier, decided to swim across the River Ouse. He was cheered on by his intoxicated companions, who failed to recognise the danger until they saw him disappear midstream.[59] Later that year, the Royal Life Saving Society began its campaign 'Don't Drink and Drown' at the University of York.[60]

Rivers flowing through industrial areas or town centres are often confined between steep banks made of brick or stone. Someone who falls into the water may be able to reach the bank and grab a projecting handhold, only to be trapped because they cannot pull themselves out of the water. Soon the cold weakens their grip, and the current wrenches them away.[61] In York, the death rate from accidental drowning was three times the national average, and in 2014, the City Council asked the Royal Society for the Prevention of Accidents (RoSPA) for safety advice.[62] On inspecting the rivers, RoSPA found 'numerous low level slip/trip/fall features that increase the likelihood of a fall into water', while at many locations, self-rescue would be difficult or impossible. RoSPA advised a number of physical safety improvements including placement of lifebelts, river ladders, grab rings and chains along the engineered vertical walls edging the river, with extended fencing, grab rails and improved street lighting along the riverbanks. In 2016, although the suggested improvements had been completed, three more people drowned in the rivers in York.[63]

In Durham between October 2013 and January 2015, three university students fell into the River Wear and drowned while walking home alone after prolonged sessions in city centre pubs. Donations by the family and friends of one of the victims enabled the Royal Lifesaving Society to provide water safety education packs to every school in County Durham, to organise demonstrations of rescue equipment by the Fire Service, and to teach lifesaving skills to more than a hundred pupils. The Police considered introducing on-the-spot fines for drunken behaviour to discourage undergraduates from putting themselves at risk.[64] The 'Don't Drink and Drown' campaign was repeated, with leaflets showing a canal-side footpath at night, captioned: 'From happy hour to nightmare in just one slip.'[65]

In Bristol between January and June 2017, the bodies of six young people were found dead in the river or the docks, three after an evening of heavy drinking.[66] During Freshers' Week in September 2017, as part of the 'Don't Drink and Drown' campaign, the Royal Life Saving Society and the Avon Fire and Rescue Service used a floating nightclub moored at the waterside in Bristol Harbour as the venue for a drowning prevention event.[67] Hundreds of students from the city's music college and the two universities were warned about the dangers of skinny-dipping and advised to take a well-lit route when walking home from waterside pubs, away from narrow paths beside rivers and canals.

WEIRS

Weirs dam back the water flowing down a river to maintain an adequate depth for boating. There are 45 weirs on the River Thames alone, [68]each with its lock to allow boats to bypass the weir and travel from one reach of the river to the next. Children often paddle and fish in the calm water above a weir or walk on the slippery structure of the weir itself. Weirs are relatively safe when there is little water spilling over the crest. However, a number of weirs are extremely dangerous, especially when the river is swollen by recent rainfall.[69] As the river pours over the crest of the weir into the stilling basin below, the plunging water forms treacherous re-circulating counter-currents – known as keepers – which trap anybody who falls in, swirling them repeatedly beneath the surface in the violent, deadly backwash.

Beside the River Coquet at Guyzance in Northumberland stands a memorial to 10 soldiers who drowned in a river-crossing exercise in 1945 when their craft was swept downstream and overturned at the weir.[70] In July 2010, a young man walked past the monument with two friends. They were discussing the wartime accident when, on an impulse, he jumped into the river just below the weir. He was caught in the counter-current and pulled underwater.[71] At first his friends thought that he was play-acting as he fought to reach the surface, but when they realised that he was in deadly earnest, they made desperate attempts to pull him to safety. He was swept out of their grasp. His body was found after a three-hour search.

In 2012, on the River Avon in Warwickshire, a family outing ended in tragedy when a rowing boat was carried over a weir and capsized, throwing a man and three children into the churning water.[72] Neighbours managed to pull two of the children to safety, but a three-year-old boy and his father drowned only yards from their new riverside home.

Cromwell Weir on the River Trent is a notorious danger point. The stilling basin beneath the weir is known locally as the Devil's Cauldron. In 1975, with the river in flood, loss of lighting left Cromwell Lock in complete darkness. During a night-time navigation exercise, ten paratroopers drowned when their boat went over the weir.[73] Only one soldier survived, pulled into a rowing boat by local policemen. In June 1981, a cabin cruiser went over the weir. Two children were rescued from the river by a Royal Air Force helicopter, but two adults drowned.[74] After this second fatal accident, a floating safety boom was placed across the river above the weir, strong enough to restrain a boat on the river, but allowing branches and other debris to slip below the boom.

In January 2012, two anglers were fishing for pike below the weir when their small boat was caught by the recirculating countercurrents and pulled upstream into the stilling basin.[75] By the time firemen trained in swift water rescue managed to pull them from the water, the force of the torrent had stripped one man of his life jacket and most of his clothes. He died later in hospital. His friend needed more than a week in hospital to recover from hypothermia and the effects of his immersion.

In October 2013, an elderly couple were making their way down the flooded River Trent in a narrowboat. As the man stepped onto the bank to moor the boat, the mooring rope slipped out of his hands. With his wheelchair-bound wife on board, the narrowboat was carried towards Cromwell Weir, where only the safety boom stopped its progress.[76] A Sea King helicopter lifted the woman to safety.

THE TIDAL RIVER THAMES

The Thames is tidal as far upstream as Teddington Lock, with more than thirty bridges crossing the river as it winds through London. Anyone who falls into the river is in the gravest danger, swept along by the fast, surging water as the tide runs in and runs out, caught in swirling eddies round every pier and buttress. Even at 'slack tide', the river is in motion, and intensely cold.

In the days when the River Thames flowed under the historic Old London Bridge, with its nineteen arches and southern drawbridge, the river level was often six feet higher on the upstream side of the bridge than the downstream side.[77] The stone pillars and the projecting stone starlings at their bases held back so much water that ferrymen gave their passengers a choice: they could either shoot the rapids under the bridge or land upriver and re-embark past the bridge. Samuel Pepys preferred to get out and walk, knowing that many of the boats were swamped and their unfortunate passengers registered as 'Drowned at the bridge'.[78]

Old London Bridge was torn down in 1832, removing the obstruction to river flow that had stood for almost seven hundred years. Victorian engineers dredged the river and constructed smooth stone embankments, so that the River Thames is now 'half as wide and twice as fast'.

In 1934, beside Tower Bridge, the stony foreshore in front of the Tower of London was converted into Tower Beach by bringing in 1,500 barge loads of sand. East End families were able to climb down a flight of stairs to picnic beside Traitor's Gate.[79] They paddled on the narrow strip of land exposed for a few hours at low tide. It was not an ideal playground. As one elderly East Ender recalled: 'There was a bit just along the edge for paddling, and we knew we mustn't step over the barrier which we could feel under the water, because on the other side it was really deep. We couldn't see the barrier, and we couldn't see our feet and ankles either, because the water was black as ink.' A lifeguard patrolled the beach, and a rowing boat patrolled the river, but danger was ever-present, from the racing current and from the inrush of the rising tide. (At Tower Bridge, the twice-daily tides on the River Thames rise and fall by as much as 23 feet.)[80] In 1971, Tower Beach was closed 'owing to pollution and the water being deemed unsafe to bathe in'. The sand has long since washed away.

In London's Dockland, warehouses once held cargoes from around the world. Now high-rise commercial buildings have taken their place, and the major docks have moved downriver. Yet the river remains busy.[81] Above Putney Bridge, rowing, boating and sailing clubs predominate. Between Putney Bridge and the Thames Barrier, passenger vessels share the river with barges, cruise ships and more than forty rowing and sailing clubs. Below the Thames Barrier, the boats are bigger, with coasters, container ships and oil tankers. Considering the currents, the tides, the bridges, and the weight of traffic on the river, major accidents are uncommon.

In August 1989, in a horrific night-time collision close to Southwark Bridge, the chartered passenger cruiser *Marchioness* was rammed by the dredger *Bowbelle*. The cruiser was torn apart and sank in less than a minute, too fast for any organised evacuation.[82] Within moments, 51 young people were drowned – 24 victims were trapped below deck, and 27 more, including the Master, were swept away by currents made more turbulent by the incoming spring tide rushing up the river. Some survivors were carried under Blackfriars Bridge, half a mile upstream.

The *Marchioness* had seven life rafts, each capable of supporting twenty people. They floated to the surface, together with seven lifebuoys, but they were inconspicuous in the dark. Only a score of the desperate people managed to grab a life raft and hold on until rescue came. Many of the 80 survivors were pulled from the water by the crew of a nearby passenger cruiser and by river police: 22 survivors were hauled onto a police launch designed for two passengers and a crew of three.[83]

Two inquests, two trials, several court hearings and a number of official enquiries followed the tragedy. To enhance river safety, long lists of recommendations were put forward and acted on.[84] Search and rescue services on the tidal Thames were boosted by the opening of four Royal National Lifeboat Stations – at Teddington, Chiswick, Tower Pier and Gravesend.[85]

In 2013, London lifeboat crews rescued 372 people from the River Thames, including 25 people who would have died within minutes but for the speed of the rescue launches and the skill of the lifeboatmen.[86] At Isleworth, a two-year-old girl fell through railings into the River Thames 12 feet below.[87] Her mother immediately climbed over the railings, jumped into the river, and succeeded in catching hold of the child. They were spotted by a passer-by who rang 999. The water was 8 feet deep and the tide was coming in. A rope was thrown, but it failed to reach them. Mother and child owed their lives to the fast response of the Chiswick RNLI Lifeboat, which was launched within 90 seconds and arrived within 7 minutes. Darkness was falling as the crew pulled them from the river, chilled but otherwise unharmed.

In sunny weather, people often swim in the Thames, at risk to their lives. In 2012, the Port of London Authority introduced a new bye-law to help people resist the temptation to enter the water: it is now an offence to swim in the busiest stretch of the tidal Thames (between Putney Bridge and Crossness, downstream from the Thames Barrier) without the prior written consent of the Harbour Master.[88] Despite the bye-law, the river continues to tempt Londoners to take the risk. Over five years – 2012 to 2016 – there were 42 fatal drowning accidents in the tidal Thames.[89]

CHAPTER EIGHTEEN

FLOODS AND STORM SURGES

In Britain, we are spared the catastrophic floods brought on by monsoons, typhoons, and tsunamis. Instead, we have rainy weather throughout the year. The prevailing wind arrives from the west, laden with water after crossing the Atlantic Ocean. As the wind blows across the mountains in the Scottish Highlands, Cumbria and Snowdonia, the air rises and cools to dewpoint. So western counties get more rain, whereas eastern counties lie in a relative rain shadow.[1]

RIVER FLOODING AND FLASH FLOODS

Rain soaks into the ground and runs off into rivers. When the land is already saturated from previous rain, rivers fills very quickly. If the river reaches 'flood stage' and overtops the riverbank, the water spills into water meadows, submerges farmland and destroys crops. When the flood stage reaches a pinch point downstream, the swollen river inundates riverside towns and villages, ruining houses, wrecking businesses and threatening lives.

In towns, rain falls onto waterproof surfaces – tiled roofs, concrete pavements and acres of tarmac. Rainwater is channelled directly into drains. In heavy downpours, drainage systems are rapidly overwhelmed.[2] Floodwater and sewage erupt from manhole covers, creating deadly pitfalls for anyone walking along the flooded pavements (including BBC reporters filming the flooding).[3]

In 2007, after the wettest June since 1882, the widespread flooding resulted in 'the country's largest peacetime emergency since World War Two',[4] causing enormous damage and claiming 13 lives. In Sheffield, the River Don burst its banks. Hundreds of people were trapped in homes and offices and had to be rescued by emergency services or airlifted to safety.[5] The railway station was evacuated and the bus service stopped running. A fourteen-year-old boy phoned his parents to let them know that his school bus had been cancelled and he was walking home with friends. Their route took them through Millhouses Park beside the River Sheaf just as the river overflowed its banks. The children splashed about in ankle-deep water and slid across the flooded grass. They could not see where the flooding ended and the river began. The boy's last slide took him over the edge of the riverbank and the water carried him away.[6] Later that evening, a man in his sixties was returning home when he found his way blocked by floodwater under Newhall Bridge. He decided to wade through, but the road dipped beneath the bridge and he was soon in water up to his neck, clinging to girders under the bridge in pitch darkness. Three men risked their lives as they rescued him, but he died in hospital the next day.[7]

Wide tracts of Gloucestershire were underwater. The rising floodwater put a large water treatment plant out of action, and troops had to be called in to protect two electricity substations supplying power to Gloucester and South Wales.[8] At Tewkesbury, built on the floodplain at the confluence of the Avon and the Severn, both rivers overflowed. Only the twelfth-century Abbey and a handful of Tudor buildings on higher ground escaped inundation.[9] After an evening in a nightclub, a young man drowned while walking home through the floods. At the Inquest, the Coroner said. 'I'm afraid that intoxication clouded his judgement and I am satisfied he didn't realise the danger he was placing himself in.'[10]

In Hull, floods affected hundreds of houses built on reclaimed marshland beside the tidal River Hull.[11] Pumping stations failed, and the drainage system was overwhelmed.[12] Floating debris blocked storm drains, preventing the escape of floodwater. A young man was clearing debris from a drain behind his workplace when his foot slipped through a metal grille guarding the mouth of a culvert. Floodwater gushed past him into the culvert and the powerful suction held him fast.[13] Neighbours, firemen and police divers fought desperately to free him. A fire engine and a police Land Rover failed to dislodge the grille. Metal-cutting equipment did not work under water. A surgeon was brought in to consider the underwater amputation of his leg, but the operation was impossible. Despite pumping, the water level rose to his chin. After a struggle for survival that lasted for four hours, he lost consciousness and died of hypothermia.[14]

Sir William Pitt chaired an independent review of the country's flood defences. Completed within a year, the 500-page report, *Learning Lessons from the 2007 Floods*, made 92 suggestions for forecasting and mapping of floods, building flood defences, maintaining power and water supplies, improving flood warnings, organising emergency rescue services, and setting limits to house-building on floodplains.[15]

◆ ◆ ◆

On 19th November 2009, a rain gauge at Seathwaite in Cumbria recorded the highest 24-hour rainfall in Britain since records began 200 years ago:[16] more than a foot of rain (314mm) fell in a single day. The rain fell on saturated farmland, and the enormous run-off exceeded the capacity of the rivers flowing through Keswick and Cockermouth.[17] In Keswick, 250 properties were flooded. In the historic centre of Cockermouth, the water was 8 feet deep, and the emergency services rescued more than 200 people from their houses and shops. The flood wave rushed down the Derwent valley, swelling the river to 25 times its usual size, causing landslides, tearing away hundreds of trees, and destroying four bridges.[18] Fifteen miles downriver, on Northside Bridge in Workington, a policeman was directing cars to safety when the sturdy sandstone arches collapsed beneath him and the flood swept him to his death.[19]

After the 2009 flooding in Keswick, six million pounds was spent on embankments, floodgates, and flood walls.[20] Reinforced glass panels preserved views of the idyllic Lakeland scenery. These traditional 'hard engineering' barriers proved inadequate. In December 2015, extreme water levels overtopped the flood walls and glass panels, and 500 homes and businesses were flooded.[21]

A different approach to flood control was taken to protect the market town of Pickering, which lies in a gorge draining water from the North Yorkshire Moors. Pickering suffered serious flooding four times between 1999 and 2007.[22] Instead of expensive concrete barriers to confine the water in Pickering Beck, local and national agencies worked together to design a sustainable drainage system.[23] They planted extensive stretches of woodland above the town. They dug out a wide, shallow reservoir capable of storing 120,000 cubic metres of floodwater and releasing it slowly. They constructed dozens of 'leaky' dams of logs and branches, which let normal flows through, but slowed the escape of water after heavy rain.[24] In recent years, despite heavy rainfall, there have been no floods in Pickering.[25]

✦ ✦ ✦

When people are threatened by flooding, much of their attention is focused on preventing floodwater from entering their homes and destroying their possessions and livelihoods. They may fail to recognise the threat to their own safety – until it is too late.

After heavy rain in 2008, youngsters were playing in a flooded meadow in Oxfordshire when a schoolboy slipped into a channel leading to a storm drain. The force of the water escaping down the underground culvert pinned him face down against the grille guarding its mouth.[26] His friend jumped into the channel to help him and was also trapped by the current. Firefighters fought to pull the boys free. Then they tried to resuscitate the schoolboy, without success. His friend was admitted to hospital, suffering from shock and hypothermia.

In Denbighshire in November 2012, a 91-year-old woman was warned about approaching floods. She told emergency workers that she wanted to remain in the warmth of her home and finish her breakfast. Later in the morning, the River Elwy burst its banks. Hundreds of homes were flooded and the 'fiercely independent' pensioner was drowned.[27] The Coroner delivered a Narrative Verdict confirming that she had twice been warned of the danger but declined the offer of help to leave. He said, 'It is likely at that time that she could have had no anticipation of the eventual extent of the flooding as no water had yet reached her door. In any event no power existed by which she could be forcibly removed from her home.' He added that lessons had been learned by the city's flood response teams, making the city a safer place than before.

After flood warnings in 2014, the Royal Life Saving Society advised householders: 'Keep an eye on weather reports for flooding in your area', and 'Prepare a flood kit (including a torch and a charged mobile phone) in case your home floods or you are trapped in a vehicle'.[28] To protect themselves and their families, people were advised not to travel in heavy rainstorms unless absolutely necessary, never to walk, swim or drive through floodwater, and never to allow children or pets to play in floodwater.[29]

Most river flooding is caused by water draining from upland catchments after heavy winter rainfall, but summer cloudbursts are also capable of causing dramatic flood damage and loss of life. On 15 August 1952, a thunderstorm poured 9 inches of rain onto Exmoor in North Devon. Floodwater escaped down steep gullies into the East and West Lyn rivers, wrenching trees and boulders from the riverbanks. The torrent plunged down the steep escarpment onto the houses, hotels and holiday camp-sites of Lynmouth. During the night, 34 people died in the flash flood.[30]

On 16 August 2004, in Boscastle on the northern coast of Cornwall, a similar localised thunderstorm at the head of a quiet valley caused a violent flash flood that uprooted trees, destroyed bridges, and demolished historic houses, shops and hotels.[31] No lives were lost in Boscastle, thanks to the heroic actions of the villagers and a major rescue operation. Seven helicopter crews airlifted 120 people stranded on rooftops. A fifteen-month-old girl was secured in a rucksack and winched to safety. A young chef pulled on his wetsuit and waded into deep water beside a car park. 'The current was so strong and they were all rolling around and being swept away,' he said. 'I managed to grab them as they came whirling past me like fish. In the end I must have saved eight or nine lives of all ages.'[32]

The danger of taking refuge in a car during a flood was demonstrated by a startling statistic. During a few hours of mayhem in Boscastle, the torrent raging down the valley swept more than sixty cars into the sea.[33]

CARS FLOAT IN FLOODS AND SINK IN RIVERS

People underestimate the risk of driving a car along a flooded road. They forget that a car engine will stall in water only a few inches deep. Worse, in two feet of water, the upthrust of buoyancy will float the car off the ground. If this happens, the car loses traction and the driver cannot steer the car.[34] It is often safer to turn back and search for another drier route.[35]

In Worcestershire in June 2007, a county court judge was driving home when floodwater lifted his car and swept it down the road. He could not escape from the car, so he used his mobile phone to alert his wife that floodwater was washing him off the road. Searchers found him the next day, drowned in his submerged car, more than 50 yards from the point where he left the road.[36] In April 2012, prompted by their sat nav, a couple tried to drive across a flooded ford in a Hampshire river. The water lifted the car from the roadway, then the current swept it 200 yards downstream. The driver managed to escape through the window and swim to safety, but her husband's attempt to rescue the family dog cost him his life.[37] He was trapped and drowned as the car overturned and sank.

Cars parked too casually by the waterside may roll into the water. In Surrey in 2000, a man was about to drive his son and his five-year-old granddaughter to the airport when he realised that he had left his spectacles in the house. He left the car engine running while he went to fetch them. The car began to move, crashed through a fence and rolled into the River Thames. His son escaped through the passenger door, but as he struggled in the water, the car floated out of his reach, turned over and sank – with the little girl still strapped into her car seat.[38] In Cardigan in 2018, a young woman parked on a slipway beside the River Teifi and left her two-year-old toddler in the car while she made a brief visit to her husband's office. When she returned a couple of minutes later, the car had disappeared. She phoned the police immediately, reporting that her car had been stolen with her child inside. In fact, the car had rolled down the slipway into the river.[39]

If a car plunges into deep water, the occupants have a couple of minutes at most to unfasten their seatbelts, open the window, and make their escape. The car will float for a few moments. Then as water enters, it will sink to the bottom engine-first, often coming to rest on its side or its roof. External water pressure holds the doors shut until the car has almost filled with water.[40] By that time, it may be impossible for the terrified, disoriented occupants to get out of the car. Recent innovations improve car safety and prevent theft, but they increase the difficulty of getting out of a sinking car. Electrically powered windows cannot be opened. Laminated glass cannot be broken. Airbags inflate, obstructing movement in front seats. Child seats and safety belts hinder escape from back seats.[41]

In 1998, two elderly sisters were imprisoned in their sinking car in a Highland river. A brave schoolboy swam out to the car, dived beneath it and tied a rope to a back wheel. A dozen rescuers pulled the car 30 feet to the riverbank. The old ladies had to be lifted to safety through the sun roof.[42]

In July 2001, a man lost control of his car as he drove round a park in Bedfordshire. The car careered over a grassy ledge beside the lake, splashed into the water, overturned, and sank to the bottom, 30 feet down. The driver and two adult passengers escaped from the sinking car, but three children were trapped in the back seat.[43] Risking their own safety, a paramedic and two firefighters dived down repeatedly, trying to reach the children. More than an hour went by before they were able to attach a rope, and a fire engine pulled the car to the surface. Despite prolonged efforts at resuscitation at the scene and in hospital, all three children died of drowning. The unlicensed driver had never had a driving lesson. He was found guilty of manslaughter and sent to prison.

On a family outing to the harbour at Buncrana in Donegal in March 2016, a car skidded down an algae-covered ramp and slid into Lough Swilly. A young man saw the accident, stripped off his clothes and swam out to the car. The driver broke the side window with his elbow and handed over his four-month-old daughter, saying, 'Save the baby.' A moment later, the car sank, taking five people to the bottom of the harbour.[44] The man who rescued the baby was awarded the Michael Heffernan Gold Medal for Marine Gallantry.[45]

In 2008, the Royal Society for the Prevention of Accidents warned that over a single year (2005), 33 vehicles plunged into deep water after traffic accidents on British roads and 28 of the occupants were drowned.[46] The low-lying farmland of East Anglia is drained by ditches, dykes and fens that lie close beside raised, narrow roads.[47] In December 2005, a father and son were drowned when their car ran off the edge of an embanked road at the side of a Cambridgeshire fen.[48] Only six weeks later, three people drowned at the same spot when their car clipped an oncoming vehicle and went into the water.[49] In August 2010, a pregnant young mother escaped through the window of her car after she crashed from a pot-holed track in Lincolnshire into a dyke 10ft deep. She dived into the water, groped for a door handle of the upturned car, released her two-year-old daughter from her car seat, and got her to the bank. But her teenage son drowned in the car.[50]

In America, as many as 350 people die every year when cars plunge into water. In 1990, researchers studied fatal car immersions in Sacramento County, California. The traumatic injuries were reported at post-mortem as 'generally minor if present at all'. Instead, 76 of the 77 victims had died of drowning, usually as the sole cause of death.[51] Yet, as the researchers pointed out, 'By convention all such deaths have been classified as motor vehicle deaths, rather than drownings.' The roads curved sharply at many of the accident sites, suggesting that lives could be saved by placing guardrails at tight bends on waterside highways to prevent cars from veering off into the river.

STORM SURGES AND COASTAL FLOODING

In fine weather, the atmospheric pressure on the surface of the sea measures around 1013 millibars.[52] If the atmospheric pressure falls (weather forecasters call this 'a depression'), the sea level rises. For each millibar fall in atmospheric pressure, the sea level rises by one centimetre.[53] In storms, the atmospheric pressure frequently falls by 30 millibars or more, so that the level of the sea rises by 30 centimetres or more, drawn into a wide dome of water, lashed into enormous waves by gale-force winds. When a storm travels from the open ocean into shallower sea near the land, this huge volume of water is thrown against the shore – as a storm surge.[54]

People fear the strength of the wind when they are menaced by hurricanes in the Caribbean or cyclones in the Far East, but most deaths are caused when giant storm surges reach the land and crash through fishing villages, seaside resorts, and coastal cities. On 7 September 1900, the first swells from an approaching hurricane rolled up the beach of Galveston, a busy port on a barrier island off the coast of Texas, where the highest ground was less than 9 feet above sea level. When the hurricane arrived a day later, a storm surge 15 feet high engulfed the island, killing more than 8000 people.[55] On 12 November 1970, half a million people perished when a cyclone in the Bay of Bengal produced a storm surge that a devastated 3,000 square miles of low-lying farmland in East Pakistan (soon to become Bangladesh).[56] On 29 August 2005, Hurricane Katrina hit the coast of Louisiana with a storm surge 28 feet high. In New Orleans, protective levees along the Mississippi gave way: 80% of the city was flooded and almost 2,000 people drowned.[57]

In Britain, it was the storm surge that made the Great Storm of 1953 so deadly. On the night of 29 January, there was a total eclipse of the Moon. The Earth lay directly between the Sun and the Moon, resulting in a high spring tide.[58] At the same time, an Atlantic storm gathered strength as it travelled past Orkney and turned south-east into the North Sea. By 31 January, the atmospheric pressure had fallen to 966 millibars, sucking seawater into an extensive mound half a metre high. Hurricane-force north winds propelled the storm surge into the shallow waters off the eastern counties of Lincolnshire, Norfolk, Suffolk, Essex and Kent. The sea rose 10 feet above its expected level. Lifted on this hill of water, 30 foot waves crashed upon the land.[59]

During the hours of darkness, sea walls were breached in more than a thousand places. Waves threw boats over harbour walls, tore houses from their foundations, smashed through factories and oil refineries and inundated 160,000 acres of land. The storm surge reached Canvey Island in the Thames Estuary after midnight, drowning 59 people, many in their beds.[60] The river walls collapsed near the Royal Victoria Docks and hundreds of low-lying houses in Canning Town and Silvertown were flooded.[61] The water level in the River Thames rose 6 feet above the predicted spring high tide level, reaching the top of the Embankment from London Bridge to Chelsea. In Britain, 307 people were drowned during the storm.[62] Across the Channel in Holland, three layers of protective dykes gave way. As the sea burst through, more than 2,000 people lost their lives.[63]

THE THAMES BARRIER PROTECTS LONDON FROM FLOODING

Since 1983, London has been protected from extensive flooding by the Thames Barrier, the brilliant result of many years of plans, designs and construction.[64] Stretching a third of a mile across the River Thames at Woolwich, ten enormous tilting gates rest in sills on the river bed. When the gates are open, river traffic, from small boats to naval warships, passes through the Thames Barrier unimpeded.[65] When tidal conditions threaten, the gates swing upwards, presenting a solid wall of steel to the incoming sea. The water level downstream from the closed Barrier often rises 12 feet higher than the water level upstream.

The Thames Barrier also protects London from river flooding after heavy rain. By closing the gates soon after low tide, the Barrier holds back the seawater that would flow upriver with the next rising tide. This allows the Thames upstream of the Barrier to act as a huge reservoir, with enough space to hold the extra water arriving from the river's rain-soaked catchment area.

The gates of the Thames Barrier first swung upwards to hold back the tide on 1 February 1st 1983. By February 2014, the gates had been raised to protect London from flooding on more than 160 occasions.[66] In years to come the Thames Barrier itself may be overtopped, since the sea level is slowly rising[67] and London is slowly sinking into its bed of clay. As increasing numbers of people settle in the Thames 'floodplain', an even more ambitious barrier will be needed, closer to the sea.[68]

CHAPTER NINETEEN

AT THE SEASIDE: TIDES, WAVES AND RIP CURRENTS

Along eleven thousand miles of British coastline,[1] the sea responds to the powerful forces of the tides, waves, winds and currents. The water surges forward and back, up and down, left and right, its incessant motion enough to overcome the most powerful swimmer. Every year, more than one hundred people drown at the seaside.[2]

THE RISE AND FALL OF THE TIDES

Twice a day the tide goes out in Morecambe Bay, exposing beaches miles wide. Twice a day the tide comes in, racing over the sands 'faster than a horse can gallop'.[3] At low tide on the Dorset coast, the beach at Lyme Regis is strewn with fossils. At high tide, the sea hurls itself against the base of the crumbling cliffs.[4] The ever-changing seashore intrigues, entices and delights us. But the incoming tide can trap and kill.

The rise and fall of the tide is governed by the gravitational attraction between the Earth and its closest neighbours, the Moon and the Sun.[5] The pull of gravity is strongest directly beneath the Moon, drawing seawater into a low, wide mound.[6] The Moon's attraction is weakest on the opposite side of the globe, allowing centrifugal force to draw seawater into a second mound. The Sun is far larger than the Moon, but it is much further away, so that its gravitational pull is weaker. Lesser mounds of seawater form directly beneath the Sun and on the opposite side of the globe.

As the Earth spins on its axis, the implacable pulls of the Moon and Sun set the oceans in motion. Yet mounds of seawater cannot move uninterrupted around the globe. The land gets in the way. In the landlocked waters of the Mediterranean, the high tide mark is rarely more than 3 feet higher than the low tide mark. In contrast, on the eastern coast of Canada, so much water pours in and out of the Bay of Fundy that the water level at high tide is 53 feet higher than the water level at low tide, the greatest tidal range in the world.[7]

Along much of the British coastline, high tide lifts the sea 10 to 20 feet above the level at low tide. In the Bristol Channel, however, the tidal range is almost as great as in the Bay of Fundy, rising and falling as much as 49 feet between one low tide and the next.[8] As the sea comes in, it advances with startling rapidity, covering the dinosaur footprints embedded on the rocks of Glamorgan to the north[9] and the sandy beaches of North Devon and Somerset to the south.

Tragic consequences follow when people forget or ignore the rise and fall of the tides. In June 2013, two teenage girls were paddling along the beach on Barry Island off the Glamorgan coast. The sea was calm, so they decided to see how far they could walk out into the sea, fully clothed. As the tide came in around them, they stood on tiptoe 50 metres from the water's edge, up to their chins in the water. Soon they were struggling to keep their heads above the waves. One girl managed to swim back to the beach, but her friend sank before she reached safety.[10] Two days later, after an intensive search, her body was recovered five miles along the coast.

Sully Island is connected to the Welsh mainland by a causeway half a mile long. Access to the island is possible for only three hours before and after low tide.[11] Over the years, hundreds of people have been stranded overnight on the tiny, windswept island. Visitors trapped on the causeway by the rising tide must be rescued by lifeboat or must swim for their lives in the cold, fast-flowing sea.[12] In May 2013, a couple were saved from drowning by two brave local men who swam out to them and dragged them ashore.[13] In July 2014, the Royal National Lifeboat Institute installed the world's first Tidal Traffic Light,[14] which allows holidaymakers to check whether they have time to get to the island and return before the causeway disappears beneath the waves. Despite the Tidal Traffic Light, lifeboats are called out to Sully Island even more often than before.[15]

Holidaymakers throng the wide beaches of North Devon every summer. Lifeguards patrol the waterline, trained and equipped to rescue swimmers struggling in the water. But lifeguards were not on duty on New Year's Eve 2013, when a family group walked out across the sands at Croyde Bay, clad in wetsuits and carrying body-boards. It was almost low tide, and the sun was shining. A witness saw them jumping the waves in water little more than knee-deep. But the tide was still going out, and within twenty minutes, two of the party were dragged out to sea. One woman battled her way back to the shore. Her sister-in-law was swept out into deeper, rougher water. She was found two hours later, floating face down in the sea.[16] She could not be revived.

Further east along the Bristol Channel, near the estuary of the River Severn, wide expanses of mudflats are exposed at low tide. Wind and sun dry the mud into a surface crust. As the tide comes in, seawater floods beneath the crust.[17] Then, when people walk along the shore, the crust breaks under their feet, trapping them in the underlying mud. Near Burnham-on-Sea in 2002, a five-year-old girl rushed towards the water's edge as the tide came in. As she neared the sea, her feet sank through the crust. Frantic rescuers, too, were hindered by the mud. Before her family could reach her, she was submerged by the waves.[18]

Within a year of the tragedy, an appeal raised enough money to buy a hovercraft able to skim over mudflats previously inaccessible from either land or sea.[19] During its first year of operation, fifty people were rescued from this dangerous stretch of coastline. Now two hovercraft manned by volunteers work closely with the inshore lifeboats of the Royal National Lifeboat Institution (RNLI) to provide emergency rescue for people and animals trapped on the shore.[20]

◆ ◆ ◆

The relative positions of the Sun and Moon change over the course of a month, affecting the rise and fall of the tide. The gravitational attraction of Moon and Sun act together at New Moon (when the Moon lies between the Sun and the Earth) and two weeks later at Full Moon (when the Moon and Sun lie on opposite sides of the Earth).[21] The combined attraction produces higher than usual high tides, and lower than usual low tides. These are the fortnightly spring tides.

One week after New Moon (at First Quarter) and one week after Full Moon (at Third Quarter), the Moon and Sun have moved out of line. The forces of their gravitational attraction now compete with each other, limiting the sea's rise at high tide and its fall at low tide. These are the intervening neap tides. (Spring tides and neap tides take their names from Anglo-Saxon words meaning 'to swell' and 'to become lower'.)[22]

This regular variation in tidal range comes as an unpleasant surprise to those holidaymakers who wrongly assume that 'spring tides' only occur in springtime, rather than every two weeks throughout the year. If they arrive at the time of a neap tide, they expect the edge of the sea to continue its same moderate advance and retreat throughout their stay. A few days later, when they set off to explore a sandy bay, they find that the faster, higher rise of the spring tide reaches the headland before they do, cutting off their return.

Coastguards, lifeboats and helicopters are frequently called out to rescue panic-stricken holidaymakers trapped by the rising tide. In April 2012, four visitors were cut off by a spring high tide in a Cornish bay. Three were rescued by lifeboat. Their companion was clinging desperately to the sheer cliff face, drenched by the waves, when a Royal Navy rescue helicopter arrived and winched her to safety.[23] The next day, four more visitors were cut off at the same spot and the rescue helicopter was summoned again. Three had to be winched from the cliff, but a woman slipped and fell into the sea. She was pulled aboard a lifeboat and taken to hospital with hypothermia. 'These incidents could have had a fatal outcome,' the coastguard commented. 'People were not aware of the state of the tides and clearly did not realise how quickly it comes in and cuts off exits.'

The Moon's orbit round the Earth is not circular, but elliptical. Its gravitational attraction is greater at its monthly perigee (when its orbit brings it closest to the Earth) than at apogee (when its orbit takes it 40,000 miles further away). Every seven months or so, the Moon reaches perigee at the time of the New Moon or the Full Moon. The result is even higher high tides and lower low tides (the perigean spring tides).[24] These are not the best days to choose for a leisurely walk along the beach,[25] or for swimming, or for coasteering,[26] an extreme sport that involves travelling along a stretch of the coastline by scrambling and climbing across rocks, leaping off ledges, and wading and swimming in the sea.

On the Dorset coast on 2 November 2013, at the time of a perigean spring tide, a coastguard helicopter winched a young man from the sea. He had been coasteering with his sister when waves swept them both into a cave in the cliffs.[27] Leaving his sister in the cave, he swam out, hoping to find help. Rough seas prevented lifeboats from reaching her. In a daring rescue attempt, a coastguard was winched down into the cave through a narrow blowhole at the top of the cliff, only to discover that the young woman had already drowned. At the Inquest, the Coroner recorded a verdict of Misadventure.[28] He said, 'What I can't ignore is that [they] put themselves in a risky situation in poor weather. They had not checked tides or anything like that. It was predictable that the weather was going to worsen. The information was available.' The coastguard was commended for his bravery.

The Earth's year-long orbit round the Sun is also elliptical, travelling closest to the Sun at 'perihelion' in early January every year. When perihelion coincides with a New Moon or Full Moon, the increased forces of gravitational attraction result in winter spring tides at their most dangerous and damaging. In March and September, spring tides rise higher and fall lower if they coincide with the spring or autumn equinox.[29] At low tide, holidaymakers can explore tempting rock pools or examine the remains of wrecks normally hidden under the sea.[30] But everybody on the beach must watch out for the turn of the tide, because the sea comes in faster and further.

◆ ◆ ◆

The Moon and Sun move round the Earth in the same direction, but at slightly different speeds. The Sun takes 24 hours to orbit the Earth. Since the Moon takes 24 hours and 50 minutes, each high tide arrives 12 hours and 25 minutes later than the one before. No two beaches have identical tides.[31] On the east coast of Britain, high tide arrives earlier in Scotland than in England, and earlier at Hartlepool than at Harwich. On the south coast, high tide arrives earlier in Cornwall than in Kent. On beaches only a mile or two apart, the time of high tide may differ by more than an hour.

Tide Tables list the time and height of high tide and low tide for each day of the year, taking account of astronomical predictions and direct measurements of local water levels, repeated over many years.[32] Tide Tables can be read on the internet, and a series of slim booklets make them widely and cheaply available in coastal towns. Yachtsmen know the value of Tide Tables. Most holidaymakers ignore them. Yet by taking a few moments to check the Tide Tables during their stay at the seaside, visitors can choose the best times for paddling, building sandcastles or beachcombing, and they can avoid the risk of being cut off by the tide.

On the flat Norfolk coast, the tide floods in very quickly. At Wells-next-the-Sea in July 2007, a fifteen-year-old boy was paddling in a channel on the salt flats between the harbour and the sea when the rising spring tide filled the channel and swept him away.[33] A huge land-air-sea search was launched. Although the boy managed to stay afloat for an hour and a half and was still alive when a rescue helicopter airlifted him to hospital, he died of hypothermia with his parents at his bedside.

In September 2008, after lifeboats had been called out a number of times to rescue visitors stranded on sandbanks along the Norfolk coast, an article in a local newspaper quoted a Hunstanton cockle-picker concerned that beach-goers remained too casual about safety. 'Tide tables are available everywhere, but people just do not seem to take notice,' he said. 'When you are on the beach and you tell people about the dangers they look at you as if you are interfering.' A local lifeboatman said, 'Most people turn up on the beach and they don't have a clue. They have no idea about the tide, wind directions or safety hazards.'[34]

In July 2009, at the Welsh resort of Tenby, a holiday group of 4 adults and 36 children linked arms and walked along a sandbank exposed by a spring low tide – just as the tide was turning.[35] Lifeguards spotted them, drove along the beach, and used a loudhailer to call them back to land. The sea came in fast, submerging the sandbank, and suddenly 40 people were out of their depth in surging water. Luckily, the lifeguards were on the spot. Using rescue boards, they pulled struggling children from the sea. One child inhaled seawater and had to be revived on the beach. Another had an acute asthma attack. Both recovered after hospital treatment, but without the lifeguards' timely, skilful rescue, a number of lives could have been lost.

In August 2016, five young men travelled from London to Camber Sands on the East Sussex coast to enjoy a sunny day by the sea. They were playing football on one of the sandbars when they were cut off by the fast rise of the incoming tide. Two of the men stepped into quicksand below the waterline and the others went to help them. All five drowned in the turbulent water swirling through the deep troughs between the sandbars.[36]

◆ ◆ ◆

The Promenade at Blackpool was widened and strengthened in 1905, using 20,000 tons of basalt. Its steep, curving sea wall stands almost a hundred yards closer to the sea than the natural position of the shoreline at high tide.[37] Holidaymakers enjoy the wide expanse of Blackpool's sandy beach, but during spring tides, waves cannon against the sea wall, and during storms, gale-driven waves wash over the Promenade. As the tide comes in, safety chains are placed across flights of steps leading down to the beach. Whether out of ignorance or recklessness, holidaymakers frequently risk their lives by climbing over the chains.

In May 2006, a young man was standing on the steps posing for a photograph when he was hit by a wave and swept into the sea. His friend dived in to rescue him. The two men made desperate attempts to climb the sea wall and a beach patrol threw them life rings, but they could not escape from the surging water.[38] The lifeboat was launched at once. Before it reached them, they had drowned. In 2011, a holidaymaker decided to go for a swim just as the rising tide reached the bottom of the access steps. He was about to enter the water when a wave dragged him from the steps. He was unconscious when he was pulled onto the lifeboat.[39] Resuscitation did not succeed.

In a BBC interview in September 2009, David Warburton, a helmsman at Blackpool's Central Promenade Lifeboat Station warned, 'You can't underestimate the dangers of the sea: to ignore them is the equivalent of playing on the motorway or on a train line.'[40] Less than a week later, a wave swept another young visitor to his death. At the inquest, the coroner commented, 'People come to Blackpool and they underestimate the awesome power of the sea and the strong undercurrents. The fatalities we have every year are people who come from out of town.' She expressed her sympathy for the bereaved family and friends. 'Those of us who live locally are aware of how powerful the sea is, but every year there is so much grief and sadness as people unwittingly think it is not a problem.'[41]

◆ ◆ ◆

On flat, wide stretches of sandy beach, the edge of the sea fades into the background, too distant to seem a threat, but when the tide is coming in, safety depends on getting off the beach well ahead of the advancing water. At nightfall on 5 February 2004, a group of illegal workers were digging for shellfish, far out on Warton Sands on the Lancaster coast. The rapid rise of the spring tide took them by surprise.[42] They were trapped by the sea as it flooded along the channels that surround the cockle beds. Their truck stalled in a water-filled gully. A few survivors managed to swim to the shore. One young man scrambled onto a sandbank in the dark. He was picked out by the searchlight beam of a rescue helicopter, and an inshore lifeboat lifted him to safety as the sea closed around him.[43] Twenty-three people were drowned.

On the coast, it is easy to become disoriented if sea fret or fog blot out the view. In 2002, a father took his nine-year-old son on a fishing trip on Ulverston Sands in Cumbria. When fog came down, he lost his bearings. The tide was coming in. Ankle-deep in water, he used his mobile phone to summon help.[44] The fog prevented a Sea King helicopter from joining the search. The rescue teams got close enough to hear his shouts, but they could not locate him in the fog. As the water rose around him, he hoisted his son onto his shoulders. His son answered the last phone call from the rescuers. 'My daddy's all right,' the boy said. Then the phone went dead.[45]

WAVES, SWELLS AND BREAKERS

Whether the tide is rising or falling, wind blowing across the surface of the sea furrows the water into an endless succession of waves.[46] The stronger the wind, the longer the storm lasts, and the greater the distance that the wind blows across the sea, the more energy is transferred from the wind to the water, and the bigger the waves. Long after the furious winds of tropical storms die down, the waves they set in motion continue to roll across the ocean. Known as swells, they often travel thousands of miles before they reach land.

When trains of waves from different storm centres meet and intermingle, they produce complex wave patterns, with some waves far bigger than the rest. According to a study by the National Institute of Oceanography, one wave in every 23 is over twice as high as its neighbours, one wave in 1,175 is over three times as high, and one wave in 3,000 rises more than four times as high as the rest.[47] This variability is well recognised by experienced surfers. They study the 'surf beat' before entering the water, and wait for the more powerful waves in a set before they try their skill.[48]

Unfortunately, non-surfers rarely notice the variability of waves. Every year, beach-goers are engulfed by so-called sleeper waves, freak waves, or rogue waves, which wash them from the shore. During a storm in 2005, two teenage boys were drowned while night-fishing high on the cliffs of Lulworth Cove in Dorset. A huge wave funnelled into the cove and swept them from their vantage point.[49] In 2018, a five-year-old girl was walking a short distance behind her mother and sister on a Dorset beach when a sleeper wave knocked her off her feet and carried her out to sea.[50] In 2019, at Freshwater on the Isle of Wight, a man climbed over a low sea wall. He was walking along the beach carrying a small boy on his back when a powerful wave crashed over them. They were saved by onlookers who rushed into the sea and pulled them to safety.[51]

The west coast of Ireland is exposed to swells that have travelled across the Atlantic. In July 2007, a celebrated rock climber had just completed an un-roped solo climb down a 600ft cliff on Valentia Island in County Kerry. He was standing on a rocky platform 10 feet above the sea when a rogue wave swept him into the water.[52] Despite a massive search, his body was never recovered. In July 2015, a wave washed a young woman from the rocks as she walked beside the sea near Baltimore Harbour in County Cork. Her boyfriend jumped into the sea to try and save her. Then his father dived in to rescue them both. All three were drowned.[53]

When winds blow across landlocked stretches of water, they often raise sizeable waves. On Loch Lomond in March 2005, a wave knocked a father and his thirteen-year-old daughter out of their boat into the frigid water.[54] Hours of searching failed to find their bodies. Then eight years later, some of the teenager's bones were washed ashore on an island in the loch.[55] Westerly winds on Lake Michigan raise waves 7 foot high, capable of washing people from breakwaters and lakeside jogging paths.[56]

Sleeper waves menace shore-based anglers standing on rocky outcrops. In August 2012, on the last day of his holiday, a fifteen-year-old boy was fishing on the Gower coast when a wave washed him into the sea.[57] His older brother swam out to him and held him at the surface. Friends phoned 999 and kept them in sight from the shore. The outgoing tide and an offshore wind carried them half a mile out to sea. They had been in the water for more than thirty minutes before a RAF helicopter and an RNLI lifeboat found them. The boy died of drowning: his brother survived. In August 2016, a family of five settled down to fish from rocks at Fistral Beach in Cornwall. A 10 foot wave washed the parents and their two-year-old daughter into the sea. The RNLI rescued the young woman, but her husband was drowned. Although the little girl was still alive when she was pulled aboard the RNLI lifeboat and airlifted to hospital, she died of drowning four days later.[58]

In New Zealand, fishing from rocky coastlines is recognised as one of the most dangerous pastimes. Between 1980 and 1995, 63 rock fishermen drowned after waves washed them into the sea.[59] In Australia, rock fishing is just as dangerous. Between 1992 and 2000, along the coast of New South Wales alone, 74 men died while rock fishing.[60] Although rock fishermen recognise their risk of being swept into the sea, very few protect themselves by wearing a life jacket.

◆ ◆ ◆

Waves carry energy as they travel across the sea, but they carry little water with them. Instead, the water simply rises into each passing crest, falls into each passing trough, then circles back to its starting point. Once waves travel across the ocean into coastal shallows, their progress is slowed by frictional drag against the seabed. When they reach water approximately half as deep as their wavelength, waves are said to 'feel bottom'.[61]

On a gently shelving seabed, frictional drag increases gradually as the waves near the shore. The waves crowd each other, each successive wave piling into the back of the wave in front as the waves begin to shoal. Their crests rise, tip forward, spill over – and the waves become breakers.[62] In the surf zone, waves have time to break and re-form again and again as they roll towards the shore, so that much of their energy is spent before they reach the land. Surfers seek out beaches where the waves shoal over sandbars, allowing them a long ride on the face of the breaker before they have to swim out to look for further action.

On reaching the shoreline, a breaker releases a swash of foaming water, which recedes as a backwash, or undertow, to be overtaken, in turn, by the swash of the next breaker.[63] The bigger the wave, the more water in the swash, and the stronger the flow of the backwash. Both the impact of the swash and the suction of the backwash may be powerful enough to knock people off their feet.[64] Injury is likely if they are caught by successive waves and thrown about on the foreshore.

Paddling in the sea and jumping in the waves are treasured joys of childhood, but young children need an adult close at hand when paddling at the seaside. Being lighter and weaker than adults, children are more easily knocked over by the swash, while the backwash can pull them down the beach and into the sea. In August 2004, two young brothers were paddling in the sea at St Bees on the Cumbrian coast. When a teenage girl urged them to go further out, they told her that they were only allowed to go knee-deep. When they tried to return to the shore, they were buffeted by the waves, then they were swept away.[65] Rescuers had to fight against a strong backwash. They saved one of the boys, but they found his brother further along the beach, face down and unconscious in the water. They tried to resuscitate him, without success.

Where the seabed shelves more steeply, the waves travel closer to the beach before they begin to shoal. The crests rise faster and higher, and the breakers deliver a huge weight of water as they plunge down onto the beach.[66] These shore breaks are recognised as a major hazard for surfers.[67] Even small waves can lift surfers bodily, tip them over the crest, and slam them face down into the sand.[68] Since waves break only once on steeply sloping beaches, they release all their energy near the water's edge, in wading depth.[69] Not only do they endanger surfers, they also injure people paddling or swimming in the shallows. The sudden arrival of a breaking wave catches them unprepared, often when they are walking out of the water or standing in less than two feet of water with their backs to the sea. After studying surf-related injuries on the steep beaches of east-coast America, researchers in Delaware advised: 'Never, ever turn your back to the waves.'[70]

On a steeply shelving beach, the impact of the waves is not the only danger. Swimmers often have to struggle against the backwash. There are many historical records of sailors wrecked off the Dorset coast who managed to reach the steep shingle bank of Chesil Beach, only to drown at the water margin.[71] The action of the waves tumbled them about, shifted the pebbles beneath them as they tried to scramble ashore, and then sucked them back into the sea. In a recent authoritative article on safety on geological field trips, Dr Ian West warned: 'Parties should not descend low on Chesil Beach except in very quiet conditions. Pebbles are washed back with the retreating wave and there is a very dangerous undertow.'[72]

On very steep beaches, with a gradient greater than 1:15, waves have no time to shoal and break. Instead, the waves surge far up the beach – at full speed – and the large volume of water in the backwash creates a powerful undertow beneath the next surging wave.[73] Holidaymakers should stay well back from the water margin on these steep beaches. If they are caught by a surging wave, they have to fight to keep their balance against the weight of the incoming water, while at the same time, the backwash pulls their feet from under them. Then, if they lose their footing, the backwash pulls them down the beach, into deep water close to the shore.

RIP CURRENTS

Backwash causes a relatively brief undertow, which ends with the next advancing wave. Rip currents are far more prolonged and treacherous.[74] They frequently catch unsuspecting swimmers in their grip, and drag them out to sea. On British beaches, rip currents drown 35 people every year.[75] In Australia, rip currents kill more people than bushfires, floods and cyclones combined.[76]

Rip currents are found wherever breakers roll towards the beach.[77] The shoaling waves transport water into the shallows, lifting the water level in the surf zone above the level of the open sea. If waves break over a sandbar, the water piles up between the sandbar and the shore. All this excess water must drain back into the sea. At a series of points along the beach, along a depression in the seabed, at the side of a jetty or through a break in a sandbar, the outflowing water rushes away from the shore as fast, powerful, dangerous rip currents. The bigger the waves, the higher the inshore water level rises. This 'wave setup' increases whenever an onshore wind is blowing. A study of 94 rip current drownings on the east coast of Florida between 1979 and 1988 confirmed that an onshore wind was blowing at the time of every fatality.[78]

Rip currents often appear as a streak of sandy discolouration, a narrow line of choppy water flecked with foam or fragments of seaweed moving away from the shore. Sometimes holidaymakers are tricked by the apparent calm of a stretch of smoother, darker water interrupting the lines of breakers. They make the mistaken assumption that they will be safer where there are no breakers – only to step straight into the pull of a rip current.[79] Even children paddling in knee-deep water can be caught in a rip current and pulled out of their depth.

When caught in a rip current, the natural reaction is to try to swim directly back to land. But even the strongest swimmers can make no headway against a flow of water moving out to sea as fast as 5 miles per hour. Despite their frantic efforts, the rip current will drag them further and further from the beach, into deeper, colder water.[80] If they struggle against a rip current, they will lose the fight. If they panic, they will drown.

Swimmers caught in a rip can often help themselves to recover their composure and conserve their energy by flipping over to float on their backs. In this position, they can take time to fill their lungs with air, check on the direction of the current, and follow the best route to escape its clutches: 'Flip, Float, and Follow'.[81] Rip currents are rarely more than 20 yards wide, so the best chance of escaping the pull of the rip is to swim diagonally across it, towards the breakers.[82] Only then should they swim towards the beach, helped on their way by the shoreward action of the breaking waves.

Swimmers who find it impossible to get out of the rip current have no option but to go with the flow. Studies from California and Australia using 'drifters' – floating devices fitted with GPSs (Global Positioning Systems) – suggest that rip currents circulate, pouring out to sea, then travelling back towards the shore.[83] So swimmers have to master their rising panic, float on their backs, and allow the water to carry them out to the seaward edge of the surf zone. At this point, the rip current slows and turns shorewards, exit becomes possible, and the shoaling waves assist them as they swim back to the beach. But they have a long, cold, demanding swim.

Far better to spot the rips and avoid them. Far better to swim, surf and bodyboard on lifeguarded beaches. Between 2006 and 2011, RNLI lifeguards went to the aid of more than 12,000 people caught in rip currents on British beaches.[84] On a single day in August 2012, lifeguards at Croyde Bay, North Devon, rescued 30 people from drowning in rip currents.[85] After studying similar mass rescues, researchers at Plymouth University, working together with the RNLI, suggest that these incidents are often triggered in calm seas at low tide, due to the sudden arrival of larger swells that shoal and break on the beach. The extra water lifts panicking bathers off their feet at the moment when strengthened rip currents catch them and drag them out to sea.[86]

A combination of factors make rip currents especially hazardous on the sandy north-west-facing beaches of Devon and Cornwall.[87] They are pounded by large waves – both Atlantic swells and waves raised by the strong onshore prevailing winds. The dramatic rise and fall of the tides become even more pronounced during the fortnightly spring tides. Inshore water is trapped when waves shoal and break over sandbars, and when the tide is going out, the receding water is channelled into strong lateral currents that feed increasingly powerful rips.[88] A spokesman for the RNLI warned, 'Rip currents occur on most beaches around the South West and can be very dangerous, especially for people who have never experienced being caught in one before, as they panic and become exhausted very quickly.'[89]

On popular surfing beaches, many lives are saved by the heroic actions of lifeguards. But lifeguards cannot be present 24 hours per day. On 7 September 2001, two men, father and son, took bodyboards out into Holywell Bay on the north coast of Cornwall. It was 7pm, an hour after the lifeguards had left the beach. The men were caught in a rip current and swept out to sea, where waves tore them from their boards.[90] Local rescuers, coastguards, two lifeboats and a naval helicopter were quickly at the scene, but both men drowned.

There was a spring tide and an onshore wind at Mawgan Porth near Newquay in October 2014, when five teenagers were caught in a rip current at low water. Two men and a woman rushed into the sea to help them. The youngsters all managed to struggle back to the beach, but the rescuers were carried away by the strengthening rip current.[91] The RNLI lifeguards who patrolled the beach throughout the summer had ended their season only three weeks earlier, and the drama took place just at the time when the instructors at the surf school had gone for lunch. Coastguards arrived within minutes, pulled the three unconscious victims from the sea, and began resuscitation on the beach. Helicopters airlifted them to hospital, but none survived.

LONGSHORE CURRENTS

Along the coast, as the tide rises and falls, powerful longshore currents flow parallel to the shore. Headlands and rocky outcrops deflect them out into deeper water further from the shore.[92] Longshore currents are particularly strong during spring tides, swirling round groynes and jetties, pouring into bays, creeks and harbours as the tide floods in, then out again as the tide ebbs away. The flow of water is often so powerful that fishing boats must wait for slack water before proceeding into harbour. And in estuaries, river currents add to the turbulence, compounding the fierce run-out during ebb tide.

Swimmers caught in these tidal currents are borne helplessly along, unable to fight against the force of the water. Thomas Hardy often swam in the sea at Weymouth in Dorset.[93] In a pivotal chapter in *Far from the Madding Crowd*, he described his anti-hero Sergeant Troy swimming in Lulworth Cove, 'a small basin of sea enclosed by the cliffs', where the water was smooth as a pond. As he swam further from the shore, he was caught in a longshore current: 'Troy found himself carried to the left and then round in a swoop out to sea'.[94] In the novel, Sergeant Troy is saved from drowning by two sailors who haul him over the stern of a rowing boat. In real life, such currents can prove too taxing for the most confident swimmer. In June 2013, a powerfully built man in his early forties got into difficulties in the sea off Beesands Beach on the South Devon coast. He was a keen supporter of 'wild swimming', engaged in an endurance swim. A canoeist found him unconscious in the water and brought him ashore.[95] Paramedics began resuscitation and he was airlifted to hospital, but he died soon afterwards.

Children can be caught in a longshore current even when paddling close to the shoreline. Shortly after arriving at Holme Beach, Norfolk on a sunny Sunday in August 1996, two young children rushed off to the water's edge. When their parents followed them a few minutes later, the children had disappeared.[96] The parents searched along the crowded beach. Police, coastguards and members of the public joined the search for the lost children, without success. It was two weeks before their bodies were found, washed onto beaches 30 miles to the west. At the Inquest, their father gave evidence. 'We simply did not believe that they had drowned and couldn't believe that two healthy children could have got into difficulties there,' he said. 'It was our impression that this was a very safe beach.'[97] A coastguard explained that the beach was undulating. Sandbars formed parallel to the shoreline, and the channels between them were often three feet deep. The children probably stepped into a drop-off. Once in a channel of deeper water, the incoming spring tide would have caught hold of them and swept them away.

Four years later, in August 2000, a five-year-old boy was swept away from a crowded beach at nearby Brancaster. He was playing at the water's edge, watched by his parents. Then his father turned to look at a plane as it passed overhead, and his mother bent to get out some sandwiches. In that moment, the boy vanished.[98] An extensive search followed. Three days later, his body was lifted from the sea.

His mother gave her support to the Maritime and Coastguard Agency (MCA) when it launched a sea safety code that urged parents to check the weather and the tides before going on the beach and to keep children within reach at all times.[99] The MCA reminded beach-goers: 'Dialling 999 and asking for the coastguard is the quickest way to get help at the Coast', yet a survey by the Sea Smart Campaign reported that few families knew what to do in an emergency.

WHEN STORM WAVES BREAK ON THE SHORE

The entire shoreline is under attack.[100] Breakers pound the rocks, grinding them into sand. Cliffs are undermined and crumble into the sea. Groynes, breakwaters, revetments and sea walls protect the coastline from the ravages of the sea, but the waves batter these shore defences too. Apart from the impact of the water itself, a surging wave acts like a compression hammer, forcing pockets of air deep into cracks in any barrier constructed to protect the beach. Then, the receding wave creates a vacuum which sucks away fragments of stone and concrete.[101]

If storm-driven onshore winds coincide with perigean spring tides, huge waves overtop breakwaters or rebound from the vertical faces of sea walls and promenades, colliding with succeeding waves in dramatic plumes of spray.[102] When a storm hit the Cornish coast at Newquay in 2008, thrill-seeking bodyboarders ignored coastguard warnings. Urged on by cheering crowds, they rode the storm surge breakers and were tossed about by rebounding waves 30 feet high.[103]

Awestruck spectators often gather to watch as waves crash over their local esplanade. Some become wave-dodgers, thrilled by the sudden violence of the waves, heedless of the risk to their lives. Yet anyone swept into such turbulent water is in the gravest danger, powerless to swim, but unable to escape from the sea. In a series of incidents in Sussex, Devon and Cornwall during storms in January 2002, four young men drowned when they were carried off by the waves.[104] Police in Brighton tried to move wave-watchers from the seafront by delivering warnings by loudhailer. At nearby St Leonards-on-Sea, a group of local people came down onto the beach in the dark to watch the breakers. A wave knocked a young woman from a groyne and dragged her into the sea. A search by helicopters and lifeboats failed to find her body.[105] A Falmouth coastguard rejected the common excuse that freak waves were to blame for the fatalities. 'One wave might be a light spray to someone on the promenade and the next could be a few hundred tonnes of water crashing down on them,' he said. 'There is a lot of time wasted dealing with situations like this where people actively put themselves in danger, as well as the lives of police and coastguards.'[106] A police spokesman repeated the warnings: 'The sea is a dangerous place at any time but with the storms we have been having it is particularly advisable not to go anywhere near the beach or exposed promenades and piers.' In August 2017, as waves overtopped the harbour wall at Portreath, three teenagers in wetsuits were photographed 'bracing themselves for impact' as a colossal wave cascaded over them, a weight of water that might easily have washed them into the sea.[107]

At an Inquest in 2005, a young man described an outing to Scarborough in stormy weather. As he walked along the seafront with his girlfriend, her two children and his own son, they were drenched by a 30ft wave. The children laughed, and on a sudden impulse, he climbed over the guardrail, jumped down onto a ledge facing the sea and stood waving his arms and shouting 'Come and get me!' at the waves. As he clambered back, his girlfriend's children ran down a nearby slipway. Within moments, waves engulfed them. Their mother tried to save them, but she too was swept away.[108]

Only two bodies were recovered from the sea. The Coroner gave a verdict of accidental death for all three victims. He said, 'This tragedy is a sharp reminder of the importance that a sea such as this is viewed from a safe distance. Young children should be kept away from the potential danger and also under tight control.' And he warned his listeners, 'The power of the sea must at no time be ignored. It can be benign, but equally its force can be unmerciful.'

CHAPTER TWENTY

PERIL ON THE SEA

Wind and waves disturb the surface of the sea. The stronger the wind, the higher the waves and the more ferocious the storms. In the days of sailing ships, seamen were at the mercy of the wind as soon as their ship left port. Sails had to catch the wind. For speed and safety, sailors had to adjust the area of unfurled sail whenever the wind strength changed.

Admiral Sir Francis Beaufort began to keep a record of wind and weather while he was still a cabin boy. In 1838, the Royal Navy adopted his Beaufort Wind Force Scale, which linked wind strength to its effect on the sails of a fully rigged frigate, the prime man-of-war of the British fleet.[1] The Beaufort Scale defined thirteen degrees of wind strength, from Force 0 (Calm) and Force 1 (Light Air – 'just sufficient to give steerage way'), through gradations of breeze, gale and storm, up to Force 12 (Hurricane – 'that which no canvas could withstand').

The Beaufort Scale is still in daily use, but now it indicates wind speeds, associated wave heights and sea conditions, with probable damage to trees and buildings.[2] Force 12 (Hurricane) gives a wind speed of 73 miles per hour or more, wave heights of 46 feet or more, adds a vivid report of sea conditions ('Huge waves. Sea is completely white with foam and spray. Air is filled with driving spray, greatly reducing visibility') and predicts damage on land ('Severe, widespread devastation to vegetation and structures. Debris and unsecured objects are hurled about').

More recently, the Saffir-Simpson Hurricane Scale defined five further degrees of hurricane-force winds, from Category 1 (wind speed 74-95 miles per hour) to Category 5 (wind speed equal to or greater than 155 miles per hour), each linked to an expected degree of structural damage.[3] Until 2009, the Saffir-Simpson Hurricane Scale also included an estimate of the lowered barometric pressure at the centre of the storm, the likely height of any associated storm surge and the extent of coastal flooding should the hurricane reached the land.

Computer-aided weather forecasts predict storms with increasing accuracy. Modern navigational aids and GPS location systems pinpoint the position of endangered ships. Satellite phones and VHF radios link ships to coastguards and search-and-rescue operations. Nevertheless, a gust of wind can capsize a yacht,[4] an extreme wave can sink a supertanker, and every week throughout the year,[5] ships are lost in storms at sea.[6]

HIGH WINDS: THE FASTNET YACHT RACE

On Saturday 11 August 1979, 303 ocean-going racing yachts set sail from Cowes on the Isle of Wight at the start of the prestigious Fastnet Yacht Race – 600 miles along the south coast, across the Irish Sea, then back to finish at Plymouth. At the start, the conditions were described as 'near perfect'. The BBC Shipping Forecast predicted, 'South-westerly winds Force 4 to 5, increasing to Force 6 or 7 for a time.' Then, at dawn on Monday 13 August, the wind died down and swells from a distant gale rolled across the smooth surface of the sea. The storm arrived as the yachts neared the lighthouse on Fastnet Rock off the southern coast of Ireland.[7] By midnight, the wind was blowing at Force 11 (Violent Storm). The atmospheric pressure on the surface of the sea (normally 1013 millibars) fell to the extremely low value of 978 millibars.

Waves more than fifty feet high smashed onto the yachts. Yachtsmen strapped themselves into safety harnesses. Many were washed overboard. Some hauled themselves back onto their vessels, but others were swept away when their harnesses tore apart. Water poured through open hatchways. Masts and rudders broke. Half of the yachts were knocked over, capsized, or rolled through 360°. Crews were trapped beneath upturned hulls. A dozen life rafts were torn away, with survivors clinging to the shredded remnants.

Over the next few hours, fifteen yachtsmen lost their lives. Helicopters, merchant ships and naval vessels rescued 136 men and women from foundering yachts, from damaged life rafts, or directly from the sea. Only 86 yachts reached Plymouth. The report of the Fastnet Race Inquiry ended with the comment, 'The sea showed that it could be a deadly enemy and those who go to sea for pleasure must do so in the full knowledge that they may encounter dangers of the highest order.'[8]

HUGE WAVES: THE CAPSIZE OF THE PRINCESS VICTORIA

In 1953, the Stanraer-to-Larne sea ferry *Princess Victoria* was the first to be feel the full strength of the Great Storm. Gale-force winds were blowing when the ship left the shelter of Loch Ryan on the west coast of Scotland for the 20 mile voyage across the Irish Sea. There were 183 people on board, 128 passengers and 55 crew. Storm waves battered open the stern doors. Water poured onto the car deck,[9] the engine room flooded and the ship began to list. Only two lifeboats got away. The rest were swamped by the waves or smashed to pieces against the hull. RAF planes dropped rubber dinghies into the sea. Braving the storm, the lifeboats from Donaghadee and Portpatrick, a trawler, cargo boats and the destroyer *HMS Contest* rescued 48 survivors. The ship sank within sight of County Down, with the loss of 133 lives, including the Captain, the Deputy Prime Minister of Northern Ireland, and the radio operator, who was awarded the George Cross.[10]

HUMAN ERROR: THE SINKING OF THE *HERALD OF FREE ENTERPRISE*

On the evening of 6 March 1987, the Roll-On/Roll-Off ferry *Herald of Free Enterprise* sailed out of Zeebrugge harbour. The bow doors had been left open, a careless mistake that led to the loss of 193 lives.[11] The assistant bo'sun should have closed the bow doors, but he was asleep in his cabin. The Chief Officer should have remained on deck to ensure that the bow doors had closed, but he assumed that the assistant bo'sun was on his way and went up to the bridge while the doors were still open. On the bridge, the Captain could not see the bow doors, and there was no warning light to alert him of the danger.

Two factors made early flooding inevitable. First, the bow of the ferry lay low in the water. (This was because two ballast tanks had been filled to allow the short loading ramp at Zeebrugge to reach the upper car deck, then the ferry sailed before the ballast tanks had been completely emptied.) Second, the ferry settled even lower in the water when it increased its speed on leaving the confines of the harbour. (This was because acceleration in shallow water results in dynamic sinkage, or 'squat', due to lowered water pressure beneath the hull.)[12] As a result, the rising crest of the bow wave topped the bulwarks and spilled over the prow. Water gushed through the open bow doors – a gaping inlet 6 metres wide – and flooded into the main car deck. In less than two minutes, the ferry tilted to port and capsized, coming to rest half-submerged on a sandbank close to the harbour mole. The crew of a dredger saw the lights go out and sounded the alarm.

As the ferry settled on the seabed, the port windows exploded. Water gushed into the passenger decks and along the tilted gangways and stairwells. With the ferry lying on its side, corridors connecting one side of the vessel to the other became deep vertical shafts. Doors swung downwards from their hinges, acting as open trapdoors above voids.[13] In the darkness, passengers and crew tumbled down into near-freezing water, bombarded by furniture and shards of broken glass. Stored life jackets broke free, floating as obstructive rafts on top of the water, but passengers could not disentangle them because their fingers were numbed by the cold. As they waited for rescue in chest-high water, many passengers lost their strength and slid under the surface.

The starboard windows remained above water level, but they were 30 feet above the panic-stricken passengers imprisoned in the upended lounges and cafeterias of the ferry. Some young men clambered up to the windows, smashed through the glass and began to lift people to safety. Dozens of ships arrived to help. Within twenty minutes of the capsize, the first rescuers boarded the ferry, and let down ropes. However, many passengers were already so chilled that they lost their grip as they climbed upwards and fell back into the water. Children had to be hauled up in canvas buckets. Helicopters brought divers to the wreck and took casualties to hospital. Over 350 survivors were rescued from the ship, including 3 men who were found in an air pocket on the car deck more than seven hours after the accident. In the New Year's Honours List, 31 people received bravery awards, including two men who were given the George Cross[14] – a chef who died while rescuing passengers from a flooded restaurant and a young father who lay across a flooded corridor as a 'human bridge' and allowed 20 people to walk across his body to a small refuge where ropes lifted them to safety.

HYPOTHERMIA IN SHIPWRECK SURVIVORS

During the Second World War, ships hit by torpedoes often sank before lifeboats could be lowered. To provide a second escape route, warships carried Carley floats stacked on their decks.[15] In an emergency, these large, light, buoyant life rafts could be lifted over the side and thrown into the sea. Survivors climbed inside the floats or clung to ropes draped round their sides. Undoubtedly, Carley floats saved the lives of many soldiers and sailors. However, Carley floats had a major disadvantage. Their floors were gratings made of wood or webbing. Survivors were held at the surface, but they remained immersed in the water. After half an hour or so, their core body temperature began to fall and they developed hypothermia.

On 16 April 1945, only three weeks before the end of the Battle of the Atlantic, a U-boat torpedoed the Canadian minesweeper *Esquimalt* off the coast of Nova Scotia. The ship sank within four minutes. The 43 men who survived the explosion hauled themselves onto four Carley floats. They were only 5 miles from land, and they could see the lights of Halifax in the distance. The survivors thought a plane had spotted them soon after the *Esquimalt* went down, but the pilot mistook the Carley floats for fishing boats and failed to report the sighting. At first, the sailors sang to keep up their spirits, then they huddled together, trying to fight the cold. When they were found 6 hours later, only 27 were still alive.[16]

In his book *Survival in Cold Water*, the physiologist William Keatinge wrote, ' The heavy loss of life at sea in the 1939–45 war led to the first general realisation that survival in water was not just a matter of having floatation equipment to prevent drowning, and that the hazards of cold immersion presented some of the most important of all practical problems of human physiology.'[17] Professor Keatinge documented the powerful reactions of the heart and lungs when cold water chills the skin, even while taking a brief cold shower.[18] Much of his research was carried out using indoor and open-air swimming pools. He demonstrated that people immersed in cold water lose more body heat when they move than when they stay still, that clothing slows the loss of body heat,[19] and that thin people lose body heat faster than fat people.[20] He highlighted the danger of sudden swimming failure when someone attempts to swim to safety in cold water.[21]

In 1963, Keatinge travelled 1,500 miles to find out why 128 people died when the liner *Lakonia* sank while cruising near Madeira with 1,022 people on board.[22] Fire broke out at 11pm on 22 December. Flames trapped a number of passengers in their cabins and killed members of the crew who tried to rescue them. At 1am, the Captain gave the order to abandon ship, but evacuation was hampered because the fire destroyed four of the lifeboats and two loaded lifeboats broke away from rusted launching gear, tipping their occupants into the sea. The remaining 18 lifeboats got away safely, but more than 150 people were left on board the burning liner. The Captain and 14 elderly passengers retreated to a glass-enclosed area at the stern.

As the fire spread, the remaining passengers and crew put on life jackets and left the liner by jumping, climbing down rope ladders, or walking down the port and starboard gangways into the sea, where they floated in water with a temperature of 18°C.[23] The Argentinian passenger ship *Salta* and the British tanker *Montcalm* responded to distress calls and arrived around 4am. Other ships joined them, including the aircraft carrier *HMS Centaur*. The *Lakonia* had drifted for several miles during the emergency, so the search for survivors lasted for hours, aided by rescue planes and helicopters.

All the people in the lifeboats survived. The Captain and his 14 companions also survived after climbing into a life raft dropped from a plane. However, many of those floating in the sea were already dead by the time rescuers reached them. Fifteen people were still alive when they were taken on board the *Montcalm* after 5 hours in the water, but they were in a state of collapse and they all died soon after rescue.[24]

Professor Keatinge asked survivors whether the cold or fear of drowning had troubled them more. Most survivors told him that they became increasingly preoccupied with the cold. Some had suffered from uncontrollable shivering just before or after being taken from the water. Others had no memory of being rescued. Two survivors told him that people floating near them became confused or delirious, then lost consciousness and drifted away. He suggested that some victims would have survived if they had been advised to put on extra layers of warm clothing before they went into the sea. Those wearing flimsy nightclothes or evening dresses would have cooled more quickly than those protected by suits and overcoats. Some survivors told him that they tried to keep warm by swimming to and fro, unaware that moving in the water increased their rate of heat loss.

He concluded: 'Immersion hypothermia was the main hazard faced by those in the water and led to most of the deaths.' More recent research has shown that shipwreck survivors are in danger of drowning well before the onset of hypothermia, even when they are held at the surface by a life jacket.[25] Some die within a few minutes from the immediate, dramatic effects of 'cold shock'.[26] Others lose their strength, gradually becoming incapable of the constant exertion needed to keep their backs turned towards the waves. Unless their life jackets have a 'splashguard', they inhale seawater whenever a wave washes over their faces, slowly drowning even at the water surface.[27]

◆ ◆ ◆

In January 1968, three deep-sea trawlers – *St Romanus*, *Kingston Peridot* and *Ross Cleveland* – left Hull for the fishing grounds off Norway and Iceland. They did not return. Caught in violent storms, with 40ft waves whipped up by hurricane-force winds, they sank with the loss of 58 men. Only one man survived.[28]

Dr Griffith Pugh had published classic studies on the endurance of Channel swimmers immersed in cold water, and on fatal exposure and hypothermia in hill walkers.[29] As a talented physiologist and an intrepid climber, he was a crucial member of the British team that conquered Everest in 1953.[30] He set off for Iceland to find out what had happened to the trawlermen and to talk to the lone survivor, Harry Eddom, the mate of the *Ross Cleveland*.

The *Ross Cleveland* had sought shelter from the storm in the freezing seas of Isafjordur, a narrow inlet on the north coast of Iceland. Two trawlermen managed to scramble into one of the life rafts.[31] The mate was chopping ice from the trawler's superstructure when a wave knocked him overboard. As the trawler capsized and sank, his two shipmates dragged him into the life raft.[32] The canopy of the life raft could not be closed. Before long, waves swamped the raft, and the bailer and survival gear were washed away.

The mate was wearing an inflatable life jacket over a waterproof smock and over-trousers, a heavy woollen sweater, seaman's moleskin trousers, thigh-length woollen stockings, woollen underwear, and thigh-length rubber boots.[33] The double protection of his windproof, waterproof outer layer and his inner layers of heat-retaining clothes gave him enough protection to survive the cold. His two companions had been below deck when the ship began to sink. Their hurried escape allowed them no time to put on waterproof clothing. They coped with their ordeal bravely, but they were wet through and chilled by the bitter cold. They both died of hypothermia within 3 hours. Harry Eddom was washed ashore at first light, after 8 hours tossed in the life raft. He could see a farmhouse in the distance, and set off towards it. It took him all day to walk 6km along the rocky snow-covered shore. Too weak to continue, he kept himself awake through the hours of darkness, standing all night in the shelter of a wall. Next morning, he reached the safety of the farmhouse.

Harry Eddom told Dr Pugh that he took several sets of outer garments and underclothes on board the trawler so that he always had dry clothes to wear. In contrast, young fishermen usually arrived on the trawler wearing jeans, shirt and jersey, short cotton underpants, string vest, and nylon socks, without a change of clothes. Only deck personnel had waterproof outer garments.

Dr Pugh suggested that all trawlermen should have their own personal waterproof 'duck suits' to wear at sea, that emergency packs containing waterproof suits should be secured inside life rafts, and that life rafts should be equipped with emergency radios. He recommended that they should be given training in the use of life rafts and the prevention and treatment of hypothermia. He wrote, 'One wonders whether the crew were fully aware of the danger of hypothermia and the fact that even those apparently dead can often be revived.' (In the circumstances of the Isafjordur tragedy, this remark indicates a degree of over-optimistic confidence in the power of resuscitation.)

COMMERCIAL FISHING: THE MOST HAZARDOUS OCCUPATION IN BRITAIN

Professor Richard Schilling was internationally known for his research on byssinosis, an occupational lung disease in cotton mill workers. In 1965, he was invited to investigate Dogger Bank Itch, an allergic dermatitis affecting Lowestoft trawlermen,[34] who became sensitised to sea chervil, a 'moss animal' that grows as gelatinous seaweed-like fronds on stones on the seabed and spills from the trawlers' nets together with the captured fish. Dr Muriel Newhouse, one of Professor Schilling's colleagues at the London School of Hygiene and Tropical Medicine, checked the crews of 55 trawlers as they returned to port. She found many seamen with the typical blistering rash on their hands, forearms and faces.[35] Before considering preventive measures, Professor Schilling decided to spend six days watching fishermen at work on a trawler at sea. 'This experience made me aware of hazards other than skin disease,' he told an audience at the Royal Society of Medicine, 'In particular, of the risk of fatal accidents as a result of ships foundering, men falling overboard or being injured by winches and fishing gear.'[36]

Professor Schilling soon realised that the available accident statistics grossly underestimated the accident rate of fishermen.[37] Until then, the most recent official figures (standardised mortality ratios or SMRs, which compared death rates of fishermen to death rates in the general population) suggested that fishermen had a lower accident mortality than 9 other occupations, including coalminers. At that time, fishermen's deaths were recorded in two separate registers – deaths on shore were recorded by local registrars and submitted to the Registrar Generals of England, Wales and Scotland, whereas deaths at sea were reported to the Port Superintendent and submitted to the Registrar General of Shipping and Seamen. The calculations of mortality rates had omitted deaths at sea, whether fishermen died when their vessel was lost (founderings, capsizings, groundings, explosions and fires) or died due to fatal accidents at sea (men killed by machinery or lost overboard). Once deaths at sea were taken into account, fishermen's fatal accident rates were shown to be more than double the rate for miners, and more than 20 times the rate for men in manufacturing industries.

In 1968, soon after the loss of *St Romanus*, *Kingston Peridot* and *Ross Cleveland*, a Committee of Inquiry into Trawler Safety (CITS) was appointed. Professor Schilling was among its members.[38] Published in 1969, the committee's report discussed accidents and deaths during the routine tasks of fishing; the extended hours of work and consequent fatigue; inadequate protective clothing on deck; and the need for better health care both at sea and between voyages.[39] The report was described as 'one of the most sweeping indictments of any British industry this century'.[40]

Many of the recommendations of the CITS were passed into law. In 1985, however, a study carried out at the University of Dundee warned that the death rate from fatal accidents involving British fishermen was even greater in the decade after the publication of the CITS report (1971–1980) than in the decade before (1961–1970).[41] The study noted: 'Compared with coalminers, fishermen were, on average, four times more likely to die from accidents at work.' A graph contrasted the gradual and sustained reduction in the mortality rate of coalminers between 1961 and 1980 with the increase in the mortality rate of fishermen.

In 1987, a further study from the University of Dundee reported that the Marine Directorate of the Department of Transport ordered fewer formal investigations into losses of fishing vessels in 1972–1982 than in 1961–1971, even though the number of losses had more than doubled.[42] This lack of inquiry into losses of fishing boats – and the associated deaths of fishermen – was contrasted with the regular publication of rigorous and detailed examinations of aircraft accidents, with their acknowledged contribution to airline safety. In 1989, after the disastrous capsize and sinking of the *Herald of Free Enterprise*, the Marine Accident Investigation Branch (MAIB) was established as the marine equivalent of the Air Accidents Investigation Branch.[43] MAIB investigates around 50 shipping accidents every year and publishes detailed analyses and recommendations.

A number of studies – by epidemiologists in Oxford and Swansea, by the Maritime and Coastguard Agency and Ministry of Transport, and by the Marine Accident Investigation Branch – have confirmed that commercial fishing is by far the most hazardous occupation in Britain.[44] Since the end of the 'cod wars' in 1976, British trawlers no longer fish in Icelandic waters. There are now fewer vessels in the British fishing fleet, and consequently there are fewer fishermen at risk. However, the dangers of distant fishing grounds have been replaced by hazards closer to home ports. In 2011, a fishing boat capsized and sank off the Cornish coast when its heavy catch of pilchards shifted in an open storage tank in the hold, creating a 'free surface effect' that changed the position of the centre of gravity and prevented the boat from righting itself as it rolled in the waves.[45]

Despite repeated attempts to improve safety by legislation, inspection and training, the fatality rate has scarcely fallen over the past 60 years. During the decade 1996–2005, 160 British fishermen lost their lives at sea: 86 died when their vessels sank and 74 suffered fatal accidents on board ship.[46] Fishing boats foundered in gales and storms, collided in the busy shipping lanes of the English Channel, capsized after snagging their nets on forgotten shipwrecks, or ran aground through faulty navigation or fatigue. Fishermen were asphyxiated by fumes below decks, struck by fishing gear or thrown against hard structures. But the commonest cause of death was falling overboard – 32 fishermen drowned after falling overboard, including 11 fishermen who were fishing alone. None of these 'men overboard' had put on a life jacket.[47]

Following Professor Schilling's revelations of the high death rate among British fishermen, the excessive danger of commercial fishing was also identified in America, Denmark, Norway,

Iceland, Canada, Australia and New Zealand.[48] Almost a third of deaths due to fatal accidents at sea followed a fall overboard.[49]

Trawlermen have a fatalistic acceptance of danger as part of the job. They know that a wave might wash them overboard, that they might lose their balance on the swaying, slippery deck and fall over the bulwark, or that they might become entangled and dragged into the sea when shooting the nets or hauling their heavy catch back on deck. Nevertheless, many choose to work on deck without wearing a life jacket, even though they know that without a life jacket, their risk of drowning is increased. Writing on fishing vessel safety in 2000, the Chief Inspector at the MAIB observed: 'The average fisherman is extraordinarily reluctant to wear a life jacket, usually because he will claim it is too bulky, is impractical for the work he does or is too expensive.'[50]

LIFE JACKETS

During the Battle of Britain, the bodies of many British airmen who parachuted into the Channel were found floating face down, their life jackets fully inflated.[51] At the Royal Air Force Research Station at Farnborough, two distinguished anaesthetists, Professor Robert Macintosh and Dr Edgar Pask, showed that the life jackets in use at that time prevented airmen from 'self-righting'. If they lost consciousness before or after they landed in the water, they drowned. Beginning in 1943, the anaesthetists conducted a series of bold experiments to improve the lifesaving potential of life jackets, studying the characteristics of 'some 20 flotation garments.'[52] A team of anxious researchers pumped air and anaesthetic gases into a volunteer, who lay unconscious as they lowered him into the deep end of a swimming pool, in fresh water, in salt water and in 3 foot waves produced by a wave machine. Then they photographed and filmed him as he floated – or sank.

Their war-time research on improving the effectiveness of life jackets remained secret until 1957, when Macintosh and Pask published a graphic account of their work in the *British Journal of Industrial Medicine*: 'The Testing of Life-jackets'. They wrote, 'The subject was deeply anaesthetised with a volatile anaesthetic. An endotracheal tube with an inflatable cuff was used to prevent water entering the subject's lungs when his head became immersed and the pharynx was packed for extra security.' They illustrated their account with 29 photographs of their life jacket trials, showing the volunteer as he risked his life, partly or wholly submerged in the water. However, they did not mention that the volunteer was actually Dr Edgar Pask himself, and that after every experiment he had to be admitted to hospital to recover from his ordeal.

Pask's war-time research also involved studies on high altitude parachute descent, on assessment of methods of artificial respiration, and on the development of survival suits.[53] After the war, as Professor of Anaesthetics at Durham University, he continued his studies on life jacket design. Instead of human subjects, however, he used test dummies.[54] By copying the size, shape and density of head, chest, abdomen and limbs in plastic, fibreglass and stainless steel, he ensured that the dummies floated in the water in a realistic manner. His work had central importance in the development of international standards for modern life jackets.[55]

Professor Edgar Pask recognised that the properties of a serviceable life jacket depend upon the particular circumstances of its use. He wrote. 'Special needs impose special requirements.'[56] Life jackets designed for the armed forces, merchant seamen, yachtsmen or passengers on cruise ships are not best suited to the needs of fishermen.

◆ ◆ ◆

Any fisherman who falls into the cold water of the open sea is in great peril.[57] His face is suddenly immersed in water colder than 15°C, triggering the immediate onset of 'cold shock'. As water seeps into his clothes and chills the muscles of his arms and legs, he develops swimming failure within a few minutes. Even if shipmates see him fall, they may not reach him in time to save his life, and if he is fishing alone, he may be incapable of hauling himself back onto his boat. If he is wearing a lifejacket, he has at least a chance of survival. And as well as the support of a life jacket, fishermen need clothing that slows the loss of body heat during immersion in cold seawater.

As long ago as 1970, the Royal Society of Medicine held a symposium on *Recent developments in personal protective clothing and equipment*.[58] Dr Muriel Newhouse (Professor Schilling's colleague in the investigation of Dogger Bank Itch) described problems with the oilskins that fishermen wore on deck at that time – usually an unlined PVC smock that reached to mid-calf, together with thigh-length sea boots: 'This smock is a bulky, heavy garment which impedes active movement; it is open at the cuffs and neck allowing penetration of water to the garments worn underneath.' And she pointed out, 'The smock and boots together are negatively buoyant, thus no protection against drowning.'[59]

Other speakers described newer materials and improved designs to protect fishermen who work for many hours on deck in wet, cold, windy conditions. Outline sketches showed zipped one-piece boiler suits, and 'duck suits' with shorter hooded jackets over waterproof bibbed trousers, worn with lighter calf-length sea boots. Test garments made of foam-backed polyurethane-coated nylon and neoprene rubber had slight positive buoyancy and provided better heat insulation, but they prevented evaporation of sweat,[60] and water seeped through the seams and zips. Black neoprene acted as an all-too-effective camouflage when worn by someone floating in the sea.[61]

Fifty years after the symposium at the Royal Society of Medicine, many fishermen still wear oilskins that give them inadequate protection if they fall overboard. To improve safety, scientists and industrial engineers working at SINTEF (the Foundation for Scientific and Industrial Research, part of the Norwegian Institute of Technology in Trondheim) decided to use 'concept engineering' to find out exactly what fishermen wanted from their protective clothing.[62] They gathered information by recording and annotating interviews with fishermen, representatives of fisheries' authorities and clothing suppliers. Using information from the interviews, they asked over a thousand fishermen and boat owners to pick their ten most important requirements.

Fishermen asked for clothes that were waterproof, light, warm, comfortable, hard-wearing, highly visible, allowed ventilation, withstood tearing by fishing hooks and were unlikely to snag on moving nets and hoists. In addition, they wanted clothes designed with built-in buoyancy. Over several years, SINTEF and the British leisure clothing firm Regatta worked together to develop prototype work suits, which were tested in SINTEF's Work Physiology Laboratory and on fishing vessels at sea. Bib-and-braces trousers were designed with large panels of closed-cell plastic foam covering the front and back of the chest. The foam panels provide 50 Newtons of in-built buoyancy (equal to the buoyancy of a near-shore life jacket), lifting the wearer into a 'self-righting' vertical floating position with head and shoulders well above the water surface. By placing the flotation in the trousers, the design provides an integrated crotch strap, and allows the fisherman to take off his jacket while at work on deck without losing his built-in personal flotation device (PFD).

In June 2006, researchers at SINTEF reported that the new work suit (Regatta Fishermen's Oilskins with Flotation) had saved the lives of two fishermen when their boat sank off the coast of northern Norway:[63] 'One of the two men on board unpacked his suit the same morning and dressed [in] it before he went fishing. After the accident he explained that he felt safe because the buoyancy aid and the upright floating position allowed him to pick up his mobile phone and call for help.' In fact, he was able to disentangle his companion from the fishing net and support him at the surface in waves 2 to 3 metres high, then contact his brother using a mobile phone kept dry inside the zipped jacket pocket.[64] His companion was wearing standard oilskins without a life jacket, and when the rescue boat arrived twenty minutes later, he had signs of hypothermia. Both men made a full recovery.

Further reports of lives saved by Regatta Fishermen's Oilskins soon followed. In January 2009, a Welsh fisherman was unloading fish from his boat when he slipped on an icy gangway, knocked his head against the quayside wall, and fell unconscious into Swansea Dock.[65] The built-in buoyancy brought him to the surface and held him there until a rescuer reached him and pulled him out of the water. In March 2011, an Irish lobster fishermen was able to swim 60 metres to the shore after his boat capsized and sank. The buoyancy panels built into his trousers held him at the surface and helped to keep his body warm in seawater at 4°C. His crew mate, wearing standard oilskins, without a life jacket, was also thrown into the sea. He tried to swim to safety, but before long, he disappeared beneath the waves.[66] By 2012, SINTEF confirmed that the oilskins developed with the help of insights gained from in-depth interviews with experienced fishermen, then tested in their laboratory in Trondheim, had saved at least 10 fishermen from drowning.[67]

In American waters, the death rate is particularly high among fishermen harvesting shellfish and king crab in the frigid stormy seas around Alaska.[68] The National Institute for Occupational Safety and Health (NIOSH) reported that from 2000 to 2014, 182 commercial fishermen drowned after falling overboard without a life jacket or personal flotation device (PFD): 'Many were within minutes of being rescued when they lost their strength, sank, and drowned. Those deaths could have been prevented if the fishermen had been wearing a PFD.'[69]

Researchers at the Alaskan Field Station of NIOSH were concerned that most fishermen found PFDs 'bulky, heavy, hot, and generally uncomfortable'. Despite the high risk of drowning if they fell overboard, they usually worked on deck without wearing a PFD.[70] The researchers selected six different designs of life jacket or PFD, and recruited 200 Alaskan commercial fishermen (crabbers, longliners, gillnetters or trawlermen). The fishermen were asked to choose their preferred design, wear it on deck for a month, then assess its comfort and acceptability.[71] At the start of the study, the majority of the fishermen admitted that they never wore a life jacket, or wore one only occasionally, even though they believed that they were likely to fall overboard at some time in their career. After wearing their PFDs for a month, most of the fishermen gave them a positive rating. In 2013, the American Centers for Disease Control and Prevention in Atlanta presented the Alaskan research, complete with colour photographs of fishermen wearing their chosen personal flotation devices (inflatable suspenders, foam work vests and Regatta Fishermen's Oilskins).[72] It is still too early to know how many fishermen can be persuaded to protect themselves by wearing life jackets or PFDs whenever they are working on deck, but the advice from Alaska and Atlanta is clear: 'Personal Flotation Devices Prevent Fishermen Deaths.'[73]

CHAPTER TWENTY-ONE

HAZARDS ON HOLIDAYS ABROAD: SWIMMING POOLS & WATER PARKS

People look forward to a foreign holiday as a high point in the year, a chance to relax, to renew their energy, to have an adventure. They set off in cheerful anticipation, their cases packed with books, cameras, sun cream, and swimming costumes. But holidays are times when people let their guard down and take risks they would not take at home. In five years – 2006 to 2010 – a grim total of 309 British holidaymakers drowned abroad, including 30 children who drowned before they reached their seventh birthday.[1]

SWIMMING POOLS AT HOTELS, HOLIDAY VILLAS AND WATER PARKS.

Foreign holidays rarely last longer than two weeks. Yet more British children drown in the swimming pools of hotels, campsites, and holiday villas abroad than drown throughout the whole year in the swimming pools in Britain.[2] And British children on holiday in America have an even higher drowning rate than those on holiday in Europe.[3] Hotel swimming pools are magnets for children. Most are prominently sited near the sunloungers on the terrace or within easy reach of the public rooms. Some elaborate pools resemble tropical lagoons, with inlets and islands that screen children from their parents' view, and with sudden slopes and steps that take children out of their depth. At larger hotel pools, an attendant may be present for part of the day, handing out towels and tidying the changing rooms, but few hotels employ lifeguards to ensure the safety of their guests. Smaller swimming pools and paddling pools are likely to be entirely unguarded. In 2002, a five-year-old boy wandered away from his grandmother while she was attending a welcome reception in a Majorcan hotel. He came upon a small deserted children's pool, where he drowned.[4] In 2008, guests at a Paris hotel discovered a child lying unconscious at the bottom of an unsupervised pirate-themed swimming pool adjacent to the play area where his mother was searching for him. He died in hospital two days later.[5] Children wake early in the bright light of summer mornings. At the Inquest on a toddler who drowned in the swimming pool of a hotel on Menorca in 2005, the Coroner described the victim as a 'very excited little boy on his first full day of holiday and wanting to use the swimming pool.' He was so eager to get to the pool that he got up before 6am, slipped out of the family's bedroom while his parents were still asleep, walked down two flights of stairs, passed through several unlocked doors, and launched himself into the deserted outdoor pool. A cleaner discovered his body floating face down in the water.[6]

Ponds, lakes, and swimming pools on campsites and caravan parks are freely accessible to children playing unsupervised outside the family's tent, cabin or caravan. An Australian study of drowning accidents in swimming pools noted the particular danger of pools at motels and caravan parks: after immersion accidents, fewer than one in five children survived.[7]

For many families who rent a holiday villa, a swimming pool in the garden is a delightful luxury. Parents seem unaware that a swimming pool presents the greatest risk of death and injury for young children on holiday abroad, especially if the pool lies close to the house, unfenced, with only a narrow terrace between the open-plan accommodation and the edge of the pool. Although Greece and Portugal boast many miles of holiday beaches, almost all of the children who drown in Greece, and more than 80% of the children who drown in the Algarve in southern Portugal, meet their deaths in swimming pools.[8]

Children are at particular risk on the first and last days of the holiday.[9] On arrival, parents are usually tired after the journey, occupied with unpacking and settling in, whereas their children are often eager to slip away and explore. In July 2003, a family flew to a resort on Tenerife, and the two boys, aged 11 and 7, set off at once to find the swimming pool in their apartment complex. Less than an hour after arriving on holiday, another guest found the younger boy drowned in the pool.[10]

In August 2016, on the first day of a family holiday, a four-year-old girl managed to climb the fence around the swimming pool of a Spanish holiday villa rented to celebrate her grandfather's sixtieth birthday. She was pulled unconscious from the pool, airlifted to hospital and placed on life support, but she died four days later.[11]

On the day the family is due to leave, with their parents absorbed in housekeeping and packing, a child may take the opportunity to wander off unsupervised or seek to prolong the fun by taking one more plunge into the swimming pool. In July 2003, while a couple loaded the family car at the end of their holiday in France, their two-year-old daughter walked through an unsecured gate in the safety fence and fell into the swimming pool unobserved.[12] Only a month later and less than five miles away, a second family suffered a similar tragedy. About midnight, a three-year-old boy woke up and slipped into the garden while his parents were packing suitcases into their car. He was found lying at the bottom of the swimming pool.[13]

In the Canary Islands in June 2019, a toddler managed to reach the swimming pool of their holiday villa while his parents packed upstairs on the last day of their holiday. He had squeezed unnoticed through a slightly open sliding door. His babysitters searched for him inside the villa for several minutes before they saw his body floating in the water.[14]

Children are also at risk when parties are held beside swimming pools. It takes only a moment for a child to slip underwater, unnoticed until too late.[15] In August 2011, a three-year-old boy drowned during a birthday party at an apartment complex on the Costa Blanca. His parents had made sure that he wore a life jacket throughout the holiday, but they let him remove it while he ate a slice of birthday cake. Moments later, as they turned to talk with friends, he fell into the pool.[16]

In August 2015, a four-year-old boy was playing with other children beside the swimming pool of a holiday villa on the Costa del Sol. His parents were busy with guests and did not see him fall into the water. He could not swim, and rescue came too late to save his life.[17]

In 2004, after an increase of 20% in the number of Britons drowning abroad, Peter Cornall, head of water and leisure safety at the Royal Society for the Prevention of Accidents (RoSPA), commented: 'Families using private villas with swimming pools must realise that they will have to be lifeguards for their children 24 hours a day. It might be safer to book a property without a pool if you have young children.'[18]

◆ ◆ ◆

France has Europe's largest market for private swimming pools – and the world's highest rate of infant drowning in private swimming pools.[19] In 2002, so many children were drowning in the pools of French holiday villas that Premier Jean-Pierre Rafferin introduced a law that obliged everyone renting property with an outdoor pool to increase safety by installing at least one of four different protective systems:[20]

- a fence round the pool 1.2 metre high with a self-closing, lockable gate
- an 'abri piscine': a retractable shelter that covers the pool
- a reinforced pool cover strong enough to support the weight of a child
- an electronic alarm, either a perimeter alarm (which sounds when someone passes through an infrared beam beside the pool) or an immersion detector (which sounds when someone jumps in, falls in, or pushes off from the side for an unauthorised swim).

Failure to comply with the law could bring a hefty fine, and if someone drowns in the pool of a rented property, the pool owner might face a manslaughter charge.[21] Yet many French pools are neither fenced nor alarmed. Some pool owners decide to avoid the expense of fencing or never get around to organising the installation. Others consider that a fence would spoil the appearance of the garden (although well-designed fences can enhance the look of a swimming pool,[22] and some fencing systems are removable, so that they can be packed away if there are no children staying at the villa).

The risk of drowning in a fenced pool is one quarter of the risk of drowning in an unfenced pool.[23] As long as the parents shut the gate, the fence continues to guard the pool when the family leaves the pool area, denying access to young children until the parents return. In contrast, a retractable pool shelter must be pushed up to the end of the pool before anyone can swim, then it must be re-extended to enclose the water before the parents leave the pool area – every time. Parents who rely on a reinforced pool cover to act as a protective barrier have the similar chore of rolling off the cover every time they return to the pool, then rolling it on again whenever they leave. If they allow the cover to remain open while they are away, there is nothing to prevent children from getting into the water. Even when closed, some pool covers offer incomplete protection. A number of children have drowned or suffered brain damage in immersion accidents when they tried to walk across a pool cover that dipped beneath them, pulled away from the edge of the pool, then pitched them into the water, trapping them underneath.

Electronic alarms put no physical barrier between the child and the pool. Once in the pool, a child who cannot swim will sink underwater within one minute. Even in daylight, it may be difficult to see a motionless child at the bottom of the pool. At dusk or after nightfall, it may be impossible to see anything, or anyone, beneath the surface.

◆ ◆ ◆

A holiday can be a good time to teach a child to swim, but if parents plan to rent a villa with a swimming pool, pre-holiday swimming lessons make more sense.[24] Whatever their swimming ability, school-age children need close supervision. Children who swim well, or think that they do, need to understand that the swimming pool is out of bounds unless a supervising adult is present. Swimming alone in an unfamiliar, unmarked pool – often chilly, even in a sunny climate – is far more taxing than swimming in company, watched by lifeguards, in the heated water of the municipal pool back home. Fatal submersion happens very quickly and quietly.

Parents with teenage children must make a number of strict rules, and insist that the rules are followed. Private swimming pools are rarely deep enough to allow safe diving – even at the 'deep end'. Since teenage boys are the most frequent victims of spinal cord injury and permanent paralysis caused by diving into shallow water,[25] the rules for getting into the water must be clear and firm. Jump in. Don't dive in. Don't push anyone into the pool. Don't leap down onto anyone in the water. No ducking. No swimming after dark. No entering the pool unobserved, for a secret swim. No swimming after drinking alcohol.[26] In 2002 in Majorca, a teenager drowned on a family holiday when he returned to the hotel after a few beers and dived fully clothed into the swimming pool to recover a lost boot.

Before their holiday begins, parents can prepare themselves to deal with an accident by brushing up their first-aid skills, including cardiopulmonary resuscitation (CPR). When they arrive at their holiday base, they need to know how to summon the local emergency services. But most importantly, they must watch over their children with an obsessive zeal, particularly if they are staying in a villa with an unfenced swimming pool.

An Australian safety campaign summed up its five-point message in a memorable jingle:[27]

Fence the Pool
Shut the Gate
Teach your kids to swim, it's great
Supervise: watch your mate
Learn how to resuscitate.

WATER PARKS, WAVE POOLS, SLIDES AND FLUMES

Water parks have become extremely popular tourist attractions. They are costly to build and costly to run. Bad publicity is bad for business, so that it is not surprising that water parks are coy about the number of injuries suffered by their visitors. Yet minor injuries are common, while severe injuries happen far more often than people realise: legal claims are usually settled out of court. As early as 1997, an article in the *New York Times* reported that at least 176 people had died by drowning in American water parks.[28] Since then, more and bigger water parks have been built around the world, and despite the presence of lifeguards, the drownings accidents continue.

Wave pools are a major draw. People are exhilarated by the action of the waves, but they are often surprised by the power of the water as it surges back and forth. They might be less surprised if they realised that wave tanks were originally developed in Germany for testing submarines during the Second World War[29] and that hydrology research laboratories in Oxfordshire are currently using wave machines to study tsunamis.[30]

Wave machines set the water in motion. In paddling pools, blasts of compressed air send ripples across the water surface. In larger wave pools, swells and breakers are created by enormous paddles or by giant holding tanks capable of dumping thousands of gallons into the deep end of the pool.[31] The artificial waves gradually increase in height and force, rolling across the pool more frequently than waves in the sea, with little time between waves for swimmers to recover their breath. For ten minutes or so, the waves keep on coming. During a short pause, the water calms and flattens – then the waves begin again.

Non-swimmers often stand waist-deep, waiting expectantly for the waves to arrive, only to be knocked off their feet by the impact. Weak swimmers are thrown against each other and against the sides of the pool, or they are swirled into deeper water and swamped. If they try to hold on to the side of the pool, the fierce up-and-down, side-to-side motion of the water breaks their grip. Children are most at risk. In 2008, after a four-year-old boy drowned while his mother waited for him at the side of the pool, the California Senate passed the Wave Pool Safety Act, which ruled that any child less than 4 feet tall must wear a life jacket in a wave pool.[32]

In some wave pools, children lie across inflated inner-tube floats. Weak swimmers feel a false sense of security when held up by a float, allowing themselves to be carried out of their depth.[33] Waves toss the floats about and slam them into each other like waterborne dodgem cars, tipping children into the pool, preventing their escape to shallow water or even their access to the surface.

Lifeguards watch the pool intently throughout the wave sequences, but they have a challenging task. They must make split-second decisions while scanning an extremely complex and confusing scene.[34] The wave pool is crowded with swimmers of all levels of skill, propelled back and forth by the action of the waves. The water surface is choppy and disturbed, its broken reflections glinting with erratic highlights. The floats block sightlines and cast shadows on the water. Shouts for help cannot be heard above the hubbub.

Lifeguards have to spot swimmers in trouble and react immediately, hauling potential victims out of the water before they lose consciousness. At some large water parks, they pull more than a hundred swimmers to safety every day. Even accomplished adult swimmers may be overwhelmed if they fail to recover from one wave before they are hit by the next one. If lifeguards are too slow to react, swimmers may suffer permanent brain damage – or they may drown.[35]

It is hard for a lifeguard to distinguish someone energetically treading water from someone desperately fighting to stay at the surface. It is even harder for a lifeguard to detect someone who is already submerged. The swimmer's shape is obscured by the shifting, fragmented reflections and image distortions on the water surface and by the partial opacity of the water.[36] Every year, victims sink underwater unobserved, only to be discovered at the bottom when the water calms between wave sessions.

When someone gets into difficulties, other swimmers are often the first to notice. In 2005, a twelve-year-old boy was taken to a Canadian wave pool only a month after he had learned to swim. Another swimmer saw him floating face down and turned him over. His face was blue. The lifeguards began cardiopulmonary resuscitation as soon as they lifted him from the water, but the boy died without regaining consciousness.[37]

Unfortunately, lifeguards do not always respond immediately when swimmers tell them that someone is floundering in the water, especially when the warning comes from a child. On a school trip to a water park in Pennsylvania in 1994, a fourteen-year-old boy sank beneath the surface in the deepest part of a wave pool.[38] His three young friends rushed over to the lifeguards, shouting, 'Someone's drowning!' The lifeguards failed to see the victim. They knew that another child had just been rescued nearby, so they assumed that the youngsters were mistaken and reassured them that the boy was safe. Nobody dived into the pool to search underwater. Nobody switched off the wave machine. It was closing time before the boy's teachers realised that he was missing. A cursory check of the pool failed to locate him. Lifeguards took part in an after-hours training session in the pool without seeing anything amiss. Ten hours after his friends told lifeguards that the boy was drowning, a maintenance man discovered his body the bottom of the pool, beside the 'drain'.

◆ ◆ ◆

In wave pools, the energy is supplied by the wave machines, and the thrill comes from the movement of the water. On a water slide, in contrast, the energy comes from the pull of gravity,[39] and the thrill comes from the movement of the swimmers themselves as they plummet down the incline to crash-land in the receiving pool. Water runs down the slides, cutting friction to a minimum, so that water slides are much faster than playground slides.[40] Once swimmers start their descent, gravity pulls them to the bottom at ever-increasing speed. In large modern water parks, the slides may drop a hundred feet from top to bottom, and some giant flumes are designed as plastic tunnels to ensure that nobody flies off while hurtling round the curves.[41]

Water slides are best avoided until children have learned to swim. If a non-swimming toddler slides down alone, a parent's fumbled catch can result in a bellyflop or a ducking. If the parent climbs up the steps with the child and they slide down together, their combined weight increases their momentum,[42] plunging them both underwater at splash down.

Older children slide and plunge on their own. At first they slide on their backs, feet first, according to the rules of the water park. After a couple of descents, however, daring youngsters may decide to break the rules and slide on their stomachs, head first.[43] Careering down a water-lubricated slide, they risk injuries to the face, head and spine.[44] The higher the slide, the faster they enter the water. An over-confident teenager who slides head first down a chute in a water park can pick up enough speed to send him crashing against the bottom of the receiving pool.

Teenage boys are heavier, and they slide faster. If they overtake another rider while speeding down the slope, the impact will send them both swerving against the sides, like luges at the Winter Olympics, crashing down onto anybody still in the drop zone. Cuts, bruises, broken bones and concussion follow collisions with other people, with the slide, and even with the opposite wall of the receiving pool. In 1998, after investigating a series of serious water slide injuries, Professor David Ball warned that there was a risk of collision even when traffic lights controlled consecutive riders, and even when everyone followed all the instructions.[45] (His research was reported in *The Times* under the headline: 'Warning: water slides are more risky than rugby.')[46]

Water parks depend on lifeguards to enforce the safety rules on water slides, both at the top of the slide where swimmers are waiting to begin their descent, and at the bottom where swimmers splash down into the receiving pool. But when a rapid succession of children plunge into the water one after another, the lifeguard's view of the pool is obscured by constant splashes and by bubbles swirling beneath the surface. Over the years, a few unfortunate victims have failed to surface after coming down a water slide, then drowned in the receiving pool, unnoticed by the lifeguard.[47] In 1997, the body of an eight-year-old boy lay underwater at the bottom of a slide for several hours. The victim was discovered only when a lifeguard dived into the pool to retrieve his sunglasses.[48] In 2011, a woman overtook a boy on a water slide and bumped into him as she landed in the water. Her apology was cut short as she sank to the bottom, 12 feet down.[49] The boy told a lifeguard what had happened, but the lifeguard said that she was taking her break, and did not go to investigate. In murky water clouded with algae, the woman's body lay undiscovered at the bottom of the pool for two days before floating to the surface.

Water parks are hard work for the parents of young children. They cannot sit resting in the shade while their children roam the water park alone. Lifeguards will be on duty beside each stretch of water, but parents bear the ultimate responsibility for their children's safety.[50] They might be wise to avoid water features with names like Typhoon Lagoon, Runaway Rapids or Tower of Power.

SUCTION ENTRAPMENT: A HIDDEN RISK IN WATER PARKS, SWIMMING POOLS, HOT TUBS AND JACUZZIS

Suction entrapment at water parks has led to many life-threatening accidents and a number of deaths.[51] A water slide has to be kept slippery. This is done by piping water from the pool up to the top of the slide, then the water cascades down the slide and returns to the pool. Strong pumps are used for this constant recycling, which means that the large drain in the wall or the floor of the pool is actually a pump inlet.[52] A specialised cover is placed across the mouth of the opening to protect swimmers from the powerful suction of the pump. If the cover is damaged or lost, the open mouth of the drainage pipe acts as a giant, deadly vacuum cleaner capable of trapping swimmers beneath the water.

Swimmers are sometimes sucked bodily into drainage pipes. In Utah in 1985, a fourteen-year-old boy was playing with his brother and two friends in a 4ft deep receiving pool that served 6 water slides. A report describes him 'hanging onto the pool edge, dangling his feet over the submerged opening of the middle of three drainage pipes, when he let go and disappeared into the pipe'. The protective cover was missing, exposing the mouth of the pump inlet. The boy was carried 93 feet along the pipe, all the way to the pump house, where he drowned.[53] In 2006, a seven-year-old Canadian boy survived a similar accident only because a lifeguard successfully restarted his heart.[54] The boy's feet were so damaged by the pump that both his legs had to be amputated below the knee.

In Lisbon in 1993, two nine-year-old Portuguese children disappeared in a water park within two days of each other. The police began a hunt for kidnappers – until the bodies of both children were discovered wedged feet first in separate drainage pipes.[55] Hundreds of enraged protesters tore down water slides and wrecked a glass-walled restaurant. Riot police were needed to restore order. The water park closed down soon afterwards.

The influential architect, Philip Johnson, designed the Water Gardens in Fort Worth, Texas, where waterfalls cascade forty feet down the rocky steps of a huge crater into the churning water of the Active Pool, a dramatic whirlpool which was meant to produce a sensation of 'pseudo-danger'.[56] Unfortunately, the strong pumps which were needed to recirculate the water from the pool up to the top of the waterfalls meant that the danger was all too real. In 2004, a young visitor slipped into the pool. She was caught in the current rotating towards the pumps. A friend who tried to save her fell into the water. Then her brother and her father jumped in to rescue them. All four drowned.[57]

◆ ◆ ◆

Whether a swimming pool is part of a major tourist attraction in a holiday resort or a family pool in the back garden, the water is kept clean and clear by continuous recirculation.[58] Every three hours or so, a volume of water equivalent to the entire contents of the pool is pumped through filters to remove dirt and debris, treated with chemicals to destroy bacteria, then returned to the pool. The main pump inlet (commonly called 'the drain') usually lies under a grille at the deepest point in the pool. Skimmer pump inlets built high on the wall at the side of the pool take water from the surface to remove leaves and floating debris.

Children often position themselves near a pump inlet, intrigued by the sensation of water flowing past their legs, but the stream of water creates a vortex which pulls them towards the grille. Without frightening them, children should be advised to keep their distance from a swimming pool drain, which probably ought to be categorised as 'an attractive nuisance'[59] – something enticing but dangerous to children, who fail to recognise the danger. Even with the cover securely in place, if a child's body covers the grille, it will be held fast by the vacuum caused by the action of the pump.

Forty years ago Virginia Hunt Newman recalled a reckless game in which a group of teenagers competed to see who could get closest to a large swimming pool drain – a game of 'Russian roulette under water' that ended with the fatal suction entrapment of a young swimming champion.[60] In 1994, the *British Medical Journal* published a photograph of an enormous livid bruise covering most of a woman's back.[61] Its size and shape matched the dimensions of the drain cover where suction had held her fast against the wall of the swimming pool.

At a resort in the Canary Islands in 2009, a British teenager drowned when he was held underwater against a 6-inch-wide grill on the bottom of a swimming pool.[62] The suction was so strong that it took the strength of six holiday-makers to pull him free. Four months later, in a water park in Thailand, another British teenager drowned when he decided to search for his lost goggles by lifting an unsecured grill that covered a drain in the floor of a shallow pool. Unaware of the risk of entrapment, he was sucked into the pump inlet.[63]

If the grill is missing, or if the screw fixings become loose, children sometimes investigate the suction by putting a hand into the mouth of an inlet pipe, only to find that they cannot pull the hand free. In Florida in 2000, a fourteen-year-old boy saw a friend dive into the deep end of the swimming pool near his apartment, move the unattached cover of the pump inlet, then replace it. Thinking that his friend had hidden something under the cover, the boy swam down and pushed his arm into the drain. He was held fast.[64] The combined efforts of several adults failed to release him. Twelve minutes passed before police broke open the door of the pump house and switched off the pump, allowing him to be lifted to the surface. Deprived of oxygen, he suffered such serious brain damage that he never regained consciousness. He remained in a coma until he died four years later.

When small pump inlets cause entrapment, an exploring hand may become so tightly jammed within the pipe that it cannot be freed, even after the pump is switched off. In London in 2005, a three-year-old girl tried to retrieve a toy by putting her hand into the drain at the bottom of a paddling pool. Fortunately, the water was too shallow to drown her, but she could not pull her hand free.[65] Firemen had to dig a trench to expose the pipe, hack through the pipe clear of her fingertips, and take her to hospital with her hand, wrist and forearm still jammed inside the offending pipe. The child needed a general anaesthetic before the combined efforts of firemen and surgeons succeeded in prising open the pipe and releasing her arm.

A swimming pool cleaner draws grit – and water – from the bottom of the pool. Its hose plugs into a 'vacuum port' fitted on the side of the pool, tapping into the suction of the pool recycling pump. This means that the pool cleaner is capable of causing suction entrapment, and it follows that swimmers should not be allowed into the pool while the pool cleaner is at work. When the pool cleaner has completed its task, it is removed from the pool, and a self-latching closure slides over the vacuum port. If the closure is missing, however, the vacuum port itself becomes a hazard. A number of inquisitive children have suffered limb entrapment while investigating the flow of water through an unprotected vacuum port.[66]

In paddling pools, suction entrapment sometimes results in the bizarre and terrible injury of evisceration.[67] Reports in medical journals in America, Australia, France, Spain and Italy have recorded the life-changing trauma suffered by unfortunate children who sat on top of an unprotected drain in a paddling pool and blocked the pump inlet with their buttocks. The suction can be strong enough to tear the child's intestines loose and draw them out of the child's bottom.[68]

On a Greek island in 2003, a six-year-old girl 'was sucked onto an uncovered drain as she climbed into a communal swimming pool'. She was held tight for several minutes, and by the time the pump was switched off, her rectum had been pulled out of her body. She was transferred to a major hospital on the mainland, where surgeons were able to return the damaged bowel to its normal position. Five more days of hospital treatment followed. Several weeks later, when she attended a Leeds hospital for review, she was still suffering pain and discomfort. The paediatricians decided 'to explore how much the travel industry is aware of this potentially devastating injury'.[69] They questioned 24 travel companies organising holidays abroad with access to swimming pools. Most travel companies knew little or nothing about suction entrapment, very few companies carried out their own safety checks on the swimming pools and water parks used by their customers, and only one company had installed accessible poolside cut-off switches that could be used, in an emergency, to shut down the recirculation pump.

◆ ◆ ◆

Suction entrapment has not been adequately recognised as a risk factor when using hot tubs, spa baths or Jacuzzis, even though the water is constantly recirculated past the heater, and the circulation pump exerts a pull of over 400 pounds. In 2002, seven-year-old Virginia Graeme Baker, granddaughter of former American Secretary of State James Baker, drowned at a friend's party when she was sucked against the drain in the bottom of a spa bath.[70] Her mother used all her strength to try to pull her out of the water, but the girl 'appeared to be attached to the bottom of the spa, as if she were tied or held down'. Two men managed to haul her off the drain, but too late to save her. Mrs Baker soon found that others had drowned in the same way. (The US Consumer Product Safety Commission recorded 74 incidents of body or limb entrapment between 1990 and 2004, including 13 deaths.)[71] She decided to change the law. She said, 'It helps me to make some sense of something that makes no sense at all.'

America is now one of the few countries which has recognised the danger of suction entrapment and has tried to limit the risk.[72] Introduced in 2008, the Virginia Graeme Baker Pool and Spa Safety Act ruled that dome-shaped anti-entrapment drain covers must replace flat drain covers in all public swimming pools, paddling pools, hot tubs, spa baths and Jacuzzis, and that only anti-entrapment drain covers could be sold for private use. (The domed shape of the drain cover prevents a vacuum from forming if the bather's body lies across the drain.) As an additional safeguard, a single drain must be fitted with an anti-entrapment device (which shuts down the pump if there is an obstruction to water flow, or breaks the vacuum by allowing the entry of air). Unless an anti-entrapment device is fitted, a single drain must be replaced by dual drains at least 3 feet apart (so avoiding the excessive suction that would follow if a single pump inlet was blocked.) Unfortunately, designing anti-entrapment drain covers has proved to be more complex than expected. In 2011, the United States Consumer Product Safety Commission recalled a million recently installed drain covers that were judged inadequate, and redrafted the standards required from manufacturers.[73]

On holiday in Queensland in 2001, a teenage girl entered a spa bath with two friends. As they dipped their heads into the water to wet their hair,[74] water flowing through the drain carried strands of her hair through the drain cover. Within a moment, water turbulence twisted the strands into a tangle behind the drain cover. She slid off the submerged seat and lay on her back at the bottom of the spa pool, held fast by her hair. Her friends could not lift her to the air, and she quickly lost consciousness. A doctor attempted mouth-to-mouth resuscitation beneath the surface while people searched frantically for a knife or scissors to cut her free. More than 10 minutes went by before she could be released. Cardiopulmonary resuscitation restored her heartbeat, but she died in an intensive care unit four days later. The US Consumer Product Safety Commission recorded 43 incidents of hair entanglement between 1990 and 2004, including 12 deaths.[75]

Since the introduction of the Virginia Graeme Baker Pool and Spa Safety Act, there have been fewer deaths due to suction entrapment in American swimming pools, hot tubs, spa baths and Jacuzzis. Yet even with approved anti-entrapment drain covers, incidents of body and limb entrapment still occur, and unlucky bathers are sometimes trapped by their hair or held underwater when their clothing or jewellery is sucked into the drain.[76]

It is highly likely that drowning deaths due to suction entrapment are under-reported.[77] If emergency staff arrive after the pump has been switched off, the victim is no longer held down by the vacuum, so that the death may be logged as a straightforward drowning fatality. Moreover, incidents of suction entrapment in water parks appear to be on the increase, probably due to the increasingly powerful pumps used to produce waves, to feed artificial rapids, and to lubricate the water slides and flumes that make a visit to a water park so exciting for holidaymakers.[78]

CHAPTER TWENTY-TWO

HAZARDS ON HOLIDAYS ABROAD: BY THE SEA

Since the 1960s, charter flights and package tours have made it possible for huge numbers of Britons to take their holidays abroad – mostly beside the sea. Freed from the routine and restraint of their working lives, people are more than usually accident-prone. When travelling to exotic destinations, tourists expect that tropical infections will be their greatest danger (and sensibly protect themselves with vaccinations and anti-malarial tablets). In fact, accidents – notably road traffic accidents and drowning – cause 25 times as many deaths as tropical infections.[1]

PROTECTING CHILDREN AT THE BEACH

Fences round a swimming pool protect children from drowning, but there are no fences along the tide line. Parents must watch over their children whenever they go to the beach. Toddlers need a grown-up beside them while they are building a sandcastle, exploring a rock pool, or paddling in the sea. And as long as children need the support of armbands or flotation vests, they must stay in the shallows under specially close supervision.

Seawater is more buoyant than fresh water, so small children often manage to launch themselves as swimmers during a seaside holiday, taking their first supervised strokes in the sheltered pools left behind as the tide goes out. However, buoyancy comes at a price. Seawater is denser and more viscous than fresh water. Swimming in the sea is harder work than swimming in a swimming pool, while children are readily tossed around by waves, and they have little strength to fight against currents.

Strict rules are needed for children who know how to swim. Stay within your depth. Swim parallel to the shore. Swim within your comfort zone, without trying to keep up with a stronger swimmer. Don't go off on your own. Come out at once if you feel cold or tired. Whether children are able to swim or not, they should not be allowed to float out of their depth while lying on air cushions or blow-up beach toys. Inflatable dinghies must be securely tethered, even in the shallows. Every year terrified children are swept out to sea clinging to their lilos.[2]

Parents have the responsibility to intervene before their children get themselves into danger. In 2007, a series of drowning accidents involving young children prompted a study of parental supervision on weekends and holidays on 18 flat-water and surfing beaches in New Zealand.[3] Most parents of children aged 0–4 years considered their children to be non-swimmers or weak swimmers and stayed close beside them. However, most parents of children aged 5–9 years considered them good swimmers, and watched them from a distance. A quarter of the youngsters were not adequately supervised while they played in the water. Their parents lay on the sand sunbathing, chatting to others, talking on mobile phones, reading or picnicking. Even parents who entered the water with their children sometimes failed to protect them sufficiently, allowing them to wander away or letting them stay in the water after the conditions became unsafe. Although few parents had any training in rescue or lifesaving, three-quarters of the men and almost half of the women felt confident in their ability to rescue their children. Men were twice as likely as women to consider their children to be at 'no risk' of drowning. The researchers commented on the tendency of parents, particularly fathers, to overrate their children's ability to cope in the water, while underrating their risk of drowning – 'a potentially fatal combination'. The report strongly recommended: 'If in doubt, keep your children out.'[4]

If adults have to rescue children from the waves, either the would-be rescuers or the endangered children may be lost. In 2007, a group of children got into difficulties while playing in the surf on a beach in southern Portugal. Six adults swam out to rescue them. All the children survived, but three of their parents were swept out to sea and drowned, while a fourth parent fought his way back to the beach before dying of a heart attack.[5] Only a month later, on the Spanish Costa Brava, a father was taking holiday snaps of his two sons as they sat on a rocky outcrop close to shore. At that moment, a wave washed the boys into the sea.[6] The father succeeded in rescuing the older boy, then he swam out to try and save the younger boy. They were seen clinging to a lifebuoy in the bay before the waves tore them away. Both father and son were drowned.

RISK TAKING BY TEENAGE BOYS AND YOUNG MEN

During holidays beside the sea, when teenage boys and young men enter the water, their energy and daring often takes them out of their depth and into danger.[7] They pay little attention to the hazards presented by sudden drop-offs in the seabed, crumbling sand spits, or water surging between jagged rocks. They don't notice the rapid advance of the rising tide or the strong undertow as the tide goes out. They pay no attention to the wind, although wind strength and wave height are linked. (Even Force 3 on the Beaufort Scale, a 'gentle breeze', means rough water, with whitecaps whipping off the waves.)[8] And in sunny weather, they expect the sea to be reasonably warm.

In a group of friends, some will be stronger swimmers than others, and this inequality often leads to a pattern of risk-taking where the stronger swimmers flaunt their prowess and the weaker swimmers struggle to keep up, each spurring the others towards more risky exploits.[9] They dive off harbour walls and collide with the seabed. They jump off cliffs and crash onto submerged rocks. They swim into submerged caves, against advice. Whether they are oblivious of the danger or aware of it (and excited by their awareness), few young men imagine that they might drown.[10]

On the Atlantic beaches of France, Spain and Portugal, they overestimate their swimming ability and underestimate the strength of the waves that surge over their heads, lift them off their feet, and cannon them against sea walls.[11] They plunge into the breakers, but forget that the water powering towards the shore must return to the open sea: rip currents drag even the strongest swimmers out to colder, deeper water offshore.[12] The water of the Mediterranean is warmer than the Atlantic Ocean and the waves less dramatic, so that young men are tempted to strike out too far from the shore, only to find their return more arduous than expected.[13] Further afield, when swimming in the tropics or the Far East, holidaymakers too often ignore the warnings of locals. Swimmers who enter the water in the early morning or at sunset have no one to warn them of dangerous currents, and no one to rescue them.[14]

In 1996, a study by American social scientists asked, 'Why are most drowning victims men?' The researchers found that fewer men than women had taken formal swimming lessons, yet significantly more men than women rated themselves as excellent or very good swimmers.[15] Men spent more time in the water, chose aquatic activities with more risk of submersion, swam more often alone or at night, and were more likely to have drunk alcohol before swimming. Indeed, risk-taking and alcohol use were often associated, placing overconfident young men in double jeopardy.

Whether swimming alone or in a group, a swimmer who mixes alcohol with seawater is more likely to get himself into trouble but less able to save himself. Alcohol impairs his judgement and cripples his stroke. It diminishes his shivering response, so that he cannot replace body heat lost to the water. It shuts down glucose production in his liver, interrupting his energy supply. It lowers his awareness of cold, so that he fails to recognise the onset of disabling hypothermia.[16] Alcohol also suppresses the protective reflexes that guard his airway (the gag reflex, the cough reflex and the control of swallowing), increasing his risk of inhaling seawater. And if the worst happens, and the swimmer is pulled unconscious from the sea, alcohol lowers his chances of responding to resuscitation.

In 2008, after a recent increase in alcohol-related drownings in New Zealand, researchers interviewed teenagers in Auckland on their attitudes to drinking alcohol before swimming.[17] Many parties, music festivals and New Year celebrations were held near the beach, and intoxicated youngsters often staggered into the sea for a swim. Both girls and boys admitted that they sometimes went 'skinny-dipping', but boys were twice as likely as girls to have combined drinking and swimming. The teenagers were aware that alcohol had blunted their judgement: 'When you're drinking, you think you're invincible.' The girls acknowledged the risks, and expressed regret ('Now I think about it, it was the stupidest thing I've ever done. But at the time it was just a whole lot of fun, but I'd never do it again.'). The boys played down the risks and appeared unlikely to change their behaviour ('Drinking at the beach is OK at 2 or 3 in the morning, so long as you have someone sober that has the ability to pull people out of the water.') The boys made the wildly optimistic assumption that another bather would notice a friend's drunken collapse into the sea at night and would succeed in a timely rescue. Yet Dr Frank Pia's eye-opening films of bathers on New York's Orchard Beach demonstrated as long ago as 1974 how easy it is to drown unnoticed, even in daylight on a gently shelving, flatwater beach protected from ocean waves.[18]

SURFING, SCUBA-DIVING AND SNORKELLING

On surfing beaches, most people in the water are able to swim, but they face constant threats to their safety from exhaustion, hypothermia,[19] concussion when struck by a surfboard,[20] spinal injury when dumped against the seabed[21] or head injury when slammed against rocks.[22] The greatest threat is from rip currents.[23] Like fast-flowing rivers gushing out to the open sea, rip currents catch unsuspecting swimmers in their grip, and drag them out beyond the surf line. More sea bathers are drowned by rip currents than by any other cause. Even a powerful swimmer will exhaust himself if he tries to swim against a rip current,[24] which will pull him out to sea faster than he can swim towards the beach. Instead of battling against a rip, he should swim diagonally across the rip current towards the nearest breakers, where the waves are running towards the beach. Only then should he aim for a familiar landmark on the shore.[25]

Along the surfing beaches of Australia, swimmers are encouraged to swim in areas patrolled by lifeguards ('Swim between the flags').[26] Swimmers who enter the water close to a surf patrol have a better chance of rescue if they get into difficulties, and a better chance of recovery if they need resuscitation.[27] Yet tourists often choose to swim from remote beaches, far from the lifeguards and paramedics who might save them from drowning.[28]

Several international airlines now screen beach safety videos on their flights to Australia in an attempt to educate their passengers about the dangers of surfing,[29] but many tourists fail to recognise that they are neither fit enough nor skilled enough to venture far from the shore. Impatient young men are eager to plunge straight into the breakers. Few choose to begin their holiday by enrolling in a class on 'surf awareness', which would teach them how to cope with the waves and how to stay out of rip currents.[30] Despite intensive efforts to protect them, overseas visitors on surfing beaches have a ninefold risk of drowning compared to Australian men.[31]

◆ ◆ ◆

Visitors to Australia also suffer a disproportionately large number of scuba-diving accidents.[32] Almost one hundred international tourists are admitted to Queensland hospitals every year after injuries in the sea – mostly scuba-related decompression injuries.[33] In 2000, a study of drowning on the Great Barrier Reef reported that 13 people died while scuba-diving within four years, 6 while visiting Australia.[34] Mastering the complexities of scuba-diving involves lengthy step-by-step instruction and practice. A brief introduction cannot cover the many technical, physical and emotional demands of scuba-diving.[35] Yet tourists offered the chance to try scuba-diving on holiday may find the opportunity hard to resist.

Ill-prepared scuba-divers with limited experience are more likely to have accidents.[36] The risks are particularly high for children and teenagers.[37] If they become anxious underwater, they take rapid, shallow breaths, which increases their anxiety. Feeling short of air, they fear that they might suffocate. Then panic sets in.[38] If a diver panics at depth, minor problems quickly escalate into life-threatening emergencies. A number of double tragedies have involved a parent and child diving together.[39] On holiday in Malta in 2006, a father and his sixteen-year-old son, both new to the sport, dived to about 20 metres. During their unexplained, hurried return to the surface, they forgot to exhale.[40] The compressed air in their lungs expanded to three times its volume, killing them both.

Scuba-divers must be physically fit, and beyond that, they must be certified as 'fit to dive'. Yet a study of one hundred consecutive scuba-diving fatalities in Australia and New Zealand found that a quarter of the victims had a history of health problems – including high blood pressure, epilepsy, diabetes and asthma, which meant that they were medically unfit to dive.[41]

Most people with asthma symptoms know better than to scuba-dive, but those who are free of symptoms when booking their holiday may be willing to gamble on their fitness to dive, unaware that plugs of stiff mucus are often present in their lungs, blocking minor airways and trapping pockets of air.[42] Some over-eager tourists are so determined to glide underwater admiring coral formations and brightly-coloured fish that they sign up for scuba diving with their asthma inhalers tucked in their pockets. Such risky behaviour can have tragic consequences. An acute asthma attack can be triggered by the exertion of the dive, by breathing in the cold, dense air delivered by the scuba regulator, or by a spray of seawater entering the diver's mouthpiece. During an asthma attack, the diver's airways are narrowed by muscle spasm and blocked by an outpouring of mucus. Air under pressure is trapped in his lungs. Then as he ascends to the surface, the trapped air expands inside his chest, ripping the delicate tissues of his lungs and endangering his life.

◆ ◆ ◆

Snorkelling is usually considered less demanding and less dangerous than scuba diving, but the a four-year study of drowning on the Great Barrier Reef reported that more people died while snorkelling than scuba-diving:[43] 18 of the 20 snorkellers who drowned were international tourists, with more victims from Britain than any other country. In Hawaii, snorkelling is the leading cause of accidental death, with 79 drownings during 2009–2013.[44] Older holidaymakers are best advised to forget about snorkelling, even in shallow water, unless they are fit enough to cope with strenuous physical exercise.[45] Children should wear a life jacket while they are learning to snorkel, and they should practise in calm, shallow water, protected from any passing watercraft.[46]

Fit young swimmers often combine snorkelling with free diving. But if they over-breathe at the surface in a misguided attempt to prolong their dive time, they risk losing consciousness under water due to shallow water blackout.[47] A letter in the *New England Journal of Medicine* points to another source of danger to snorkellers: scuba-divers sometimes offer snorkellers a puff of compressed air from the scuba mouthpiece to extend their time beneath the surface.[48] However, snorkellers must breathe out the expanding air before they rise to the surface, otherwise they risk catastrophic injuries to their lungs.

JELLYFISH CAUSE MORE DEATHS THAN SHARKS

In subtropical and tropical seas, swimmers run a slight but definite risk of attack by large predators such as sharks,[49] and saltwater crocodiles,[50] and by venomous creatures such as the blue-ringed octopus.[51] But they run a far greater risk from deadly species of jellyfish. The trailing, toxin-laden tentacles of box jellyfish floating off the beaches of tourist destinations in the Pacific and Indian Oceans, the Caribbean islands and the Gulf of Mexico kill far more swimmers than shark attacks. As many as 40 deaths annually are reported in the Philippines alone.[52] Some victims drown because the agonising pain of the stings makes swimming impossible. Others are poisoned by the venom or overwhelmed by their own allergic reaction to the stings. Children are specially sensitive, sometimes dying within minutes of being stung.[53]

On the Great Barrier Reef and on leeward beaches of Hawaii, including Waikiki, sea bathing is out of bounds during the seasonal blooms of jellyfish: swimming is confined to swimming pools. People with an allergy to bee stings are likely to be highly allergic to jellyfish stings, and they should take their EpiPens to the beach.[54] Children should be warned never to touch jellyfish. Their sting cells are potent whether they are floating in the sea or lying dead and drying on the sand.[55]

Vinegar inactivates jellyfish sting cells that have not yet discharged their venom, whereas contact with fresh water, ice or alcohol causes further firing of sting cells and the injection of more venom.[56] Anyone who develops breathing difficulties after a jellyfish sting should be treated as a medical emergency. Victims may need immediate, prolonged cardiopulmonary resuscitation.

Anti-venom has been developed against box jellyfish toxin,[57] but, as yet, there is no available anti-venom to counter the toxin of Irukandji jellyfish, a hyper-venomous box jellyfish first identified fifty years ago on the Queensland coast of Australia.[58] Transparent and tiny, they are invisible in the water. Their sting is almost painless, but within half an hour the toxin causes life-threatening symptoms of Irukandji syndrome: painful muscle cramps, vomiting, feelings of terror, and extremely high blood pressure sometimes ending in fatal brain haemorrhage. Doctors in Queensland now advise swimmers, 'Always have vinegar on you.'[59] Anyone stung by an Irukandji jellyfish needs hospital treatment fast.

The Portuguese Man-of-War is a worldwide threat to swimmers, especially when their blue-crested sails are blown onto a windward shore.[60] Their intensely painful stings are designed to paralyse small fish by injecting a neurotoxin which affects the heart and the muscles of respiration. With adhesive tentacles up to 50 feet long, they are a danger to swimmers and scuba-divers who are not protected by wetsuits or stinger suits. Fragments of the tentacles clinging to a swimmer's skin should be doused with vinegar before gently removing them, without touching them with bare fingers. Swimmers in pain or affected by the toxin or an allergic reaction need urgent medical help.

WINDSURFING, WATERSKIING AND WATER-TUBING

During seaside holidays abroad, visitors have time to play and cash to spend on aquatic sports, each with its own attractions, its own quirks, and its own risks. Water sport is a growth industry. Some holidaymakers will already be familiar with their chosen sport, but most will be trying something new, which makes them very dependent on the safety-consciousness of the operators who rent out the equipment. Not surprisingly, the rate of injury is higher among tourists than among local residents.[61]

Windsurfing is fairly simple to learn, but beginners spend much of their time losing their balance, falling into the sea, treading water and hauling themselves back onto the board. They should wear wetsuits and buoyancy vests because they soon become cold and tired. Rescue boats are sometimes needed when offshore breezes blow exhausted windsurfers out to sea.[62] Kitesurfing is far more demanding than windsurfing, and training is essential. Even after training, accidents are frequent, especially during launching or landing.[63]

When waterskiing, safety depends on the skill and common sense of the person operating the tow boat, but a careful observer is also needed in the tow boat.[64] The driver powers the boat back and forth parallel to the beach at speeds of up to 30 miles per hour: the observer looks out for obstacles in the water and protects waterskiers by noting their location when they fall into the sea. Waterskiers should wear a certified buoyancy jacket to cushion the impact of a fall and to support them in the water while they wait for the tow boat to return to pick them up. When the boat arrives, they must keep well away from the propeller, which can slice through flesh like a bandsaw. Propeller strikes are one of the most frequent causes of serious injury to waterskiers.[65]

At many resorts water-tubing is on offer, a variation of waterskiing. Instead of balancing on waterskis, youngsters are towed through the water clinging to inflatables shaped like doughnuts or bananas. At first sight, water-tubing looks more amusing, easier and safer than waterskiing, but the riders have no control over their inflatables or their own trajectories. If the tow boat makes a sharp turn, the inflatable skims across the surface in a wide arc,[66] dragging riders through the water, smashing them against swimmers and obstructions, launching them into the air, or throwing them onto the shore. In 2018, a young man was thrown from a banana boat when it overturned in the Santa Clara Reservoir in Portugal. Divers searched for three weeks before his body was found.[67]

PEDALOS, DINGHIES, JET SKIS, SPEED BOATS AND MOTORBOATS

Pedalos and inflatable dinghies are so buoyant and appear so stable that holidaymakers put to sea with small children, never thinking to wear a life jacket, but the sea has to be calm. While on holiday in Cyprus in 2001, four friends hired a pedalo. When it capsized in choppy seas, three managed to swim to safety, but the fourth man drowned.[68] In 2003, a family bought a dinghy from a kiosk on a Cyprus beach. The mother and her nine-year-old son set off to reach a small offshore island, but the dinghy overturned a hundred yards from the shore, throwing them into the sea. Leaving his daughter on the beach, the father swam out to save them. Rescuers pulled the boy unconscious from the water. He recovered in hospital, but both of his parents drowned.[69]

Lifeguards make good use of jet skis (often called personal watercraft or PWCs) to reach endangered surfers with the minimum delay. Jet skis are also widely available for hire to holidaymakers, who are usually unaware of how powerful they are and how difficult to handle. Drivers get a minimum of instruction before they set off, and they are soon speeding through the water and changing directions in a series of tight turns, their sight lines obscured by water spray. But a jet ski has no brakes and no rudder. This means that if the power is cut to avoid a collision, the jet ski careers onwards for a hundred yards or more 'like a car on ice'.[70] Accidents are common, one-quarter of them within an hour of hiring the jet ski.[71] Inexperienced teenagers frequently lose control of their high-horsepower machines and plough through groups of unprotected swimmers or smash into other vessels, causing a rising toll of deaths from blunt injury and drowning.[72]

Speedboats power through the waves, bouncing like bucking broncos. Sometimes the person at the wheel is unseated and thrown into the sea. With no one at the controls, the spinning action of the propeller forces the speedboat into a tight turn, so that the boat circles back at full speed towards the ejected driver, who is struggling in the water, unable to get out of the way. This danger can be avoided by using a 'kill cord'. Before the boat leaves the mooring, one end of the kill cord is fastened to the driver's leg and the other end is clipped into an emergency stop switch on the controls. Then if the driver falls overboard, the speedboat will come to a rapid halt.

Boats without a functioning kill cord have caused many serious accidents, killing and maiming victims in British waters and in holiday destinations round the world.[73] In May 2013, lack of a kill cord led to a horrific accident in calm seas off Padstow in Cornwall.[74] A family of six were thrown from their speedboat, and the boat locked into a series of tight clockwise circles, repeatedly running over the defenceless swimmers. Two were fatally injured, and the propeller blades wounded all four survivors. Only the skill of a local waterskiing instructor prevented an even worse outcome. Pulling his own speedboat alongside, he slowed the runaway boat by entangling its propeller with a rope. Then he jumped on board and stopped the engine.[75]

◆ ◆ ◆

Most holidaymakers realise that they need skilled instructors while they learn how to harness the power of the wind in sail-powered yachts. They assume, however, that handling a hired motorboat is as straightforward as driving their car, and they are often remarkably casual about the weather forecast and the state of the sea beyond the harbour wall. Yet compared to larger commercial boats, the small boats hired by tourists are more vulnerable to damage in a collision, easier to swamp when overloaded, more likely to capsize due to the poorly balanced weight of passengers and gear, and more unstable in rough seas.[76] In 2009, the United States Coast Guard Accident Report on boating accidents recorded 543 deaths by drowning, and noted that most fatal boating accidents involved boats smaller than 26 feet, most happened close to shore, and three-quarters of the victims had not taken a boating course of any kind.[77]

Alcohol is involved in one-third of all marine fatalities.[78] Young men taking a boat out to sea are warned not to drink alcohol and advised to wear life jackets. Unfortunately, too many break open their packs of beer and leave their life jackets stowed in lockers. Whether in a car or a boat, alcohol impairs the judgement and coordination of the person at the wheel, slowing his reactions when fast responses are needed. Common sense and the law demand that the skipper must remain sober, but tipsy passengers are also accident-prone.[79] They become at once more boisterous and less adept at keeping their balance on the pitching, slippery deck, increasing the risk of someone falling into the sea, and decreasing the survival chances of any 'Man Overboard'.[80]

Even if the fall is witnessed and the alarm sounded immediately, a considerable time elapses before the skipper can change course and return to rescue the casualty.[81] With only his head bobbing above the surface, he will be hard to locate unless someone on board acts as a spotter, pointing at the 'man overboard' and keeping him constantly in view. At night or in fog, the casualty may be impossible to find unless he carries a mobile phone in a waterproof pouch, or his life jacket is fitted with a strobe light, a Personal Locator Beacon (PLB) or an Automatic Identification System Man Overboard Device (AIS).[82]

His would-be rescuers expect him to splash, wave, and call for help. If he does not signal his distress, they may assume that he is comfortably treading water, when actually he is desperately trying to stay at the surface, exhausted and fighting for every breath.[83] He may be too chilled to assist his rescuers, who then have the difficult task of lifting the dead weight of a helpless or unconscious casualty from the water onto the deck.[84] If he is not wearing a life jacket, he may sink before the boat can reach him.

The sudden lurching of drunken passengers can capsize a small boat,[85] throwing everybody on board into the sea. The 2009 USCG Accident Report listed capsizing as 'the biggest single danger' for boats under 26 feet in length.[86] Whether victims fall overboard or are thrown into the sea during a capsize, they will suffer the sudden onset of cold shock as they enter the water, then the rapid progressive loss of strength and coordination as their muscles chill. The colder the water, the greater their peril. A US Coast Guard report pointed out that nine out of ten people who drown in boating accidents are not wearing a life jacket, and gave the emphatic advice: 'The most important survival technique is putting on your life jacket before you need it.'[87]

◆ ◆ ◆

Travellers to the wilder shores of the world face unfamiliar dangers far from home, with less likelihood of speedy rescue or adequate resuscitation.[88] They trust themselves to overloaded ferries.[89] They contend with monsoon rains or catastrophic floods,[90] with hurricanes in the West Indies[91] and typhoons in the East Indies.[92] There were 230,000 victims of the South Asia tsunami on Boxing Day 2004. Among them were more than 9,000 European tourists.[93]

CHAPTER TWENTY-THREE

SHALLOW WATER BLACKOUT

We breathe automatically, usually without thinking about it. We capture oxygen when we breathe in and expel carbon dioxide when we breathe out. The rate and depth of our breathing is controlled by the respiratory centre in the brain. If we try to hold our breath, within half a minute the respiratory centre prompts us, then forces us, to breathe in. Surprisingly, it is the level of carbon dioxide in the blood, much more than the level of oxygen, which regulates respiratory drive.[1] When the carbon dioxide level rises, as it does during exercise, the respiratory centre stimulates faster, deeper breathing. If the level of carbon dioxide in the blood is low, the respiratory drive weakens, delaying the urge to breathe. Such a fall in respiratory drive has been responsible for many drownings due to shallow water blackout.[2]

SHALLOW WATER BLACKOUT IN COMPETENT SWIMMERS WHO OVER-BREATHE BEFORE SWIMMING UNDERWATER

The most frequent victims of these tragic accidents are confident, determined boys and young men who swim well.[3] They get into difficulties when they try to achieve a personal best for underwater breath-holding, or when they compete to see who can swim further underwater.[4]

Typically, a swimmer discovers that if he takes a series of rapid, deep breaths as he prepares to enter the water, he can stay below the surface for longer or swim further before he has to come up for air. He is under the mistaken impression that his over-breathing stores more oxygen in his body, allowing him extra time underwater. In reality, he hardly changes his oxygenation, since normal breathing is enough to ensure that his circulating blood leaves his lungs with 97% oxygen saturation of the haemoglobin in his red cells.[5] Instead of capturing more oxygen, his over-breathing expels carbon dioxide from his body. The more he practises, the more carbon dioxide he breathes away, the longer he can hold his breath and the further he can swim underwater.

While he remains underwater, he cannot capture oxygen from the air, but his tissues constantly remove oxygen from his bloodstream. And here lies the danger. As soon as the level of oxygen in his arterial blood falls below a critical level – from 97% down to 55% oxygen saturation – he will lose consciousness suddenly and without warning.[6] If he is lucky, someone will see him lying prone and inert at the bottom of the pool, will realise that he has lost consciousness, and will pull him out of the water in time to revive him.[7] Unfortunately, he may be unlucky. Shallow water blackout claims a number of lives every year.[8]

In 1993, a young American spearfishing champion thought that he could improve his skin-diving if he practised holding his breath under water in the shallow end of his local swimming pool. He bought a new stopwatch, swam a length or two, slid down to a sitting position with his back against the tiled wall, and began to time himself. He lost consciousness. Then he inhaled water.[9] Another swimmer found him submerged and motionless, still clutching the stopwatch. He could not be revived.

During the 2012 Christmas holidays, a twelve-year-old Australian boy took part in several underwater breath-holding contests with his friends. A month later, alone while swimming underwater laps, he lost consciousness and drowned in the shallow end of the family swimming pool. His mother was nearby, but she heard nothing to alert her. 'When I saw him he was at the bottom of the pool.[10] At first I thought, is he just playing, mucking around? It was so quick.'[11]

Teenagers may be so intent on beating friendly rivals in an underwater swimming contest that they over-breathe, sometimes to the point of dizziness, before diving in. Once under way, they concentrate on their progress through the water. If they notice an urge to breathe, they choose to ignore the warning. Then they black out. Writing in 1999, a prominent American swimming coach described a 'near drowning incident' in his own backyard pool.[12] His three sons were competing to see who could swim furthest underwater. His oldest son, an Olympic water polo player, took a number of deep breaths before he set off to beat the combined distance swum by his younger twin brothers. It took him about a minute to cover the distance, but then he failed to come up for air. Instead, he lay underwater, scarcely moving. The younger boys shouted, 'Dad… He's pretending to drown'. Recognising the danger, the coach dived into the pool, rescued his blue, unconscious son and gave him immediate lifesaving mouth-to-mouth resuscitation.

Thirty years ago, a British doctor realised that her son had had a lucky escape after suffering a shallow water blackout while swimming 50 metres underwater across an open-air swimming pool. He was saved from drowning by three friends who found him lying face down on the bottom. She decided to record the episode for a medical magazine. 'He can remember approaching the end of the pool and thinking that he could make it, and then he regained consciousness in the first-aid room,' she wrote. 'I feel that parents, teachers and swimming pool personnel should be made more aware of the extremely dangerous practice of over-breathing before swimming underwater – more publicity would avert some tragedies.'[13]

There are still too few people aware of shallow water blackout, and the tragedies continue. Moreover, the underlying cause of these fatal accidents is often missed because the victims usually inhale water before they die, so that the unexpected deaths of skilled swimmers may be incorrectly recorded as due to drowning rather than shallow water blackout. As a result, neither swimmers nor lifeguards are sufficiently alerted to the risks of underwater breath-holding.

Lifeguards tend to focus on the weaker swimmers splashing about at the surface of the water. They do not monitor a strong swimmer so closely.[14] And since a shallow water blackout happens instantly, without a struggle, when the swimmer is already underwater or slumped against the wall in the shallow end, he may not be discovered until it is too late.

Other swimmers in the pool may be too involved with their own swimming to recognise that someone needs help. In 1999, a teenager blacked out in a college pool while trying to hold his breath underwater. Nearby, a woman was swimming laps. She swam past the unconscious teenager twice as he floated face-down and motionless. She only registered that something was wrong when she passed him a third time, still face-down and motionless, and now lying deeply submerged.[15] She pulled him to the surface, but he did not respond to resuscitation.

Normally, blood has an oxygen saturation of 75% when it enters the lungs after circulating round the body. Then, within moments, blood is re-oxygenated and leaves the lungs with an oxygen saturation of 97%.[16] Once prolonged breath-holding has resulted in shallow water blackout, however, blood enters and leaves the lungs with an oxygen saturation as low as 45%.[17] The swimmer's body tissues are starved of oxygen, with far too little oxygen reaching his brain.[18] Even if the swimmer is noticed soon after losing consciousness, and given prompt resuscitation, he may not survive.[19]

◆ ◆ ◆

Although most of the victims are male, girls also fall victim to shallow water blackout, provided that they swim well enough and are sufficiently competitive. In Texas in 2003, during a training session of a school swimming team, a sixteen-year-old girl swam 50 metres underwater. Her fifteen-year-old teammate was determined to do as well, but her attempt almost killed her.[20] Before she dived in, she breathed rapidly and deeply a number of times, hoping to improve her chances of success. She managed to complete her 50 metre underwater swim, but instead of climbing out of the pool, she floated face-down in the water. At first, her friends did not realise that she was in trouble. They assumed that she was 'alive and kicking' because her arms and legs were twitching to and fro. When they jumped in to help her, they were shocked to find that she was blue and deeply unconscious. She was given skilled resuscitation, but three minutes went by before she began to respond. Her rhythmic jerking was a sign of loss of motor control due to cerebral hypoxia (caused by the low level of oxygen in the blood supplying her brain).[21]

In breath-hold diving competitions, judges disqualify contestants who show signs of cerebral hypoxia. (Approximately 10% of contestants lose consciousness or suffer loss of motor control – and are duly disqualified.)[22] The involuntary rhythmic jerking of cerebral hypoxia bears a slight resemblance to South American dance moves, and breath-hold divers have given these warning twitches the inappropriately light-hearted nickname – 'samba'.[23]

BLACKOUT IN SYNCHRONISED SWIMMERS

During synchronised swimming, swimmers have to cope with a demanding combination of breath-holding and vigorous exercise under water. In competition, they submerge repeatedly for 45 seconds, or longer, while performing the strenuous muscular manoeuvres needed to complete four set figures followed by a taxing free programme. Synchronised swimmers often become dizzy or disoriented. Some even black out during practice sessions and in competition.

In 1995, British physiologists reported that the judges at a regional synchronised swimming championship gave higher marks to teams with longer submersion times, thereby encouraging undesirable levels of hypoxia.[24] They studied nine members of the Great Britain Synchronised Swimming Team, including the entire 1992 Olympics squad, measuring the level of oxygen and carbon dioxide in breath samples while the swimmers performed the underwater sequences of their competition programme. The swimmers were aware of the dangers of over-breathing. They breathed normally before their swim, and all showed the increased levels of carbon dioxide in their lungs that would be expected after their swim. Yet as they completed the free programme, all the swimmers were described as cyanosed (blue in the face) and mildly confused. And all were found to have 'significant hypoxia', including three who developed 'a degree of hypoxia often associated with loss of consciousness'.

In Seattle in 2008, six members of an elite American synchronised swimming team were involved in a hypoxic incident while taking part in demanding practice drills in preparation for an upcoming competition. The girls, aged 13 to 15 years old, developed symptoms of marked oxygen deprivation, including disorientation, tunnel vision, memory loss, numbness of the arms and legs, muscle weakness and cyanosis. Four girls lost consciousness and sank.[25] Their coaches and lifeguards pulled them to the surface and lifted them out of the water. Two girls needed mouth-to-mouth resuscitation. One girl needed cardiopulmonary resuscitation. They recovered consciousness at the poolside, and were given oxygen on their way to hospital. All the girls made a full recovery.

BLACKOUT IN ELITE SWIMMERS DURING TRAINING

Elite swimmers learn to cope with discomfort, even pain, during training, in the hope that their endurance will pay dividends during competition. Swimming 200 metres of fast freestyle is enough to lower their arterial oxygen saturation from 97% to 94%.[26] As they prepare for the aerobic and anaerobic demands of racing, sessions of vigorous interval training reduce their oxygen saturation even further. Some training schedules are designed to push swimmers into hypoxia, with instructions to take five, seven or nine strokes between breaths, breathing out slowly into the water to get rid of the accumulating carbon dioxide that would prompt them to take an extra breath.[27] By the end of each training session, swimmers have built up a massive oxygen debt.

Hypoxic training is dangerous if it is pushed to extremes in extended underwater swims or in 'lung buster 25 repeats' which encourage determined swimmers to ignore the urge to breathe. In 2009, a Canadian champion blacked out briefly during a training session after completing a length of underwater breaststroke.[28] He told himself that a chest infection and compression from his new full-body swimsuit had limited his breathing during his lengths on the surface. Nevertheless, he reported the episode to his coach, who arranged several days of hospital tests to exclude a physical abnormality. The tests were normal, and he was cleared to resume training and competing.

In Florida in 2015, an elite swimmer pushed himself too far. At the end of a long workout, watched by his family, he attempted to swim a 'hundred' (four 25 metre laps underwater without coming up for air). Before he completed the distance, his sister realised that he had stopped moving and alerted the lifeguards. He was pulled from the water deeply unconscious and could not be revived.[29]

◆ ◆ ◆

Swimming underwater minimises wave drag. Before 1957, racing rules allowed breaststroke swimmers to decide for themselves how far they would swim underwater before they returned to the surface. As a consequence, a number of swimmers passed out underwater.[30] The rules were changed after the 1956 Melbourne Olympics, when the winner of the Men's 200 metre Breaststroke swam most of the race underwater.[31] Since then, breaststroke swimmers are only allowed to complete one stroke cycle underwater at the start of the race and after every turn.[32] For the major part of every length, their heads must break the surface during every pull of the arms.

Backstroker Dave Berkoff won two medals at the 1988 Olympics in Seoul using the dolphin kick at the start of each length. His innovation was called 'the Berkoff blastoff'.[33] Elite swimmers soon added the dolphin kick to the underwater glide in freestyle and butterfly events.[34] In competition, the dolphin kick is limited to the first 15 yards at the start and after each turn.[35] In training sessions, swimmers practice 'dolphin kicking out of starts and turns', but swimmers who are determined to master the vigorous up-and-down undulations of the dolphin kick often push on much further than the 15 metre mark, risking hypoxia because of the demands of the stroke.[36] At North Baltimore Aquatic Club in 2012, a talented fourteen-year-old stayed in the pool after a Sunday morning training session and set off to swim 50 metres underwater, powered by the dolphin kick. He was spotted lying motionless at the bottom of the pool.[37] He was given resuscitation, taken to hospital and placed on life support, but he had suffered such severe hypoxic brain damage that he died three days later.

After the intense activity of their training sessions, swimmers usually cool down with 15 minutes of leisurely swimming. The gentler pace of exercise, combined with deep, unhurried breathing, returns their raised blood lactate to near-normal levels more quickly than passive resting.[38] Some elite swimmers chose to cool down by swimming underwater, unaware that their exertion will have left them with a huge oxygen debt and that every stroke extracts further oxygen from their blood.[39] In 2006, a seventeen-year-old swimmer with hopes of a place on the British Olympic team was discovered on the bottom of the pool at the end of a coaching session.[40] He trained hard as usual on the night he drowned, then he took a couple of deep breaths and dived in from the deep end, intending to swim two lengths underwater before heading for the changing room. He swam one and a half lengths without difficulty before he suffered a shallow water blackout. As he lay unconscious, he inhaled water. Vigorous efforts at resuscitation failed to save his life.[41]

BLACKOUT IN SNORKELLERS AND SKIN-DIVERS

Shallow water blackout also endangers another group of skilled swimmers – experienced snorkellers and skin-divers, particularly those who hunt fish with spear guns or cameras. Because their time beneath the surface is so limited, they often give themselves extra moments of dive time by over-breathing just before they dive, a doubly dangerous action in these adventurous young men. Not only does over-breathing delay the urge to breathe by lowering the level of carbon dioxide in the blood, but it also brings on feelings of euphoria and overconfidence, encouraging risk-taking.[42]

The skin-diver's energetic underwater swimming quickly uses up the meagre store of oxygen in his body, but as long as the level of carbon dioxide in his blood remains too low to prompt him to breathe, he goes on with his dive. But he will lose consciousness as soon his brain is deprived of oxygen, suffering shallow water blackout with little chance of successful rescue or resuscitation.[43]

As well as being at risk from shallow water blackout, a skin-diver is also at risk from deep water blackout (or 'hypoxia of ascent'). Deep water blackout is not related to lowered carbon dioxide levels in the diver's blood, but to the pressure changes in his lungs during descent and ascent.[44] At the surface, the air in the diver's lungs is at atmospheric pressure, and the partial pressure of oxygen is around 104mm Hg.[45] When a skin-diver is skilled enough to dive to a depth of 10 metres, the air in his lungs is compressed to half its volume and the air pressure is doubled, including the partial pressure of oxygen. As a result, the blood flowing through his lungs can capture adequate amounts of oxygen for longer, allowing the skin-diver to continue his hunt.

If he focuses on the thrill of the chase too long, however, these extra moments of enjoyment may be dearly bought. His misjudgement has trapped him at depth. As he rises towards the surface, the air in his lungs re-expands, the air pressure in his lungs decreases, and the partial pressure of whatever oxygen remains in the lungs falls even more. (The blackout threshold lies around 30mm Hg.)[46] With so little oxygen available in his lungs, his blood captures too little oxygen, his brain receives too little oxygen, and he will pass out before he reaches the surface. In a fatal dive, pre-dive over-breathing and the fall of partial pressure of oxygen during ascent may both be implicated – shallow water blackout and deep water blackout in deadly combination.[47]

If a skin-diver loses consciousness under water, he usually develops laryngospasm, a protective reflex which closes off the entrance to the trachea and stops water entering the lungs.[48] However, as the hypoxia becomes more severe, the laryngospasm relaxes. Then, even if another diver pulls him up to the surface, his chances of recovery are poor. As he rises to the surface and his compressed lungs re-expand, the resulting syphon effect sucks water through his windpipe down into the depths of his lungs.

A skin-diver has to make sure that he surfaces before he runs out of oxygen. Any skin-diver who surfaces safely, but has symptoms of trembling or tunnel vision, who sees stars or has a brief memory loss, should realise that he escaped death by a whisker and that he must change his skin-diving technique.[49] Unlike a scuba-diver, who has to breathe out as he ascends to avoid decompression injury, a skin-diver should never breathe out underwater. If he does so, he loses precious oxygen from his lungs (and he loses buoyancy.) Moreover, he is likely to trigger a reflex (the Hering-Breuer Deflation Reflex) which responds to exhalation by compelling him to inhale.[50]

A skin-diver must also make sure that he surfaces before carbon dioxide builds up in his bloodstream to a level which forces him to breathe in. The breath-hold breakpoint is reached when the partial pressure of carbon dioxide in the blood climbs to 55mm Hg (well above the normal level of 40mm Hg).[51] A skin-diver who reaches the breath-hold breakpoint while he is still underwater will feel an overwhelming urge to inhale. Even if he realises that a breath will probably kill him, he will be unable to resist the compulsion to breathe in. But instead of breathing air into his lungs, he will breathe in a jet of pressurised water.

CHAPTER TWENTY-FOUR

HEALTH WARNINGS

A number of medical conditions carry an increased risk of drowning. Those at risk should be warned that special precautions are needed whenever they swim. And they are better protected if their families also know about the risk.

A small number of people are found to carry a gene mutation that greatly increases their risk of drowning. They must be warned to stop swimming entirely. Genetic screening of their relatives may identify potential drowning victims as yet unaware of their danger.

Competitive athletes with undiagnosed heart disease are at risk of sudden death. Cardiac screening may allow treatment in good time. Older swimmers should avoid over-exertion.

EPILEPSY

People with epilepsy have an increased risk of drowning, even when they are taking full anti-epileptic medication.[1] Researchers at the Institute of Neurology in London calculated: 'The risk of drowning in people with epilepsy is raised 15- to 19-fold compared to people in the general population.'[2] An American study estimated that children with epilepsy are 20 times more likely to drown while swimming than children without epilepsy, and they are almost a hundred times more likely to drown in the bathtub.[3] They remain at increased risk of drowning even if their seizures appear well controlled. A Canadian study of drowning deaths in people with epilepsy included a victim who had been free of seizures for 15 years.[4]

Swimming exposes people with epilepsy to a number of stimuli which can trigger seizures. A sudden ducking or a dive into cold water may bring about a 'startle' seizure.[5] Light patterns on the rippling surface of the water close to the swimmer's eyes can initiate seizures in people with photosensitive epilepsy.[6] Gasping for breath during a swimming race and exhaustion at the end of the race have both been linked to the onset of petit mal absences and seizures.[7] Swimming underwater demands sustained breath-holding and muscular effort, increasing the risk of a seizure.

Professor John Pearn has advised: 'If an epileptic child is mentally normal, well controlled with anticonvulsants, and supervised in the water, then the risk of drowning is very small.'[8]

But he warned: 'Epileptics should never swim or bathe alone.'[9] Parents must ensure that their child is accompanied by a skilled, attentive adult swimmer who knows the diagnosis, who can recognise the onset of a seizure, and who is strong enough to hold the child's face above the water surface during any jerking.[10] Support and rescue is relatively easy at the shallow end of a swimming pool, where the companion swimmer can stand while holding the child. Adequate support and rescue is harder to achieve if a child has a seizure in the deep end. A lifeguard beside the pool should be informed of the diagnosis and asked to keep the child and the companion swimmer under close observation.[11] Then, if the child has a seizure, the lifeguard can help the companion swimmer to hold the child's head above the water surface and to lift the child out of the water as soon as the jerking has ceased.

In Britain, the Epilepsy Association encourages children to take part in a wide range of sporting activities. Swimming is included, with the advice: 'Extra safety precautions are needed' and 'Adequate supervision is essential'. Unfortunately, safety advice for teachers is softened in a misguided attempt to avoid embarrassment to the child. An advisory leaflet comments: 'One-to-one supervision may be necessary. Some schools use a "buddy system" which pairs up pupils so that everyone has someone to look out for them in the water. This may help a child feel they are being treated the same as the other children, as well as increasing everyone's safety in the water.'[12] If a child with epilepsy has a seizure while swimming, however, the youngster acting as buddy is an inadequate substitute for an adult companion swimmer. Moreover, the child's safety becomes the responsibility of the buddy, who would be unfairly burdened with guilt should anything go wrong.

In 2001, Professor Frank Besag pointed out: 'Any seizure that involves impaired awareness or notable loss of control of motor function must be considered to place the person at great risk in the water'. He described a fourteen-year-old boy with epilepsy who drowned in a lake on a school outing while swimming with 11 children and 15 teachers.[13] He was seen playing happily in the water, but nobody saw him sink below the surface. He was missed only when the children were leaving the water. Professor Besag stressed the particular risk of drowning in people who suffer from tonic 'grand mal' seizures. He explained: 'During a tonic seizure the muscles of the chest wall contract and much of the air from the lungs may be expelled.' The swimmer immediately loses buoyancy and sinks to the bottom so rapidly that companions fail to notice his sudden disappearance. Once the initial spasms subside, his chest will re-expand, but because he is submerged, he will inhale water instead of air.

Adults with epilepsy need to be made aware that they have an increased risk of drowning, and, however well they swim, they should be advised that they need close supervision when they are in the water.[14] Their families, too, should be told of the increased risk and the need for close supervision. Weak swimmers are safer if their companion swimmers swim alongside them. Strong swimmers may be better protected if their companion swimmers watch from the side of the pool, ready to dive in at the first sign of a seizure.[15]

In 2006, the *British Medical Journal* published a case report as their Lesson of the Week which showed that even a brief submersion is fraught with danger when someone with epilepsy swims alone.[16] A 44-year-old woman with epilepsy had a seizure while swimming by herself in her local swimming pool. She was submerged for only a minute before she was rescued. The seizure soon stopped, and she insisted that she was well enough to return home. Once home, she became breathless and began to cough up frothy pink mucus. She called an ambulance, but she had another seizure on the way to hospital, and arrived deeply unconscious, apparently suffering pulmonary oedema due to heart failure. She was placed on a ventilator at once, but the correct diagnosis of secondary drowning was not made for several hours. She recovered after 8 days in intensive care.

Swimming in open water or in the sea, even with a trained companion, is very unwise. A talented young teacher was determined to continue surfing after he developed epilepsy in his twenties. In September 2010, he was surfing with his brother and several friends. His brother lost sight of him and reported his disappearance to the coastguard at once.[17] He was found face down in the water, given immediate cardiopulmonary resuscitation and airlifted to hospital, where he died of drowning. An Australian study noted that several people with epilepsy drowned while snorkelling.[18]

When people with epilepsy go fishing or boating, or take part in water sports such as canoeing, it is essential that they wear a life jacket at all times.[19] An opportunity to emphasise the importance of life jackets was missed when the Department of Health published a slim booklet, *'Epilepsy in later life'*, in 1998. In the full-colour photograph on the front cover, an elderly man and a young boy are fishing from a rowing boat beside a reed bed. Neither is wearing a life jacket.[20]

Scuba-diving is entirely inappropriate for people with epilepsy, however well controlled their seizures.[21] If a scuba-diver has a seizure under water, he will probably dislodge his mouthpiece and lose his air supply. Whether he returns to the surface by his own efforts or with the assistance of rescuers, he will suffer decompression injuries if he fails to breathe out the air expanding in his chest. And he will drown if he inhales seawater underwater. Moreover, his seizure threatens the safety of any diver who tries to save him.

ASTHMA

Asthma causes wheezing and shortness of breath due to bronchospasm, a narrowing of the airways. The severity of bronchospasm can be assessed by measuring how much air someone can breathe out in a second after taking a deep inhalation. (This is called the forced expiratory volume in the first second or FEV1.)[22]

In 1971, Australian researchers measured FEV1 in 50 people (40 asthmatic patients and 10 controls) before and after 8 minutes of running, cycling and swimming.[23] After these brief exertions, most of the asthmatics had significantly lowered FEV1, indicating a degree of bronchospasm, whereas the controls had unchanged FEV1. The researchers noticed that the fall in FEV1 in asthmatics was more marked after running or cycling than after swimming. They concluded: 'Swimming provokes less exercise-induced asthma than either running or cycling.' They suggested: 'Swimming should be recommended in preference to running or cycling as an exercise programme for adults and children with asthma.'

In a follow-up study, the researchers found that children with asthma had less troublesome wheezing after taking part in a 5-month-long swimming regime. They also had better posture, increased fitness and used less asthma medication. And, of course, their swimming improved.[24] Nevertheless, treadmill tests showed that the severity of their exercise-induced asthma remained unchanged. Even if asthmatics choose swimming in preference to other sports, and increase their muscular strength and coordination by taking part in regular swimming workouts, they remain asthmatic. They still have to contend with exercise-induced narrowing of their airways.

The prevalence of asthma in the general population is approximately 8%.[25] Among competitive swimmers, however, a larger percentage suffer from asthma, with worse symptoms during and after practice sessions in the pool. At the 2008 Olympic Games at Beijing, 20% of racers and synchronised swimmers needed medication to control their asthma symptoms.[26]

Until recently, the high incidence of asthma symptoms in elite swimmers was thought to be because people with asthma choose swimming as a sport, rather than running or cycling. But there may be another explanation. Elite swimmers draw huge volumes of air into their lungs during many hours of endurance training in the chlorinated water of swimming pools. Chlorine in the water combines with body proteins and sweat to produce irritant chemicals, including chloramines and trihalomethane (commonly known as chloroform).[27] There is now increasing evidence that exercise-induced asthma in elite swimmers is related to exposure to chlorine derivatives in the water and in the air just above the water surface.

In striking contrast to the high incidence of asthma in Olympic swimmers, only 5% of Olympic high divers have asthma.[28] Their training involves similarly intensive, prolonged training sessions, but despite the frequency of their practice dives, Olympic high divers have only fleeting contact with chlorine and chlorine derivatives in the water of the diving pool and in the air above it.

Whatever their level of aquatic skill, when swimmers with asthma leave the relative safety of a swimming pool to swim in open water or in the sea, they must remember their limitations. Exercise-induced asthma narrows their airways and obstructs their breathing, yet they have to inhale large volumes of air in order to supply their muscles with adequate oxygen. When swimming in rough water or when surfing, water droplets fill the air, aggravating their wheezing and chest tightness, even triggering a full-blown asthma attack that prevents a return to safety.[29] And if an asthmatic swimmer needs resuscitation, the rescuers' efforts will be obstructed by the presence of asthma-related bronchospasm and an outpouring of bronchial mucus.

✦ ✦ ✦

Scuba-diving is an exceptionally dangerous sport for people with asthma. They cannot use inhalers underwater, yet the effort needed to swim against a current can trigger exercise-induced bronchospasm. The dense, dry, cold air issuing from the scuba regulator valve is likely to irritate an asthmatic diver's airway, and seawater spraying into an ill-fitting mask can bring on an acute asthma attack, with sticky mucus filling narrowed airways. The anxiety produced by the onset of an asthma attack may quickly turn into panic, followed by a dangerously rapid ascent to the surface.

At depth, scuba-divers breathe compressed air. During their return to the surface, their lives depend on their ability to expel the air expanding in their lungs. They must breathe out constantly throughout their ascent. If their breathing is restricted because of an asthma attack, or if they forget to breathe out because of panic, the expanding air causes 'pulmonary barotrauma of ascent', including burst lung and air embolism.[30]

Even if the dive goes smoothly and the diver remembers to breathe out as he rises to the surface, asthmatics are at risk due to the presence of plugs of mucus in many small bronchi. On land, they are unaware of these minor blockages, which cause no symptoms, even though pockets of air are trapped behind each plug.[31] While ascending at the end of the dive, however, the trapped air expands, rupturing the delicate tissues of the lungs. Since the relative changes of air pressure and volume are most marked close to the water surface, scuba-divers with asthma are at risk even in shallow dives. Indeed, scuba-divers who denied or hid their asthma diagnosis have developed serious symptoms, including burst lung and cerebral air embolism, while ascending from a practice dive in a swimming pool.[32]

Experts are divided on whether a 'certificate of fitness to dive' should be given to someone with asthma. Opinions vary from extreme caution to surprising permissiveness. Some consider that any history of asthma is an absolute contraindication.[33] Others argue that there is little added risk if a diver has been free of bronchospasm over the previous five years, or two years or one year.[34] A few advise asthmatic divers 'not to dive within 48 hours of wheezing'.[35]

Whatever they have heard about the risks, many asthmatics present themselves for sport diving medicals. A review of the records of 200 would-be scuba-divers who failed their sport diving medicals found that 80 (40%) had been failed because of asthma,[36] including a man who claimed that he had not had any problems with his asthma for 16 years – but still carried an inhaler. Half of those who admitted to having only 'childhood asthma' still had symptoms, signs or abnormal lung function tests indicating continuing asthma – and were pronounced unfit to dive.

Since many asthmatics rule themselves out because of their symptoms, and many are found medically unfit to dive, a lower percentage of asthmatics would be expected among scuba-divers than among the general population. Unfortunately, people with asthma are well represented among the victims of fatal scuba accidents.[37] In a study of a hundred consecutive scuba-diving fatalities in Australia and New Zealand during the 1980s, no past medical history was available for more than half of the victims. Nevertheless, asthma was definitely implicated in at least nine deaths.

Asthmatic scuba-divers involved in a diving accident are often very difficult to treat.[38] Their tissues are starved of oxygen because both water inhalation and asthma compromise their breathing. They may be deeply unconscious due to the effects of near-drowning and/or pulmonary barotrauma leading to air embolism. Air bubbles in the brain may cause repeated grand mal seizures. They urgently need both decompression therapy and intensive care. If they survive, they may be left with severe and permanent brain damage.

GENES THAT GREATLY INCREASE THE RISK OF DROWNING

The sudden death of a fit young person is sometimes linked to an inherited disturbance of heart rhythm. The most common inherited disturbance is Long QT Syndrome, which affects 1 in 2,000 of the population.[39] The heart appears anatomically normal, but electrocardiogram tracings reveal a subtle abnormality: after each contraction of the cardiac muscle cells, the ventricles take longer than usual to prepare for the next beat. People with Long QT Syndrome are prone to sudden episodes of excessively rapid heartbeat, palpitations, dizziness and fainting, sometimes ending in cardiac arrest. The episodes are triggered by stress, by a number of medications (including amiodarone and erythromycin), by sudden noise – and by swimming.[40]

Importantly, people with Long QT Syndrome appear to be at greater risk of developing an abnormal heart rhythm in the water than on land.[41] Life-threatening episodes may be triggered by sudden immersion of the face, by entering cold water or by the exertion of swimming. And clearly, any swimmer who faints while swimming is in immediate danger of drowning.

Long QT Syndrome often runs in families. In 1997, a ten-year-old boy was rescued unconscious from the bottom of a public swimming pool in Minnesota.[42] Paramedics noticed an unusual rapid ventricular rhythm – torsades de pointes[43] – on the boy's electrocardiogram. They treated him with a defibrillator at the side of the pool before admitting him to an Emergency Room at the Mayo Clinic. As the boy recovered, his electrocardiogram tracings showed changes typical of Long QT Syndrome. A research team at the Mayo Clinic screened 26 of the boy's first degree relatives: 12 of them had Long QT Syndrome.[44] The team identified a genetic mutation of the KVLQT1 gene in the boy and in 9 of his relatives.[45]

In the absence of previous episodes of fainting or palpitations, a diagnosis of Long QT Syndrome may be impossible before the fatal drowning incident. In 1998, a nineteen-year-old American woman was transferred to the Mayo Clinic after she was found face down at the bottom of a swimming pool in Iowa. She never regained consciousness and died in hospital twelve days later.[46] The provisional cause of death was 'anoxic encephalopathy after near-drowning' (brain damage due to cerebral hypoxia during submersion), but further investigations were carried out because she was a good swimmer, the pool was only 4 feet deep, and electrocardiographic tracings taken between her admission to hospital and her subsequent death raised the possibility of Long QT Syndrome. Post-mortem molecular genetic tests revealed that the victim had a mutation in the KVLQT1 gene. Her family were offered genetic screening, and the same mutation was identified in her grandfather, mother, and younger sister.

The researchers at the Mayo Clinic embarked on a larger study of blood or stored tissue samples from 35 unrelated subjects with Long QT Syndrome.[47] They found genetic mutations of the KVLQT1 gene in 9 subjects, including all 6 subjects who had a personal history or a family history of drowning or near-drowning. One of these genetic mutations was discovered in a small paraffin-embedded tissue specimen from a young girl who had drowned more than 20 years before. Commentators wrote about a 'gene for drowning'.[48]

✦ ✦ ✦

Similar life-threatening episodes of rapid heartbeat, palpitations, dizziness, fainting and cardiac arrest occur in young people with Catecholaminergic Polymorphic Ventricular Tachycardia (CPVT), an arrhythmia found in about 1 in 10,000 people. Those affected usually have a genetic mutation of the RyR2 gene.[49] Most of the time, their heartbeat is controlled and regular, but bursts of extremely rapid heartbeats follow the release of adrenaline into the bloodstream in response to stress, fear, or physical exertion – including swimming. They too are at increased risk of drowning.[50]

One drowning victim with CPVT was a promising young footballer who had been rescued from a swimming pool on two occasions after losing consciousness in the water. Between episodes, his electrocardiogram tracings were normal, and the diagnosis was missed. He died in a Welsh river after his canoe capsized, tipping him into cold swirling water. The diagnosis was made only after his death.[51]

Both Long QT Syndrome and CPVT are now regarded as 'cardiac channelopathies', a group of molecular malfunctions that disturb the rhythmic changes of electrical impulses on the surface of cardiac muscle fibres during contraction and relaxation.[52] Once the dangerous nature of the palpitations is recognised, the arrhythmia can be suppressed with medication, or it can be detected and terminated by a small battery-powered implantable cardioverter-defibrillator (ICD) placed under the skin on the chest wall.[53] Patients should be warned not to swim or scuba-dive because, even after the implantation of an ICD, swimming may trigger a dangerous arrhythmia.[54]

If the victim of an unexplained drowning is found to have a genetic marker for Long QT Syndrome or CPVT, family members should be offered screening for the telltale gene mutations.[55] Those found to have inherited this susceptibility to sudden death can then be protected by medication or by the implantation of an ICD. They should also be warned not to swim, and not to scuba-dive.[56]

These molecular malfunctions are relatively rare, but diagnosis is important because they have a high death rate, and unexplained faints may precede a fatal collapse. A leading paediatric cardiologist advised:'Any child who has a serious faint or cardiac arrest whilst swimming should be considered to have a channelopathy until proven otherwise.'[57] Post-mortem diagnosis is always worthwhile, even if it comes too late to save the victim.[58] The guilt felt by companions after an unsuccessful attempt at rescue and resuscitation may be lessened once they know the reason for the fatal accident. Grieving relatives understand why their loved one died. And on occasions, unfair accusations directed against lifeguards can be challenged and dismissed.[59]

In 2001, a thirteen-year-old girl drowned in the swimming pool of a leisure centre. The Magistrates in Bishop Auckland imposed a maximum fine of £20,000 on the District Council because only one lifeguard had been on duty, instead of the two required by the regulations.[60] Before she died, the girl had seen doctors on three occasions because of fainting. She won medals for sprinting, but during a presentation at her local running club, her grandmother witnessed her collapse in what seemed to be a fit.[61] Two years after she drowned, her thirteen-year-old brother died during a supervised swimming lesson.[62] At his inquest, the pathologist gave his opinion that both children had Long QT Syndrome, and that both had died of natural causes.

DISEASES OF CARDIAC MUSCLE FIBRES

In young adults, sudden cardiac death is sometimes due to cardiomyopathy, a group of diseases that affect the cardiac muscle itself.[63] In hypertrophic cardiomyopathy, the individual cardiac muscle fibres become progressively enlarged and jumbled, and the walls of the left ventricle thicken, obstructing blood flow.[64] In right ventricular dysplasia/cardiomyopathy, the muscle cells of the wall of the right ventricle are replaced by fat and scar tissue.[65]

A young family man remained fit enough to run his own business while receiving medical treatment for hypertrophic cardiomyopathy. In 2010, he drowned in the swimming pool of his hotel in Tenerife moments after playing a game of water polo.[66] The exertion of swimming had over-taxed his heart. An acclaimed photographer with cardiomyopathy had episodes of irregular heartbeats and was warned by her cardiologist not to swim in open water. In August 2013, she went swimming alone in the Ladies' Bathing Pond on Hampstead Heath, a deep, secluded pool shaded by a grove of trees. Nobody noticed her slip beneath the surface.[67] At closing time, lifeguards found clothing and a mobile phone in the changing rooms. Police divers retrieved her body from the bottom of the pond.

In an Italian study of 60 young people who died suddenly, researchers identified 12 victims with advanced right ventricular dysplasia/cardiomyopathy, unsuspected and undiagnosed during life,[68] even though some of the victims had a history of palpitations, episodes of fainting, or abnormal electrocardiogram tracings. Italian researchers noted that the death rate due to sudden cardiac death was twice as high in young athletes as in non-athletes, with cardiomyopathy identified as the leading cause of sudden cardiac death.[69] In 1982, compulsory cardiac screening was introduced for all young Italian athletes, including swimmers, before they are allowed to take part in sports training and competition. Between 1982 and 2004, 42,386 young athletes were screened, and almost 900 were disqualified from participation in competitive sport because of their family history, previous symptoms, medical findings or abnormal electrocardiogram tracings. By 2001–2004, the death rate from sudden cardiac death in the screened athletic population in the Veneto region had fallen to one-tenth of its rate in 1979–1982. In contrast, the death rate from sudden cardiac death in the unscreened 'non-athlete' population remained unchanged.[70]

In Britain, unsuspected cardiomyopathy is the leading cause of sudden death in athletes.[71] In a policy statement on cardiac screening for professional athletes, the British Heart Foundation gave its support to 'targeted expert assessment of families where there is a high risk of inherited cardiac disease, or where there has been a sudden unexplained death'.[72] In the House of Commons in 2004, Dari Taylor, MP for Stockton, introduced her private members' bill: 'Cardiac Risk in the Young (Screening) Bill'.[73] Although the bill did not become law, the debate increased awareness of cardiac arrhythmia and sudden cardiac death. The charity Cardiac Risk in the Young (CRY) continues to recommend targeted screening for athletes and competitive swimmers, for children and young adults with warning symptoms such as fainting, palpitations, chest pain, or seizures during exercise, for those who survive an episode of near-drowning, and for families who have lost a young relative due to unexplained drowning or sudden cardiac death before the age of 50.[74]

HEART DISEASE IN LATER LIFE

Older swimmers have older hearts. A pensioner who decides to take an unaccustomed swim must be able to cope with the extra muscular effort involved. The heart has to beat harder and faster to supply blood to the muscles that propel him through the water and move air in and out of his lungs. Blood flow through the coronary arteries must double or treble to supply the extra oxygen needed for the powerful contractions of his heart.[75] If his coronary arteries have narrowed over the years, his heart muscle will be starved of oxygenated blood. Some older men who drown while swimming are found to have suffered a heart attack before sinking below the water surface.[76]

In 1999, an Australian study of 60 snorkelling deaths during the decade 1987–1996 reported that 18 of the victims died from cardiac causes, mostly men in late middle age.[77] Some had chosen to snorkel despite taking medication for high blood pressure or irregular heartbeats, including three who had suffered a previous heart attack, two who were known to have cardiomyopathy, and one who had had triple coronary artery bypass surgery. These victims must have imagined that swimming with a snorkel would be a gentle form of exercise, but their hearts failed to cope with the extra effort involved. In 2012, the follow-up study of snorkelling-related deaths in Australia reported even more fatalities in 1994–2006. Over thirteen years, 140 people died while snorkelling: 60 of these deaths were attributed to cardiac causes, including 34 victims with a known history of cardiac disease.[78] The researchers recommended: 'Intending snorkellers with pre-existing cardiac disease should have a prior medical consultation and be prudent in their physical exertion.'

◆ ◆ ◆

Older people who wish to improve their general level of day-to-day fitness are often advised to take up swimming. A gentle regime of unhurried lengths in their local swimming pool is considered an extremely safe way to build up their exercise tolerance. But the emphasis must be on 'gentle' and 'unhurried'. Most benefit is gained from short, relaxed swims. There are limited benefits to be gained by straining to increase speed or by ploughing through too many lengths. Laboured breathing and awkward posture increase the stresses and strains on the swimmer's body instead of decreasing them. Lane swimming is best avoided when the pool is busy.[79] The lanes are likely to be occupied by regulars counting their lengths, impatient if their progress is hindered by a slower swimmer who has to take a breather in the shallow end.

Swimming, usually lane swimming, is sometimes advised as part of an exercise programme during convalescence after heart surgery, including coronary artery bypass surgery. Fit, sporty advisers commonly underestimate the effort involved in swimming, with its high demand on coronary blood flow. A quote in a widely distributed Health Education Council booklet, 'Beating Heart Disease', typically underplays the strain on a damaged heart: 'I used to play a lot of squash before my heart attack, but for the moment I'm happy with cycling and swimming until I feel fit enough for something more strenuous.'[80]

Recuperating swimmers should be warned that aching in the jaw (angina) or pain in the chest, the left shoulder or down the left arm are all warning signs of coronary insufficiency. They should be firmly advised never to ignore these warnings. Instead they must stop swimming at once, climb slowly out of the swimming pool, tell the lifeguard, and seek early medical advice. Angina usually settles with rest because the heart's oxygen-deprived muscle fibres recover once their workload eases. If symptoms of angina continue after a five-minute rest, an ambulance should be summoned to the swimming pool.

Angina and other symptoms of coronary insufficiency are brought on by reversible chemical changes in the muscle fibres of the heart. Swimmers are courting disaster if they ignore the pain and swim on regardless. Reversible changes quickly become irreversible. Some of the cardiac muscle fibres may die, or their coordinated contractions may splinter into uncoordinated fibrillation. Within seconds, the vital pumping action of the heart may be lost.

Yet swimmers have a chance of surviving even this cardiac emergency if they are spotted early by a lifeguard, who has been trained in both rescue techniques and cardiopulmonary resuscitation. In a Croydon swimming pool in 2012, a lifeguard made his 'first full rescue' when he saw a 73-year-old swimmer floating face down, unconscious after a heart attack.[81] The lifeguard dived into the water, lifted the pensioner from the pool, gave him cardiopulmonary resuscitation – and got him breathing again by the time the ambulance arrived to rush him to hospital.

Another elderly swimmer swam three times a week in his local pool in Romford although he knew that he had a blocked coronary artery. In 2006, he escaped death by a narrow margin when he suffered cardiac arrest 20 lengths into his customary hundred lengths swim.[82] He was pulseless when pulled from the pool, and his lips were blue. The lifeguards began cardiopulmonary resuscitation and used a defibrillator to shock his heart back into action. He regained consciousness two days later in an intensive care unit. After a coronary bypass operation and six weeks in a cardiac ward, he made a good recovery. At the age of 71, he began to swim again. But only twice a week.

CHAPTER TWENTY-FIVE

DROWNING IN THE BATH

In 2001, a study commissioned by the British government, *Drownings in the Home and Garden*, reported that 443 people died from drowning in the bath during the seven years 1993–1999. That is, more than 60 victims drowned in the bath every year. [1] The number of bathtub fatalities is somewhat lower now, but the familiar domestic bath can be a dangerous place, particularly for the very young and for people with epilepsy. [2]

BATHTUB DROWNING OF BABIES, TODDLERS AND PRESCHOOL CHILDREN

Babies are defenceless. At bath time, they are totally dependent on whoever undresses them and puts them into the water – usually the baby's mother. So it is shocking to realise that one of the most frequent causes of accidental death in babies is drowning in the bath. [3]

When babies are sitting in their bath, they make small movements all the time. They keep their balance by bending their bodies this way and that, wobbling their heads, waving their arms, jerking their legs. They grab at bath toys floating out of reach. And wet babies are slippery. Soaped and buffed, they slide around on the smooth enamelled surface of the bath. They need to have someone sitting beside the bath to support and steady them. Otherwise they topple over and lie spread-eagled, helpless in the water.

From the age of eight months onwards, a babies can sit unsupported, so their mothers may assume that it safe to leave them alone in the bath for a short time. But at this age, babies are learning to stand, and if they try to stand in the bath, they are likely to tumble into the water.

Even small amounts of water can be lethal. Babies who fall face down can drown in two inches of water. They cannot lift their faces above the water surface because their heads are too heavy for the weak muscles of their necks. They cannot push themselves up with their puny arms. Their legs slide away from under them. They lack the coordination to roll over onto their backs.

At bath time, a baby should not be left alone for a moment. Towels, creams, nappies and night-clothes must be collected and placed ready to hand before the baby is lowered into the water. If the telephone rings in the hall or someone knocks at the front door, the mother has a choice of ignoring the summons or taking the baby with her, wrapped in a towel, when she answers the phone or the door.

In New Zealand, nurses take the opportunity to discuss water safety with mothers when they bring their babies for a routine health check at the age of five months. To strengthen the safety message, new mothers are given a blue non-slip bathmat printed in bold white letters: 'Always supervise children around water – always.'[4] The advice is repeated in Māori. The timing is key. For the mother of a five-month-old baby, the safety message is given just before her baby learns to sit unsupported, so that it helps her to resist the temptation to leave her baby alone in the bath, however briefly. Since the scheme began in 2010, thousands of non-slip bathmats have found a useful place in New Zealand bathtubs.[5]

◆ ◆ ◆

Bath seats for babies were introduced in America in 1981.[6] They were designed for babies 5 to 10 months old, to encircle and support them. At this age, babies are learning to sit up, but they are still too weak and unstable to be able to sit without their mothers' helping hands. Bath seats soon became valued bath aids because they made it easier for mothers to soap and rinse their babies and to play with them at bath time.

But a baby bath seat has a serious drawback. Since it takes a little time to lift the baby out of the bath seat, the mother may be tempted to leave the baby playing in the bath for a few minutes, or longer, while she copes with a brief task elsewhere. The bath seat gives her a false sense of security because the baby seems to be protected from danger, held safely while sitting in the water.[7]

A bath seat is safe enough if the mother stays beside her baby all the time. But it is the mother's presence and attention that keeps the baby safe – not the bath seat. As long as she remains in view, the baby will sit happily in the bath seat, but if she suddenly disappears from sight, the baby will twist around searching for her. Then the bath seat may fail to keep the baby safe.

Most accidents to babies in bath seats follow one of three patterns:[8]

- the bath seat tips over, pitching the baby into the water
- the baby succeeds in climbing out of the bath seat, only to tumble into the water
- the design of the bath seat allows the baby to slide into the water. (Babies are meant to sit with legs straddling a central bar at the front of the seat. If they squirm both legs through the space on one side of the central bar, there is nothing to stop them from slipping downwards into the water, an accident frequent enough to have a nickname – 'submarining.')[9]

The first record of a bath seat being implicated in the drowning of a baby was in 1983, only two years after bath seats appeared on the market. In America between 1983 and 2001, 78 babies drowned while using bath seats.[10] Various models were withdrawn as unsafe, but the new models were just as accident-prone. By 2009, 174 babies had drowned while using bath seats, and a further 300 babies had suffered life-threatening submersions, leaving a number with permanent brain damage.[11]

In 2005, a British study described 7 drowning incidents associated with baby bath seats – one baby suffered serious brain damage and six babies died of drowning.[12] But parents continue to view bath seats as safety devices, so that these accidents continue to happen. Indeed, bath seats are involved in one-third of accidental drowning deaths in children aged two years or younger.[13] In Gloucestershire in 2009, a nine-month-old baby managed to get out of her bath seat after her mother left the bathroom. Shortly afterwards, the mother returned to find her baby face down in 6 inches of water. She was pronounced dead in hospital later that evening.[14] The baby's mother told the Inquest: 'I believed that once sat in the seat she would not be able to get out. In fact we would find it difficult to get her out.' In his closing remarks, the Coroner commented, 'This is not the first case in which a child – thought to have been safely positioned – has been left unattended for a short period of time only for tragedy to strike.'

In Scotland in 2013, an eleven-month-old baby tipped backwards in his bath seat and drowned when his mother left him, for seconds, to fetch his pyjamas.[15] His heartbroken parents set up a Facebook site, 'Karson's Story', to warn others of the danger, urging parents not to leave their child unattended in the bath, no matter what the circumstances.

Even with their mothers beside them, babies have drowned wedged in bath seats that tipped over.[16] In Northern Ireland in 2009, a mother narrowly avoided a tragedy while bathing two of her children together. While she was washing the toddler's hair, the ten-month-old baby turned round in her bath seat, slid into the water, and lost consciousness.[17] Luckily, the baby responded to her mother's immediate rescue breaths. 'I had her wee blue body in my arms,' the mother said, 'and I was blowing into her mouth to get her to breathe.' Thanks to the mother's fast reaction and her first-aid training, the baby made a full recovery after a brief stay in hospital.

◆ ◆ ◆

Toddlers are almost as likely as babies to drown in the bath. They topple over when they are sitting in the bath. They stand up, lose their balance, and fall into the water. They try to right themselves, but the smooth curve of the bath's enamelled surface has no handholds. If there is no one in the bathroom to steady and protect them, they slide under water.

Toddlers are at risk from the moment that the bath begins to fill until the water is drained away.[18] A few are so fascinated by water that they climb into the bath unobserved.[19] Others lean over the rim to dabble their fingers in the water or to play with the bubbles, but the weight of their heads makes them top-heavy, so that they see-saw into the water.[20] In 2001, the British study *Drownings in the Home and Garden* reported that, between 1993–1999, 58 children drowned in the bath before their fourth birthday: 43 of them were less than two years old.[21]

Babies and toddlers who drown in the bath have almost always been left on their own before the fatal accident,[22] yet many parents fail to recognise the risk. In a large survey in Atlanta in 2003, parents who brought their children to hospital emergency departments were asked about bathtub supervision. Almost a third of parents confirmed that they sometimes left children aged 5 years or younger unsupervised in the bath, while 15% of babies less than one year old had been left alone in the bath, usually while the parent fetched a towel or a nappy or answered the telephone.[23] One mother told the researchers that she had left her eight-month-old baby sitting alone in the bath for more than five minutes while she cooked a meal.

A mother who leaves the bathroom when her child is sitting in the bath expects to hear warning noises if the child gets into difficulties – cries, splashes or calls for help. Silence should worry her more. Drowning happens too suddenly, quickly and quietly for any useful warning. A child capsizes in a moment, and once submerged, cannot call for help. At an inquest in 2002, a mother told the coroner how she had turned on her baby monitor so that she could listen to her eleven-month-old twins while she went to bring them clean clothes. The sounds stopped. She ran upstairs. But she was already too late.[24] One of the babies was lying unconscious in water only a few inches deep. He could not be revived.

Bath water is warm. There is no protective slowing of brain chemistry that may preserve the life of a child submerged in icy water. Rather the reverse. Even a short submersion speeds up the child's metabolism. Deprived of oxygen, the baby may suffer fatal brain damage within a minute or two. Attempts at immediate resuscitation are always worthwhile, but they often fail.

◆ ◆ ◆

Although the risk of drowning in the bath is greatest for babies and toddlers, preschool children still need bath-time supervision to keep them safe. An American paediatric emergency clinic reported that more than 200 children needed treatment for bath-time injuries in the course of three years, mostly gashes on the face or scalp caused when they slipped in the bath and fell against a hard object such as a tap or the rim of the bathtub, often when the responsible adult was out of the bathroom.[25] During 1996–1999, the United States Consumer Products Safety Commission (CPSC) received reports of 292 children who drowned in bathtubs before their fifth birthday: 51 of the victims were 2 to 4 years old when they died.[26]

Some accidents happen when parents over-estimate the competence of older children and expect them to supervise younger brothers or sisters at bath-time.[27] Children given such a responsibility may fail to recognise that a struggling toddler is in danger, or they may lack the strength needed to lift the toddler to safety. Two children together in a bath are likely to play more boisterously than one child alone. In an Australian study of 17 fatal accidents where an unsupervised child drowned while left in the bath with a sibling, the victims were all younger than the survivors[28] (10 of the victims and 12 of the survivors were boys). A prank may turn into a tragedy if an older child puts a toddler into the bath without their parents' knowledge. Sibling rivalry may play a part in some incidents. An American study of submersion injuries noted that several children were rescued by parents who briefly left the bathroom, then returned to find the older child sitting on the younger one or holding a resented toddler underwater.[29]

Although most young children who drown in the bath come to grief during brief lapses in parental supervision, there are no clear-cut boundaries separating less-than-ideal care from neglect, or neglect from child abuse.[30] Some children drown in the over-casual care of a drunken parent or childminder.[31] Others are found to have fading bruises and healed fractures, telltale marks of a cruel upbringing, with drowning as the last, fatal incident in a succession of non-accidental injuries.[32] A few victims are deliberately pushed beneath the surface and held there by a depressed or psychotic parent.[33] The emergency services are sometimes called to an unconscious child who survived a deliberate attack only because the attacker was seized by panic or remorse and snatched the child from the water at the last moment.[34] In the course of two years – 1988–1989 – the Coroner pronounced a verdict of unlawful killing or culpable homicide at five Inquests of British children who drowned in the bath.[35]

Experts at Cardiff University suggested, 'Child abuse should be considered in incidents where the story is other than that of the "typical accident description".[36] That there should be such a 'typical accident description' emphasises how frequently this totally avoidable tragedy occurs. New parents need to understand how easily young children can drown in the bath. Then they won't leave their baby sitting alone in the bath while they fetch pyjamas from the bedroom.

BATHTUB DROWNING IN PEOPLE WITH EPILEPSY

After the age of five years, children are very unlikely to drown in the bath – unless they have epilepsy. In school-age children and in young adults, an epileptic seizure is the commonest cause of drowning in the bath. In 1985, an American study under the auspices of the Centers for Disease Control in Atlanta analysed all 710 bathtub-related drownings that took place over two years (1979–1980).[37] A history of a seizure disorder was found in 144 of the victims. Importantly, more than 40% of victims aged 10 to 39 years had a history of a seizure disorder.[38] The authors commented, 'The risk of drowning in bathtubs by persons with a history of a seizure disorder has not been emphasized in the epilepsy literature; the problem was referred to in only one of the 14 texts on epilepsy reviewed. Such drownings, however, have been described in a number of case reports.'

The British 2001 study *Drownings in the Home and Garden* reported that 118 people with epilepsy drowned in the bath in 1993–1999. The great majority of the victims were between 10 years and 50 years of age when they died; girls and young women were the most frequent victims.[39]

In 1993, researchers at the University of Washington, Seattle, reviewed the records of all King County residents aged 0–19 years who suffered a submersion incident during the years 1974–1990. They reported that children with epilepsy were 47 times more likely than children without epilepsy to suffer a submersion injury in the bath and almost one hundred times more likely to drown in the bath.[40]

In a Canadian study of 25 seizure-related drownings, 8 people with epilepsy drowned in swimming pools, lakes or rivers, but 15 drowned while taking a bath, unsupervised. In addition, one victim had a seizure while taking a shower in a bathtub and drowned because his body obstructed the escape of water from the drain, and one victim had a seizure in a Jacuzzi and drowned because his companion was unable to lift him to the surface of the water.[41] In all except one of the bathtub drownings, other people were elsewhere in the house at the time of the accident, unaware of the tragedy until it was too late. Low levels of anticonvulsant medication were found in most of those who died, suggesting that a higher dose or better control might have avoided some of the deaths. However, 'therapeutic' levels of anticonvulsant medication were found in a number of the victims, so their epilepsy appeared to be well controlled at the time they were found drowned in the bathtub.

Someone who has a seizure at work or in the street is likely to recover consciousness surrounded by concerned and helpful witnesses. In contrast, someone who has a seizure in the bath is likely to drown unless another person sees the drama and manages to pull the victim's head out of the water.[42] After the sudden death of a student or a young parent, newspapers sometimes quote the comment of the coroner, or another authority figure, that the onset of a fit while the victim was having a bath was 'a freak accident' or 'a very awful coincidence'.[43] These comments are kindly meant, no doubt, but they are misleading. They overlook the fact that bathtub drowning is distressingly common in people with epilepsy. An opportunity is lost to alert those with a history of seizures that they are at increased risk of drowning while taking a bath, especially if they have a seizure while they are alone, with nobody to come to their rescue.[44] Moreover, the families of people with epilepsy are not always warned that their loved ones are at increased risk of having a seizure while taking a bath. And since they remain unaware of the drowning risk, they take no effective safety precautions.[45]

A young woman with epilepsy had as many as five epileptic fits a day. Uninformed of any drowning risk, with her husband in another room, she was taking an unsupervised bath when she had an epileptic fit and sank below the surface. She was saved from drowning by her five-year-old daughter.[46] She explained, 'Lucy was in the bathroom at the time and grabbed me by the hair. She was also slapping my face and shouting for her dad who was busy with the youngest at the time. He helped me out of the bath and when I came round I had no idea who anybody was, although I was okay after fifteen minutes.'

Another young woman with epilepsy was not so lucky. She had not had a fit for three years when her daughter found her unresponsive in the bath. The little girl told her father, 'Mummy's asleep in the bath.' He rushed upstairs, but she had already drowned.[47]

People with epilepsy should be told that taking a bath may sometimes trigger an epileptic fit, and that baths are best avoided altogether. Light-sensitivity (also called photosensitivity) affects one in every twenty people with epilepsy, most commonly teenagers and young adults.[48] Strobe lighting is now well recognised as a trigger for seizures in people with light-sensitive epilepsy, but seizures can also be triggered when more subtle contrasts of swirling lines and grid patterns are viewed at close quarters.[49] Bathroom strip lights often flicker, while rippling fragmented reflections on the surface of the water, such as bright overhead bathroom lighting, broken streaks of sunlight shining through Venetian blinds or checkerboard patterns made by tiles on the bathroom walls, are all capable of inducing seizures.

In susceptible young people, seizures may be triggered as they look down on the surface of the water while they are preparing to climb into the bath, or when they are leaning across the bath to turn off the tap. A number of victims have been discovered collapsed across the side of the bath, with their heads submerged. A fifteen-year-old girl had been symptom-free for several years, but she had a seizure and drowned while kneeling over the side of the bath to wash her hair under a running tap.[50] A young mother was giving her child a bath when she suffered an epileptic seizure. Her head slumped into the water and she drowned.[51]

The temperature of the bath water may be the critical factor in some drownings. Hot-water epilepsy is well documented, especially in southern India,[52] where it is linked to the local custom of pouring hot water (at 40–50°C) over the head while taking a bath. Sometimes hot-water epilepsy can be controlled simply by replacing hot water with lukewarm water. However, as many as a third of people with established hot-water epilepsy continue to have seizures when they take a bath, even when the water is relatively cool.

Hot-water epilepsy is not confined to India. Neurologists in Birmingham documented two patients who had a number of seizures over several years, always when taking hot baths:[53] a seventeen-year-old girl had eight epileptic fits in the bath, despite treatment with anticonvulsant medicine, and a 34-year-old man had six epileptic fits in the bath, biting his tongue on two occasions. They were advised to stop taking baths, and at follow-up they had suffered no further seizures. The neurologists wrote, 'Avoiding the stimulus may be a more effective treatment than anticonvulsant medication.' They stressed the particular risk when seizures are triggered by taking a bath: 'Avoidance of the precipitant seems especially important in this circumstance as there must be real risk of death by drowning.'

It is curious that people who have already survived a bathtub seizure continue to take unsupervised baths, at risk to their lives. Scientists studying both light-sensitive epilepsy and hot-water epilepsy have noticed that a number of their patients speak of intense feelings of pleasure at the moment when abnormal activity appears on their electro-encephalograph tracings and their seizures begin. Some people with epilepsy even provoke the attacks, for instance by waving their fingers back and forth in front of their eyes,[54] or by deliberately pouring hot water onto their heads.[55] A few are able to induce the feelings of pleasure while stopping short of provoking full-blown fits: in order to experience the aura, they are willing to risk a seizure. All those who have suffered a seizure in the bath should be firmly advised never to take another bath, whether they find baths particularly pleasurable or not.

Showers are much safer than baths for anyone with epilepsy.[56] Yet showers are not entirely without risk. Someone who has a seizure while taking a shower may be injured by striking metal shower fittings or by falling onto the shower tray. Worse, the shower stall itself may become a trap. If the victim's unconscious body blocks the drain, enough water may collect in the base of the shower to cause death by drowning.[57] In 2013, a conscientious young man missed an important day at work. His colleagues grew alarmed and went to his flat. They heard the shower running, saw water flowing under the bathroom door, and found his body slumped on the floor of the shower stall.[58]

The danger of a seizure-related fall can be limited by taking a shower while sitting in the bath, with the bath plug removed so that the water drains away, and as an additional safeguard, by using a handheld shower head which releases water only when the button is held down. And for safety's sake, people with epilepsy should not take a bath or a shower unless someone else is in the bathroom, or paying attention nearby with the bathroom door unlocked.

BATHTUB DROWNING WHILE RELAXING WITH A DRINK OF ALCOHOL

After the age of 40 years, alcohol replaces epilepsy as the most frequent personal risk factor associated with drowning in the bath.[59] Many people find a warm bath is soothing after a hectic working day. Some combine a bath before bedtime with a glass – or two glasses – of wine. But, according to Jonathan Howland, an American epidemiologist, 'You get into a hot tub to relax and you drink to relax. When you put those two things together, you get more than you bargain for.'[60] Some people relax so much that they go to sleep and slide gently under the water. Others become so confused and uncoordinated that they fall into the water while trying to get out of the bath. A few become so incapacitated by alcohol (and/or drugs) that they pass out in the bath and drown.[61]

Excessive alcohol intake over a number of years is toxic to the heart, damaging the cardiac muscle and triggering irregular heartbeats.[62] Anyone whose heart skips a beat while they are soaking in a warm bath with a glass of wine should take this as a clear warning to get out of the bath and to cut down their alcohol intake. Less common than skipped beats, but much more serious, is the sudden onset of atrial fibrillation,[63] when the muscle fibres of the two upper chambers of the heart lose their ability to pull together, and each fibre contracts at its own fast, uncoordinated pace, impairing the pumping action of the heart. Holiday heart syndrome was first described thirty years ago in people admitted to hospital after developing palpitations and atrial fibrillation during bouts of drinking at the weekend, at Christmas, or at New Year.[64]

When people relax in the bath, the water should be tepid, rather than hot. If the temperature of the water approaches blood heat, their skin flushes and they begin to sweat.[65] Flushing and sweating are responses designed to increase the rate of cooling on land, but they have no effect when people are lying in hot bathwater. Indeed, the dilated blood vessels in the skin absorb more heat from the bathwater, instead of losing heat to the air, and sweat has no cooling effect unless it evaporates from the skin. A glass of wine also speeds the absorption of heat from the water because alcohol causes blood vessels to dilate. The combined effects of alcohol intake and immersion in hot water may lead to a faint when the bather stands up to get out of the bath.

The temperature of the water in a Jacuzzi or a hot tub is often as high as 40°C,[66] maintained by constant recirculation through a heater. Bathers can become dangerously overheated in a matter of minutes. Normal body temperature is 37°C, and a rise in core temperature of only 2°C tips the bather into hyperthermia, a life-threatening medical emergency. As long ago as 1979, the American Consumer Product Safety Commission (CPSC) warned about the risk of overheating and drowning in hot tubs, with the added caution, 'The risk of drowning is significantly heightened if individuals consume alcoholic beverages while, or prior to, soaking in hot water.'[67]

Hot tubs, like garden ponds, are dangerously attractive to young children. Toddlers who climb into hot tubs unobserved are extremely vulnerable to heat stroke and drowning.[68] As early as 1990, researchers in California warned that 74 children had already drowned in hot tubs, spas and whirlpool baths,[69] and as more domestic hot tubs were installed, more children are at risk. In Scotland in 2012, an eighteen-month-old girl was playing in the garden when she fell into the family hot tub. She died in hospital the next day.[70]

BATHTUB DROWNING IN PENSIONERS WHO FALL WHILE GETTING INTO OR OUT OF THE BATH

Falls can happen at any age, but the risk of falling is highest in young children and in pensioners,[71] especially elderly people who take tablets to lower their blood pressure or help them to sleep, who have a failing memory, impaired balance, weakened hand grip or impaired sight.[72] 'Evidence of having fallen' is a common personal risk indicator in Americans over 60 years old who drown in the bath.[73]

Getting into the bath, so quickly and easily accomplished by a younger, fitter person, becomes increasingly problematic for old people with limited mobility.[74] They have to balance on one foot while lifting the other foot and swinging it over the side of the bath, planting it on the bottom, then transferring their weight from one foot to the other. They regain their balance only when they are standing firmly with both feet in the bath. They will fall if either leg gives way, if either foot catches on the rim of the bath, or if they lose their balance at any moment during the transfer. When pensioners sit down in the bath, they have to lower their hips as if to floor level, a feat of gymnastic control that would be taxing enough when carried out in their living room. The slippery surface of the bath is an added complication.[75]

Getting out of the bath is just as tricky. Some have difficulty when rising to their feet. Others trip when coping with the complex step-over to exit from the bathtub. They may pitch forward against hard surfaces or projecting bathroom fittings. If the bathroom floor is wet and slippery, and the bathmat skids away, they may tumble backwards into the bath.[76] Well-sited grab bars make balancing easier when they enter and leave the bath or the shower,[77] but swinging glass shower doors are not designed to support their weight.[78]

Studies of potential hazards in the homes of elderly people have identified the bathroom as the most hazardous room in the house.[79] As bath-taking becomes more onerous, pensioners are probably wise to limit the number of baths they take, even if they have someone to help them.

◆ ◆ ◆

In Japan, most elderly people take daily baths that involve lengthy soaking in the bathtub, immersed to the neck in very hot water (40–42°C)[80] – and a startling number of Japanese pensioners die in the bath. Researchers at Kagoshima University recorded that more Japanese die in the bath than in road traffic accidents.[81] Between October 1999 and March 2000, the Tokyo Metropolitan Fire Department dealt with 623 cases of sudden death during bathing. From these figures, the Tokyo Metropolitan Institute of Gerontology calculated that over the course of 1999, as many as 14,000 elderly Japanese people died suddenly while taking a bath.[82]

In *The Road to Wigan Pier*, George Orwell wrote, 'Less than half the houses in England have bathrooms.' He looked ahead to better times: 'But the English are growing visibly cleaner, and we may hope that in a hundred years they will be almost as clean as the Japanese.'[83] He was writing in 1937. In those days, miners had to wash themselves in tin baths in front of the kitchen fire. Tin baths are antiquated curiosities now, whereas plumbed-in bathtubs have their valued place in almost every home. Nevertheless, the bathtub has been identified as a domestic danger spot.

DROWNING IN HOSPITAL BATHTUBS

Frail, confused and vulnerable patients are at risk of drowning in hospital bathtubs if their nurses are too busy or too casual to supervise them properly.

In June 2007, a young woman was admitted in labour to a Nottinghamshire maternity ward. She had a recorded history of fainting attacks, more frequent since the recent death of her mother. A midwife suggested she have a bath to ease the labour pains. Although a friend reminded the midwife of the fainting attacks and asked that she be kept under observation, she was left unattended in the bath for 45 minutes.[84] At this point, midwives found her submerged and unconscious. Doctors delivered her baby by Caesarian section. The baby survived, but the first-time mother died in intensive care eight days later. The Coroner recorded an Open Verdict.

In March 2013, an autistic teenager with epilepsy and learning difficulties was admitted to a specialist NHS learning disability unit for assessment. He had several seizures while in the unit. A nurse was asked on two occasions to complete a care plan, but she had no experience of doing care plans for people with epilepsy and failed to carry out the request. The young man enjoyed taking baths. In July 2013, after running water into a deep, steep-sided bath, a nurse left him unsupervised, with the bathroom door locked. Twenty minutes later, another staff member checked on him – and found him lying underwater, deeply unconscious.[85] He did not respond to resuscitation. At the Inquest, the Coroner gave a Narrative Verdict, noting that neglect contributed to the young man's death. The learning disability unit was transferred to new management.[86]

A 78-year-old widow with dementia was admitted to a Norwich hospital in September 2014. On 15 October 15, an occupational therapist discovered her in an unlocked bathroom on the ward, trying to run herself a bath. This incident was reported to one of the nurses, but the bathroom remained unlocked. The next day, she went missing, and a search found her drowned in the bath.[87] The Inquest verdict was Accidental Death, but the Norfolk and Suffolk NHS Trust was fined £366,000 for breaching health and safety regulations.[88]

A 58-year-old woman suffered from paranoid schizophrenia for many years. In February 2014, she was admitted to a Middlesbrough mental health hospital after cutting her arm with a broken teacup in her residential home. She told the psychiatrist that she planned to drown herself. Nevertheless, she was given permission to have a bath – with a nurse in attendance. While she was still in the bath, the nurse went off to attend to another patient, leaving her alone for 17 minutes. The nurse returned to find the woman drowned.[89] She had been in the hospital for less than an hour.

◆ ◆ ◆

As well as offering warm baths to women in labour, maternity departments now have midwife-led Birth Centres where women can choose to have a water birth, spending part or all of their labour in an oversized bathtub called a 'birthing pool'. Studies suggest that women who sit in the warm water of a 'birthing pool' have a shorter labour, require less pain relief,[90] and benefit from the constant close attention of the midwife who cared for them throughout their pregnancy.[91]

Studies also suggest that babies born in a 'birthing pool' do as well as babies of low-risk mothers born in the conventional way.[92] The National Childbirth Trust reassures mothers-to-be who decide to have a water birth, 'Don't worry, your baby won't drown. If your baby is born in water, they are brought gently to the surface by the mother or the midwife. The baby will not breathe until they meet the air, and they continue to get oxygen through the umbilical cord.' (sic)[93]

Nevertheless, some babies inhale water from the birthing pool. Reports by obstetricians in Britain, America, New Zealand, Italy and Switzerland describe babies who needed urgent transfer to a neonatal intensive care unit because they developed respiratory distress caused by water aspiration, sometimes followed by the development of pulmonary oedema, pneumonia or hypoxic brain damage.[94] Although fatalities are rare, babies have died from drowning or from florid pneumonia and sepsis after inhaling contaminated water during a water birth.[95]

CHAPTER TWENTY-SIX

CHILDREN DROWNING ON SCHOOL TRIPS

In July 1985, a party of 50 pupils from a Buckinghamshire school travelled to Land's End with their headmaster, two teachers and two parents. Some of the children were allowed to wander along the beach on their own. Four boys were drowned when waves washed them off the rocks at the base of a cliff. The Coroner brought in a verdict of Death by Misadventure.[1]

Between 1985 and 2017, 35 children drowned on school trips run by English schools:

- 5 children drowned in swimming pools
- 4 children drowned in lakes
- 1 child drowned while swimming across a flooded quarry
- 1 child drowned in a pool beneath a waterfall
- 1 child drowned in the artificial rapids at a theme park
- 10 children drowned in rivers
- 1 child drowned in a flooded cave
- 12 children drowned in the sea.

Certain accident scenarios recurred. Parents' written warnings to the school – that their child could not swim, or had a history of epilepsy, or had suffered blackouts while swimming – went unheeded.[2] Teachers made infrequent head counts, so that they failed to notice that a child was missing until the group returned to their bus,[3] or until they got back to their hotel.[4] In two incidents, a child was first noticed lying at the bottom of a swimming pool only after the other children in the group had left the pool, and had dried and dressed, at which point a small pile of folded clothes was discovered and a frantic search began.[5] On at least five occasions, a child drowned while the teacher's attention was focused on other children who needed help.[6]

Children were shepherded to the brink of waterfalls or 'plunge pools' carved by mountain torrents. Some fell in and others were encouraged to jump into the frigid, turbulent water, where they suffered cold shock, panicked and sank.[7] Teachers got into difficulties themselves when they jumped into plunge pools or fell into the sea, so that they were unable to save children who needed rescue.[8] Indeed, a student drowned in a plunge pool in a swollen river in the Brecon Beacons while attempting to rescue his teacher and another student, who both survived.[9]

Groups of children were taken 'river walking' after heavy rain, when the force of the current meant that anyone who slipped might be swept away.[10] Teenage girls were sent on cross-country treks in stormy weather. Weighed down by heavy backpacks, they were expected to ford mountain streams in spate.[11] Teenage boys on a rugby trip to Canada were taken to swim in a frigid lake, where their coach took fifty photographs on his mobile phone but failed to intervene when they started pushing each other off a jetty.[12] Six young girls were allowed to ride, unsupervised, down the 'Splash Canyon' rapids at a theme park. They stood up in the bobbing, spinning dinghy, leaned over the side and dipped their hands into the water. When they hit a barrier, a girl was 'propelled' out of the dinghy and fell into a holding reservoir 12 feet deep. She could not swim.[13]

Children with poor swimming skills were taken to the seaside and allowed to paddle – with such inadequate supervision that a ten-year-old girl was discovered floating unconscious after waves washed her into the sea,[14] and the body of a thirteen-year-old girl was found on the beach by a fisherman hours after she stepped out of her depth, sank and drowned.[15] Four youngsters drowned in Lyme Bay when their kayaks were blown out to sea.[16] A nine-year-old girl was trapped unnoticed under a capsized boat in Portsmouth harbour, hampered in her attempts to escape by the buoyancy of her life jacket.[17] A teenage boy drowned during a sea survival exercise in his school swimming pool, unable to extricate himself after righting an upturned life raft, while his friends and supervisors continued with safety exercises nearby.[18]

◆ ◆ ◆

At the ensuing Inquests, surviving children gave graphic, harrowing accounts of the unfolding tragedies.[19] A sixth-former described an unreported near-miss during a morning river walk, not reported to the group about to encounter the treacherous water conditions that afternoon.[20] A witness remarked on the inattention displayed by teachers supervising a group of children on their first trip abroad.[21] A mountain rescue expert spoke of a foolhardy trip in atrocious weather conditions.[22] Marine accident investigators reported that an overloaded boat had been leaking before it capsized.[23] Colleagues recalled that their warning advice was ignored.[24] A teacher confirmed that he had left vital rescue equipment in the boot of his distant car.[25] Teachers admitted that they allowed children to enter the sea from steeply shelving or rocky beaches, despite breaking waves, strong currents, fading light, and official hazard warning notices.[26]

At the conclusion of these inquests, the usual verdicts were Accidental Death (sometimes expanded to 'Accidental Death, contributed to by neglect') or 'Misadventure'. On occasions, the

Coroner gave a Narrative Verdict, a brief summary of what happened, noting mistakes that contributed to the death. Coroners recorded seven verdicts of Manslaughter, and referred a number of cases to the Crown Prosecution Service. A decision to prosecute was very unusual. Only two people were jailed after a Manslaughter verdict in a British court: the managing director of an adventure activity centre[27] and a teacher both served brief prison terms.[28] In addition, a French court found a teacher guilty of negligent homicide (manslaughter) after a pupil drowned in the sea during a school trip to Le Touquet.[29] He was given a six-month suspended prison sentence, which was cleared on appeal.[30] Several schools and local authorities were fined under the Health and Safety at Work Act 1974. Understandably, some bereaved parents took legal advice on private prosecutions.

DROWNING OF CHILDREN ON TRIPS RUN BY THEIR PRIMARY SCHOOLS

Between 1985 and 2001, 11 children drowned during trips run by their primary schools. Most were non-swimmers or weak swimmers. The regulations allow teacher–pupil ratios of 1:15 for children aged 9 to 11 years, and 1:6 for younger children.[31] It is a demanding task to supervise so many active young children after they have entered the water. Teachers cannot give their constant, undivided attention to every child in their group. They cannot keep track of 15 juniors who may, or may not, be able to swim. They cannot stay within grabbing distance of 6 young children frisking about in the water.

Descriptions of most of these fatal accidents suggest that the teachers involved failed to realise how quickly and quietly a child can drown. They approached their task of supervision as if they were keeping a general eye on children during playground duty, rather than performing the intense visual scanning that is expected of a lifeguard.[32]

In July 2001, six primary school staff members took 41 pupils on a trip to France. During an outing to a theme park, 16 of the children were taken to paddle and swim from a crowded bathing beach on the shore of a lake. The adults in charge failed to notice that an eleven-year-old girl had disappeared. She could swim 10 yards at most.[33] Her teachers only realised that she was missing when her clothes were found unclaimed on the beach at the end of the visit. The teachers feared that she had been abducted, and the police were called in. After a manhunt and three days of searching, her body was found floating among reeds near the edge of the lake, a few yards from where she had last been seen.

The French authorities summoned four teachers to appear before an investigating magistrate in an involuntary homicide (manslaughter) inquiry that lasted for six years.[34] No prosecution followed. Nevertheless, the wide press coverage of the accident, and the legal repercussions, alerted teachers to the danger to their pupils' lives, and to their own career prospects. Between 2001 and 2017, no children drowned on trips run by their primary schools.

Soon after this tragic incident, the second-largest teachers' union, the National Association of Schoolmasters and Union of Women Teachers (NASUWT), advised its members not to take part in any further school trips. The general secretary of the union commented, 'Parents are increasingly reluctant to accept the concept of a genuine accident.'[35] More recently, Sir Ranulph Fiennes, the intrepid adventurer, derided teachers who wished to avoid risk on school trips as 'scared stiff of the blame-claim culture.'[36] These patronising comments bring to mind the authoritarian attitudes mocked in Stanley Holloway's famous monologue *The Lion and Albert*.[37] Yet most parents, and most teachers, see the value in school trips. They want the school trips to continue, but they want every child to return home alive.[38]

DROWNING OF CHILDREN ON TRIPS RUN BY THEIR SECONDARY SCHOOLS

Secondary schools often enrol their pupils in outdoor pursuits organised by commercial adventure activity centres. It took the deaths of four daring youngsters and an Act of Parliament to ensure that the staff at adventure activity centres are suitably trained for their responsibilities.

In March 1993, eight Plymouth teenagers, their teacher and two inexperienced instructors from an adventure activity centre set off in kayaks to paddle two miles along the coast from Lyme Regis to Charmouth. The falling tide and an offshore wind pulled them out into rough water in Lyme Bay. They had no radios or flares. Waves swamped the kayaks, which capsized, pitching everyone into the sea. The teenagers were wearing life jackets that were not fully inflated. Waves broke over their uncovered heads, and their thin wetsuits could not protect them from hypothermia. A captain of a fishing boat spotted an upturned kayak floating in the sea and raised the alarm. Before rescue came, 4 of the teenagers had drowned.[39] The managing director of the adventure activity centre was found guilty of manslaughter and given a three-year prison sentence.[40]

The mother of one of the victims declared, 'I was determined to make sure that no other adventure company was allowed to do the same thing in future.' She joined with other bereaved parents in pressing for a change in the law. David Jamieson, MP for Plymouth, supported them. He tabled a Private Member's Bill, which became the *Activity Centres (Young Person's Safety) Act 1995*.[41] The creation of the Adventure Activities Licensing Authority (AALA) followed in 1996.

The AALA was given the duty to inspect, regulate and license all adventure activity centres that offered courses in caving, climbing, trekking and water sports,[42] the four areas of outdoor pursuit that were considered most in need of robust safety management. The aim was to 'allow young people to experience exciting and stimulating activities outdoors without being exposed to avoidable risks of death or disabling injury'.[43]

AALA inspectors found high standards at the great majority of adventure activity centres, and they were duly licensed. However, at least sixty centres either discontinued adventure activity courses for which they would have needed a licence or were refused licences after AALA inspection.[44] Initial fears that the legislation would deter the provision of outdoor activity proved unwarranted.[45] Instead, the legislation improved the safety record and enhanced the credibility of adventure activity centres that provide licensed courses in caving, climbing, trekking and water sports. Over the years, more adventure activity centres were granted licences for activities covered by the AALA, including sea kayaking, the activity involved in the Lyme Bay tragedy.[46]

The National Union of Teachers (NUT), the largest teachers' union, urged that schools should be required to seek a licence if they took their own pupils on outdoor activities for which adventure activity centres would have needed a licence. However, the advice from the NUT was ignored, and schools and voluntary organisations were excluded from the legislation.[47] Since 1996, no child has died while taking part in a licensed course at an approved adventure activity centre, but 15 secondary school children drowned during adventurous outdoor activities led by their teachers.

◆ ◆ ◆

Most secondary school children know how to swim, but their teachers know too little about water – its weight, its coldness, the immense force of water on the move, the damage water can inflict on the human body. From one moment to the next, a well-ordered school trip can change into a frightening, desperate, muddled drama where the rescue of an endangered child rapidly becomes impossible.

In October 2000, a large party of teenagers gathered in the Yorkshire Dales for a week-long residential adventure activity course. Groups of pupils were to be taken 'river walking' along the picturesque watercourse of Stainforth Beck.[48] Days of rain had filled the underground caverns in its catchment area.[49] Continuing rain, unable to soak into the sodden ground, ran off into the beck, raising its level and increasing its flow.

On the morning of 10 October, rain was falling heavily as two teachers set off with their group for their river walk. They were wearing helmets but not life jackets.[50] The water reached their knees, then rose to their waists. To get the youngsters across the beck, a sixth-form helper stood in midstream and handed them to a teacher one by one. One of the boys slipped and clung to an older pupil for support. The force of the current carried another boy fifteen feet downstream before he regained his footing.[51] The group returned to base in good spirits and spent the afternoon taking part in an indoor climbing activity.

No warning of the depth, speed and force of the water was sent to the two teachers setting out for the afternoon river walk with their group of fifteen pupils. The water was 'shin-deep' when they stepped into the beck, but before they had progressed 70 yards upstream, the group was struggling through tumbling floodwater 3 feet deep.[52] Two boys refused to get into the beck, saying that the water was 'too fast'.[53] A girl stumbled and was allowed to climb out. Then three teenagers lost their footing and were carried downstream. A boy managed to save himself by grabbing hold of a boulder, but two girls were swept away.[54]

One of the girls laughed and waved for a few moments as the water spun her round, but she was powerless to withstand the power of the current. In an attempt to reach her, a teacher allowed herself to float some distance downstream,[55] but the girl was swept along in the faster current on the opposite side of the beck. The other teacher raced down the bank, and managed to grab the girl by the shoulder. He was dragged into the beck, and when he caught hold of an overhanging branch, the water tore the girl from his grasp.[56] Her body was found the next day, five miles downstream. After almost three weeks of intensive searching by volunteers, police and army divers and helicopters equipped with heat-seeking sensors, the body of the second girl was recovered even further from the accident site.[57]

Over several weeks in February and March 2002, national newspapers carried daily reports of the evidence given at the inquest on the two drowned schoolgirls. Although Stainforth Beck was 'prone to rapid and irregular fluctuations in level' after heavy rainfall in its catchment area, no local advice had been sought, no notice was taken of the recent rain, and no morning weather forecasts had been consulted before two groups of children were taken into the rushing water.[58] A police officer filmed the beck two hours after the accident, and told the inquest that it would have been too dangerous for him to cross. A local instructor with an AALA licence for gorge-walking said that he would have been quite happy to lead the children as long as the water was less than 6 inches deep: 'Anything near the knees is far too deep.' A water bailiff gave his opinion, 'It was highly irresponsible to let children walk in the river that day.' He commented, 'I have seen some stupid things, but walking through a river in full flood defies belief.'[59]

River walking was not regulated by the Adventure Activity Licensing Authority, but Marcus Bailie, the chief inspector of the AALA, gave evidence at the Inquest. While he praised the heroic actions of the teachers in their efforts to rescue the two girls from the beck, he said, 'Whether they should have entered is another matter, and I believe it was an error of judgement.' He suggested that the dangers of river walking had not been recognised before the fatal accident, and commented, 'I don't believe this outcome was foreseen by anyone, which was why the control measures to prevent it were not in place.'[60] In fact, in June 1998, two years before the girls lost their lives, an experienced outdoor pursuits instructor was swept away and drowned while shepherding children across a river in Wales.[61] The Coroner directed the jury to bring in verdicts of Accidental Death, adding that he would contact the Educational Authority and the Health and Safety Executive to call for action to prevent similar tragedies. The Education Authority was heavily fined.

In May 2002, only two months after the announcement of the Inquest verdicts on the fatalities at Stainforth Beck, two teachers arrived at a Lakeland youth hostel together with 15 high school pupils. A ten-year-old boy had been allowed to come along because his mother, an Educational Support Worker, was helping to supervise the group.[62] Among their outdoor activities over the weekend, they planned to jump into a 'plunge pool' fed by a waterfall in Glenridding Beck. On their way from the hostel to the plunge pool, they met teachers leading another school party who had intended to jump but had decided against it, and who warned them not to jump. The beck was in spate, flooded with water from the rain-soaked heights of Helvellyn. The waterfall thundered into the plunge pool. The churning water in the pool was intensely cold (8°C) and 8 feet deep.[63]

Ignoring the warning, the teacher led the group up to a flat ledge above the pool, jumped into white water 13 feet below, and climbed out of the pool. Next, a thirteen-year-old boy jumped and got out safely. Then the ten-year-old boy jumped. He suffered immediate cold shock and panicked,[64] unable to climb out of the pool.

The teacher jumped into the pool again, but he could not cope with the struggling boy, who repeatedly pushed him under the water. The teacher was soon weakened by the cold. 'All of a sudden my legs had gone and I just couldn't move,' he told the police later. 'He was drowning me and I just let go of him.'[65]

The boy's mother leaped down into the pool and managed to push the boy to the side of the pool, but she could not lift him out of the water, and no one had brought a rope. For a time she was able to hold his head above the surface, but as the water chilled her, she lost her strength, her grip loosened, and the boy drifted from her arms. The current swept him over the lip of the plunge pool and a hundred yards down the flooded beck. His mother was saved when another thirteen-year-old pupil jumped into the pool and pushed her up onto a ledge. Mountain rescue volunteers on exercise nearby rushed to help the group. They recovered the body of the boy, who was declared dead at the scene. His mother and the pupil who saved her were airlifted to hospital, where they were treated for hypothermia.

The teacher was prosecuted, and pleaded guilty to Manslaughter, the first teacher to admit to manslaughter on a school trip. He was sent to prison for a year. When the Judge watched a film of the plunge pool, taken just after the incident, he said, "It struck me as unbelievably foolhardy and negligent that anyone would venture into the beck when it was in that state of full spate, or allow any child to plunge from the rocks into the pool below."[66]

Three years after the boy's death, the Health and Safety Executive issued a lengthy report 'published to help prevent further tragedies, not to blame.'[67] The report should be required reading for anyone intending to lead a group of children on similar outdoor activities. '*Glenridding Beck – Investigation Report*' gives a moment by moment description of the drama, with a critical commentary on the succession of blunders which culminated in the boy's death. It points out that combined water/rock activities such as plunge pooling should always be treated as 'high hazard', and it suggests that an adequately prepared teacher should hold qualifications in life-saving and/or white water rescue and in first-aid, including training in resuscitation and the management of fractures and hypothermia.

Nowhere, however, does the report question whether the activity of plunge pooling is appropriate for groups of children on school trips. Back in 1996, Marcus Bailie, as chief inspector of the Health and Safety Executive, wrote: 'There are only three things which will cause death or disabling injury during an activity session; Drowning, Impact with something solid (which either falls onto you or onto which you fall), Exposure/Hypothermia.'[68] It seems perverse that organised school parties allow children to launch themselves from narrow rocky ledges into deep, cold water, whereas police officers and rescue organisations are busily engaged in discouraging youngsters from jumping off bridges, leaping into quarries, and tombstoning from cliffs.

◆ ◆ ◆

A number of children have died of drowning during activities which their teachers considered to be 'low-risk', including activities run for several years without a hitch. Teachers forget that sudden, unforeseen changes in the weather may quickly turn 'low-risk' into 'high-risk'. If the wind blows harder, the waves will be higher. After heavy rainfall, streams turn into torrents and flood-prone caves are flooded.

In November 2005, a school party set off from an outdoor education centre in the Yorkshire Dales to give 11 teenagers their first experience of caving.[69] The caves and caverns of limestone country are leached out by underground rivers fed by rain. Caving experts describe the chosen cave as 'extremely flood-prone', with a heightened risk of flooding whenever the nearby Scar House Reservoir is full, especially if a westerly wind blows water over the lip of the reservoir dam.[70] Moreover, cavers are warned not to enter the cave when water from the dam overspill fills the normally dry bed of the River Nidd beneath the dam and flows past the mouth of the cave.[71]

No rain was falling on the morning of the school trip, but two weeks of recent heavy rainfall had filled the dam, and a westerly wind was blowing so much water over the lip of the dam that a passerby stopped and filmed the spectacle on his mobile phone.[72] A teacher, a caving instructor and a seventeen-year-old trainee were in charge of the schoolchildren, but they made no inspection of the dam en route to the cave, and when they found that the river was too deep to cross near the cave mouth, they made a detour to find a safe place to cross, then continued into the cave.[73]

The teenagers would have had a satisfactory introduction to caving, and there would have been no accident, had their exploration stopped at the end of a huge, boulder-strewn cavern a quarter of a mile from the entrance. Instead, they were taken deeper into the cave, creeping on hands and knees through a low, narrow passage called the Crawl, 40 feet long and in places only 27 inches high, a passage which was known to fill with water if the cave flooded.[74] (Cavers call this 'sumping off'.) They were well beyond the Crawl when the cave system began to fill with water.[75]

By the time the group made their hurried return to the Crawl – their only exit – water was within four inches of the roof of the passage, and rising. The instructor told them, 'We are going to have to get through the tunnel now, or we won't get out alive.'[76] Four terrified youngsters managed to make their way through the Crawl before the water reached its roof. In panic and confusion, the rest of the party had to submerge themselves in cold, dark, muddy water and struggle halfway along the Crawl to a point where the roof rose enough to allow them to take a breath in an air pocket. Then they had to plunge underwater a second time. In his last brave act, a fourteen-year-old boy offered to let a school friend escape before him. His friend got through, but the boy failed to emerge from the Crawl.[77] The flood had receded by the time rescuers reached him. They found him lying chilled and deeply unconscious, face down in a pool of water, the light on his helmet still shining. He was taken to hospital, but he died from drowning a few hours later.[78]

It was two years before the Coroner delivered a narrative verdict. The Health and Safety Executive brought a legal action against North Yorkshire County Council, which owns and runs the outdoor education centre.[79] In 2010, after a brief hearing at Harrowgate Magistrate's Court and a month-long trial at Leeds Crown Court, the County Council was found 'Not Guilty' of breaching health and safety laws.[80] No school groups have entered the caving system since the fatal accident.

DISCOUNTING THE SIGNIFICANCE OF DROWNING ACCIDENTS ON SCHOOL TRIPS

Over the years, a number of comments by those involved in the provision of outdoor adventure activities for schoolchildren have down-played the risk of fatal accidents. In 2005, the Parliamentary Select Committee on Education and Skills published the report 'Education Outside the Classroom'.[81] Among the Written Evidence was a Memorandum submitted by the Adventure Activities Licensing Authority (AALA). The Memorandum included an Appendix: 'Fatal Accidents on School Visits', which listed the deaths of 5 adults and 52 children between May 1985 and June 2004.[82] The most frequent cause of death was drowning: 2 adults and 22 children died of drowning on school visits. The list was incomplete: 5 children who drowned on school visits between May 1985 and June 2004 should have been added to the list. And the deaths continued. Children drowned on school trips in 2005, 2006, 2007, 2008, 2015 (two fatalities) and 2017 (two fatalities).

The Adventure Activities Licensing Authority (AALA) was created in response to the drowning of four teenagers while kayaking in Lyme Bay in 1993, and its raison d'être is to protect young people from 'avoidable risks of death and disabling injury'. So it is disappointing that the AALA Memorandum discounted the significance of its own list of 52 children who died on school visits:[83] 'Whilst these figures may at first seem alarmingly high they need to be set in the context of other causes of death amongst young people and the population in general.'

Under a subheading 'Lives lost and lives saved', the memorandum makes a completely baseless claim: 'Trying to reduce accidents outside the classroom will make it more difficult for young people to develop a fit and healthy lifestyle, with a corresponding dramatic increase in the risk to their future physical as well as emotional well-being' – as if the loss of these young lives on school trips might be considered an unavoidable cull to ensure the survival of their classmates.

Some outdoor enthusiasts, as adventurous types themselves, consider parents overanxious, overprotective, too ready to deny their children opportunities for adventure and excitement. Quoted soon after the drowning of two schoolgirls in Stainforth Beck in 2000, Sir Chris Bonington, the celebrated mountaineer, asked, 'After every crisis there are screams to make things safer, but at what cost to education?'[84] He was chairman of the risk management committee of Outward Bound at the time.

Interviewed in 2009, Sir Ranulph Fiennes was clear that youngsters should be encouraged to go on 'adventure outings'. He said, 'What concerns me is the growing negative connotation of the word risk – we have got to make a clear distinction between recklessness and taking risk. In the right place, taking risks can be very positive. It is vital that children are taught to understand and manage risk from an early age.'[85] However, teachers are often surprisingly lax in their risk assessment, underestimating the danger of water and overestimating their pupils' swimming ability. Since teachers themselves often misjudge the risks on school trips involving water, they may be unable to teach their pupils 'to understand and manage the risks from an early age.' Moreover, children, especially teenage boys, are prone to test boundaries and push their luck, sometimes with tragic consequences.[86]

Tim Gill, campaigner and prolific writer on children's play, ended a recent article in the *Guardian* with an irresponsible plea. He wrote, 'What is needed is nothing less than the wholesale rejection of the philosophy of protection. In its place, what we need to adopt is a philosophy of resilience that truly embraces risk, uncertainty and real challenge – even real danger – as essential ingredients of a rounded childhood.' Surely, few parents would let their children go on school trips that deliberately exposed them to 'risk, uncertainty and real challenge – even real danger'.[87]

A more balanced view – that it is reasonable to question whether the benefits justify the risks – is taken by Andrew Brookes, an Australian expert on outdoor education and safety management. In the first of a series of papers on fatal accidents, he wrote: 'I contend, as a researcher, outdoor educator, and parent, that fatality prevention must be approached from the standpoint that there is no acceptable rate of accidental deaths in outdoor education.'[88]

More than forty years ago, James Hogan (who established the first Outward Bound School in 1941,[89] and later became Chairman of the Duke of Edinburgh Award Expedition Advice Panel) expressed his uneasiness about the motivation of some enthusiasts who push for more arduous, action-oriented trips.[90] He questioned the truth of their claims that when children are 'stretched' by taking part in inherently risky outdoor activities, they become more self-reliant, more committed to pressing on when they encounter difficulty, more determined to overcome fear, more ready to help each other. He pointed out that children taking part in Outward Bound courses suffered few serious accidents: 'Why? Because professional educators – especially those who wish to continue in practice – must reject with finality the deliberate "courting" of risk.'

He acknowledged: 'Of course there may be great value in allowing young people to assess for themselves their competence in face of hostile weather, terrain, or simple unexpected circumstance.' But he emphasised the psychological and physical dangers involved: 'Failure in the same conditions may be disastrous. It may deepen insecurity and lack of confidence – or, at worst, result in loss of life.'[91]

Writing in 2002, after many years of practical experience in outdoor education in South Australia, Rob Hogan suggested that teachers often approach pre-trip risk management (and the documentation required nowadays) by paying too much attention to routine preparation tasks, while paying too little attention to the unlikely possibility of serious injury.[92] He advised teachers to consider possible dangers arising from environmental factors (such as weather, terrain, remoteness), from human factors (the age variation, the different levels of fitness, experience and coping skills of group leaders and students), and the choice or lack of equipment (clothing, ropes, buoyancy aids). When teachers focus their attention on how they might prevent or manage sudden emergencies, they minimise the risk of death and disabling injury.

In their publications 'Glenridding Beck – Investigation Report' and 'Glendinning Beck – Conclusions', the Health and Safety Executive (HSE) advised teachers to provide 'sufficient information on hazards and risks to allow parents to make informed decisions about their child's participation', to obtain 'relevant medical information', and to secure parental signatures on consent forms.[93] However, the HSE revealed a chilly level of calculation behind this advice: 'A parent who has knowingly accepted a risk (for instance that a child might trip on a mountain path and break a limb) when giving consent will have few grounds for complaint against a diligent leader if such an incident occurs.'[94] The emphasis here is on risk management as a means of avoiding litigation rather than safety management to prevent injury.

The Parliamentary report 'Education Outside the Classroom' of 2005, and the later report 'Transforming Education Outside the Classroom' of 2010, both called on the Department of Education to work with teachers' unions to ensure that teachers did not feel vulnerable to 'vexatious litigation'.[95] In fact, very few of the bereaved parents resorted to the law, although newspapers record their heartbroken comments after the announcement of the inquest verdict and their expressions of disbelief or anger at the end of a Crown Court trial when they realised that no one was to be held responsible for actions which resulted in the death of their child. 'How can a little girl go to school and die during a swimming lesson, and no one is to blame?' asked the mother of a four-year-old girl who drowned in 2001, discovered at the bottom of a swimming pool as her supervisors were about to leave the building with the rest of her class.[96]

Often parents hope that the evidence heard in court by the coroner or judge will be put to use to protect children in the future. In 2003, after hearing the guilty verdict on the teacher in charge when his son jumped into a Cumbrian plunge pool and drowned, the boy's father said, 'What we have got to remember here today is a ten-year-old boy has lost his life and we have to learn lessons over this so that it does not happen again.'[97]

The parents and grandparents of the fourteen-year-old boy who drowned while caving in 2005 were present in Harrogate Magistrate's Court when they heard that the case brought by the Health and Safety Executive against North Yorkshire County Council was to be transferred to the York Crown Court. The boy's mother said, 'The county council has always maintained that it was a freak accident, but there was evidence at the inquest to suggest otherwise.' She supported a further trial: 'We need to make sure that all adequate precautions are being taken to prevent another tragedy, as this would be the only good to come out of all this.'[98]

Almost five years after the death of her son, the County Council was found not guilty of breaching health and safety laws. In 2010, at the end of the legal process, although no one was officially blamed, the judge said, 'I trust that this will not be regarded as a reason for any complacency or self-congratulation by those involved in organising and delivering these activities.'[99] Speaking after the verdict, the boy's father said, 'It may be that the detailed examination in this court of the actions of North Yorkshire County Council will result in measures being taken to ensure that other young people are not exposed to the terrible and traumatic experiences described to the jury by those that were underground on that fateful day. We fervently hope so.'[100]

OFFICIAL ADVICE TO SCHOOLS CONCERNING SAFETY DURING SCHOOL TRIPS

In 2003, the Central Council for Physical Education, under the auspices of the Department of Education and Skills, published the leaflet: 'Group Safety at Water Margins'.[101] It was aimed at teachers and others who would be taking children on 'low-risk' outdoor trips 'near or in water – such as a walk along a riverbank or seashore, collecting samples from ponds and streams, or paddling or walking in gentle shallow water'. The leaflet was disappointingly vague about which activities were suitable for children at primary school, which activities should be reserved for children at secondary school, and which activities were best avoided as too risky.

Teachers were told: 'If you do lead the visit yourself you should take a number of steps to identify the foreseeable hazards', but with little indication of what the steps or the hazards might be. They were advised, several times, to have a 'Plan B' at the ready, 'as an alternative, not as an emergency procedure', since they might need to change their original plan 'for any number of reasons' – presumably because the original plan was no longer appropriate.

Even though they would be taking charge of schoolchildren who might step into, wade into, jump into, dive into, or fall into water, the leaflet did not query whether teachers knew how to swim or suggest that they should have lifesaving or first-aid qualifications. Indeed, on its opening page the leaflet made clear that it did not cover 'swimming and other activities that require water safety or rescue qualifications and equipment, or water-going craft'. Nor were teachers advised to find out if the children could swim, and if so, how well. (Well-prepared teachers should surely be advised to ask parents for this information beforehand, in writing, so that they do not have to ask the children on the day.) Teachers were, however, advised to take along 'an adequate first aid box' and to check the whereabouts of the nearest hospital. And they were invited, rather tentatively: 'If you have been trained, and are currently practised in the use of throwlines, you may wish to take one with you.'

The leaflet included some helpful warnings:

+ 'Remember that fast moving water above knee height is likely to knock people off their feet.'
+ 'The tide may advance more quickly than your group can retreat.'
+ 'Remember that sudden and unexpected immersion in cold water has a rapid and dramatic effect on the body's systems and will impair people's ability to reach safety.'

However, the leaflet contained not so much as a hint that any child had drowned on any school trip, even though 13 children had drowned on school trips in the previous five years.

+ + +

In 2008, the Council for Learning Outside the Classroom(CLOtC) introduced a national accreditation scheme that awards the LOtC Quality Badge to 'organisations that provide good quality educational experiences and manage risk effectively'.[102] Teachers planning school trips are encouraged to make use of organisations awarded the LOtC Quality Badge. The Outdoors Education Advisers' Panel (OEAP) organises training courses for leaders of outdoor learning and educational visits.[103] OEAP issues helpful guidance for schools, including specialist advice on group management and supervision, risk management, adventure activities, safe use of swimming pools during off-site visits, swimming and paddling in natural water (river, canal, sea or lake), and discusses what the law requires, how to comply with it, and what may happen following an accident and incident.[104] In 2017, OEAP published a much improved version of the 2003 leaflet 'Group Safety at Water Margins'.[105]

In 2013, the Royal Society for the Prevention of Accidents (RoSPA) published its own guidelines: *'Planning and Leading Visits and Adventurous Activities – Guidance for schools and colleges teaching children and young people from 5 to 18 years'.*[106] The emphasis of the publication was on pre-trip planning and checklists: defining legal responsibilities of school governors, head teachers, teachers, non-teaching staff and volunteers; giving parents information about the trip and obtaining their written consent; selecting staff/pupil ratios; ensuring adequate insurance cover for personal injury, public liability and medical expenses; recording that significant risks were assessed and that 'all reasonably practicable precautions' were taken.

School trips abroad involved extra pre-trip planning and checklists involving travel and accommodation, valid passports and visas, and relevant immunisations. When abroad, teachers are expected to act as 'reasonably careful parents', coping if a child has an illness or an accident, supervising any medication, giving advice on avoiding sunburn, dehydration and food poisoning, and ensuring that pupils comply with local laws and customs.

The guide confirmed that 'swimming outdoors' was a high-risk activity, and warned: 'Statistically, this is the most risky activity for children. An accident in any depth of water can be fatal unless prompt action is taken. Being able to swim well in a swimming pool does not guarantee safety in outdoor water. The cold water temperature is always a potential hazard.' The guide advised teachers that effective water safety training should be included in all water activity programmes.

For several years, the Health and Safety Executive considered winding down the Adventure Activity Licensing Scheme, but no agreement was reached on what to put in its place.[107] In 2018, the Department of Education published 'Health and safety on educational visits', which commends the Council for Learning Outside the Classroom, the LOtC Quality Badge scheme, the Outdoors Educator Advisers' Panel and the Adventure Activities Licensing Authority.[108]

FAILURE TO LEARN FROM FATAL ACCIDENTS, INJURIES AND NEAR MISSES ON SCHOOL TRIPS

Police investigate every fatal accident. At the Inquest, the Coroner seeks to establish exactly what happened, and why. Once the Inquest verdict is announced, however, the learning process often seems to peter out. From time to time, brief notes (called Infologs) appear on the AALA website about 'incidents, accidents and near misses',[109] but there is no defined mechanism to keep head teachers and their staff informed about deaths or injuries on school trips. No British studies have been published on this clearly defined category of fatal accident, or on near misses where injury was narrowly averted. As a result, children drown in carbon copies of accidents that drowned children years before.

The investigation of an aircraft accident is very different.[110] It involves two stages. First, the circumstances and causes of the accident are defined as exactly as possible. Second, safety recommendations are made 'with a view to the preservation of life and the avoidance of accidents in the future.' These recommendations often lead to changes in aircraft design or pilot training which are widely accepted and applied throughout the industry.

In July 2015, a talented seventeen-year-old schoolboy from Surrey drowned on the first day of a rugby tour to Canada. After two morning rugby training sessions, 25 teenagers were taken to swim at a nearby lake. He was standing at the end of a jetty when a friend pushed him into the water. He surfaced briefly, then sank out of sight. It was fifteen minutes before he was located under the jetty and lifted from the lake, and he did not respond to resuscitation. The temperature of the lake was 16°C at the surface (with colder water below the surface layer), whereas his previous experience of swimming had been limited to the warmed water of swimming pools. It was thought likely that he developed cold shock before he drowned. He also had asthma.[111]

At the schoolboy's Inquest in 2018, an expert witness was critical of the adequacy of the risk assessment conducted when planning the trip and at the scene of the accident. There was no lifeguard presence, and the supervision was inadequate, preventing consideration of the safest site of entry to the water, and of the risks of unacclimatised swimmers entering cold open natural waters. He pointed out that the rate of drowning was particularly high in the teenage years because of ignorance of dangers, bravado and peer pressure. He said that schools do not teach children to swim to a sufficiently high standard to be safe, and there is no requirement for any swimming at secondary school.

At the end of the Inquest, the Coroner recorded a verdict of Accidental Death, and he also sent a Regulation 28 Report to Prevent Future Deaths to the Secretary of State for Education.[112] He supported the recommendations of the expert witness: that the minister review the curriculum and introduce swimming instruction with an ambitious target of competence; that Department of Education guidance to schools should include the risks of inexperienced or weaker swimmers swimming in open and cold waters; and that schools should adopt, as routine, prior to school trips, a proper test of the swimming ability of each participant, and should use the test results in the risk assessments of the activities planned for the trip.

◆ ◆ ◆

Fatal accidents on school trips are, by definition, unintended, but they are not necessarily unpredictable – even if they are not predicted by the teachers involved.[113] Drowning is the commonest cause of death in fatal accidents to children on school trips, usually because their teachers fail to realise how vulnerable their pupils are or how destructive water can be. As a spokeswoman for the Royal Life Saving Society pointed out in 2001, 'When children swim on school trips it is often in the sea or open water – the most dangerous places to swim.'[114] Perhaps schools could be advised to choose safer venues for education outside the classroom, and stay away from water. Certainly, teachers should take note of the cautionary entry in Ambrose Bierce's dictionary: 'ACCIDENT: An inevitable occurrence due to the action of immutable natural laws.'[115]

CHAPTER TWENTY-SEVEN

THE GLOBAL DEATH TOLL OF DROWNING

The Asian tsunami of 2004 killed 230,000 people in one devastating event.[1] It is a shock to discover that an even greater number of people die of drowning every single year. Yet this epidemic of drowning deaths has been overlooked until recently. The World Health Organisation (WHO) published its first report on drowning only in 2014. The report stated: 'Drowning is a serious and neglected public health threat claiming the lives of 372,000 people a year worldwide. More than 90% of these deaths occur in low- and middle-income countries.'[2]

The WHO report admitted that even these alarming figures are an underestimate, given that official statistics frequently 'exclude intentional drowning deaths (suicide and homicide) and drowning deaths caused by flood disasters and water transport incidents (including those where vessels carrying migrants, refugees and stateless people capsize during so-called irregular transport on water).'[3] Moreover, many drowning deaths go unrecorded. Victims are pulled lifeless from village ponds. Their deaths are reported to the village elders. Burials follow soon after. Since the victims are not brought to hospital, they do not appear in hospital statistics. The inevitable result is serious under-reporting of national death rates.

Around the world, immunisation programmes protect infants from infectious diseases that once caused many early deaths. In low-income countries, however, children who survive infancy are still menaced by the 'silent epidemic' of drowning.[4] The United Nations Children's Fund (Unicef) highlighted drowning as 'the major killer of children over the age of one year.'[5]

Epidemiologists in Bangladesh were the first to recognise drowning as the commonest cause of death in children who survive infancy. Careful data collection over a number of years in one rural area was followed by a nationwide survey of more than 400,000 adults and their children.[6] These studies indicated that almost 17,000 children drowned in Bangladesh every year[7] – that is, 46 children drowned every day. Children aged one to four years old are the most vulnerable, with an annual mortality rate due to drowning of 86.3 per 100,000,[8] more than 40 times higher than the equivalent rate in Britain.

Many factors contribute to this loss of life, including inadequate supervision of the younger children in large families, the presence of numerous wells, ponds, irrigation ditches and rivers close to homes, water-filled paddy fields, mangrove swamps, unsafe bridges and fords, defective fishing boats and overloaded ferries.

In addition to these frequent individual drowning tragedies, Bangladesh suffers from repeated disasters linked to geography and climate. A combination of water run-off from the Himalayan snows, monsoon rains, cyclones, storm surges and the slow rise in sea level in the Bay of Bengal threaten the heavily populated Ganges-Brahmaputra Delta, the largest river delta in the world.[9] Floods engulf entire villages, sweeping hundreds of people to their deaths, together with the turfed embankments that were meant to protect them. In 1998, floods submerged Dhaka, the capital, under 6 feet of water, inundated two-thirds of the country, and drowned more than a thousand people.[10] In 2007, a cyclone lifted a tidal surge 20 feet high, which killed 15,000 people.[11]

In Vietnam, too, death by drowning was recognised only recently as a major killer. In 2003, an injury survey for Unicef reported: 'From infancy to puberty, drowning is the overwhelming cause of death in every age group and far outstrips other causes.' More than 14,000 children drowned in 2001 – that is, 38 children drowned every day. Three-quarters of the victims were between 5 to 14 years old.[12] The survey commented: 'Many Vietnamese parents refer to the ending of the school year and the beginning of the summer vacation period as "the drowning season".'[13]

In 2011, a National Injury Survey in Thailand reported a similar loss of life from childhood drowning,[14] with children in rural areas almost five times as likely to drown as children in towns. Two-year-old boys had the highest death rate due to drowning (106.8 per 100,000). A Sri Lanka doctor told an international conference on drowning prevention that houses in rural areas often relied on water from open, uncovered wells: 'A toddler who crawls to an unprotected well sees his own reflection in the water and it usually ends in a tragedy'.[15]

Drowning is a frequent cause of death in the Caribbean and in Central and South America, but according to James Vaughan, International Director of the Royal National Lifeboat Institution: 'Specific data is often unavailable or too limited to provide an accurate figure.'[16] In 2004, at least 2,000 people drowned when floods cascaded down from mountainsides in Haiti and the Dominican Republic, where most of the trees had been chopped down to make charcoal. A report in the New York Times noted: 'Death counts remain estimates with officials citing conflicting and sometimes second-hand information.'[17] Brazilian doctors confirmed 85,540 deaths from drowning in the years 1996–2007, but according to Dr David Szpilman, a respected authority on drowning and lifesaving, only half of the death certificates recorded whether the fatality took place in a bathtub, pool, river, lake or the sea: many fatal drowning accidents go unreported, and deaths from drowning in natural disasters are completely excluded from official statistics.[18]

In Africa, reliable information on drowning deaths is extremely scanty. In 2007, a conference paper under the auspices of the International Life Saving Federation, began: 'The task of compiling statistics on drowning fatalities in South Africa is currently a challenging one.'[19] The task is just as challenging today. In 2013, Job Kania, the President of the Kenya Lifesaving Federation, expressed his concern: 'Death by drowning occurs far more often than generally perceived in Kenya.'[20] It was not possible to establish reliable data on age, gender or activity of drowning victims or to map the exact locations of fatal incidents due to the lack of proper records. Witnesses of drowning accidents failed to inform the police. Rescuers did not record incidents. Bereaved families concealed the circumstances of drowning deaths. Few people could swim, and even fewer had any lifesaving skills.

In Uganda, the River Nile flows north from Lake Victoria to Lake Albert via the extensive swamps and wetlands of Lake Kyoga, which vary in extent and depth according to the season of the year. Newspapers carry reports of water-related accidents if there is a major loss of life, but very many fatal drowning accidents involve one or two victims, and those deaths go unreported and unrecorded.[21] Malaria and HIV are seen as more pressing problems than drowning. Tanzania controls the southern half of Lake Victoria, the largest lake in Africa, and it also has a coastline stretching for 500 miles on the western shores of the Indian Ocean. Yet neither the central government nor the local authorities collect data on drowning. Researchers admit: 'Establishing surveillance where it has not existed is difficult in resource poor settings. Without information derived in this way, motivation to implement intervention strategies is difficult to generate.'[22] They describe the vicious circle succinctly:

+ 'No drowning surveillance – no statistics.
+ No statistics – no evidence of need.
+ No evidence of need – no attention.
+ No attention – no intervention.'

SUCCESSFUL DROWNING INTERVENTION IN BANGLADESH, VIETNAM AND SRI LANKA

In 2005, the researchers who drew attention to the high rate of drowning in Bangladesh founded the Centre for Injury Prevention and Research, Bangladesh (CIPRB) in Dhaka.[23] They realised that the preventive strategies tried and tested in high-income countries would have to be adapted for the different circumstances in the tropics, and introduced gradually to ensure acceptance. In 2006, they set up PRECISE (Prevention of Child Injury through Social-intervention and Education),[24] a community-based project 'to identify the effective interventions which can be replicated and scaled up to the rest of the country.' In 2009, with the help and support of The Alliance for Safe Children (TASC) and the Royal Life Saving Society of Australia, CIPRB established the International Drowning Research Centre – Bangladesh, where a number of different projects are planned and assessed.[25]

Most young children who drown are discovered in ponds, ditches and rivers within 20 metres of their homes.[26] Many had been in the care of another child who was only slightly older. More than a third were alone after wandering out of the house, leaving their mothers busily engaged in work or food preparation. A team from Baltimore's Johns Hopkins University introduced villagers to simple and cost-effective ways of keeping babies and toddlers under observation – by fitting gates and barriers across doorways, by using locally made playpens, and by running community day care centres or crèches called Anchals, safe havens where children aged 18 months to 5 years could play during the morning (when most drownings happen).[27] The drowning rate in children who attend Anchals is one fifth of the drowning rate in children who do not attend Anchals.[28]

The staff of the CIPRB attend many of the community meetings held in rural villages after a child has died of drowning. They help the child's parents, relatives, neighbours and the village elders to discuss the fatal incident. They encourage villagers to suggest practical ways to prevent drowning by identifying danger spots and dealing with them,[29] for instance, by putting childproof covers on wells, filling in superfluous drains and ditches, and erecting fences round particularly dangerous ponds. Audiences of hundreds of villagers learn about drowning risks and injury prevention as they watch video docudramas or attend performances of travelling theatre groups.[30]

◆ ◆ ◆

Despite the high risk of drowning in Bangladesh, researchers realised that few adults and very few children knew how to swim. SwimSafe programmes (funded by the Royal Life Saving Society of Australia and The Alliance for Safe Children) recruited and trained local swimming instructors to teach survival swimming and water safety to children aged 4 to 12 years.[31] In Dhaka, swimming lessons are held in four large portable plastic pools. In rural areas, children are taught to swim in local ponds, where a fenced underwater platform provides a defined area of water about 2 feet deep. This protected corral is surrounded by a deeper area, also guarded with a perimeter fence.[32]

Three weeks of lessons are usually enough for children to learn how to float on their backs, tread water for 30 seconds and swim 25 metres. 'When we started most of the instructors were men,' one of the organisers explained, 'but we found that women were much more reliable, and were available to do the lessons. So now most of the instructors are women.' After women became instructors, families sent their daughters to the swimming lessons as well as their sons.[33] By 2013, SwimSafe instructors had taught over 300,000 children to swim.[34] The drowning rate in children who took part in SwimSafe programmes is only one-tenth of the drowning rate in children who did not attend SwimSafe.[35]

There was concern that teaching young children to swim might encourage them to enter the water on their own, to spend more time in the water or take risks while swimming. In fact, children taught survival swimming spend less time in the water and they take fewer risks than children who learn to swim by copying the self-taught 'natural swimming' of their friends.[36]

SwimSafe teaches children land-based, non-contact 'Reach and Throw' techniques for rescuing a drowning victim – stretching out a hand, holding out a stick, or throwing a rope or something buoyant that could serve as a life ring. Children are advised to stay at the water's edge and to enter the water only as a last resort. It was only when researchers questioned children about actual rescues that they realised that only 10% of the rescuers actually used the safer, land-based, non-contact rescue techniques they had been taught.[37] Sticks and branches are routinely collected for firewood, so there were few sticks available for rescuers to hold out to children struggling in the water. The great majority of rescuers entered the water and made direct contact with drowning children. Indeed, half of the rescuers had to swim to reach the child they saved. SwimSafe trainers have now adapted their advice to include in-water approaches that minimise contact, such as suggesting that a rescuer might strip off a T-shirt or a scarf, and hold it towards the drowning victim.

SwimSafe teaches mouth-to-mouth resuscitation and cardiopulmonary resuscitation (CPR) to children as young as 9 years.[38] Youngsters with differing levels of literacy were able to perform CPR competently after two days of training.[39] But in the villages of Bangladesh, rescuers must cope with drowning children without any back-up from paramedics bringing oxygen and defibrillators, so that even competent CPR may not be enough.[40]

Few adults in Bangladesh know how to resuscitate drowning victims using rescue breaths and cardiac massage.[41] Even when children are found floating in a nearby pond soon after falling into the water, rescuers often make no attempt at resuscitation, believing that 'drowning is a natural phenomenon, or "God's wish", and can't be prevented'.[42] Rescuers sometimes resort to ineffective traditional methods of resuscitation. They try to induce vomiting by putting salt or ashes into the child's mouth, by whirling the unconscious child overhead, or by spinning the child around as he lies face down on the ground.[43] Many villagers feel uneasy at the thought of giving CPR to their elders or to someone of the opposite sex. They may be more willing to consider giving mouth-to-mouth resuscitation to immediate family members than to distant relations or strangers.[44]

In 2012, the Royal National Lifeboat Institute (RNLI) became involved in training 15 local lifeguard volunteers at Cox's Bazar, a long sandy beach on the south-east coast of Bangladesh. Since then, RNLI lifeguards have returned to Bangladesh a number of times,[45] assisting local lifeguards to develop their lifesaving organisation, teaching them how to train their own lifeguards, testing a prototype rescue board made out of jute, teaching flood rescue techniques to members of the Bangladesh Fire and Civil Defence service, giving search and rescue training to the Bangladesh Coast Guard, and working directly with the International Drowning Research Centre in Dhaka. In 2016, five flood rescue instructors from the Bangladesh Fire

Service and Civil Defence travelled to Scotland to take part in advanced training in swift water rescue and incident command.[46]

◆ ◆ ◆

In 1997, more than twenty years after the Vietnam War came to an end, Congressman Pete Peterson was appointed the first post-war United States Ambassador to Vietnam.[47] He had been a prisoner-of-war in Hanoi for six years, from 1966 until 1973. Three months after his return to Vietnam, he told his former captors, 'Reconciliation is not only possible but absolutely the way to reach out.' Visiting overcrowded hospitals round the country, he noticed that half the beds were occupied by victims of injury. He soon recognised the importance of injury prevention, especially the prevention of drowning. He was instrumental in setting up The Alliance for Safe Children (TASC) and SwimSafe.

Instead of building expensive permanent swimming pools, SwimSafe sited large portable plastic swimming pools in the grounds of primary schools in the coastal city of Da Nang.[48] SwimSafe also runs survival swimming classes in a standard in-ground swimming pool at the rehabilitation hospital and on two of Da Nang's beaches, and trains and certifies swimming teachers, lifeguards and instructor trainers. Thousands of children aged 6 to 12 years have already taken part in specially developed courses in survival swimming. After 20 sessions they can swim 25 metres, float, tread water and carry out non-contact rescues. Children aged 9 years or more are given training in cardiopulmonary resuscitation. SwimSafe also uses portable pools to teach survival swimming to children in Thailand.[49]

◆ ◆ ◆

For children in low-income countries who live within a few feet of ponds and drainage ditches, knowing how to swim is a vital skill that gives them life-long protection. Even when adults were swept away by giant waves in the Asian tsunami of 2004, knowing how to swim tripled their chances of survival.[50] Moreover, children whose parents survived the tsunami were 15 to 20 times more likely to survive than children whose parents lost their lives.[51]

In Sri Lanka, 35,000 people were killed by the tsunami.[52] Far more women died than men.[53] A Buckinghamshire swimming teacher, Christine Fonfe, was so moved by the horrifying TV coverage of the tsunami that she decided to travel to Sri Lanka to help.[54] When she first arrived, she taught swimming to children living in a camp for tsunami survivors. She soon discovered that none of their mothers could swim. Cultural expectations meant that women were not taught to swim in childhood. Once girls reached puberty, they were excluded from swimming pools, and they were expected to wear tight, wrap-around saris while washing at the well or on the beach. She realised: 'The drowning survival odds are stacked against women by culture and couture.'[55] In an emergency, mothers could neither rescue their drowning children nor save themselves from drowning.

Christina Fonfe decided to teach women to swim. She refurbished a dilapidated swimming pool on an abandoned coconut plantation 'where women could learn to swim out of sight of men'. She persuaded women out of their saris into swimsuits.[56] She reasoned, 'People do not drown because they cannot swim, but drown because they cannot breathe.' So the women were first taught to 'Float-and-Breathe' on their backs. Next, they practised relaxed gliding. Then they learned backstroke and crawl. Once they could jump into deep water, float on their backs for ten minutes, swim any stroke for a hundred metres, and climb out of deep water unaided, they were awarded their 'icanswimcanyou' certificate.[57] Students were then invited to bring a friend and teach her what they had learned. This reinforced their own progress and prepared them for teaching their own children to swim. It also identified potential swimming teachers, who were given further training.

In 2012, the British High Commission donated a portable pool to her project,[58] and in 2015, Christina Fonfe was awarded the British Empire Medal. [59]After teaching more than 5,000 women and teenage girls to swim over eleven years, she was able to hand over the management of the Sri Lanka Women's Swimming Project to a woman from Sri Lanka.

CHAPTER TWENTY-EIGHT

A DROWNING ACCIDENT EXPLAINED

Chapters 13 to 27 introduced a series of drowning scenarios, looking at typical accidents involving victims from infancy to old age. Now it is time to look more closely at the tragedy of drowning as it affects one unlucky individual.

After work on a hot summer evening, a 25-year-old carpenter decides to cool down by swimming in a nearby lake. Within a few minutes he will be at the point of death. To understand how his body reacts in those few minutes, we will follow his ordeal from the first moments of danger to the moment when his heart stops beating.

THE FIRST MOMENTS OF DANGER

The young man usually swims in his local swimming pool. The water is far colder in the lake, and it becomes colder as he swims away from the shore. As his muscles chill, his strokes lose strength and coordination.[1] He makes slow progress through the water. Swimming is harder work than he expected, but he cannot wade to safety because he is out of his depth.

As the swimmer tires in deep water, his predicament comes as a terrifying surprise. When he first realises that he is in difficulty, he shouts for help. No help comes, so he forces himself to swim on. His straining muscles demand more oxygen and produce more carbon dioxide, prompting him to take deep, hurried breaths. His heart beats harder and faster. Blood flow increases sevenfold round his body, while the entire network of pulmonary blood vessels opens up, easing the transit of so much extra blood through his lungs.[2]

The swimmer is soon exhausted, and now he cannot shout for help because he needs all his lung power for breathing. He lies lower in the water. His mouth begins to dip below the surface. He has to struggle to reach the air. His muscles run short of oxygen for aerobic contraction, so they become more dependent on anaerobic energy production. Loaded with lactic acid, they lose their power to support him in the water.[3] He sinks under the surface.

THE FIGHT TO REGAIN THE SURFACE

As soon as the swimmer recognises that his life is in danger, an involuntary, uncontrollable 'fight or flight' reaction releases adrenaline and other stress hormones into his bloodstream.[4] His heart races. His blood pressure rises. His airways widen. His liver releases extra glucose into his bloodstream. He feels a surge of energy and a renewal of strength. He manages to pull himself back to the surface and gasps in air. His struggle to survive continues for a little while longer.

He sinks back under water. For a time, he manages to hold his breath and his desperate efforts take him to the surface briefly for a second time, even for a third, fourth or fifth time, but he sinks again after each attempt to save himself. He finds it harder and harder to lift his mouth clear of the water. Then as he breathes in, he takes in a little water with the air. He coughs, but coughing expels air from his lungs. He loses buoyancy and has to work even harder to rise to the surface. Sooner rather than later, his strength fails him. He cannot manage to pull himself up to the air. Now when he sinks, he goes on sinking.

THE OVERWHELMING URGE TO BREATHE

While the swimmer struggles underwater, he holds his breath for as long as he possibly can. But now he cannot rid his body of carbon dioxide, which accumulates in his bloodstream. The respiratory centre in his brain responds to the increasing level of carbon dioxide by prompting him to breathe.[5] His diaphragm begins to jerk up and down, its contractions stimulated by the build-up of carbon dioxide.[6] The urge to breathe becomes increasingly difficult to resist, and he counters by swallowing repeatedly, gulping mouthfuls of water.[7] Nevertheless, however hard he tries to prolong his breath-holding, within one and a half minutes at the most, and usually much sooner, he will be compelled to take a breath.[8] Then he will inhale water instead of air.

THE PROTECTIVE REFLEX OF LARYNGOSPASM

The larynx sits at the top of the windpipe, guarding its entrance.[9] Nicknamed the Adam's Apple because of its bulging shape at the front of the neck, the larynx is normally open, allowing air to flow in and out of the lungs for breathing and speaking. During swallowing, the larynx is closed off to prevent saliva, food or liquid from 'going down the wrong way' into the lungs. Muscles lift the larynx upwards and forwards, sealing the top of the windpipe against a stiffened flap at the back of the tongue. At the same time, the soft palate is raised, blocking the back of the nose, and the edges of the vocal cords are pressed tightly together.[10] These coordinated movements ensure that anything swallowed is forced to slide down the gullet into the stomach.[11] A reflex triggered by the respiratory centre stops people from breathing while they swallow.[12] And they swallow, usually without thinking about it, hundreds of times every day.

As soon as water enters the swimmer's airway, it triggers a powerful protective reflex. The muscles of his throat contract in a prolonged, intense swallowing action called laryngospasm.[13] Laryngospasm stops further water from entering the swimmer's airway,[14] but it does not provide him with air. He still tries to breathe, driven by the rising level of carbon dioxide in his blood. Yet as long as the laryngospasm continues, nothing can enter or leave his lungs. The contractions of his diaphragm create a bellows effect that sucks water down his gullet, so that his stomach becomes distended with water.

Frederick Banting, who won a Nobel Prize in 1923 for the isolation of insulin, studied the physiology of drowning in a series of animal experiments during the 1930s. He found that laryngospasm could be triggered by putting as little as 2ml of water into the larynx.[15]

THE VITAL SUPPLY OF OXYGEN IS CUT OFF, AND THE SWIMMER LOSES CONSCIOUSNESS

Normally, every breath renews the air in the lungs and, as blood flows through the lungs, haemoglobin in the red cells forms oxyhaemoglobin, capturing its full quota of oxygen within seconds. When blood leaves the lungs and returns to the left side of the heart, it has an oxygen saturation of 97%.[16] The presence of large amounts of oxyhaemoglobin gives arterial blood its bright crimson colour. As blood is pumped round the body, the tissues extract the oxygen they need from the red cells. The colour of the blood gradually changes from crimson to dark purple-red, and the oxygen saturation falls to 75%. This normal 'de-oxygenated' venous blood returns to the right side of the heart to be pumped again through the lungs.[17]

When the swimmer sinks below the surface of the water, he cannot renew his vital supply of oxygen. During the stages of breath-holding and laryngospasm, the blood flowing through his lungs extracts much of the oxygen that remains in the alveoli. With less and less oxygen available for capture, the blood leaves his lungs without its normal cargo of oxyhaemoglobin and carries less oxygen round his body. Once the oxygen saturation of arterial blood falls to 55%, so little oxygen reaches the swimmer's brain that he loses consciousness.[18] He has begun to suffocate, even before an appreciable amount of water has entered his lungs.

Once the swimmer lies unconscious in the water, he is totally unable to help himself. Now only minutes are left before his drowning accident becomes a drowning fatality. Nevertheless, if he is rescued at this point, his lungs will have suffered little damage, and he stands a good chance of making a full recovery.

WATER ENTERS THE SWIMMER'S LUNGS

As unconsciousness deepens, the laryngospasm relaxes, exposing the entrance to the larynx.

The swimmer continues his efforts to breathe, and without the protection of laryngospasm, he draws water through his windpipe, deep into his lungs. He also inhales fragments of waterweed and aquatic flora and fauna floating in the water,[19] including diatoms,[20] a group of algae with distinctive silica-impregnated cell walls. If he sinks to the bottom, he may inhale mud[21] or sand.[22]

The sudden inrush of water into the swimmer's lungs triggers a reaction similar to an acute asthma attack.[23] Muscle fibres contract in the walls of his airways. Mucus pours into the narrowed bronchi, blocking some of them completely. The entry of as little as half a cup of water causes abrupt closure of many terminal bronchioles, the last and smallest branches of the bronchial tree,[24] trapping air in the alveoli they supply. Despite the narrowing of the bronchi and widespread closure of terminal bronchioles, water penetrates the depth of his lungs.[25] His lungs still contain a lot of air, but the entry of a modest volume of water is enough to cause life-threatening injury.[26]

WATER IS ABSORBED INTO THE SWIMMER'S BLOODSTREAM

The thinnest of barriers separates the water entering the swimmer's lungs from the blood flowing through his alveolar capillaries. Much of the water he inhales is rapidly absorbed into his pulmonary circulation, increasing his blood volume and lowering the concentration of sodium, chloride and calcium.[27] The more water he inhales, the more enters his circulation, and the more marked the changes in his blood. His blood may become so diluted that some of the red cells swell and burst (a process called haemolysis), releasing haemoglobin and potassium into his bloodstream and turning his urine red (haemoglobinuria).[28] Scattered blood clots may form in small blood vessels (a complication called disseminated intravascular coagulation),[29] depriving his tissues of their vital blood supply and causing extensive bruising or internal bleeding.[30]

The immediate effects of seawater drowning are rather different. Seawater contains three times as much salt as blood plasma. When seawater enters the lungs, some of it is absorbed into the pulmonary circulation, but the strong osmotic pull of the salty water draws a greater volume of fluid out of the bloodstream into the lungs. The blood becomes more salty and more concentrated, and the lungs become increasingly waterlogged. So much fluid may accumulate in seawater-damaged alveoli that bloodstained froth pushes its way into neighbouring, unaffected alveoli, damaging them in their turn.[31]

In a series of influential animal studies on drowning in the 1940s and 1950s, physiologists filled the lungs of laboratory animals with fresh water or seawater and recorded a number of dramatic changes in the volume and chemistry of the circulating blood.[32] However, human victims of fatal drowning are usually dead long before enough water enters their lungs to cause such severe dilution or concentration of their blood.[33]

INHALED WATER INACTIVATES OR DESTROYS SURFACTANT

Surfactant is a detergent-like substance that coats the alveoli. It reduces the pull of surface tension, allowing the alveoli to expand when air enters the lungs, and allowing the alveoli to contract without collapsing when air leaves the lungs.[34] Surfactant appears in the developing lungs of babies while they are still in the womb. Its presence reduces the work of breathing after they are born. Premature babies, born before their lungs have produced enough surfactant, have to struggle for every breath.[35] Without skilled treatment, they develop life-threatening respiratory distress syndrome of the newborn.

Fresh water inactivates surfactant, while seawater washes surfactant away and destroys the cells that produce surfactant and store it.[36] Without surfactant, the full force of surface tension acts on the moist tissues of the swimmer's lungs. The alveoli shrink and their walls stick together, losing much of the surface area available for gas exchange.[37] In places, the alveoli collapse entirely and the open lacy structure of the lungs congeals into airless masses with the consistency of liver.

During the swimmer's desperate attempts to breathe, a mixture of surfactant, water, mucus and air is whipped into a stiff, tenacious froth[38] that fills his airways and bubbles from his nose and mouth in mushrooms of sticky foam.

INHALED WATER DAMAGES THE SWIMMER'S LUNGS

The swimmer's involuntary attempts to breathe continue. Water fills or over-fills some of the alveoli, while many others are distorted and over-distended by trapped de-oxygenated air. The delicate cells lining the alveoli are damaged by lack of oxygen.[39] The alveolar capillaries are either compressed or stretched to breaking point.[40] Blood still flows through the swimmer's lungs, but its red cells cannot capture oxygen from alveoli filled with water or with de-oxygenated air.

PLASMA FLUID LEAKS INTO THE SWIMMER'S LUNGS FROM DAMAGED CAPILLARIES

During the swimmer's struggle to survive, his heart pumps huge volumes of blood through his lungs. At first, his thin-walled alveolar capillaries accommodate the extra blood flow. But inhaled water acts as an irritant. Just as blisters form when the skin is burned, sticky tissue fluid leaks from damaged alveolar capillaries into the alveoli and into the flimsy lung tissues between them.[41] The swimmer's lungs become clogged and stiffened by a double load of liquid – by the water that he inhaled and by widespread oedema fluid leaking from the damaged capillaries in his lungs.[42]

The damage to his lungs is now so severe that even if he is rescued, and begins to breathe air again, he will be unable to capture adequate amounts of oxygen without skilled hospital treatment.

ONCE OXYGEN SATURATION FALLS TO 45%, THE SWIMMER NO LONGER TRIES TO BREATHE

Normally, there is a minute-by-minute balance between the amount of air inhaled into the lungs and the amount of blood flowing through the lungs.[43] Enough oxygen reaches the alveoli to ensure an oxygen saturation of 97%, and enough blood flows through the alveolar capillaries to capture sufficient oxygen for the metabolic activities of the body.

Once the swimmer lies unconscious in the water, this moment-by-moment balance is lost. For a time, he continues his efforts to breathe, but he cannot renew the air – or the oxygen – in his lungs. As a result, blood is 'shunted' through his lungs, entering and leaving the lungs in an increasingly de-oxygenated state.[44] If no rescue comes, his oxygen-starved tissues use up whatever oxygen remains in his bloodstream. Within a minute or two of submersion, his arterial blood has an oxygen saturation of only 45% – far less oxygen than would be found in normal venous blood.[45] The colour of his arterial blood changes from bright crimson to dark maroon. His lips turn blue. Deprived of oxygen, the respiratory centre in his brain shuts down. He no longer tries to breathe.[46]

THE SWIMMER'S BLOOD BECOMES TOO ACIDIC

Carbon dioxide is constantly formed in the tissues. Normally, it is breathed away (exhaled air contains one hundred times more carbon dioxide than inhaled air).[47] By ridding the body of excess carbon dioxide, the blood is kept at its near-neutral pH of 7.4, a level that suits the complex chemical processes of the body.[48]

As soon as the swimmer sinks beneath the surface, carbon dioxide accumulates in his body,[49] mostly as carbonic acid in his bloodstream. In addition, his muscles produced large amounts of lactic acid during his desperate struggle to reach the air. Moreover, his oxygen-starved tissues now depend on anaerobic energy production, and the partial breakdown of glucose releases even more lactic acid.[50] The increasing acidity of his blood (acidosis) disturbs the chemical reactions in every organ, especially his heart, his kidneys and his brain.[51]

THE PULMONARY BLOOD VESSELS CONSTRICT, OVERBURDENING THE SWIMMER'S HEART

Normally, the partial pressure of oxygen in alveolar air is about 107mm Hg.[52] A protective reflex is triggered if the partial pressure of oxygen falls below 70mm Hg in a localised area of the lung. The reflex improves oxygen capture by constricting the pulmonary vessels that supply the affected alveoli, which ensures that blood is rerouted through well-aerated lung tissue nearby.[53]

In a drowning accident, however, the sudden fall of the partial pressure of oxygen is so widespread and extreme that this 'protective' reflex fails to protect. Instead, it causes constriction of the pulmonary blood vessels throughout the lungs.[54] Such extensive constriction obstructs pulmonary blood flow and causes a sudden dramatic rise in pulmonary blood pressure.[55]

Normally, the blood pressure in the pulmonary arteries is only one tenth of the blood pressure in the aorta,[56] allowing the right ventricle to pump blood against very little resistance. The walls of the right ventricle are thinner and less muscular than the walls of the left ventricle,[57] so the sudden dramatic rise in pulmonary blood pressure puts the right ventricle under immense strain.[58] Its muscular contractions are too weak to overcome the raised pressure in the pulmonary circulation. Too little blood is pumped through his lungs, whereas blood arrives as fast as ever from its journey round his body, over-filling the right ventricle and stretching its walls.

The tricuspid valve sits between the right atrium and right ventricle, ensuring the one-way flow of venous blood from the atrium into the ventricle.[59] Normally, the valve opens while the right ventricle is filling and closes when the right ventricle contracts. As the right ventricle balloons, however, the leaflets of the tricuspid valve are pulled apart, so that blood floods backwards into the right atrium with every contraction of the right ventricle.[60] The right atrium then balloons in turn, followed by the vena cava. Soon, the chambers of the swimmer's right heart and the great veins become engorged with dark, de-oxygenated blood.[61] With the onset of right heart failure, the two sides of his heart can no longer pump in tandem, and the balanced action of the heart is lost.

THE SWIMMER'S HEART STOPS BEATING

The beating heart needs a constant supply of oxygen to provide enough energy for its ceaseless contractions.[62] Two coronary arteries branch directly from the aorta. A branch of the right coronary artery supplies oxygenated blood to the sinoatrial node,[63] where specialised cardiac pacemaker cells control the rate and rhythm of the heartbeat. Normally, during strenuous exercise, almost a litre of oxygenated blood flows through the coronary arteries every minute.[64]

After the swimmer sinks below the water surface, increasingly oxygen-depleted blood flows through his coronary arteries. The beating of his heart slows and weakens.[65] From the start of the drowning accident, the cardiac pacemaker cells have been bombarded by adrenaline and other 'fight and flight' hormones. Once the sinoatrial node is deprived of oxygen, the control of the pacemaker cells is lost. Instead of the normal orderly spread of muscle contraction throughout the heart, the ventricles beat more slowly than the atria, and muscle fibres mistime their contractions.[66]

For a short time, the heart continues to beat, but if the heart's coordinated contraction splinters into fibrillation, each muscle fibre beats at its own pace.[67] Once this happens, the heart becomes useless as a pump and no blood is propelled round the swimmer's body. Soon, usually within one or two minutes of inhaling water deep into his lungs, his heart stops beating altogether.[68]

LACK OF OXYGEN CAUSES PROGRESSIVE DAMAGE TO THE BRAIN

Every cell in the body needs oxygen, but the brain is particularly sensitive to oxygen lack. The swimmer loses consciousness early in the drowning process, and unconsciousness deepens with each moment that passes. Lack of oxygen affects the most metabolically active regions of the brain first, and damages them the most.[69] Nerve cells sicken and die on the surface of the cerebral hemispheres (especially in areas dealing with movement and vision)[70] and in the cerebellum, the hippocampus and the basal ganglia (especially in areas involved with memory and the coordination of movement). The metabolism of his brain is also impaired by the increasing acidity of his blood. As the minutes pass, he suffers progressive, irreversible damage to his brain. Sometimes convulsions jerk his body. Soon he ceases to struggle and enters a deepening coma. Brain death will follow before long.

It is important to remember that the swimmer's life may be saved by timely rescue, rapidly followed by skilled resuscitation. He may recover after he has ceased to breathe, even after his heart has stopped beating. After brain death, however, there is no possibility of recovery.

PART THREE

COPING WITH A DROWNING ACCIDENT

In a drowning accident, every minute counts. The sooner the rescue, the better the outcome. A struggling child who is pulled from the water before sinking below the surface will suffer no more than a fright. An unconscious child who is rescued before inhaling water will make a full recovery. Once the child breathes in water instead of air, lung damage is inevitable.

When a drowning victim loses consciousness, lack of oxygen has already affected brain function, at least temporarily. If the victim stops breathing, permanent brain damage will soon follow, unless someone restores the oxygen supply by giving mouth-to-mouth resuscitation. If the victim's heart stops beating, death is moments away unless someone at the scene begins immediate cardiopulmonary resuscitation (heart massage combined with mouth-to-mouth resuscitation). Whenever resuscitation is needed, it should be started at once, while the ambulance is on its way.

All drowning victims who lose consciousness, however briefly, and all who inhale water, however small an amount, must be admitted to hospital for urgent assessment and treatment. The lucky ones will be fit for discharge within hours. Others may need mechanical ventilation and intensive care. Despite the best available treatment, some will suffer permanent brain damage due to oxygen deprivation, and some will die.

An Inquest is held after every drowning death. The Royal Society for the Prevention of Accidents publishes the number of 'Accidental drowning deaths' every year, which includes only those victims given an Inquest verdict of Accidental Death or Death by Natural Causes. Since this excludes almost 40% of drowning deaths in the United Kingdom, the size of the drowning problem is seriously understated.

Too many families have to cope with the tragic aftermath of a drowning accident. If their loved one dies, they may be overwhelmed by feelings of sadness and guilt. If their loved one survives with severe permanent brain damage, they face a prolonged nightmare.

CHAPTER TWENTY-NINE

RESCUE

Successful rescues often depend on the fast reaction by a parent, a friend or an untrained bystander. Unfortunately, rescue may be delayed because the danger is recognised too late.[1] Parents leave young children unattended. Onlookers pay no attention when a girl paddling close beside them steps out of her depth. A group of boys assume that their friend is larking about as usual, not realising that he is fighting to stay at the surface. Although lifeguards watch swimmers for signs of distress, they may be slower to notice someone who slips quietly under the water.

When swimmers first realise that they are in danger, they are usually capable of keeping themselves afloat for a limited time. So witnesses must cheer them on, encourage them to float face up, to fill their lungs with air, and to continue their struggle for survival as long as they possibly can. Once drowning victims lose consciousness or sink below the surface, rescues become even more urgent and much more testing.

Whenever possible, rescuers should avoid getting into the water themselves.[2] Rather than jumping in at once, they are wise to look for other ways to help the victim. Safer methods can be listed briefly – Reach, Throw, Wade, Row.[3] While the rescue is underway, an early 999 call ensures that the emergency services arrive without delay.

REACH

If someone is struggling close to the water's edge, the quickest method of rescue is also the safest – the rescuer simply lies flat on the ground and stretches out to catch hold of the victim's hand, hair, or clothes. If the victim is just out of reach, the rescuer can hold out a belt, a scarf, a branch, an oar or a 'shepherd's crook' for the victim to grab,[4] while taking care to avoid being pulled into the water. In 2001, a seven-year-old boy rescued his little sister from a canal by holding out the bamboo pole of his fishing net and dragging her safely to the bank.[5] Neither child could swim.

THROW

If the victim is too far from the edge to be reached directly, the rescuer can throw a lifeline by fastening one end of a rope to a firm support, tossing the free end to the swimmer, then slowly hauling in the rope. (Several casts may be needed before the victim manages to catch the rope.) Even children can learn to throw a lifeline to someone 30 feet away. A hundred years ago, Boy Scouts practised throwing life-lines at a target representing a drowning man, a camp-fire game called 'Flinging the Squaler'.[6] (A squaler is a short piece of cane weighted with lead, attached to the end of a lifeline.)[7] At life-saving championships in Australia, lifeguards continue the tradition, winning prizes for the speed and accuracy of their line throwing.

A rescuer can help an exhausted swimmer to stay at the surface by throwing a lifebelt. If a tether-line is fastened to the lifebelt, bystanders can help the rescuer to pull the swimmer to safety.[8] However, non-swimmers or weak swimmers may be unable to tread water long enough to grab a lifebelt or a lifeline. Indeed, they may be so focused on their frantic struggle to survive that they fail to notice a lifebelt that lands within easy reach.[9] And even when swimmers manage to catch hold of a lifeline, they soon lose grip strength as the cold water chills their hands. Then the lifeline may slip from their grasp.[10]

Hauling someone to safety from turbulent water is not without risk. In 2012, a young man was swept into the sea while fishing from rocks on the Cornish coast. His father threw him a lifebelt and began to drag him to safety. Then violent swells washed the father into the sea. They both clung to the lifebelt as waves tossed them about under the cliff, but the father died in his son's arms shortly before the lifeboat arrived to rescue them.[11]

Failing a lifebelt, a rescuer can throw a blow-up beach toy into the water, or a football, a picnic cooler, a foam cushion, a spare tyre – anything that will float. Even an air-filled plastic bottle will give the swimmer a little more buoyancy.[12]

WADE

When the water is shallow, a rescuer may be able to wade out to the victim, as long as the current is slight and the surface is firm enough to walk on. A wading stick helps with balance and allows the rescuer to probe the bottom for sudden changes in depth.[13] A wading stick also extends the rescuer's reach, so that a terrified drowning victim grabs the wading stick rather than the rescuer.[14]

ROW

When the victim is further from the water's edge, a rescuer may choose to put on a life jacket and row out to help. It is safer to lift the victim over the transom at the stern, because hauling someone over the side of a rowing boat can cause a capsize. Sometimes a rescuer is unable to hoist a heavy victim into the boat.[15] Nevertheless, the rescuer should never leave the boat to assist someone struggling in the water since the boat is likely to drift away from both rescuer and victim.[16] In 2012, a ten-year-old boy rowed out to rescue two people who were thrown into the sea when their sailing boat capsized half a mile out from the Devon coast. He was not strong enough to lift them from the sea, so he tied a rope to their boat and they held on as he towed them back to safety, chilled but otherwise unhurt.[17]

In February 2019, a quick-thinking kayaker saved the life of a tourist who fell 18 feet into the River Thames after climbing a lamppost on Westminster Bridge to take a photograph. The tourist hit the water and went under, but the rescuer paddled across, grabbed him as soon as he came back to the surface, and hauled him into a lifebelt thrown by an onlooker.[18] The Metropolitan Police Marine Unit and the RNLI arrived minutes later and lifted the tourist from the river. Then he was taken to hospital for treatment.

ENTERING THE WATER TO RESCUE SOMEONE IN DANGER OF DROWNING

Whatever their swimming ability, would-be rescuers should take a few moments to consider their own safety before they enter the water. However, rather than watch someone drown, a parent, a friend, or a bystander may feel impelled to go to their aid. In Oxford in 2001, a young chef in a riverside restaurant saw a woman fall out of a rowing boat. He dived into the river from the restaurant window and brought her safely to the bank.[19] (Neither the woman in the water nor her husband in the rowing boat could swim.)

In 2009, a pensioner spotted the body of a young man floating face down in a Surrey canal. A trio of onlookers were already phoning the emergency services, but the pensioner jumped in at once.[20] 'It was a no-brainer really,' he explained afterwards, 'I could see someone in the water and they were going to be dead in a few minutes so I just got on with it.' He lifted the unconscious victim's head out of the water, whereupon the young man took a huge gasp of air and began to breathe again. Then the elderly rescuer kept him at the surface until the Fire Brigade arrived. After a brief hospital stay, the young man returned home unharmed. A police spokesman applauded the pensioner's heroic action: 'There could have been a very different ending to this incident. It is to his credit that he took matters into his own hands and saved the victim and we cannot praise his efforts highly enough.'

In 2010, a courageous young man managed to rescue a toddler from a Leicester canal although he himself was unable to swim. 'I wasn't expecting the water to be so deep,' he said. 'It went over my head. I managed to bounce along the bottom until I reached the boy and held him over my head, holding my breath under the water until I reached the side.'[21] He then gave the child mouth-to-mouth resuscitation, following the telephoned instructions of the

999 operator. Four minutes went by before the toddler began to breathe again, but he was pronounced 'fit and well' when he left hospital two days later.

Jumping into a swimming pool is usually safe enough. There are a number of reports of holidaymakers taking immediate action and saving a child's life. In 2000, near Tarragona in eastern Spain, two Scottish teenagers saw a seven-year-old boy floating in a hotel swimming pool, submerged and unconscious. While one youngster jumped in to rescue the boy, the other rushed off to find a friend with paramedic training, who revived the boy with mouth-to-mouth resuscitation. [22] A year later in the Canary Islands, a man on honeymoon dived into the deep end of a swimming pool and saved the life of a three-year-old boy who fell in and 'started sinking as soon as he hit the water.'[23]

In 2010 in Yorkshire, an anxious father vaulted down from the spectator gallery overlooking his local swimming pool and leapt into the water fully clothed to save his four-year-old son who was floundering and sinking in the deep end during a swimming lesson.[24] The instructor had thrown a float into the water and was holding out a rescue pole, but the boy did not have the strength to reach them. The instructor and two lifeguards were still watching the boy when his father grabbed him, lifted him to the surface, and succeeded in rescuing him before he took water into his lungs.

In January 2013, a gust of wind blew a baby into a Devon harbour, still strapped into his pushchair. The pushchair drifted away on the outgoing tide, with the baby lying face down in the water. The Dock Master jumped into the sea, swam to the pushchair, turned it right side up, and towed it back to the harbour wall. Someone let down a rope and hauled the pushchair up to the quayside, with the unconscious baby still strapped inside.[25] An off-duty nurse ran across and gave the baby cardiopulmonary resuscitation on the spot. An air ambulance took him to hospital, where he made a full recovery.

In December 2016, a four-year-old boy fell into the Thames near London Bridge. Alerted by the mother's screams, an office worker dived in and managed to grab the boy before he was swept away.[26] 'I just acted on instinct without thinking,' he said. 'I do a lot of triathlons and open swimming so I thought I could help.' Onlookers threw an emergency ring to the rescuer and dragged him to safety, clutching the fortunate boy.

◆ ◆ ◆

A rescuer may swim out to help, only to arrive after the drowning victim has disappeared beneath the surface. Few drowning victims stay at the surface for long. Once they begin to sink, they usually drift down to the bottom within a few seconds.[27] Young children sink even more quickly than adults, often head first. Lying prone and unmoving, the body of an unconscious victim is often surprisingly difficult to see, especially when hidden among fronds of waterweed.

Once the victim has been located, the physical demands on the rescuer are extreme. The 'dead weight' of an unconscious body must be lifted to the surface, slowed by the resistance of the water. If the water is more than a few feet deep, it may be impossible for an untrained rescuer to hold his breath long enough to dive down and recover the victim.

On a hot Saturday in May 2010, five teenagers decided to cool off in a Thames-side dock at Rotherhithe where the water was more than 20 feet deep. Only four of them could swim.[28] Despite a warning from the other boys not to jump in, the non-swimmer leapt into the water, struggled briefly, then began to sink. 'I dived in and tried to grab him by the side. I was trying to pull him up but he was too heavy. He was struggling and pulling me down,' his friend told the Inquest. 'He was slipping away. I tried grabbing his hair, but I could not hold on. I had to go up to get some air. When I dived back I could not find him.' A policeman arrived quickly and made several dives, but without success. He described the water as shockingly cold and very murky: 'Just a few feet down I could not see the end of my arm in the water.' Police divers were needed to recover the victim's body.

SELF-RESCUE: MINIMISING THE EFFECTS OF 'COLD SHOCK' AND HYPOTHERMIA IN OPEN WATER AND THE SEA

Most people assume that if they trip, slip, or fall into cold water fully clothed, their sodden clothes will pull them down. In fact, when people first enter the water, the air trapped in their clothes makes them surprisingly buoyant for a short but critical time.[29] Researchers at the University of Portsmouth found that bulky winter clothes increase buoyancy even more than thin summer cottons. Modern trainers and hiking boots, with their foam padding and air pockets, also act as buoyancy aids.[30]

Sudden entry into cold water triggers 'cold shock', with irregular heartbeats, marked increase in blood pressure,[31] and a powerful gasp reflex followed by several minutes of uncontrollable panting.[32] In 2015, while rowing with his family on the Loch of Aboyne near Aberdeen, a young man lost his grip on the oars and they floated away. He jumped into the water to retrieve them, suffered immediate 'cold shock' and could not swim back to the boat. Fearing that he might drown, his mother jumped in to save him, only to become helpless in her turn.[33] Two campers heard the family's screams. They swam out and pulled both victims of 'cold shock' 20 yards to the shore. The young man had lost consciousness, but he recovered in hospital.

Even skilled swimmers suffer from 'cold shock' if their experience of swimming is limited to the warmed water of swimming pools. The Portsmouth researchers suggest that both adults and children could increase their chances of survival after falling into cold open water if they 'Float First' for two or three minutes before they begin to swim. By giving themselves time to get their breathing under control, they limit the danger of water inhalation during the early phase of 'cold shock'.[34] And nearby rescuers are given more time to reach out and pull them from the water.

For a short swim to the margin of a pond or the bank of a river, swimmers usually use head-up crawl or breaststroke. But whatever their level of skill, moving through the water is harder for those who have fallen in fully clothed. Crawl is particularly taxing because every arm stroke involves lifting the extra weight of water held in the sleeves of their clothes. Breaststroke is a better choice, but if they tire, they will be able to breathe even more freely if they switch to survival backstroke[35] (sculling by the arms and whip-kicks by the legs, alternating with relaxed glides), which lifts the mouth well clear of the water. When caught by the current in a river, it is safer to float on the back, feet first, steering for slack water on the inside of a bend, where the swimmer has an increased chance of running aground and climbing onto the bank.[36]

During the Second World War, many ships were torpedoed in the warm water of the South Pacific. Survivors who struck out for a distant shore often drowned on the way, due to the effects of injury, shark attack, strong currents, cramp and fatigue. Fred Lanoue introduced the technique of 'drownproofing', which emphasised the importance of economy of effort to preserve body heat, conserve energy and avoid exhaustion.[37] Troops were trained to travel slowly through the water using a relaxed 'kick-and-glide' stroke. When they tired, they floated vertically, with their lungs fully inflated to increase buoyancy. Between each breath, they allowed themselves to sink slightly, so that their heads were supported by the water rather than their own efforts. 'Drownproofing' remains part of the basic training of the American maritime special forces, the Navy Seals.[38] However, the technique is inappropriate in colder seas. The rich network of blood vessels in the scalp does not constrict when cooled, so survivors rapidly lose heat as their heads dip underwater.[39]

Survivors of boating accidents were once advised to strip off their outer clothes to make swimming easier. American servicemen were trained to remove their uniforms and transform them into buoyancy aids by blowing air inside the sleeves and trouser legs.[40] Yet swimmers are well advised to keep on their clothes and headgear. Thick protective clothing and padded life jackets may be lifesaving in the colder waters of the North Atlantic or the North Sea.[41] Even thin fabric provides insulation by retaining a layer of warm water next to the skin (water that has been warmed by absorbing body heat from the survivor).[42]

For yachtsmen and boaters, the most important rescue strategy is the decision to put on their life jacket before they leave the jetty.[43] If they fall into the sea, their skin becomes chilled and numb within a few minutes. As the water continues to extract heat from their bodies, they lose power in the muscles of their hands, arms and legs. Survivors of a capsize are advised to stay with the boat, and to slow the rate of heat loss by climbing out of the water onto the upturned hull.[44] But the longer they wait, the colder they grow.

They have to decide whether to wait for rescue or swim for the shore. If they decide to wait, they will develop hypothermia over the course of half an hour, making them increasingly dependent on a rescue that may never come. If they decide to swim for it, they risk the onset of swimming failure before they reach safety.[45] Canadians experts now advise that, even in water as cold as at 10°C, self-rescue swimming may be an appropriate choice for a fit survivor (with or without a life jacket) if the likelihood of rescue is low and the shore is less than half a mile away.[46]

In 1984, after his boat capsized and sank 3 miles from land, a 22-year-old Icelandic fisherman swam for 6 hours in near-freezing water (5°C) before he reached the shore.[47] In hospital in Iceland, his deep-body temperature was below 34°C, but he had 'almost no symptoms of hypothermia or vasodilatation, only of dehydration'. He agreed to submit to investigation in the water tank of the London Hospital's hypothermia laboratory. Professor Keatinge confirmed his subject's remarkable ability to maintain his body temperature in frigid water, and suggested that the thickness of his subcutaneous fat provided enough insulation to prevent 'peripheral heat loss and progressive body cooling'.[48]

Yet this brief experiment did not mention the strength, courage and determination that helped the Icelander to survive when his four shipmates drowned.[49] He swam alone towards the lights of a distant village. He kept his mind focussed by talking to the seabirds flying around him. When he reached the shore, his way was barred by a cliff, so he re-entered the sea and swam on until he found a cove where he managed to struggle ashore. Then he walked barefoot for two miles across a frozen lava field, his feet cut and bleeding from shards of volcanic glass. Desperately thirsty, he smashed an inch-thick layer of ice covering a water butt with his fist. It seems that he never gave himself up for lost.

Determination also played a part in the survival of a British woman who fell from a cruise ship in the Adriatic in August 2018.[50] The alarm was raised, and the ship doubled back to search for her, but failed to find her in the dark. In mid-summer, the temperature at the surface of the Adriatic was as warm as a swimming pool (25°C or more), and the sea was very calm.[51] As she floated alone in the sea at midnight, 60 miles off the coast of Croatia, she coped with her predicament by singing. The search resumed at first light, and helicopter crew spotted her swimming only a mile from where she fell from the ship. A Croatian navy patrol ship reached her 10 hours after she fell into the sea. A sailor dived into the water and hauled her to safety, exhausted and chilled, but 'very glad to be alive'.

SUPERVISION AND RESCUE BY LIFEGUARDS IN SWIMMING POOLS AND BY THE SEA

Lifeguards on duty in public swimming pools are trained to watch for the earliest signs that a rescue might be needed. They notice toddlers about to wander out of their depth and guide them back to shallower water. They pay special attention to children who keep their faces tipped back while they hop about in chest-deep water.[52] (Timid beginners are more likely to need rescue than confident swimmers who are happy to get their faces wet.) They look out for youngsters who are clinging onto the water gulley in the deep end, pulling themselves along hand over hand. They follow the progress of anyone whose swimming is limited to dog-paddle.

Lifeguards jump into the swimming pool to catch hold of struggling children before they go under. They react immediately if beginners show signs of quiet panic (lying low in the water with head tilted back, eyes wide with alarm or tightly closed, legs dangling, arms pummelling the water without forward movement).[53] They rescue frightened swimmers who are waving or calling for help. They check on lane swimmers who halt mid-length and float motionless for more than ten seconds.

Swimming pools are safer places when there is firm supervision from lifeguards. They limit risky behaviour of youngsters who push others into the pool, jump down onto swimming friends, or play-fight in the water. They prevent diving into the shallow end of the pool (which risks a broken neck) and put a stop to over-breathing before underwater swimming (which risks shallow water blackout). Modern leisure centres and water parks need extra lifeguards to patrol the added attractions of slides, fountains, flumes, waves and water jets. During in-service sessions to refresh and improve their rescue skills, lifeguards practise on manikins and on each other.[54] Most lifeguard rescues are completed before the process of drowning has even begun, with the happy result that there are very few deaths from drowning in public swimming pools in Britain.

◆ ◆ ◆

Lifeguards cannot watch every swimmer all the time. Instead, they scan the pool to identify potential trouble before anyone comes to grief, sweeping their gaze across the surface of the water, moving their heads to look directly at each area in turn.[55] While their sharp central vision takes in most detail, scanning also includes awareness of movements at the corner of the eye.

Lifeguards must pay full attention while they are on duty. They cannot do their job properly if they are assigned other tasks which interrupt their scanning duties or if they lose concentration while they chat to a colleague.[56] To stay vigilant, they need frequent scheduled breaks, handing over to a colleague every half hour or so. To stay ready for action, they need frequent changes in position, walking round the pool or moving from their elevated lifeguard's chair down to the poolside and back again.

Scanning the surface of the water is not enough. Lifeguards must scan the water under the surface as well. Yet their vision is tricked by the constant image distortions and dancing reflections from the water surface. To add to their difficulty, water is translucent rather than transparent, scattering what little light is reflected from the bottom of the swimming pool. As a result, it is surprisingly difficult to detect someone lying at the bottom of a swimming pool.[57]

In 2007, Canadian researchers highlighted the problem that lifeguards face when scanning below the water surface. They used a plastic manikin for their tests and assumed that lifeguards who could see the lane markings on the bottom of the pool would be able to see a submerged swimmer. 'But the reality was very different,' they reported. Lifeguards may see no more than a distorted, indistinct shadow underwater,[58] scarcely enough to alert them that immediate rescue is needed: 'The lifeguard must be very close to the victim – less than 10 metres in calm water and as close as 2 metres if there is water turbulence.' They were surprised to find that scanning from a raised lifeguard's chair increased the difficulty. Instead, a system of walking lifeguard patrols increased the lifeguard's chance of noticing a submerged victim.

Other pool users often have a clearer view of a submerged swimmer and may be the first to raise the alarm and attempt a rescue.[59] A lifeguard is wise to respond at once to any warning that someone is lying on the bottom. If a necessary rescue is delayed and the victim dies of drowning, the lifeguard will be summoned to appear in court. In Scarborough in 2000, a teenager with epilepsy dived into deep water at a seaside water park, but remained submerged. An inexperienced lifeguard ignored several warnings that someone was lying at the bottom of the pool.[60] (His view was obscured by streams of bubbles piped into the water under the diving board.) Five minutes went by before a schoolgirl dived down and pulled the unconscious teenager to the surface. A second lifeguard lifted him from the water and began resuscitation at once, but the boy died in hospital the next day. At the Inquest, the Coroner brought in a verdict of Accidental Death, but he criticised the lifeguard for ignoring the children's warnings. Three years later, after prosecution by the Health and Safety Executive for failing to safeguard the swimmer, both the lifeguard and Scarborough Council were found guilty, and fined.[61]

In a crowded swimming pool in Wales in August 2005, a ten-year-old girl jumped in at the deep end, swam a few strokes, then sank without a struggle and lay motionless on the bottom, 12 feet down. With so many swimmers splashing in the water, the lifeguards did not see her sink, but they were alerted within seconds by a warning signal from an underwater camera,[62] fitted only two years before because of the great depth of the pool. Her life was save by the fast response of the nearest lifeguard, who dived in and brought her to the surface, unconscious. She responded to mouth-to-mouth resuscitation and needed only a brief stay in hospital.

◆ ◆ ◆

Along the British coastline, thanks largely to the Royal National Lifeboat Institution (RNLI), lifeguards protect holidaymakers on popular beaches. Lifeguards swim well, and they are trained to cope with dangerous water conditions.[63] They often use rescue boards and jet skis to reach drowning victims quickly, bring them to land successfully, and ensure their own safety throughout the rescue.[64] They are ready to spring into action. Drowning happens fast, so rescues must happen faster. In 2011, RNLI beach lifeguards saved 84 lives.[65]

Lifeguards take special care when rescuing a swimmer who dived into shallow water and struck his head on the seabed,[66] or a surfer who was 'dumped' head first by a plunging wave.[67] After such an forceful impact, there is a risk that the casualty may have broken his neck. To avoid worsening the injury, lifeguards restrict the movement of his head and neck as they lift him from the water. Cervical collars, backboards and head-blocks ensure that the casualty remains immobilised until a full assessment is possible.[68] These precautions prevent his head from flopping backwards, which might crush his spinal cord against a spur of bone, condemning him to spend the rest of his life as a paralysed invalid.

Whether they patrol a flatwater beach or pounding surf, lifeguards are aware that every rescue of 'an active drowning victim' may turn into a high-risk encounter. A trained lifeguard swims to a rescue carrying a buoyant rescue float attached by a cord to a shoulder strap. Nearing the victim, the lifeguard pushes the rescue float forwards and down, underneath the victim's hands. By using the rescue float, the lifeguard avoids the clutches of the struggling victim, who grabs the float instead of the rescuer.[69] Two types of float, both developed by Californian lifeguards, are now used worldwide for rescues in swimming pools and on beaches.[70] The Peterson rescue tube is a flexible foam-rubber band, 4 feet long and 6 inches wide, which can be clipped round the body of an exhausted or unconscious victim. The Burnside rescue buoy is a rigid, hollow plastic moulding, 28 inches long and 9 inches wide, with handholds along the sides for one or more victims to grab. An unconscious adult is an awkward burden, but lifeguards become adept at using the added buoyancy of their rescue float to lift submerged drowning victims to the surface and keep them at the surface as they are towed to safety.[71]

Surf lifesaving began in Australia more than a hundred years ago.[72] In 1956, after the Melbourne Olympic Games, the Australians organised an international surf lifesaving competition, and Californian surfers introduced their Australian hosts to rescue floats and surfboards.[73] Australian surf lifesavers rescue more than 13,000 swimmers from rip currents every year.[74] Tourists are urged to 'swim between the flags', where lifeguards watch over them and use surf boards or jet skis to reach them quickly if rescue is needed.[75] Despite efforts to keep people safe on Australian beaches, swimmers drown in rip currents every year, the great majority more than a kilometre from beaches patrolled and guarded by lifeguards.[76]

Lifeguards rescued almost 95,000 people from American beaches in 2015, the majority from rip currents. Although there were 108 fatal seaside drowning accidents, only 13 occurred on beaches where lifeguards were on duty.[77] In 1989, county officials in Florida decided to limit taxpayers' costs by dispensing with lifeguards at American Beach, near Jacksonville, Florida. Then, in June 1994, five swimmers drowned on a single day.[78] The lifeguards were reinstated at once.[79]

Lifeguards protect tourists on some of the idyllic holiday beaches in South-East Asia, but most beaches are unguarded. Swimmers have no one to warn them of dangerous currents if they enter the water in the early morning or at sunset, and they have no one to rescue them if the currents pull them away from the shore.[80]

◆ ◆ ◆

Lifeboats manned by the volunteer crews of the Royal National Lifeboat Institution (RNLI) are launched hundreds of times every year, and RNLI lifeboatmen rescue more than twenty people from the sea every day.[81] They save families cut off by the rising tide, holidaymakers floating out to sea on airbeds, sailors clinging to capsized yachts, fishermen in peril on foundering trawlers and workers on offshore oil rigs in the North Sea.[82]

Helicopters from the Navy, the Maritime and Coastguard Agency and the Royal Air Force search for survivors of marine emergencies and inshore boating and swimming accidents, and winch casualties from the sea.[83] Whenever possible, after lengthy immersion, chilled survivors are lifted horizontally to avoid sudden collapse due to a life-threatening fall in blood pressure.[84] In the helicopter, on-board paramedics provide skilled resuscitation during the flight to hospital.

RESCUE BY FIRE CREWS TRAINED IN SWIFT WATER AND ICE RESCUE

Specialist knowledge and training is needed for some rescues because of the extreme danger of the water. When a child is caught in the vicious recirculating current beneath a weir or a kayaker is trapped in a white-water torrent, anguished witnesses on the riverbank have no alternative but to wait for the arrival of fire crews trained in swift water rescue.[85]

Fire crews should also be summoned if someone falls through the ice.[86] A few victims manage to rescue themselves by stretching their arms across the surface of the ice, kicking their legs so that they lie horizontal in the water, then clawing their way onto the ice and rolling to safety. Unfortunately, once the ice has broken, it is likely to break again under the added weight of their water-soaked clothes, so that they plunge into the freezing water a second time.

With so little time to act before a child chills and sinks, people who witness the accident might attempt a rescue by holding out a pole, by throwing a rope from the bank (a pre-tied loop at the end of the rope will help the victim hold on with fingers numbed by the freezing water).[87] A daring rescuer may attempt to grab the child's arm after inching a short way along a ladder extended over the pond, stabilised by the weight of people sitting on the landward end, but by the time that a ladder has been found and positioned, and enough neighbours have arrived to provide an adequate counterweight, the rescue may well be too late.

The emergency services are clear in their advice. No matter how desperate would-be rescuers may feel, they should not walk out onto ice which has already given way. If a child has fallen through the ice, it will certainly not support the weight of an adult.[88] In the harsh freeze of January 2009, the RoSPA issued a warning:[89]

'We are urging people not to walk out on the ice because you never know whether it will hold your weight and by the time you find out that it will not, it is normally too late. We are encouraging parents to talk to their children about the hazards of ice-covered lakes, rivers, canals and ponds and the reasons to stay off the ice.

We also advise dog owners to keep their pets on a lead when they are near frozen water and not to throw sticks or balls on to the ice for them to retrieve.'

Adults and children put themselves at risk if they venture onto the ice to rescue a dog. The Fire Brigade accept emergency calls to rescue dogs from frozen ponds. The episode is used as a training exercise for rescuing people. In December 2009, after a team of firemen rescued a dog that had fallen through the thin covering of ice on a frozen pond, the Fire Brigade's station commander commented, 'The people in this case did exactly the right thing by phoning us, and not attempting to rescue it themselves. Too many people have drowned trying to rescue their dogs.'[90] Indeed, more than half of the people who die after falling through the ice are drowned while attempting to rescue another person or a dog from the freezing water.[91]

UNTRAINED RESCUERS OFTEN PUT THEMSELVES IN DANGER

Some experts recently suggested that the terms 'active drowning' and 'passive drowning' should be abandoned, dismissing them as confusing and imprecise.[92] Actually, the distinction is very important – for the safety of the rescuer. In an 'active drowning' incident, the victim is conscious and struggling to stay at the surface, whereas in a 'passive drowning' incident, the victim has already lost consciousness and is therefore unaware of the rescuer's approach.

For someone untrained in rescue techniques, swimming to a terrified drowning victim is a very perilous undertaking.[93] Rescuers often fail to pace themselves, swimming out as fast as possible, then arriving too tired and breathless to cope with the rescue or the strenuous task of towing the victim back to safety.[94] Indeed, they may so overtax their strength that they become casualties themselves. Not only do they frequently overestimate their own swimming ability, but too often they forget that the apparently helpless victim struggling in the water is capable of overpowering and drowning them.[95]

Drowning people do, indeed, 'clutch at straws': waterweed or broken twigs are often found grasped in the victim's hand after a fatal drowning. Panic releases adrenaline, and adrenaline gives the victim enormous strength. Crazed with fear, he will try to throw his arms in a stranglehold round the rescuer's neck, or climb on top of the rescuer's body to reach the air. Even small children can surprise unprepared rescuers and pull them under the water.[96] David Smith, a retired Commander in the US Coast Guard warned: 'Drowning humans cannot see, hear or reason. They have only one driving need and that is for air. They will kill you to get it.'[97]

People attending lifesaving classes are shown methods of breaking wrist-grips and head-holds. These releases work well enough when practised on well-mannered teammates during training sessions in a swimming pool. The familiar releases are less than fail-safe when needed to fend off the frantic embrace of a terrified individual who believes that he is about to drown. Manuals advise rescuers to approach a drowning swimmer from behind, but fail to warn that a desperate victim who retains any swimming ability can easily twist round in the water. Rescuers have to be ready to move backwards, out of the way, if victims lunge towards them.[98]

Even trained lifeguards cannot always avoid a swimmer's clutches. A survey of almost 500 Canadian and American lifeguards (all trained in the use of rescue buoys to avoid direct contact with drowning swimmers during rescue) reported that more than a third had been grasped around the head, neck, arms, wrists, or body by panicking swimmers.[99] The lifeguards responded by shoving the victims away with a hand, knee or foot, by ducking them underwater, by breaking their grip, or, in two instances, by striking them (although neither blow had much effect). Then they successfully completed the rescues.

Lacking the protection of a rescue buoy, an untrained rescuer can keep a drowning victim at a safe distance by swimming out with buoyancy aid, a towel, a T-shirt, even a length of stick. The victim then grabs the object instead of the rescuer's arm or neck. A terrified drowning victim usually regains a degree of self-control once supported by a rescue buoy or another buoyant object. Brief words of reassurance from the untrained rescuer usually have a calming effect:[100] 'You'll be fine,' or 'I'm going to help you,' followed by 'Rest for a moment and recover your breath.'

The towing techniques described in textbooks and taught in lifesaving classes rely on a degree of buoyancy and self-support by the victim. An untrained rescuer may have taken part in lifesaving classes and may have practised towing cooperative classmates across a swimming pool. But the effort is far greater when towing a real-life drowning victim through water, which is always cold and often in motion.[101] Whatever stroke is used, the rescuer is soon exhausted. Sometimes a grateful victim is able to assist their rescuer by contributing an occasional leg kick during the tow. However, a conscious frightened victim is capable of attacking the rescuer at any moment, while an unconscious victim or a non-swimmer must be held at the surface throughout the tow, legs trailing downwards, slowing progress due to drag.

◆ ◆ ◆

When untrained rescuers go to the aid of someone drowning in open water or the sea, they put themselves at risk from the effects of 'cold shock' and 'swimming failure'.[102] Rather than jumping into the water, a slower, gradual entry is safer, and almost as fast.[103] Surf lifeguards become cold-hardened, but when holidaymakers attempt a rescue, even in calm waters off a sandy beach, their muscles rapidly become chilled and stiffened, hampering their efforts to bring the victim to safety.

Rescuers often forget the enormous power of moving water.[104] In a river, the current swirls with the same strength against the rescuer and the victim he is trying to reach. In the sea, waves may overwhelm the rescuer as quickly as they overwhelmed the swimmer needing rescue. Rip currents carry both victim and rescuer away from the beach. In 2012, a woman saw her two sons caught in a rip current while bodyboarding on a Cornish beach. She swam out to try and save them, but she too was caught in the rip current. Two people on the beach rushed into the sea to help. Together with lifeguards and the Bude inshore lifeboat, they succeeded in rescuing the boys, but they were unable to reach their mother, who drowned.[105] After battling with waves that blocked his view of the beach, one of the rescuers, a confident swimmer, 6ft 4inches tall, described his ordeal as 'a near-death experience'.[106]

In 2000, 15 people drowned in Britain while attempting to rescue someone else from drowning,[107] and lives are lost at home and abroad in similar tragedies every year. In Majorca in 2012, on their first holiday abroad, a young father saw his seven-year-old son washed off a rocky outcrop into the sea. He tried to save the boy by jumping into the waves – even though he could not swim. Both father and son were drowned.[108] In Tenerife in 2014, two British doctors were drowned as they tried to rescue their children who had been swept into the sea.[109]

A study in Turkey highlighted the danger of drowning for rescuers with inadequate swimming and rescue skills. Within four years (2005 to 2008), 114 rescuers drowned together with 60 of the drowning victims they were trying to save.[110] Two-thirds of these incidents took place in open water away from the sea. Most of the rescuers who drowned were teenage boys.

In 2010, researchers in Australia reported that 98 people had died over 17 years (1992–2009) while attempting to rescue a drowning victim.[111] A related study looked more closely at 15 incidents in which 17 rescuers drowned in an 'altruistic and instinctive impulse to save a drowning child's life'. Apart from one bystander, the rescuers who drowned were all parents or close relatives of the children they were trying to save. All but one of the children survived. The researchers wrote about an 'aquatic victim-instead-of-rescuer syndrome'.[112] They concluded that many deaths from drowning could be avoided if parents supervised their children more closely, paid more attention to drowning risks, learned how to throw a lifeline, and prepared themselves to cope with accidents in the water by mastering 'the simple and basic lifesaving skills of non-contact rescue'.

In New Zealand in 1980–2012, 81 people drowned while attempting to rescue others.[113] In a water safety survey at the Pasifika Festival in Auckland in 2013, people were asked how far they could swim and how they would respond to seeing someone in trouble in the water. Almost two-thirds of the respondents could swim less than a hundred metres, yet almost half said that they would jump in and save the victim. These results suggested that 'few respondents would meet the demands of supporting and transporting a struggling victim to safety in the water without putting themselves at considerable risk of drowning'.

◆ ◆ ◆

It is understandable that rescuers willingly endanger their own lives to save the lives of drowning children. Dog owners, however, are well advised to resist the urge to throw themselves into the water to save their pets. Most dogs are strong swimmers. They use the same muscles for running and for 'dog-paddle'. They usually cope with turbulent water rather better than their owners.[114]

Dogs are best kept under close control when walking along riverbanks or seaside promenades. If a dog plunges into the river or chases a ball into the sea, it looks so vulnerable – with its wet fur plastered to its skull and its nose just visible above the water surface – that its owner may feel duty-bound to attempt a rescue, forgetting the risks from the current, the waves and the chill of the water. Inevitably, some dog owners need to be rescued themselves. But rescue is not always possible.[115]

Several drowning accidents in Blackpool have involved holidaymakers trying to rescue their dogs from the sea. In the Jubilee Gardens on the North Shore a memorial plaque and a recent sculpture commemorate a fatal accident in January 1983 when four people died. A young man was engulfed by a wave at high tide as he waded in to save his dog. Four police officers went to his aid. They roped themselves together as they entered the sea, but the waves lifted them off their feet and smashed them against the sea wall. The dog owner and three of the police officers were drowned.[116]

During spring tides, waves cannon against the Promenade at high tide. Anyone entering the sea is dragged violently back and forth by crashing rebounding waves. In May 2005, two teenage friends climbed over the safety chains that barred access to the beach as they attempted to rescue their dog from the sea, but they both drowned.[117] Nearby, in July 2007, a man entered the sea in a vain attempt to rescue his two dogs. As he struggled in the water, someone threw him a lifebelt, but it fell short.[118] A witness said, 'There were a lot of people looking on but nobody could have gone in to get him. The sea is too dangerous there.' Two lifeboats were launched. He was soon found, floating face down and unconscious. The crew of the inshore lifeboat began resuscitation at once, helped by paramedics as soon as they got him ashore, but he could not be revived.

In 2009, a young couple were so anxious to rescue their dogs from a rain-swollen river that they abandoned their baby girl in her pushchair on the riverbank. Both parents were swept away and drowned, leaving their orphaned daughter to grow up without them.[119]

PROBLEMS WITH SAFETY GUIDELINES FOR FIREMEN AND POLICE OFFICERS

Firemen and police officers save many people from drowning. But the main purpose of the Fire Brigade is to fight fires, and the main purpose of the Police Force is to fight crime. Time given for training in water rescue is limited. Firemen and police officers have drowned during rescue attempts after entering the water weighed down by bulky uniforms, carrying inadequate equipment.

In Lancashire in 1999, a fifteen-year-old boy disappeared after plunging into a lake from a rope swing. His friends raised the alarm, and in an urgent rescue attempt, three firemen tied tether lines round their waists, waded into the lake, and swam towards an area where someone had reported seeing bubbles in the water. A tether line snagged on a submerged branch, pulling one of the firemen underwater.[120] He had no quick-release harness, and although his colleagues repeatedly dragged him to the surface, they could not free him. He developed hypothermia and died in the water. The teenager's body was recovered some time later. The fireman's death prompted the Health and Safety Executive to prosecute the Greater Manchester Fire Authority for a 'breach of duty of care' under Section 2 of the Health and Safety at Work Act. In 2004, the Crown Court trial ended in acquittal.[121]

For several years before the accident, there had been a number of inconclusive discussions about training for water rescues, and the need for life jackets, buoyancy aids and safety harnesses to protect firemen who entered the water.[122] After the accident, training and equipment were improved, and rescue procedures were defined and announced.

In 2007, a fireman put on a life jacket and safety harness before he plunged into the River Tay to rescue a 20-year-old woman from certain death.[123] In the strong current, his safety line snapped. Nevertheless, he managed to reach the woman and drag her across to steps in the riverbank. Both the victim and her rescuer survived, although they needed hospital treatment for hypothermia.

The fireman's action breached Tayside Fire and Rescue policy, and he faced the possibility of disciplinary action because he had broken standing instructions which forbade entry into the water, and restricted him to the use of poles and ropes. He protested, 'I was expected to watch that young girl die in front of me. As a father and a caring human being, I couldn't live with myself if I'd had to do that.'[124] Two months later, he was told that no disciplinary action would be taken. He said, 'We're very relieved that common sense has won the day but the regulations haven't changed. I'd still face disciplinary action because I'd still have to do the same thing tomorrow if I saw a drowning person.' The Royal Humane Society honoured him with an award for bravery.[125]

◆ ◆ ◆

Until 2002, all police recruits were trained in swimming and lifesaving. Then, in July 2002, two teenage boys died from drowning in the swimming pool at the Metropolitan Police Training College at Hendon, North London.[126] The boys were in a group of 35 schoolchildren supervised by 6 youth workers as part of a summer activity scheme organised by Barnet Council. Although the regulations required the presence of two lifeguards, only one police officer was on lifeguard duty at the time. He was attending to a boy with a gashed knee when the two teenagers jumped together from the top of the diving stage and failed to surface. The police officer plunged into the pool at once, lifted them from the water and began resuscitation.[127] They were rushed to hospital. One boy was pronounced dead within a few hours. The other boy died in an Intensive Care Unit six days later.

The police officer was charged on two counts of manslaughter and with contravening the 1974 Health and Safety at Work Act. In 2006, after an eight-week trial at the Old Bailey, he was cleared of manslaughter. Soon afterwards, he was cleared of the health and safety charges.[128] The Health and Safety Executive also prosecuted the Office of the Commissioner of the Metropolitan Police and Barnet Council. Both were heavily fined.[129] After the accident, the swimming pool at Hendon Police College was closed. It was later filled in.

In 2007, a seven-year-old boy died of drowning after he was pulled from a Lancashire pond.[130] Newspapers carried the poignant and troubling story. The boy's younger sister was hunting for tadpoles when she fell into the water. Then the boy jumped into the pond to save her from drowning. Although her weight pushed him underwater, he held her up, and two elderly fishermen lifted her to safety.[131] At the Inquest it emerged that two Police Community Support Officers (PCSOs) had arrived at the scene only minutes after the boy sank below the surface, but they did not enter the water.[132] When a Police Sergeant arrived soon after, he stripped off his body armour, got into the water, and lifted the boy onto the bank, unconscious but still alive. He began resuscitation at once.[133] The boy was airlifted to Preston Royal Hospital and transferred to Royal Manchester Children's Hospital, where he died a day later.[134]

The Detective Chief Inspector who led the police inquiry into the boy's death gave public support to the PCSOs. He said, 'It would have been inappropriate for PCSOs, who are not trained in water rescue, to enter the pond.'[135] However, in an interview after the inquest, the Chairman of the Greater Manchester Police Federation commented, 'Every single officer that went to training school with me 30 years ago left with a lifesaving certificate of some sort.'[136]

We expect police officers (and PCSOs) to be able to swim. We also expect them to have been trained in lifesaving, by attending classes during working hours or in their own time. It is disappointing to realise that many have had no such training, and, as a consequence, some decide that a simple rescue of a drowning child is beyond them.

The Irish Police (Garda Siochana) still have a dedicated swimming pool at their training college, and swimming remains on the syllabus. Between 2000 and 2006, Garda officers – men and women – rescued ninety people from drowning, mostly from rivers, usually at night.[137] Many casualties were drunk, suicidal, or both. Some jumped from the midpoint of bridges into dark, fast-running water out of reach of the bank, others were unable or unwilling to grab life rings thrown within their reach. Nevertheless, the police officers risked their lives in icy water, swimming out against the current to reach the casualties and haul them back to land. They worked as teams, using their collective knowledge, experience and swimming ability, with a clear chain of command. No Garda officers drowned.

✦ ✦ ✦

When Health and Safety regulations and in-service guidelines are applied too rigidly and punitively, they stifle initiative and sideline experience. In March 2011, a man was feeding swans beside a shallow ornamental lake in a Hampshire park. As he leaned over to pick a plastic bag from the water, he had an epileptic fit, and fell into the lake.[138] He floated face down, 20 feet from the edge of the lake, in water approximately 3ft deep. A woman saw him fall and was on the point of wading out to him herself, but she was in charge of her little grandson, so she thought it best to call 999.[139] A fire engine arrived within three minutes, then another fire engine, two police cars, two ambulances and an air ambulance. No one entered the lake. Although all the firemen could swim, their water rescue training barred them from entering water more than ankle-deep. A policeman prepared to wade in, but his control room ordered him not to enter the water. A paramedic with swift water training also offered to go over to the man, only to be told that he was not properly equipped. A 'specialised water rescue unit' arrived half an hour later, by which time the man had floated further from the edge of the lake. Two firemen wearing waterproof clothing and life jackets had no difficulty in walking across to the victim and removing him from the water. He was taken to hospital, where he was pronounced dead.

The fire station watch officer told the Inquest that when he arrived with the first fire crew the victim was not moving. 'The witnesses told me that the body had been in the water for five or ten minutes,' he said. 'There were no obvious signs of life so from that I made an assessment that this was a body retrieval and not a rescue.'[140] However, a pathologist said, 'If he had been taken out of the water after ten minutes there is a slim chance he could have been resuscitated.'[141]

Firemen and police officers demonstrate their courage every day, and it would be wrong to suggest that the firemen and police officers who stood beside this calm, shallow boating lake in a municipal park were risk-averse out of cowardice. But their overcautious risk assessment and rule-following made them seem incompetent and callous.[142]

The Crown Prosecution Service now recognises that fireman and police officers acting heroically in the course of a rescue 'may decide to act in a way which puts their safety at significant risk in order to perform their duty'. New guidelines make it unlikely that they would be prosecuted.[143] Managers, on the other hand, have a 'duty of care' for the safety of emergency crews. They know that the death of a fireman or police officer during a rescue might result in charges of 'corporate manslaughter'. So regulations and guidelines remain – with constant updates to reflect improvements in rescue techniques.

◆ ◆ ◆

Whether victims of drowning accidents are pulled from the water by desperate relatives, concerned bystanders or trained professionals, their rescue must be followed by speedy first aid, early resuscitation, and transfer to hospital without delay.

CHAPTER THIRTY

FIRST AID AND RESUSCITATION AT THE SCENE

A drowning accident is a life-or-death drama. Even after a successful, timely rescue, the victim is likely to be exhausted, chilled and overwrought. Once on dry land, he should be encouraged to rest and recuperate,[1] preferably in the recovery position, lying on his right side, right leg straight, both arms and the left leg bent forward at right angles to his body. His head should be gently tilted back to keep his airway open, then his left hand is placed under his right cheek to lift his face from the ground and prevent his head from slumping forward. He should be wrapped in dry clothes, towels or blankets to trap body warmth and to protect him from cold air (wind chill) and from the cooling effect of evaporation from wet clothes and wet hair.[2] A plastic space blanket is helpful as an outer wrapping. And an ambulance should be called to the scene.

If an unconscious drowning victim is still breathing, he should also be placed in the recovery position. Lying semi-prone, he is protected from three dangers. First, his tongue cannot fall backwards, blocking his windpipe. Second, mucus, blood or water can drain from his mouth and nose instead of choking him. Third, if he vomits, he is less likely to inhale regurgitated water or stomach contents. During the anxious wait for the ambulance, the rescuer must check the victim's breathing and pulse repeatedly.

Everyone who lost consciousness, however briefly, and everyone who inhaled water, no matter how small the amount, must be admitted to hospital for assessment,[3] and kept under observation for a few hours at least. Inhaled water damages the lungs, although the extent of the damage may not be obvious at first. It may take 48 hours before any abnormal signs appear on chest X-rays.[4] Seawater, in particular, causes progressive lung damage. The salty water pulls liquid out of the bloodstream into increasingly stiff and waterlogged lungs.[5]

Sometimes a swimmer takes some water into his lungs but revives quickly and declines further attention.[6] Thinking he has had a lucky escape, possibly feeling embarrassed at the fuss he caused, he leaves the scene of the accident and returns home. However, the slightest cough, wheeze or chest tightness may be early signals of worse to come. Several hours after his apparent recovery from near-drowning, he begins to feel breathless and coughs up frothy, pink-tinged fluid. Now he needs urgent hospital admission because further deterioration may be sudden and life-threatening. This frightening and avoidable scenario was once classed as 'secondary drowning'.[7] In USA, it is often called 'parking lot drowning'.[8]

◆ ◆ ◆

If an unconscious drowning victim has stopped breathing, he will die unless someone gives him mouth-to-mouth resuscitation. If his heart has stopped beating, he will die unless someone gives him cardiopulmonary resuscitation – mouth-to-mouth resuscitation combined with heart massage. His chances of survival are hugely increased if the rescuer remembers the ABC of basic life support – A for Airway, B for Breathing, C for Circulation – and begins resuscitation immediately, without waiting for paramedics to arrive at the accident scene.[9]

MOUTH-TO-MOUTH RESUSCITATION

If drowning victims are not breathing for themselves, the rescuer, a lifeguard or a willing bystander must do the work of breathing for them. Getting air into their lungs is crucial to their survival. They need mouth-to-mouth resuscitation – at once.

On Australian surfing beaches, lifeguards are trained and equipped to begin mouth-to-mouth resuscitation as soon as they have the victim secure at the surface, even in deep water.[10] Brazilian research confirms that drowning victims have a better chance of survival if lifeguards begin mouth-to-mouth resuscitation in the sea instead of waiting until they reach the shore.[11] An untrained rescuer should concentrate on getting the victim to the water's edge: mouth-to-mouth resuscitation must wait until the victim is lifted from the water and placed face-up on the pool surround, the riverbank, or the beach. On a sloping beach, the victim should be positioned parallel to the shoreline, out of reach of the waves.[12] (Formerly, the rescuer was advised to place the victim's head lower down the slope to allow the escape of water from the lungs, but this position increases the risk of vomiting.)

When an unconscious drowning victim is lying on his back, gravity pulls the jaw downwards and the tongue falls backwards, blocking the airway. To keep the airway open, the victim's head must be tilted backwards by pressing firmly down on the forehead and lifting the chin.[13] (If there is the possibility of a spinal injury, neck movements must be kept to an absolute minimum, so the airway is kept open by a constant upward pull under the angles of the jaw.)

The rescuer kneels beside the victim's head and begins mouth-to-mouth resuscitation by pinching the nostrils and blowing two deep breaths into the victim's mouth.[14] Sometimes the mouth is clamped tightly shut.[15] If so, rescue breaths can be given through the nose until the jaw muscles relax. (When the victim is a young child, resuscitation is often easier if the rescuer's mouth covers the child's nose and mouth.[16] This is the method of choice with babies, since they are natural nose-breathers, with relatively large tongues.)

As the rescuer's expired air enters the victim's lungs, the victim's chest rises. Then, as the rescuer takes in the next breath of fresh air, the victim's chest falls. If the chest fails to rise and fall, the position of the head and chin must be adjusted to ensure that the airway is open. Sometimes regurgitated food blocks the victim's throat, and the obstructing debris must be cleared from the airway before rescue breaths can reach the lungs.

After giving two breaths of exhaled air, the rescuer should feel for the victim's pulse in the carotid artery on either side of the Adam's apple. (In a baby, it may be easier to find the pulse in the brachial artery on the inside of the upper arm, or the 'apex beat' of the heart, to the left of the breastbone.) If a pulse is present, even if it is weak and slow, the rescuer should concentrate on continuing to give mouth-to-mouth resuscitation. Twelve times per minute – every 5 seconds or so. The rescuer breathes enough air into the victim's mouth to make the chest rise. (Rescue breaths should be smaller for a child, and faster: twenty times per minute – every 3 seconds). Every minute or so, the rescuer should check that the carotid pulse is still beating in the victim's neck.

Gentle rescue breaths are usually enough to fill the lungs of a victim rescued before any water was inhaled. Once the victim has inhaled water, however, the lungs rapidly become stiff and unyielding, so that the rescuer must breathe into the victim's mouth with a fair amount of force.[17] The rescuer should end each rescue breath as soon as the victim's chest begins to rise. If rescue breaths are too forceful and prolonged, the increased pressure inside the victim's chest slows the return of blood to the heart,[18] and some of the rescuer's expired air may travel down the victim's gullet, inflating his stomach and increasing the risk of vomiting. Moreover, the rescuer may become faint or dizzy due to over-breathing.[19]

Atmospheric air contains 21% oxygen, whereas the rescuer's expired air usually contains around 16% oxygen. If the rescuer takes a shallow breath immediately before each mouth-to-mouth exhalation, the expired air may contain almost 18% oxygen.[20] A rescuer who has access to an oxygen supply (for instance, on a dive boat) can take breaths of oxygen instead of air, greatly increasing the percentage of oxygen in the air blown into the victim's lungs.[21]

When victims begin to breathe for themselves after receiving mouth-to-mouth resuscitation, their early spontaneous breaths are often shallow and irregular. If so, rescuers should continue with mouth-to-mouth resuscitation, timing their rescue breaths to coincide with the victims' faltering attempts to inhale. After longer submersions, drowning victims may respond only after prolonged mouth-to-mouth resuscitation. Nevertheless, rescuers should persevere with their task until the paramedics arrive to take over. The victim's life depends on the oxygen in the rescuer's expired air.

♦ ♦ ♦

For hundreds of years, midwives have breathed life into the mouths of newborn babies who failed to take their first gasp. In the eighteenth century, lifesaving organisations in Holland and Britain advised rescuers to give mouth-to-mouth resuscitation to victims of drowning.[22] A guide for swimmers and skaters advised rescuers to 'put the pipe of a pair of bellows into one of the nostrils… and blow gently till the breast be a little raised… If no bellows are at hand, let the assistant blow into the nostrils of the drowned person with his breath, through a small pipe.'[23]

During the nineteenth century, mouth-to-mouth resuscitation fell out of favour. Chemical analysis revealed that there was less oxygen in expired air than in the air around us (16% compared to 21%), so expired air was considered to have been 'devitalised' by its passage through someone else's lungs.[24] Worse, chemists found very much more carbon dioxide in expired air than in atmospheric air (4% compared to 0.04%), so expired air was thought to be decidedly harmful.[25] In 1829, the French physiologist Leroy d'Etoilles warned that bellows damaged the lungs by over-inflating them,[26] so bellows were abandoned. And once bacteriologists had discovered germs, rescuers became unwilling to risk possible exposure to tuberculosis or diphtheria.[27]

In place of mouth-to-mouth resuscitation, rescuers turned to mechanical methods of artificial respiration (notably the methods introduced by Marshall Hall and Silvester in the nineteenth century, and by Schafer, Holger Nielsen and Eve in the twentieth century).[28] The benefits of mouth-to-mouth resuscitation were forgotten until 1946, when Dr James Elam arrived in Minneapolis to work as an anaesthetist in the middle of a polio epidemic.[29] Patients with respiratory paralysis due to polio were dependant on mechanical respirators ('iron lungs'). When the machines malfunctioned, Dr Elam saved patients from suffocation by breathing into their noses.

In 1954 he published research showing that expired air contains enough oxygen to keep a patient alive.[30] His colleague, Dr Peter Safar, discovered that tilting back the head of an unconscious patient prevented the tongue from blocking the airway.[31] Together they established that mouth-to-mouth resuscitation was superior to mechanical methods of artificial respiration.[32]

Whether the new treatment was called mouth-to-mouth resuscitation, expired air ventilation, rescue breathing or artificial respiration, it was quickly recognised as 'an easily-learned, lifesaving procedure.'[33] No professional knowledge or equipment was necessary. Mouth-to-mouth resuscitation could be carried out by anyone, wherever and whenever it was needed, after the briefest of training. The *Journal of the American Medical Association* (*JAMA*) gave it an early seal of approval: 'It has revived many victims unresponsive to other methods and has been proved in real emergencies under field conditions.'

In the 1960s, soon after its rediscovery, mouth-to-mouth resuscitation became known as 'the kiss of life', and 'the kiss of life' it remains, despite the objections of a few pedantic doctors. (A letter in the *British Medical Journal* in 1965 asked, 'Is it too late to expunge that wretched expression "the kiss of life" from our language? It is most unpleasant that the necessary physical contact for an urgent therapeutic procedure should be associated with an act of affection, in order to try and induce a sense of drama.')[34] Yet it is the association between lifesaving and loving kindness which makes the term 'kiss of life' so apt. And surely nothing could be more dramatic than a victim saved from death by the touch of a rescuer's lips.

Mouth-to-mouth resuscitation is certainly dramatic, always nerve-racking, and often downright unpleasant.[35] Mushrooms of slimy, sticky foam commonly bubble out of the victim's mouth and nose, constantly reappearing after they are wiped away. False teeth slide out of place and have to be fished out of the victim's mouth. Fluid often accumulates in the victim's throat, sloshing and gurgling in time to the rescue breaths, so that resuscitation must be interrupted while the victim is rolled into a semi-prone position to allow the fluid to drain away. And most victims vomit.

Australian lifeguards and British paramedics confirm that vomiting and regurgitation of stomach contents are major problems during resuscitation.[36] Drowning victims often swallow large quantities of water while fighting desperately to hold their breath underwater.[37] The moment when the victim begins to breathe again is often marked by a sudden spewing of pints of swallowed water. A doctor who gave successful mouth-to-mouth resuscitation to his drowning son recalled, 'Four or five breaths later, he gave a cough and vomited all over the poolside. There is no adequate description for the relief I felt on seeing this.'[38]

Vomiting may be seen as a positive sign, but a drowning victim often vomits water mixed with stomach acids, digestive juices and half-digested food. If this corrosive slurry enters the victim's lungs, it causes an intense reaction.[39] Chemical pneumonia, bacterial infection and abscess cavities are frequent. At post-mortem, inhaled vomit is a common finding in the lungs of those who die of drowning.[40]

A drowning victim who loses consciousness also loses the protection of the gastro-oesophageal sphincter, a ring of muscles at the junction between the gullet and the stomach. Normally, the muscle acts as a gatekeeper, allowing food and drink to enter the stomach while preventing the contents of the stomach from re-entering the gullet. When the gastro-oesophageal sphincter relaxes, caustic stomach contents are free to travel backwards, seeping or surging up the gullet and into the victim's throat.[41] Unfortunately, the vigorous manoeuvres of cardiopulmonary resuscitation frequently aggravate this regurgitation of stomach contents. The protection of the gag reflex and the cough reflex are also lost, increasing the likelihood that regurgitated stomach contents will enter the victim's lungs.

In the operating theatre, anaesthetists guard against the aspiration of regurgitated stomach contents by pressing downwards on the neck below the Adam's apple, closing off the gullet by pinching it between the windpipe and the cervical vertebrae (Selleck's cricoid compression).[42]

Rescuers can only watch for the first signs of retching or abdominal heaving, turn drowning victims rapidly over to one side, clear their mouths, then continue with mouth-to-mouth resuscitation.

Clearly, someone trained in mouth-to-mouth resuscitation is likely to have a better technique than an untrained rescuer. However, as long as the victim's chest rises and falls in rhythm with the rescue breaths, an untrained rescuer is providing a supply of oxygen to preserve the delicate chemistry of the victim's brain and keep the victim's heart beating. Since any delay to the start of mouth-to-mouth resuscitation is likely to have serious consequences, untrained rescuers should not feel too self-critical. Even if they have never attended a first-aid class, they should begin mouth-to-mouth resuscitation at once, and continue, for more than an hour if need be, until the victim is breathing strongly, until another rescuer takes their place, or until paramedics arrive to relieve them.

CARDIOPULMONARY RESUSCITATION (CPR)

A drowning victim who has no pulse or heartbeat is on the point of death, and the rescuer is faced with two tasks instead of one. In addition to breathing for the victim, the rescuer must propel blood around the victim's body. Cardiopulmonary resuscitation combines mouth-to-mouth resuscitation with repeated sequences of vigorous chest compressions (cardiac massage). Trained lifeguards sometimes start mouth-to-mouth resuscitation in deep water, but chest compressions cannot begin until the victim is lying on a firm, flat surface, so there is no need to check for a pulse or a heartbeat during a long tow to the beach.

Before starting cardiac massage, it is important to be sure that the victim's heart has definitely stopped beating. After a long immersion in cold water, a drowning victim may have an exceedingly slow pulse, and chest compression applied to a cold, still-beating heart can trigger ventricular fibrillation.[43]

The rescuer begins CPR by kneeling beside the victim's head and giving two deep rescue breaths, then feels for a pulse at the wrist and neck and listens for a heartbeat on the left side of the chest. Very occasionally, if the victim's heart stopped after sudden entry into cold water, two rescue breaths are enough to restore the heartbeat.[44]

For cardiac massage, the rescuer changes position and kneels beside the victim's chest. Placing one hand over the other on the victim's sternum (breastbone) at the level of the nipples, with shoulders directly above the hands and elbows straight, the rescuer bends from the hips, using the weight of the body rather than the muscles of the arms, to compress the victim's chest at the fast and furious rate of a hundred compressions per minute.[45] (Fingertip pressure is enough for a young child.)

After each compression, the rescuer releases the pressure on the victim's chest while keeping the hands in position on the sternum. As the ribcage springs back to its normal shape, the sudden increase in chest volume creates a partial vacuum which draws blood from the major veins (the pulmonary veins and the vena cava) into the heart, ready for the next compression.[46] Both compression and release are equally important. They should be given equal time.

By pushing the sternum 2 inches downwards at each chest compression (only 1 inch in a young child), the rescuer squeezes the heart between the ribcage and the spine.[47] The one-way valves inside the heart ensure that the blood follows its normal route. Blood in the right ventricle is pushed along the pulmonary arteries into the lungs, where it takes up the oxygen supplied by the rescuer's mouth-to-mouth resuscitation. Blood in the left ventricle is pushed into the aorta, then round the body, supplying oxygen to the tissues that need it – most importantly, to the brain.

After every 30 chest compressions, the rescuer returns to kneel beside the victim's head, tilts the head backwards to reopen the airway, and gives 2 further vigorous mouth-to-mouth breaths.[48] These efforts allow the rescuer to continue mouth-to-mouth resuscitation while giving approximately 70 chest compressions per minute. When giving CPR, as in mouth-to-mouth resuscitation, the rescuer should stop exhaling as soon as the victim's chest rises,[49] so avoiding excessive pressure inside the victim's chest, which would slow the return of blood to the heart.

CPR is exhausting, particularly when carried out by a single rescuer. A force of 100–125 pounds is needed to compress the chest by 2 inches.[50] (Despite the springy resistance of the ribcage, ribs are often broken.)[51] CPR becomes more effective if two rescuers share the work, one giving mouth-to-mouth respiration, and the other giving chest compressions, swapping places from time to time to rest the rescuer who is giving heart massage.[52] By maintaining an uninterrupted rhythm, usually 2 rescue breaths alternating with 15 chest compressions, two rescuers can increase the supply of oxygenated blood to the victim's brain. Counting aloud helps to coordinate their efforts, and avoids rescue breaths from clashing with chest compressions.

The heart normally pumps 5 litres of blood per minute through the lungs and 5 litres of blood per minute round the body, with every heartbeat pumping about 70 millilitres of blood to the lungs and 70 millilitres round the body.[53] Studies suggest that CPR succeeds in pumping only a third of this volume through the lungs and around the body.[54] Yet this is enough to change the colour of the blood from the dark purple of de-oxygenated haemoglobin to the scarlet of oxyhaemoglobin. And the victim's blue lips become pink again.

Every few minutes, the rescuer should pause briefly to check for a pulse. If the victim's heart starts to beat again, the chest compressions should be stopped. But the rescuer must check the pulse repeatedly, so that chest compressions can be restarted at once if the heartbeats falter and cease.

As well as beginning CPR early, rescuers need to find the strength and determination to continue with CPR until the ambulance arrives. Researchers in New Zealand surveyed almost

2,000 parents with toddlers enrolled in playschools or swimming classes. Although two-thirds of the parents had attended formal training in resuscitation, the majority were under the mistaken impression that CPR could be abandoned after a trial lasting only five minutes.[55] But rescuers should not be tempted to give up hope so soon. Even if the victim shows absolutely no response to resuscitation, they should continue CPR – for an hour or more – until paramedics can take over and continue CPR while the victim is rushed to hospital.[56]

◆ ◆ ◆

Before 1960, cardiac massage was a high-risk technique restricted to the operating theatre, usually performed by cardiothoracic surgeons during open-chest surgery.[57] Then, in July 1960, a team at Johns Hopkins University in Maryland reported successful closed-chest cardiac massage in a series of surgical patients whose hearts had stopped beating while they were being anaesthetised.[58] Within a year, mouth-to-mouth resuscitation had been combined with cardiac massage. The modern era of cardiopulmonary resuscitation had begun.[59]

Hospital staff enthusiastically adopted CPR for patients in cardiac arrest. Initially, however, relatively few of the staff could be trained in the technique because it was unsafe to demonstrate CPR on healthy volunteers. Luckily, the Norwegian toymaker Åsmund Laerdal designed an adequate substitute to use in training classes, a life-sized plastic manikin that he named Resusci Anne.[60] (Curiously, the face of the Resusci Anne was modelled on the likeness of a young woman who was said to have drowned in the River Seine.)[61]

In 1960, 200 Norwegian schoolchildren were the first to be taught mouth-to-mouth resuscitation using the manikin.[62] The Resusci Anne was introduced in America to train hospital staff dealing with cardiac arrest, but the manikin soon proved invaluable for lifeguards and paramedics training in the resuscitation of drowning victims. Over the years, the design has become increasingly sophisticated, with electronic displays to measure performance and provide feedback. Laerdal went on to produce child-sized manikins that float on the surface or sink to the bottom of the pool, allowing lifeguards to practice rescue drills and in-water rescue breathing.[63]

Training in CPR is now a standard part of first-aid courses in hospitals. Organisations such as the St John Ambulance and the Red Cross run CPR training courses open to the public. The Royal Life Saving Society teaches survival, rescue and resuscitation skills in numerous swimming clubs, both inland and by the sea. A number of schools teach CPR to their pupils, with nine-year-old children taking turns to give rescue breaths and chest compressions to a Resusci Anne manikin.[64]

During a family party in 2015, a seven-year-old American girl spotted a young relative floating unconscious in the family swimming pool. She pulled him from the water, gave him a couple of rescue breaths and began heart massage.[65] He vomited quantities of water before the adults realised what was happening and took over the CPR. He began to breathe again and, after a brief hospitalisation, he made a rapid recovery. His young rescuer had no training in CPR apart from watching videos with her mother, an emergency room nurse.

◆ ◆ ◆

More than twenty years ago, Professor John Pearn suggested that skilled resuscitation given immediately after rescue might save the lives of 30% of drowning children who would otherwise die.[66] He advised that 'every parent should be a first aider.' Certainly, all parents with a swimming pool in their garden should have training in CPR.

In 2007, an Australian report clearly demonstrated that early resuscitation improves the prospects of survival.[67] Within a span of eleven days, 8 children were admitted to Sydney hospitals after drowning accidents. Four children received mouth-to-mouth resuscitation or CPR by parents or bystanders within 5 minutes of submersion, and all four survived without brain damage. In contrast, four children received delayed resuscitation (one child was submerged for 20 minutes before receiving CPR, and three were given CPR only after the arrival of emergency paramedics): all four children died, two in the emergency department and two after a diagnosis of brain death was made and intensive care was withdrawn.

The official guidelines of the American Heart Association emphasise that drowning victims who need CPR should be given mouth-to-mouth resuscitation to blow air into their lungs as well as heart massage to pump the oxygenated blood round their bodies.[68] After the appearance of AIDS, bystanders became less willing to give mouth-to-mouth resuscitation to strangers who collapsed in the street.[69] Cardiologists promoted 'Compression-only CPR' (cardiac massage without mouth-to-mouth resuscitation) in the hope that more people would try to resuscitate heart attack victims.[70] In 2010, the results of successful trials of 'Compression-only CPR' for victims of out-of-hospital cardiac arrest were published in the *Journal of the American Medical Association* and the *Lancet*.[71] Unfortunately, some commentators failed to recognise that 'Compression-only CPR' is totally inadequate for drowning victims. A national newspaper ran an article with the headline 'Skip the "kiss" when giving the kiss of life doctors recommend'. The accompanying photograph did not show a middle-aged man collapsed on a city pavement, but an attractive young woman in a swimming costume receiving 'Compression-only CPR' while lying on a beach.[72]

There is no record of anyone contracting AIDS from a drowning victim during mouth-to-mouth resuscitation,[73] and the likelihood of catching other infections is extremely low.[74] Lifeguards and paramedics use disposable masks to avoid coming into direct contact with a victim's mouth while giving mouth-to-mouth resuscitation or standard CPR. For parents called upon to give vital mouth-to-mouth resuscitation to their own child or their child's friend, the risk of picking up a harmful germ will be the last thing on their mind.

When people attend classes in CPR, they concentrate on learning a passable technique. The only emotion they are likely to feel is faint embarrassment when taking their turn to inflate the plastic lungs of the Resusci Anne. If they are called on to put their training into action, however, they are suddenly faced with giving mouth-to-mouth resuscitation and heart massage to wet, cold, blue, apparently lifeless human beings – perhaps to their own child.

For rescuers, the horrifying drama of a drowning accident releases intense emotions of fright, panic, confusion, and guilt. A surge of adrenaline sets their hearts pounding and their hands trembling. As they begin CPR, they have difficulty in remembering the drill.[75] They are often surprised and disturbed by their reactions. They may decide that they have failed a crucial test of character. They may judge themselves feeble, clumsy and inadequate. But this judgement is too severe. It is important for rescuers to realise that these 'fight and flight' reactions are entirely normal – at first. After a few minutes, it is usually possible for rescuers to bring their emotions under control. Their focus changes, moving away from their unexpected reaction to adrenaline overload. By counting every rescue breath and every chest compression, they can concentrate on the task of resuscitation. And as they do so, they give the drowning victim the best possible chance of survival.

Some rescuers have to give CPR while following instructions given over the phone by an emergency call handler.[76] Others must attempt to give CPR without any training at all. If so, they should remember that even imperfect resuscitation is very much better than no resuscitation at all.

DEFIBRILLATION

When a drowning victim is starved of oxygen or develops hypothermia, the regular rhythm of the heartbeat may change to ventricular fibrillation, with uncoordinated contractions of the individual muscle fibres of the heart. Blood no longer flows round his body. His heart will not regain a useful pumping action unless it is shocked by a bolt of electricity. Defibrillation briefly stops all the uncoordinated electrical activity in the cardiac muscle fibres, giving the heart's specialised pacemaker cells a chance to re-establish a coordinated rhythm.[77]

A drowning victim with an abnormal heart rhythm has a better chance of survival if necessary defibrillation can be administered without delay by a lifeguard at the swimming pool or on the beach, instead of waiting for the arrival of the ambulance. More than one shock from a defibrillator may be needed to restart the regular beating of the heart.[78] In Cornwall in 2015, a former Olympic swimmer in her sixties suffered a cardiac arrest while lane swimming in her local pool. While one of the lifeguards lifted her from the water, another grabbed a defibrillator, and 'after three shocks her pulse returned'.[79]

After prolonged sea immersion, the life jacketed survivor of a boating accident is usually profoundly hypothermic, with an extremely slow pulse. If rescuers start heart massage, they are likely to trigger ventricular fibrillation. Moreover, once the victim's core temperature falls below 28°C, his heart may not respond to defibrillation. He then becomes entirely dependent on vigorous CPR until he arrives in hospital, where he can receive advanced life support. There he will be gradually rewarmed, and as his core temperature rises towards 32°C, a further trial of defibrillation may be successful in re-establishing the regular beating of his heart.[80]

✦ ✦ ✦

Before 1947, ventricular fibrillation could not be treated. The chaotic quivering muscle activity rendered the heart useless as a pump, and the victim died at once. Early successes in defibrillation, led by Professor Claude Beck, a cardiac surgeon at Western Reserve University Hospital in Cleveland, Ohio, involved cumbersome machines plugged into the mains supply, with silver paddles placed directly onto the surface of the heart.[81] This open-chest defibrillation could be attempted only in a well-equipped operating theatres, and only in a patient whose heart went into fibrillation during a scheduled operation.

Then, in 1955, a doctor with coronary heart disease collapsed just outside Western Reserve University Hospital in Cleveland, Ohio. He was immediately brought into an operating theatre, his chest was opened, and defibrillator paddles were pressed against the writhing muscle of his heart. After several electric shocks, his heart began to beat again.[82] The cardiac surgeon, Claude Beck, had photographs taken of the doctor in his street clothes during the operation, to show that the cardiac arrest had happened outside hospital. Professor Beck wrote, 'The heart wants to beat, and often it needs only a second chance.'[83] The doctor survived for another 28 years.

Electrical engineers and doctors at Johns Hopkins University (the same team who first recognised the value of closed-chest cardiac massage) were partly financed by grants from electricity companies troubled by the deaths of workers who developed cardiac fibrillation after apparently trivial electric shocks.[84] By 1957, the team had designed a defibrillator that delivered a 'counter-shock' through the chest wall and used it to treat hospital patients who developed ventricular fibrillation in the operating room, emergency ward or cardiac catheterisation laboratory.[85]

For a number of years, defibrillation was only possible in hospitals. Then compact automated defibrillators were developed, which could be programmed to recognise ventricular fibrillation and deliver an appropriate electric shock. The Australian airline Quantas pioneered the training of flight crews in effective use of defibrillators during flights and at airline terminals.[86] By 2000, security officers in a number of American casinos were using defibrillators to restart the hearts of people who collapsed in the gaming rooms.[87]

Now every ambulance is equipped with a defibrillator, and every paramedic is trained to use one. And many swimming pools, water parks and public beaches have defibrillators on site, allowing lifeguards to treat life-threatening ventricular fibrillation before the ambulance arrives.[88]

REWARMING

The longer the immersion and the colder the water, the more heat is drained from the body of a drowning victim. A conscious survivor suffering from mild hypothermia has a core temperature above 34°C and needs only a supervised bath, a change of clothes, a warm drink, and bed rest.[89] A survivor with a core temperature below 34°C is suffering from moderate hypothermia. At the scene of the accident, passive rewarming is safer than active rewarming. Avoiding rough handling, the victim should be stripped of wet clothes, wrapped in blankets, allowed to generate body heat from the muscular activity of shivering, then rapidly transported to hospital. Once there, the victim can be rewarmed under the skilled supervision of trained staff.[90]

The core body temperature of a hypothermic survivor will continue to fall after removal from the water, sometimes by 2°C or more. This 'afterdrop' is due to heat conduction from the warmer body core to the chilled outer shell of muscle, fat and skin, and to the gradual re-opening of constricted blood vessels.[91] Rubbing the arms and legs only serves to increase the 'afterdrop', and may sometimes trigger ventricular fibrillation.

A survivor whose core temperature falls below 30°C has developed severe hypothermia. He loses consciousness. His shivering stops, and since he cannot generate body heat through muscular activity, some gentle external warmth is indicated. Tepid hot-water bottles (wrapped in towels to avoid burning the skin) can be placed under the survivor's armpits or over the femoral blood vessels in the groin.[92] The cooling of his body slows the chemical reactions in the vital organs, so that his sluggish circulation may suffice until he reaches an emergency centre, where rewarming can proceed under close supervision of trained staff, linked closely with the many other measures needed to save his life.

◆ ◆ ◆

Open water is always cold, and it is usually considerably colder a few feet down. Once submerged, the bodies of drowning victims lose heat very quickly. Hypothermia puts people in mortal danger, yet, paradoxically, rapid cooling may occasionally provide victims with brief protection. As their temperature falls, their metabolic processes slow down and their tissues consume less oxygen. Most people die of drowning if they are submerged for 10 minutes, but cold water extends the post-submersion survival time in a lucky few.

These unexpected survivors are more often children than adults, perhaps protected by the mammalian diving response, an oxygen-conserving reaction found in seals and other mammals that pursue their prey underwater.[93] (The mammalian diving response allows them to hold their breath for longer than usual, slows their heartbeat to half its normal rate, constricts the blood vessels that supply their skin, arms, legs and abdomen, but preserves the blood flow to the chest and head – making the last remnants of circulating oxygen available to the heart and the brain.)

Children cool far more rapidly than adults.[94] There is usually little fat beneath their skin to act as insulation, and cold water soon extracts heat from their small bodies. In the early stages of drowning, children often swallow repeatedly, filling the stomach with cold water. Cold water fills their nasal sinuses, separated from the brain by only paper-thin plates of bone. The scalp and skull are also thin, allowing fast cooling of the brain. Once the brain is chilled, the diminishing supply of oxygen in the blood may be enough to preserve vital metabolic activity for half an hour or more. A few children appear to enter a state of suspended animation after submersion in cold water, especially in water colder than 10°C. Even in Florida, open water may be sufficiently cold to offer some protection. In 2003, a toddler was pulled from a freshwater creek after lying at the bottom for 20 minutes. His core temperature registered less than 27°C. He survived with normal brain function.[95]

In Britain, cold winter weather allows some victims to survive prolonged submersions in open water or in the sea.[96] On Boxing Day 2003, a two-year-old boy was found lying face down in a garden pond in Birmingham. He appeared to be dead. His father rang 999 and the operator explained how to give CPR over the phone. His mother immediately started mouth-to-mouth resuscitation and heart massage.[97] Paramedics arrived, and continued CPR en route to hospital. His core temperature was 26°C at hospital admission. A respirator supplied him with oxygen-enriched air, and a team of doctors and nurses continued the heart massage while his cold body was slowly warmed by flushing warm liquid in and out of his bladder and stomach. After five hours, a heartbeat was detected. After six hours, his pupils began to react to light. After seven and a half hours he was stable enough for transfer to an Intensive Care Unit, where he spent two weeks on a life-support machine. Five weeks after the accident, he returned home, fully recovered.

Sometimes there appears to be no hope of recovery when a young man is pulled from cold water. He lies on the ground, inert, blue, pulseless, his dilated pupils unresponsive to light, his body icy to the touch, his muscles stiffened as in rigor mortis. Yet there remains a possibility that he could be saved by immediate CPR, with advanced life support en route to hospital,[98] followed by skilled intensive care. Instead of assuming that the drowning victim is dead, the rescuer or bystanders should begin CPR at once and keep it going strongly until the paramedics arrive. Only rarely should a drowning victim be pronounced dead at the scene. The formal diagnosis of death is best deferred until the victim can be examined in hospital.

On some occasions, rescuers decide too readily that it would be pointless to attempt CPR, or they abandon CPR after only a brief trial. In a Scottish study of 93 drowning deaths between 1991 and 1997, 28 of the victims were recovered from the water less than an hour after they were seen alive. The emergency services made no attempt to resuscitate 6 of the victims, and a further 5 victims were declared dead at the scene after only brief attempts at resuscitation.[99]

The authors concluded that the emergency services were failing to provide appropriate cardiopulmonary resuscitation, or failing to continue their efforts until the victims arrived in hospital. In their judgement: 'There appears to be a potential to reduce the drowning death rate by improving resuscitation. The emergency services and the public should be educated about the need to resuscitate those found in water.'

The resuscitation of a young Scotsman in the cold winter of 1995 was more determined – and succeeded in saving his life.[100] He fell through the ice covering a reservoir while trying to rescue his dog. He was able to keep his head above the surface, but he had been immersed in icy water for more than an hour before the Fire Brigade succeeded in extricating him, after a hazardous rescue. He was given oxygen in the ambulance. En route to hospital, he lost consciousness, stopped breathing and developed ventricular fibrillation, which failed to respond to three attempts at defibrillation. Vigorous CPR began at once and continued after hospital admission. By this time his core temperature had fallen to 28.3°C. Several hours of complex resuscitation followed, including cardiopulmonary bypass, extra-corporeal rewarming and successful defibrillation. After three days of mechanical ventilation, he was out of danger. He returned home fit in body and mind. The surgeons wrote, 'With a significant drop in the core temperature the immersion hypothermic casualty may appear clinically dead (without palpable pulse, blood pressure, or respiration), but may still be successfully resuscitated with little or no neurological deficit.' They concluded, 'The only certain diagnosis of death is failure to recover on rewarming.'

In 2014, doctors working at the University Hospital of North Norway in Tromsø, 200 miles north of the Arctic Circle, reported the outcome for 34 victims admitted over the past 28 years suffering from severe hypothermia with cardiac arrest.[101] After early basic life support and CPR, they were airlifted to hospital with advanced life support, then rewarmed in hospital by extracorporeal circulation. Ten were treated before 1999, but none survived: 24 were treated during 1999–2013, with 9 survivors, including 8 victims of drowning accidents.

TRANSFER TO HOSPITAL

Victims of drowning accidents who lost consciousness or inhaled water are best transferred to hospital for urgent assessment. The journey is usually made by ambulance, but a helicopter may be summoned if a victim is rescued in a critical condition or has to be winched up from the sea.

Paramedics make a major contribution during the journey to hospital. As soon as the ambulance arrives, a paramedic places a 'bag valve mask' (a BVM or Ambu bag) over the victim's nose and mouth and gently squeezes the self-inflating bag 12 times per minute, filling the victim's lungs with 100% oxygen.[102] (Before the arrival of the ambulance, mouth-to-mouth resuscitation supplied the victim with expired air containing 18% oxygen at most.)

An unconscious drowning victim is at risk of vomiting swallowed water mixed with half-digested food, then inhaling the corrosive liquid into his lungs. Paramedics prevent this dangerous complication by sliding a cuffed endotracheal tube down his windpipe and slipping a nasogastric tube down his nose into his stomach to empty its contents.[103]

If the victim has no pulse, paramedics continue and intensify the heart massage. They record the victim's electrocardiogram (ECG) throughout the ambulance journey to the hospital. When necessary, they use a defibrillator to shock the heart into a regular rhythm. If there is no response to defibrillation, they continue vigorous heart massage until hospital emergency staff take over. They set up an intravenous drip. And they warn the hospital receiving staff that the victim of a drowning accident is about to arrive,[104] ensuring that, as soon as the ambulance arrives at the Accident and Emergency Department, or the helicopter touches down, a team is ready to continue the skilled assessment and treatment that the victim will require.

CHAPTER THIRTY-ONE

HOSPITAL TREATMENT AND INTENSIVE CARE

Most people who survive a drowning accident long enough to be admitted to hospital will respond to treatment and return home in reasonable health. Those who are conscious and able to breathe for themselves when they are wheeled into the Accident and Emergency Department usually make a quick recovery. The outlook is much worse for drowning victims who arrive unconscious and still dependent on cardiopulmonary resuscitation (CPR).[1] Some die soon after admission. Others die after days of complex treatment in the Intensive Care Unit, no matter how skilled the hospital staff and how high-tech the available equipment. A number are left with devastating brain damage caused by lack of oxygen during their submersion.

Even after prompt rescue and immediate resuscitation, many victims of drowning accidents fail to reach hospital alive. Instead of being admitted for treatment, they are confirmed as 'Dead On Arrival'. In 1992, Professor John Pearn pointed out, 'The size of the problem is partly masked as most of the immersion victims bypass the hospital and go straight to the mortuary, and are not seen in the ambience of the intensive care ward.'[2] He noted 'the very high fatality:survivor ratios for immersion accidents'. In an earlier study, he reported that only half of the children who lose consciousness in fresh water survive.[3] The survival rates have improved somewhat since then, but they still make grim reading.

A British study included all 330 children aged 14 years or younger who lost consciousness in drowning accidents during 1988 and 1989.[4] Almost half of the children died: 142 children died before hospital admission, and 16 children died up to 5 days after hospital admission without regaining consciousness. There were 172 survivors: 162 made a full recovery (including all 125 children who regained consciousness before they arrived in hospital), but 10 developed spastic quadriplegia requiring permanent, full-time nursing care. When they were wheeled into the Accident and Emergency Department, 33 children appeared beyond help – deeply unconscious, without breathing or heartbeat, their pupils fixed and dilated, their core body temperature as low as 28°C. After advanced life support, rewarming and intensive care, 10 of these desperately ill children made a full recovery. The researchers concluded that resuscitation should not be abandoned unless victims fail to respond after they have been rewarmed.

Between 1984 and 1992, the emergency and intensive care departments of a large Californian hospital admitted 166 children aged 14 years or younger after drowning accidents, often in the relatively warmer water of domestic swimming pools.[5] Children with a 'good' outcome were almost five times as likely to have received immediate resuscitation at the scene than children with a 'poor' outcome. Among the 136 children with a 'good' outcome, however, the researchers included 12 survivors who had suffered brain damage that left them with ataxia (impaired balance and coordination) or spasticity (excessive muscle tightness and stiffness). In the group with a 'poor' outcome, 22 children died after hospital admission and 8 survivors had brain damage so severe that they were left in a persistent vegetative state,[6] unable to help themselves or respond to their families. Thus, despite advanced life support and intensive care, a quarter of the children who survived long enough to be admitted to hospital either died or were left with incapacitating brain damage.

ASSESSMENT AND TREATMENT ON ARRIVAL IN HOSPITAL

When the ambulance arrives at the Accident and Emergency Department, one of the paramedics hands over to the medical staff with a brief description of the drowning accident, the rescue, and treatment already given. Drowning victims who are conscious on admission will be able to describe what happened in more detail.

Survivors need rapid assessment. Those who did not lose consciousness or inhale water will be allowed to rest in bed under close observation. As long as they are breathing normally and have normal chest sounds and chest X-rays, they will be given high-flow oxygen to breathe via a facemask.[7] A pulse oximeter is clipped onto a finger to measure the oxygen saturation of haemoglobin in their arterial blood.[8] If repeated oximeter readings are normal (95–98% oxygen saturation), the red cells are capturing a full load of oxygen in their lungs and carrying normal amounts of oxygen to their tissues.[9] The oxygen enrichment can be phased out gradually until they are breathing room air. All being well, the fortunate survivors will be able to return home within 24 hours of admission.

If a survivor lost unconscious or inhaled water during a drowning accident, a member of the emergency team covers the victim's nose and mouth with a close-fitting 'bag valve mask' (a BVM or Ambu bag) and squeezes its self-inflating bag every 5 seconds to ventilate the lungs with 100% oxygen.[10] Later, 100% oxygen will be replaced with oxygen-enriched air (usually 50% oxygen).

After inhaling water, a survivor is likely to be coughing, wheezing or complaining of pain in the chest. The inhaled water damages the lungs, and oedema fluid oozes into the affected alveoli.[11] The 'bag valve mask' allows the operator to maintain the air pressure in the lungs at a level slightly above atmospheric pressure throughout the breathing cycle.[12] This 'continuous positive airway pressure' (CPAP) helps to re-open collapsed alveoli and speeds the absorption of oedema fluid leaking from damaged alveolar capillaries.[13]

Distressed or semi-conscious drowning victims are often overwrought. They are surcharged with adrenaline and frightened by the mask enclosing their mouth and nose. Their breathing difficulties are aggravated if they fight against attempts to push air into their lungs.[14] Simple explanations and whispered words of encouragement will help them to stay calm. The comments can be brief, the words simple, but the positive messages from rescuers, paramedics, doctors and nurses are potent: 'I am going to help you,' 'You are doing well,' 'These results look good.' Some victims may seem unresponsive, or may be unable to reply, but they often hear the voices of rescuers, the emergency crew and hospital staff, and react to their intonation.

During resuscitation, drowning victims are at risk of vomiting swallowed water and half-digested food and inhaling it into their lungs.[15] To protect against this dangerous complication, a cuffed endotracheal tube is inserted into the airway en route to hospital or soon after arrival at the Accident and Emergency Department.[16] The emergency team organises blood tests and chest X-rays (X-rays of the cervical spine are also taken if the victim bruised or grazed his forehead when diving into shallow water.)[17] Blood gas analysis usually shows a reduced level of oxygen in the bloodstream, confirming damage to the lungs. Continuous ECG monitoring allows fast response to any abnormal heart rhythm. An intravenous drip is set up, and fluids are given to counter the excessive acidity of the blood, boost blood sugar levels and restore falling blood volume.[18] The rectal temperature is taken to assess the degree of hypothermia. A catheter is passed into the bladder.

Liquid often wells up into the trachea, a mixture of inhaled water and oedema fluid that interferes with efforts to supply the victim with oxygen. Ventilation with the 'bag valve mask' has to be interrupted, often repeatedly, while a suction device extracts the offending liquid from the victim's airway.[19] (On occasions, as much as 2 litres of liquid has to be sucked away.) Only when frequent suction is no longer needed can the drowning victim be connected to a mechanical ventilator and moved to the Intensive Care Unit.

TREATMENT IN THE INTENSIVE CARE UNIT

The victim of a drowning accident often suffers complex damage to many systems of the body. A team of doctors, nurses and technicians work together, coping with a series of life-threatening dramas that need immediate assessment and treatment.[20]

Damaged lungs. Normally, breathing involves little effort. Once water enters the lungs, however, breathing gradually becomes more laborious. Early chest X-rays may look normal, only for abnormal shadows to appear a few hours later.[21] The damaging effects of water inhalation are not uniform throughout the lungs. Where surfactant has been destroyed or washed away, the walls of the alveoli crumple and stick together, resisting expansion. The airways are choked with foam. The delicate, pliant tissues of the lungs become bloated and stiffened as oedema fluid oozes from alveolar capillaries.[22] As the hours go by, breathing becomes more and more exhausting. Prolonged forceful contractions of the diaphragm and intercostal muscles are needed to draw air into increasingly rigid, waterlogged lungs. Blood flowing in the walls of collapsed or fluid-filled alveoli is 'shunted' through the lungs without capturing oxygen. As lung function deteriorates, some victims develop adult respiratory distress syndrome (ARDS),[23] with effects similar to those of bird flu or poisoned gas.

The foremost task of treatment is to ensure adequate oxygenation. Drowning victims who inhaled water often need the assistance of a mechanical ventilator to ensure adequate oxygen capture by the red cells in the pulmonary capillaries.[24] Since 100% oxygen is toxic long-term, the ventilator usually supplies air containing 30–50% oxygen,[25] inflating the lungs during each inspiration through a mask, an endotracheal tube, or a tracheotomy in the front of the neck.

The volume of oxygen-enriched air delivered to the lungs is carefully controlled, both to provide adequate lung filling and to minimise injury due to over-distension. Importantly, the ventilator maintains the air pressure slightly above atmospheric pressure between breathing cycles. This 'positive end-expiratory pressure' (PEEP) combats the damage caused by water inhalation due to loss of surfactant:[26] once the alveoli are re-opened, slight positive pressure prevents them from collapsing once more. PEEP also limits the formation of pulmonary oedema, so that less fluid leaks from damaged capillaries into the alveoli and the surrounding lung tissues. With gradual 'alveolar recruitment', more air remains in the lungs at the end of expiration, and the blood is more fully oxygenated as it leaves the lungs, returns to the heart, and makes its circuit round the body.

Medication is often needed to treat the asthma-like bronchial constriction which obstructs breathing, and which causes air trapping during mechanical ventilation. Some experts prescribe intravenous steroids in high doses to counter the development of ARDS.[27] Surfactant replacement therapy has improved lung function in some drowning victims, but it is not given routinely.[28]

Sometimes the lungs of drowning victims are damaged so severely that mechanical ventilation fails to provide sufficient oxygenation – even if 100% oxygen is used. Specialised heart-lung machines may be the only means of keeping them alive – oxygenating their blood and sustaining their circulation until they can breathe for themselves again.[29]

Cardiac arrest. If the drowning victim has no heartbeat, hospital staff continue heart massage during the transfer from Accident and Emergency Department to the Intensive Care Unit. A combination of ventilation with 100% oxygen and vigorous heart massage may succeed in pumping enough oxygenated blood through the victim's coronary arteries to restart the beating of the heart.

Sometimes heart massage succeeds only after hours of effort.[30] In 2003, a two-year-old boy was pulled from a pond apparently dead. His mother began cardiopulmonary resuscitation at once. Paramedics and hospital staff continued heart massage for more than seven hours before his heart responded with an adequate beat. The boy made a full recovery.

The muscular exertions of the hospital team are sometimes replaced by a mechanical compressor. The Lucas device (Lund University Cardiac Assistance System) acts like a sophisticated sink plunger. Its pressure pad pushes down on the sternum, compressing the heart, then its suction cup pulls up the chest wall, producing negative pressure inside the chest and ensuring that blood flows from the great veins into the heart, ready for the next compression.[31]

Rewarming. All drowning victims are chilled when they are rescued from open water or the sea.[32] Their core body temperature is often below 35°C, and may fall below 30°C, a state of severe hypothermia. Once in hospital, a variety of rewarming methods are used to bring the core temperature back towards normal, usually aiming to raise the core temperature by 1°C per hour. Some drowning victims are placed on heated blankets or under a radiant heater. Some are given warm, moist air to breathe (inhalation rewarming). Warmed liquid may be run into the veins (intravenous infusion), into the bladder (bladder irrigation), or into the stomach or abdominal cavity (gastric or peritoneal lavage). Drowning victims with a core temperature below 28°C are often treated by warming their blood as it circulates through a heart-lung machine (cardiopulmonary bypass with extracorporeal rewarming). Hypothermic patients are usually warmed slowly to 32°C, then kept mildly hypothermic for at least 12 hours, to continue the protection that a slower rate of metabolism affords the brain.[33]

For decades, patients have been routinely cooled before open heart surgery, then rewarmed at the end of the operation.[34] Mountaineers buried by avalanches have been successfully resuscitated, even after their hearts were 'frozen stiff' under the snow.[35] Nevertheless, the re-warming of hypothermic drowning victims is fraught with danger, and whatever method is used, over-hasty treatment has resulted in many deaths.[36]

If a victim's core temperature has fallen below 28°C, severe hypothermia may be complicated by ventricular fibrillation that fails to respond to defibrillation.[37] Heart massage must be used to keep the victim alive until the core temperature rises to 32°C. At this point, a further attempt at defibrillation may be successful. Then for several hours after the return of a normal heartbeat, the victim's temperature is usually maintained around 32°C.

The chosen method of rewarming may influence the extent of recovery in survivors. In 2004, twin toddlers were involved in the same drowning accident after falling into a pond. Both were given skilled intensive care, but they were admitted to different hospitals with different treatment regimes. One of the twins had a far better outcome.[38] She was rewarmed to 32°C, then kept mildly hypothermic for 72 hours. Although she developed ARDS, lung infection and multiple organ failure, and needed eleven days of mechanical ventilation, she made a full recovery, with no evidence of brain damage. The other twin was rewarmed to normal temperature within 5 hours. He developed aspiration-related pneumonia, and needed six days of mechanical ventilation. A month later, a cranial CT scan showed marked brain atrophy, and he was left permanently spastic and semi-conscious, imprisoned in a persistent vegetative state.

The colder the water, the faster heat is extracted from the body. As the vital organs cool, their metabolism slows and their consumption of oxygen falls. Unless the water is extremely cold, however, cooling takes too long to have a protective effect. The water in swimming pools is usually warmed to a temperature around 26°. Studies in America and Saudi Arabia suggest that few children escape death or serious brain damage after lying at the bottom of a swimming pool for 10 minutes.[39] The water in a baby's bath, with its temperature at a comfortable 'blood heat', offers none of the protective slowing of metabolism or lowering of oxygen requirements afforded by ice-cold water. This may explain why bathtub drownings are often fatal after a very short submersion, even if efforts at resuscitation begin almost immediately.

In water colder than 10°C, drowning victims cool rapidly. Their metabolic processes soon shut down, allowing the unexpected survival of some drowning victims (usually children), even after prolonged submersion.[40] Great knowledge, skill, determination and teamwork is needed to return these drowning victims to life and health. The case reports of their remarkable recoveries make exciting reading.

In 1986, a toddler in Utah fell into a creek filled with run-off from melting snow. She was submerged for more than an hour before she was rescued.[41] Her core temperature fell to 19°C, she had no heartbeat, and her pupils were fixed and dilated. Pre-hospital CPR was followed by advanced life support, including rewarming with oxygen-enriched mist at 40°C via an endotracheal tube, intravenous fluids warmed to 40°C, and continuous 40°C gastric lavage. Chest compression went on for three hours before she was connected to a heart-lung machine with a membrane oxygenator and a paediatric heat exchanger. As her core temperature rose to 25°C, her heart began to fibrillate, then a few minutes later, it converted to a regular beat. Her temperature returned to normal after 53 minutes of extra-corporeal rewarming, and she was disconnected from the heart-lung machine. She needed mechanical ventilation for six days. After eight stormy weeks in hospital, she made a good recovery, apart from mild impairment of her long-term memory.[42]

Few adults survive prolonged submersion. In northern Norway in 1999, a 29-year-old doctor was skiing with two friends when she fell head first into an ice-covered mountain stream. The current swept her under a layer of ice 8 inches thick. She was trapped between rocks, breathing in a pocket of air. Unable to release her, her friends used a mobile phone to summon a rescue helicopter. She struggled for 40 minutes, wedged beneath the ice, but as her body chilled, her movements slowed and stopped. Then she lost consciousness and sank into the icy stream.[43]

She lay submerged for a further 40 minutes before her friends were able to release her. One recalled, 'I thought we were taking a friend, dead, out of the water.' Both her friends were doctors. They began vigorous CPR, the first response in an impressive series of life support measures, including ventilation with 100% oxygen, and continued cardiac massage during an hour-long helicopter flight, rewarming by cardiopulmonary bypass, 35 days on a respirator, and 2 months in the intensive care unit.

On the helicopter, her electrocardiogram tracing showed a flat line, and there was no response to defibrillation. Her core temperature was 13.7°C when she arrived in hospital. She was rewarmed over the next 4 hours. When her temperature rose to 31°C, her heart began to beat again. Five months after the accident, following prolonged rehabilitation, she returned to work as a doctor. Dr Mads Gilbert, the consultant in charge of her resuscitation, commented, 'Her body had time to cool down completely before the heart stopped. Her brain was so cold when the heart stopped that the brain cells needed very little oxygen, so the brain could survive for quite a prolonged time.'[44]

Lung infections. Lung infections are common after drowning accidents. Even the filtered, chlorinated water in swimming pools is not germ-free. Most water samples contain bacteria such as staphylococci (which can cause boils)[45] and coliforms (which can cause diarrhoea). Open water teems with microscopic creatures.[46] After heavy rain, streams are contaminated by the run-off from manured fields and rivers are polluted by the release of raw sewage.[47] Coastal waters contain both marine microorganisms and germs linked to domestic sewage outflows.[48] The author of a highly regarded treatment review recommended that all drowning victims with evidence of water inhalation should be given intravenous antibiotics,[49] with the explanation: 'It is difficult to accept that the aspiration of water into the lungs is a sterile process or that secondary infection will not occur in those who have inhaled water from swimming pools, the sea or rivers.'

As well as inhaling water, drowning victims often inhale fragments of waterweed floating in the water.[50] Others inhale sand from the seabed[51] or mud from the bottom of a pond.[52] Using a bronchoscope, it may be possible to wash out much of the inhaled sand or mud, but some waterborne microorganisms are inevitably left behind in the nasal sinuses and the lungs.[53]

Victims of drowning accidents often develop pneumonia in the days following their admission to hospital.[54] Those who need the help of a mechanical ventilator are particularly susceptible.[55] Bacteria already resident in the victim's nose or throat may cause an 'opportunistic' infection in their damaged lungs, and hospitals have their own menacing bacterial populations. Diagnosis is not straightforward, because symptoms, signs and X-ray appearances related to drowning often obscure early indications of lung infection, or mimic infection when none is present. If lung abscesses develop, the offending microorganisms are often unusual species of bacteria and fungi, difficult to identify and difficult to treat.[56]

Inhaled microorganisms often spread beyond the lungs. Diatoms (algae with distinctive rigid cell walls impregnated with silica) frequently enter the pulmonary bloodstream during drowning, and they can be identified at post-mortem in distant parts of the body.[57] In 2006, Dutch paediatricians reported invasive aspergillosis in the lungs and brain of a child who died in an intensive care unit some days after she was rescued from a pond.[58] Two severely injured survivors of the 2004 Indian Ocean tsunami were repatriated to Switzerland. They required prolonged treatment for pneumonia and abscesses in the lung and brain due to infection by obscure antibiotic-resistant organisms inhaled while lying in warm, stagnant water.[59]

Other signs of injury. Medication for seizure control is important both for victims known to have epilepsy and for victims who develop fits due to drowning-related brain damage.[60] Some victims of freshwater drowning develop extensive bruising, bleeding and abnormal clotting of the blood, only controlled by transfusions of platelets or fresh frozen plasma.[61] An Australian study reported acute renal impairment in half of the victims of drowning accidents admitted to hospital from Sydney beaches.[62] Some had such serious kidney failure that renal dialysis was necessary.

Some drowning accidents follow a traumatic injury or an acute illness that leaves the victim helpless in the water. Lacerations and broken bones need appropriate attention. A precipitating illness must be diagnosed and treated. In Germany in 2011, a 55-year-old doctor was found floating in a swimming pool, apparently dead after a drowning accident. He was given immediate poolside resuscitation and defibrillation. After he arrived in hospital, his blocked coronary artery was identified and successfully dilated.[63] Then he was admitted to the Intensive Care Unit for treatment of his almost-fatal drowning. He needed 10 days of mechanical ventilation, but he recovered and returned to work.

PREDICTING THE OUTCOME AFTER A DROWNING ACCIDENT

Drowning victims have a good chance of full recovery if they arrive in hospital aware of their surroundings. However, those who are admitted to the Intensive Care Unit after lengthy submersion and prolonged cardiovascular resuscitation may have suffered such severe oxygen deprivation that their brains are damaged beyond repair. American researchers observed: 'It is easier to restart the heart than it is to restart the head.'[64]

Brain damage often gets worse after admission to hospital.[65] A mechanical ventilator inflates the victim's lungs with oxygen-enriched air, but oxygen capture is inefficient or impossible in the lung tissues most affected by inhaled water. Even after successful defibrillation of the victim's heart, the blood pressure is likely to remain lower than normal, reducing the volume of blood flowing through the brain. When the circulation is restored after a cardiac arrest, chemicals are released from damaged brain tissue, which cause further brain damage.[66] And swelling of the brain increases pressure within the rigid skull, restricts blood flow and limits oxygen supply to the brain.

The most active areas of the brain are the worst affected by lack of oxygen[67] – the cerebral cortex, the hippocampus, the basal ganglia and the cerebellum. A number of tests are used to assess brain function soon after hospital admission and in the following days, weeks and months. Straightforward checks such as the Glasgow Coma Score record eye, speech and motor responses.[68] More specialised diagnostic tests may be needed,[69] including electroencephalography (EEG), computerised tomography (CT scans), magnetic resonance imaging (MRI scans), intracranial pressure measurements, brain stem audio-evoked responses and blood tests for biomarkers such as neurone-specific enolase.

No early tests can give a reliable prediction of whether an unconscious drowning victim will live or die, or whether a survivor will make a good recovery or will be condemned to a vegetative existence.[70] This means that all drowning victims should be given the chance to benefit from skilled intensive care. As advanced life support systems become increasingly sophisticated, however, some doctors have warned that prolonged heroic treatment may save a few drowning victims from death, only to swell the number of severely brain-damaged survivors with no quality of life.[71]

Nevertheless, even unconscious survivors admitted to hospital with obvious signs of severe damage to the brain and brain stem may show steady improvement with the passage of time.[72] The toddler who survived after submersion for 66 minutes in an icy creek suffered such severe oxygen deprivation that when she recovered consciousness, she was blind. But over the next seven weeks, her cortical blindness slowly resolved and she gradually regained her sight.[73]

The doctor trapped beneath the ice for 80 minutes regained consciousness 14 days after her accident to find herself paralysed from the neck down. She reproached her colleagues, telling them, 'You should have let me die.'[74] As she began to recover from the paralysis, she apologised for her despairing words. Eight months after the accident, her only problem was a slight tingling in her hands.[75] Moreover, she was able to return to work in the hospital where her life was saved.

CHAPTER THIRTY-TWO

THE PROBLEM OF WATER IN THE LUNGS

In 1963, Dr R. K. Haugen, a pathologist in Florida, documented a series of nine deaths due to sudden asphyxiation in people who went out for a celebratory meal, choked on their food, and died clutching their throats.[1] Their companions, including doctors who tried to help them, wrongly assumed that they had died of heart attacks. At post-mortem, Dr Haugen found sizeable pieces of steak or fish stuck fast between the vocal cords or wedged in the larynx or trachea, inhaled too deeply to be seen, and impossible to remove by hand. He suggested that 'the only effective means of treatment is an emergency, on the scene, tracheotomy.' His special report was published in the *Journal of the American Medical Association*, with the title 'The Café Coronary: Sudden Deaths in Restaurants.'

In 1972, Dr Henry Heimlich read an article in the *Sunday New York Times Magazine* about accidental deaths. His attention was caught by the disturbingly high number of choking fatalities – 3,000 Americans died from choking every year.[2] As a thoracic surgeon with a special interest in swallowing disorders, Dr Heimlich decided to search for a simple, fast method that a bystander could use to save someone from choking to death. (When someone is choking, there is no time to send for a doctor.)

He carried out a number of experiments in his laboratory in a Cincinnati hospital. By June 1974, he was satisfied that he had found a practical way to deal with choking caused by food stuck in the throat. Writing in *Emergency Medicine*, a medical newspaper, he suggested that someone who recognises the emergency should stand behind the choking victim, place his fists in the victim's midriff, and pull upwards and backwards with a sudden powerful jerk, several times if needed. This action pushes up the diaphragm, compresses the victim's lungs, forces a jet of air up the windpipe, and expels the obstructing food. His article acknowledged Dr Haugen's report in its title: 'Pop Goes the Café Coronary.'[3]

A syndicated journalist read his article and wrote about the method in his column. Within a week, a retired restaurateur had saved a neighbour who was choking on a piece of chicken. There were reports of similar rescues across the country.[4] The choking remedy worked so well that the editors of the *Journal of the American Medical Association* invited Dr Heimlich to write about his method of subdiaphragmatic compression, which they called 'the Heimlich maneuver'.

In 1975, the *Journal of the American Medical Association* published his paper 'A Life-Saving Maneuver to Prevent Food-Choking' as a Special Communication.[5] He described his first subjects – four anaesthetised beagle dogs. He slipped a cuffed endotracheal tube into a dog's airway, plugged its opening with a rubber stopper, and distended the cuff with enough air to cause total obstruction. Then he tried to expel the tube from the dog's airway. He had no success when he pressed on the animal's chest. The ribs prevented sufficient compression of the lungs, while any rise in pressure inside the lungs simply led to downward movement of the diaphragm. When he pressed beneath the diaphragm, however, the endotracheal tube 'popped out of the trachea'. Next, he packed a dog's throat with raw hamburger meat from the hospital cafeteria. Nothing happened when he pressed on the animal's chest, but when he pressed upwards on the diaphragm, 'The meat shot out of the dog's mouth!'[6]

He then measured the volume and pressure of air expelled from the mouth during subdiaphragmatic compression in ten conscious, non-choking human volunteers. Almost a litre of air (twice the volume of a normal breath) was forcefully expelled within a quarter of a second. In line drawings, he showed how subdiaphragmatic compression could be applied to choking victims whether they remained upright or had collapsed on the floor.

Heimlich added that since the publication of his article in *Emergency Medicine* nine months before, he had received 162 reports of people saved from choking to death, including 41 children. Rescuers succeeded in expelling food, pills, sweets, and even a wad of foam rubber gouged out of a mattress cover. Six people saved themselves from choking to death while they were alone, four by pressing their fist into their own abdomen, one by pressing hard against the back of a kitchen chair, and another against the edge of a sink. Almost as an afterthought, he added: 'Five reports tell of using the maneuver on drowning victims, an application that we did not anticipate.'[7] He wrote, 'Each rescuer stated that water "gushed" from the victim's mouth.' Subdiaphragmatic thrusts appeared to expel inhaled water from the trachea in the same way as they expelled inhaled food.

In 1981, Heimlich published a paper in the *Annals of Emergency Medicine* with the title 'Subdiaphragmatic Pressure to Expel Water from the Lungs of Drowning Persons'.[8] He suggested that subdiaphragmatic pressure might be useful as an initial first-aid measure for drowning victims before rescuers begin mouth-to-mouth resuscitation or cardiopulmonary resuscitation (CPR). He saw the two approaches as complementary, allowing even more lives to be saved.

His suggestion was not taken as an interesting possibility that deserved further investigation, but as a direct attack on mouth-to-mouth resuscitation and CPR. Experts in resuscitation decided that he was a dangerous crank who championed the use of untried subdiaphragmatic pressure in place of proven methods validated by careful research. They were particularly scathing about his claim that some unconscious drowning victims had been resuscitated by subdiaphragmatic pressure alone. These cases were dismissed as 'worthless anecdotal evidence'.[9] The quarrelling escalated into 'the Heimlich controversy', which continues to this day.

◆ ◆ ◆

Some of the doctors who reacted so strongly against Heimlich's suggestion had taken part in what amounted to a medical revolution in the late 1950s and early 1960s when the dramatic arrival of mouth-to-mouth resuscitation and cardiopulmonary resuscitation relegated mechanical methods of 'artificial respiration' to the pages of medical history. All the research that led to the introduction of these new lifesaving techniques took place in hospital wards and laboratories. Between 1954 and 1961, studies were published on mouth-to-mouth resuscitation in patients suffering from poliomyelitis, children admitted to hospital for minor operations, adult volunteers in laboratories in Buffalo, Baltimore and the Army Chemical Center in Edgewood, Maryland, medical patients in cardiac arrest, and psychiatric patients during electroconvulsive therapy.[10] Meanwhile, heart massage was successfully tested on hospital patients who developed cardiac arrest or ventricular fibrillation in the operating theatre or the recovery room.[11] An early trial of cardiopulmonary resuscitation (combining mouth-to-mouth resuscitation with cardiac massage) involved hospital patients in cardiac arrest and medical student volunteers.[12] None of the hospital patients and volunteers involved in any of these studies had inhaled water.

In April 1958, in the *New England Journal of Medicine*, Drs Safar, Escarraga and Elam published an influential study that compared mouth-to-mouth resuscitation with two methods of manual artificial respiration (the chest-pressure arm-lift methods of Silvester and Holger Nielsen) which were widely used in hospitals at that time.[13] Hospital patients and volunteers were deeply anaesthetised, then paralysed with muscle relaxants (the modern equivalents of curare) for up to three hours. While they lay unconscious and unable to breathe for themselves, they were given mouth-to-mouth resuscitation by firefighters, policemen, medical students, doctors, nurses, housewives and Boy Scouts. The study reported that mouth-to-mouth resuscitation was far more effective than manual artificial respiration in getting air into the lungs. Although no drowning victims were involved, the authors included the categorical advice: '*In drowning victims*, who often swallow large amounts of water, the operator should not waste valuable seconds by preliminary turning of the patient. When the patient is removed from the water, he should be placed supine, and the rescuer should start mouth-to-mouth or mouth-to-airway breathing *immediately*.' [Emphases in the original article.]

By 1962, several short instructional films had been made. Raymond Massey, star of *Dr Kildare*, a popular TV show of the day, was chosen as narrator for 'The Pulse of Life',[14] which demonstrated 'procedures for restoring heartbeat and breathing in the crucial first minutes after a heart attack by re-enacting typical cases and showing correct methods of action.' The same methods of resuscitation were then applied to drowning victims.

The firm official advice has continued unchanged for more than fifty years:[15] 'Even if water is aspirated, there is no need to clear the airway of aspirated water.' Hypoxia (low oxygen intake), hypoxaemia (too little oxygen in the blood), hypercarbia (too much carbon dioxide) and ventricular fibrillation are seen as the major dangers in drowning. The danger of water inhalation *per se* is considered unimportant, and has been largely ignored.

This ruling explains why drowning victims are almost always laid face up immediately after rescue, and why attempts are rarely made to remove inhaled water before the start of mouth-to-mouth resuscitation or CPR. Yet during resuscitation at the accident scene, rescuers often have to roll victims to the side again and again to clear the inhaled water that gurgles into their throats.[16] When lifeguards and paramedics take over from rescuers, they often use suction apparatus to clear water and foam from the flooded airways of drowning victims. And after arrival in hospital, emergency staff often have to interrupt urgent resuscitation repeatedly to allow suction of the water/oedema fluid that wells up from the depths of the victims' lungs.[17]

◆ ◆ ◆

In 1982, a year after Heimlich's paper was published in *Annals of Emergency Medicine*, a team of researchers in Pittsburgh embarked on an animal study 'to determine whether gravity drainage or abdominal thrusts would alter the clinical course and recovery of pulmonary function after experimental near drowning with sea water.'[18] The team included Dr Peter Safar, one of the trailblazers of mouth-to-mouth resuscitation and CPR.

They anaesthetised fifteen dogs, asphyxiated them by cutting off their air supply, then poured seawater down endotracheal tubes into their lungs (30ml of water per kilo bodyweight, a lethal amount).[19] Five dogs lay horizontally, five were tipped head-down at an angle of 30°, and five were given several abdominal thrusts. After 2 minutes of 'drowning', water was allowed to drain from their endotracheal tubes.

Within one minute, significantly more water drained from the dogs tipped head down or given abdominal thrusts than the dogs that lay horizontally. These results showed that the Heimlich manoeuvre could, indeed, quickly expel an appreciable amount of the water inhaled during drowning – although the authors did not draw that conclusion. Over 10 minutes, more water drained from the dogs' lungs than the researchers had poured in, indicating that inhaled seawater, with its high salt content, was already pulling fluid from the blood into the alveoli. Again, more water drained from the dogs tipped head down or given abdominal thrusts than the dogs laid horizontally.

The dogs all received a similar sequence of resuscitation procedures 'to simulate on-site basic life support... advanced life support... and hospital care'. They were closely monitored, recording changes in blood oxygenation and a number of physiological tests. Overall, the dogs given initial abdominal thrusts did no better, and no worse, than the others. Six hours after water was poured into their lungs, the dogs were killed, and the ratio between the wet weight and dry weight of each lung was calculated. The dogs given abdominal thrusts had the lowest average wet/dry ratios, suggesting that their lungs were less waterlogged or less congested than the others, but the difference was not found to be statistically significant, and the lungs were not examined under the microscope.

The authors concluded: 'It appears that passive horizontal drainage during emergency resuscitation is sufficient, and that therefore intensive respiratory therapy is what determines oxygenation and shunting more than initial fluid drainage.' One might argue, however, that a number of studies had already shown that dogs did not survive the rapidly deteriorating injury caused by flooding their lungs with such large volumes of water. The experiment was not repeated using smaller volumes of water.

In 1986, a committee at the University of Florida approved further research on resuscitation after drowning, using 22 dogs. Mouth-to-mouth resuscitation was to be used on some of the dogs and the Heimlich manoeuvre on others. Water dyed in contrasting colours would be run into the dogs' lungs and stomachs, to measure the volume expelled during resuscitation. However, at a news conference following the committee's decision, the anaesthetist leading the research said that he was 'concerned that using the Heimlich maneuver prior to mouth-to-mouth ventilation may be inappropriate and possibly dangerous in the treatment of near drowning.'[20] He told reporters, 'Research has shown that oxygen is needed immediately to prevent brain injury or death.' Claiming that 70,000 people survive near-drowning each year using current techniques, he remarked, 'The Heimlich maneuver can also cause people to inhale stomach materials and in essence "drown again". Some people have been injured by the treatment.'

In response, Heimlich wrote to a Florida newspaper dismissing the planned experiments as biased.[21] (This would seem to be fair comment, but was unlikely to improve his chances of friendly collaboration.) Animal rights campaigners held demonstrations in the University campus. Heimlich said that the experiments were unnecessary and cruel. He took a cocker spaniel to a televised press conference and thrust its head into a fish tank to prove his point.[22] The researchers said that they had proposed the experiments to show that the Heimlich manoeuvre was ineffective, and put the research on hold.[23]

◆ ◆ ◆

Although Heimlich's opponents agreed that subdiaphragmatic compression might succeed in expelling water from the mouth of a drowning victim, they argued that the water was more likely to have come from the stomach than from the lungs. They referred to a study by two anaesthetists involved in the early trials of mouth-to-mouth resuscitation twenty years before, published in the *Lancet* in 1962. The anaesthetists noticed: 'When expired-air respiration is used in a case of drowning, water is often seen to run out of the patient's mouth after the first inflations.'[24] A brief description of their experiment seems appropriate because the study has been repeatedly cited as an authoritative reference by those arguing against the use of the Heimlich manoeuvre to expel water from the airways of drowning victims before starting mouth-to-mouth resuscitation.

The anaesthetists studied 'nine bodies soon after death'. They placed cuffed tubes into the trachea and oesophagus and poured a litre of salty water through the tracheal tube into the lungs. They noted that a small volume of fluid (5–40ml, on average 16ml) drained from the lungs after pressure on the front of the chest and abdomen. The bodies were 'turned face-down in a jack-knife position'. Rather larger volumes of fluid (5–600ml, on average 170ml) drained from the lungs after pressure on the back. The bodies were turned face up once more. The lungs were 'ventilated with 1 to 2 litres of air' via the tracheal tube, but no more fluid drained from the lungs. The entire sequence was repeated, with similar results.

Next, the anaesthetists poured salty water through the oesophageal tube into the stomach and measured the volume of fluid (250 to 1,000ml) that was retained in the stomach. The lungs were again inflated with 1 to 2 litres of air, but no fluid drained from either the lungs or the stomach. Finally, 'the stomach was distended with 1 to 2 litres of air through the oesophageal tube', and they noted that 50–500ml of fluid 'came back' from the oesophageal tube.

In their summary, the anaesthetists wrote, 'We conclude that mechanical efforts to drain the lungs are of no practical use'. And they declared, 'When more than a small amount of water flows from the mouth of a drowning patient, subjected to artificial respiration, it comes from the stomach.'

However, the anaesthetists were hampered by their experimental design, and their results did not support these rather dogmatic assertions. First, they had actually succeeded in removing substantial quantities of water from the lungs by brief postural drainage when the bodies were lying face down. (Before the introduction of mouth-to-mouth resuscitation, rescuers laid drowning victims face down while giving 'artificial respiration', and their repeated forceful efforts to pull air into the lungs and push it out again also squeezed water out of the lungs – water that drained, by gravity, onto the ground.)[25]

Second, little or no water would be expected to drain from the lungs of bodies placed on their backs after both face-up and face-down postural drainage. In this position, any remaining water would have settled below the level of the trachea and the tracheal drainage tube. During ventilation or re-ventilation, the small amount of remaining water would have been propelled deeper into the tissues of the lungs. Lungs normally contain more than 5 litres of air after a full inspiration. Lungs deflate after death, but they are easily reflated before rigor mortis sets in, allowing plenty of potential space to accommodate 1 to 2 litres of air.

Third, the gastro-oesophageal sphincter, the ring of muscles which normally prevents stomach contents from flowing back into the oesophagus, relaxes after death.[26] Therefore, it should have been no surprise that water drained from the oesophageal tube after blowing 1 to 2 litres of air into the water-filled stomach of a corpse.

In 1981, when Heimlich suggested that expelling inhaled water from the airways of a drowning victim might improve the effectiveness of resuscitation, his critics ruled against him, citing the 1962 study that claimed that 'mechanical efforts to drain the lungs are of no practical use' and that water flowing from the victim's mouth 'comes from the stomach'.[27]

◆ ◆ ◆

As early as 1959, the risk from regurgitation and vomiting during mouth-to-mouth resuscitation of drowning victims was discussed at a symposium on artificial respiration held at the Royal Society of Medicine in London.[28] Professor Ronald Woolner, director of the Research Department of Anaesthetics at the Royal College of Surgeons, said, 'It could be argued that the supine position required for expired air resuscitation makes airway obstruction – whether by soft tissues, foreign matter or water – more likely, and less easily rectified than does the prone position.' But he supposed that 'obstruction by water is probably not of great practical importance: unless there is no pulmonary circulation, water is absorbed rapidly from the lungs.' A spokesman for the British Red Cross Society, Major-General Hilton-Sergeant, considered that much more investigation was required. He said, 'In the case of casualties with water, food or debris blocking the air passages, the first step to be taken by the first-aider would seem to be to force air out of the lungs, not force air into the lungs.'

In 1960, Dr James Elam was an author of the first published study of successful mouth-to-mouth resuscitation in drowning victims.[29] Its emphasis was on how acceptable the rescuers found their experience of mouth-to-mouth resuscitation, given that 16 of the 21 asphyxiated victims vomited during resuscitation, 2 had convulsions, and 2 died after a brief recovery. In 1988, an Australian study recorded that lifeguards had to deal with vomiting and regurgitation in 68% of victims given mouth-to-mouth resuscitation and 86% of victims given CPR.[30]

In his 1981 article in the *Annals of Emergency Surgery*, Heimlich described several ways of applying subdiaphragmatic pressure to drowning victims. He wrote, 'In shallow water, it is possible to perform the thrusts from behind as in the primary position for choking victims. The rescuer stands or kneels behind the victim, places his arms round the victim's abdomen slightly above the navel and below the rib cage, grasps his own fist with his other hand, and presses into the abdomen with a quick upward thrust.' However, he gave most emphasis to subdiaphragmatic pressure applied when the drowning victim was lying face up 'already in position for mouth-to-mouth resuscitation.'[31] A line drawing shows the rescuer kneeling astride the victim, with his hands about to push down on the victim's midriff (where the stomach is likely to be full of swallowed water and/or food). He conceded, 'There is a possibility of aspiration of fluid when applying subdiaphragmatic pressure in the supine position', but he considered that turning the drowning victim's head to the side would be enough to allow water or gastric contents to flow out of the mouth without causing harm.

Knowing that many victims vomit during resuscitation, Heimlich underestimated the degree of adverse criticism that followed whenever the use of subdiaphragmatic pressure resulted in vomiting before mouth-to-mouth resuscitation had even begun. In 1987, Dr James Orlowski, a specialist in intensive care, published a case report of a ten-year-old boy with epilepsy who was rescued from a swimming pool after a brief submersion. The Heimlich manoeuvre was followed by vomiting, which interfered with cardiopulmonary resuscitation.[32] The boy suffered severe brain damage and remained totally helpless in a vegetative state until his death seven years later. Since then, vomiting has been viewed as a dangerous complication of the Heimlich manoeuvre, even though vomiting of drowning victims during cardiopulmonary resuscitation is accepted as common and unavoidable.

◆ ◆ ◆

Heimlich and the medical opinion-formers locked horns. Heimlich argued stubbornly, and incorrectly, that flooding of the lungs occurs routinely in drowning victims,[33] and that water in the airways prevents air from reaching the alveoli.[34] His opponents insisted, just as stubbornly and perhaps with more than a little sophistry, that water in the lungs is not part of the problem. 'Some victims aspirate no water because they develop laryngospasm or breath-holding,' they maintained, 'Only a modest amount of water is aspirated by the majority of drowning victims, and aspirated water is rapidly absorbed into the central circulation.'[35]

Yet it has been clear from the earliest days of mouth-to-mouth resuscitation that the entry of water into the lungs causes serious injury over and above the effect of asphyxiation. Indeed, in 1960, Dr Safar and his colleagues used dogs in an experiment 'designed to simulate human victims of submersion'.[36] The dogs' air supply was cut off until they stopped trying to breathe. Then either they were deprived of air for an additional 30 seconds or their lungs were flooded with water for 30 seconds. Attempts were made to resuscitate the dogs, using intermittent positive pressure with room air. The dogs with water in their lungs all died, whereas the asphyxiated controls all survived.

It is illogical to deny the importance of water inhalation considering how much research on drowning has involved pouring measured quantities of water into the airways of rats, rabbits, sheep and dogs. Professor Jerome Modell, whose research on drowning is internationally recognised, found a close correlation between the volume of water introduced and the damaging effect on the animal.[37] An authoritative text on post-mortem findings in human drowning victims states that changes seen in the lungs 'are related to the penetration of drowning liquid into the airways.'[38] A recent review by an international group of experts comments, 'If the person is rescued alive, the clinical picture is determined predominantly by the amount of water that has been aspirated and its effects.'[39]

Professor John Pearn, another major figure in drowning research, pointed out that 'within minutes after inhalation of small quantities of water (2.5ml/kg)' as much as 75% of the blood flowing through the pulmonary circulation leaves the lungs without capturing oxygen.[40] And it is precisely because the inhalation of such a relatively small volume of water (175ml in a 70kg man, only enough to fill a teacup) may result in such a serious loss of lung function that initial subdiaphragmatic thrusts might allow more resuscitations to succeed.

Normally, the trachea and bronchi contain about 150ml of air.[41] The protective reflex of laryngospasm is triggered as soon as a drowning victim begins to inhale water. If the victim is rescued while his throat is still clamped shut by active laryngospasm, only a small volume of water will have penetrated further than his vocal cords. Yet, when mouth-to-mouth resuscitation is begun at this early stage in the drowning process, the first rescue breaths will force any water in his throat, trachea and bronchi deeper into his lungs. The presence of inhaled water will not prevent the entry of the rescuer's expired air,[42] but every rescue breath will propel the inhaled water along the bronchioles into the alveoli, damaging the delicate surfaces where oxygen-capture takes place and stimulating the release of oedema fluid into the lungs. It seems reasonable, therefore, to suppose that it might be a good idea to expel any water present in the throat, trachea and bronchi before the start of mouth-to-mouth resuscitation.

Once the protection of laryngospasm is lost, the drowning victim draws water deeper into his lungs. A brief series of subdiaphragmatic thrusts would not remove all the water from the lungs before the start of mouth-to-mouth resuscitation. Nevertheless, the expulsion of water still in the trachea and bronchi, together with some of the watery foam clogging the airways, might allow the incoming rescue breaths to reach more alveoli still capable of capturing oxygen.

◆ ◆ ◆

With the opening of water parks and wave pools across America in the 1970s, there was a marked increase in the number of drownings in lifeguarded swimming pools. To protect their customers, water parks hired lifeguards in large numbers and the training and accreditation of lifeguards became a serious business. John Hunsucker, Professor Emeritus of Engineering at the University of Houston, served as a member of the Lifeguarding/Bather Supervision Technical Committee that gives advice on lifeguard training and cardiopulmonary resuscitation to the US Centers for Disease Control and Prevention in Atlanta.[43] After working as a lifeguard in his youth, he became an expert trainer of lifeguards for the American Red Cross. In 1974, he formed a company to provide lifeguards for water parks,[44] training them to scan the water intently, to dive to the rescue as soon as they saw a swimmer in trouble, and to give immediate resuscitation whenever it was needed.

In 1995, John Hunsucker began to train lifeguards to give a modified version of the Heimlich manoeuvre before they started mouth-to-mouth resuscitation.[45] In Hunsucker's version, as soon as the lifeguard reaches an unconscious swimmer, or while towing an unconscious swimmer to the edge of the pool, he gives the victim 5 brief subdiaphragmatic thrusts to expel water from the airway.[46] In shallow water, the lifeguard stands, supports the victim from behind, lifts him into a sitting position with his head clear of the water surface and tips the victim slightly forward. In deep water, the lifeguard maintains his buoyancy by wedging his rescue tube between his chest and the victim's back. Sometimes the lifeguard follows the subdiaphragmatic thrusts with one or two rescue breaths while still in the water. If the victim has not started to breathe again by the time he is lifted from the pool, the lifeguard immediately begins conventional mouth-to-mouth resuscitation or cardiopulmonary resuscitation (CPR).

It takes a rescuing lifeguard only 6 seconds to give 5 in-water Heimlich manoeuvres to an unconscious non-breathing drowning victim,[47] so that the start of mouth-to-mouth resuscitation is scarcely delayed. The subdiaphragmatic thrusts force the victim's diaphragm upwards, compressing the lungs. The sudden increase in pressure in the airway opens the throat if it is still clamped shut by laryngospasm, and ejects any water in the airway. (Witnesses report that as much as a cupful of water is expelled.)[48] By holding the drowning victim upright with his head forward, the lifeguard ensures that ejected water splashes harmlessly into the pool,[49] and even if the victim vomits, he is unlikely to inhale stomach contents into his lungs. After each subdiaphragmatic thrust, the pressure on the abdomen is released, and, as the diaphragm descends, some air is pulled into the lungs.

John Hunsucker agreed that some of the water may, indeed, come from the stomach rather than the lungs, but he suggested that partial emptying of the stomach might allow the diaphragm to move more freely.[50] In an interview in 2007, he dismissed the complaint that subdiaphragmatic thrusts cause vomiting, pointing out that vomiting is common during mouth-to-mouth resuscitation and CPR: 'Either way they're going to vomit, so what's the difference.'[51]

In 1995, another prominent lifeguard-training company, Ellis & Associates, Inc., introduced the Heimlich manoeuvre as the first intervention during resuscitation. They were able to compare results before and after the change in protocols. They found 'similar and successful outcomes' in casualties given immediate CPR or initial Heimlich manoeuvre followed by CPR, but they noted, that casualties given immediate CPR stayed in hospital for an average of four days, whereas casualties given the Heimlich manoeuvre before CPR went home after only a few hours. In 2000, however, Ellis changed its protocols 'to align its training procedures with the standard approach.'[52]

In 2011, John Hunsucker and his colleague Scott Davison published the outcome of 56,000 rescues over a decade in water parks staffed by lifeguards trained by his company.[53] (There were 63,800,000 admissions to the parks). In the great majority of rescues, the lifeguards arrived in time to rescue swimmers before they lost consciousness. However, there were 32 rescues of unconscious, non-breathing drowning victims. The lifeguards revived 14 of these unconscious 'nonbreathers' using in-water Heimlich manoeuvres alone, without the need for mouth-to-mouth resuscitation. A further 14 victims of near-drowning recovered after poolside mouth-to-mouth resuscitation or CPR, but 4 victims failed to respond to all efforts at resuscitation and died of drowning.

The majority of the victims were rescued from wave pools and activity pools in large water parks. Thanks to the vigilance of the lifeguards and their rapid interventions, the victims suffered relatively brief submersions. Early rescue made a major contribution to the lifeguards' success. Yet in-water Heimlich manoeuvres seem to have played a useful part in resuscitation, since almost half of the unconscious non-breathing victims began to breathe for themselves as soon as they were given 5 subdiaphragmatic thrusts, and three-quarters of the victims who needed mouth-to-mouth resuscitation or CPR after rescue were rated as 'Alert' at the poolside.

◆ ◆ ◆

Lifeguards who followed John Hunsucker's protocol had an impressive safety record, with one fatality per 15,950,000 water park visitors, yet he has been the subject of repeated newspaper attacks. The usual complaint is that the use of initial in-water Heimlich manoeuvres does not conform to the *American Heart Association Guidelines for Cardiopulmonary Resuscitation and Emergency Cardiovascular Care*.[54] The Guidelines, published in 2010, fill more than 300 pages and cover the entire spectrum of emergency resuscitation, with the emphasis on resuscitation of patients with cardiovascular disease and stroke.[55] Drowning is covered very briefly, with three paragraphs in Part 5: 'Basic Life Support',[56] two paragraphs in Part 13: 'Pediatric Basic Life Support',[57] and less than two pages in Part 12: 'Cardiac Arrest in Special Situations' (between 'Cardiac Arrest in Avalanche Victims' and 'Cardiac Arrest Associated With Electric Shock and Lightning Strikes').[58]

The Guidelines support the use of the Heimlich manoeuvre for choking, but advise against its use for drowning.[59] In 'Basic Life Support' the Guidelines suggest: 'There is no evidence that water acts as an obstructive foreign body. Maneuvers to relieve foreign-body airway obstruction are not recommended for drowning victims because such maneuvers are not necessary and they can cause injury, vomiting, aspiration, and delay of CPR.' In 'Cardiac Arrest in Special Situations. Part 12.11: Drowning' the Guidelines comment, 'Attempts to remove water from the breathing passages by any means other than suction (e.g. abdominal thrusts or the Heimlich maneuver) are unnecessary and potentially dangerous. The routine use of abdominal thrusts or the Heimlich maneuver for drowning victims is not recommended.' The updated AHA Guidelines published in 2015 and 2018 did not change this advice.

Most lifesaving organisations take their cue directly or indirectly from the Guidelines. If resuscitation of a drowning victim is unsuccessful, the lifeguard is unlikely to be censured provided that the Guidelines were followed. Orthodoxy demands that mouth-to-mouth resuscitation should begin immediately. Any delay is liable to be criticised. If a lifeguard deviates from the Guidelines and the casualty dies, the lifeguard and the swimming pool management is likely to face serious consequences, including job loss, legal costs, denial of insurance payments and criminal trials.[60]

Failure to follow the Guidelines may result in censure even when the drowning victim survives. An interesting example of this mindset can be found in a case report of a successful resuscitation after a near-fatal drowning accident on the north-east coast of England.[61] While attempting to rescue a swimmer from the cold water of the North Sea in 1990, a fit young lifeguard caught his clothing on a hidden obstacle, which trapped him underwater. When his colleagues reached him, he appeared to be dead. Twenty minutes went by before they were able to bring him ashore and begin resuscitation. They laid him face down and gave him artificial respiration using the Holger Nielsen method. Then he was turned onto his back and an off-duty ambulance woman organised standard cardiopulmonary resuscitation. After 15 minutes, he gasped, a weak pulse was felt, and he vomited a large quantity of water. He arrived at hospital in a coma, and was still unresponsive when he was admitted to the intensive care unit. After a day of mechanical ventilation on a respirator, he was able to breathe for himself. He needed treatment for a chest infection, but he was discharged home without brain damage.

The authors correctly pointed out, 'This case illustrates the value of continuing resuscitative efforts even in the apparently dead drowning victim.' Yet even though the drowning victim made a full recovery, they complained, 'When landed on the beach obsolete and ineffectual methods of resuscitation were tried.' Then they complained once more, 'The lifeguards involved in this case were ignorant of modern methods of resuscitation.' It seems not to have been considered that the initial artificial respiration by the Holger Nielsen method might have contributed to the successful outcome by forcing much of the inhaled seawater out of the victim's lungs before mouth-to-mouth resuscitation began.

In their discussion, they referred to 'a similarly submerged, trapped and apparently drowned casualty'. It is clear from reading this second case report that cardiopulmonary resuscitation began as soon as rescuers brought the young man to land, but this time there was no successful outcome.[62] In hospital, the victim's heartbeat returned, and he began to breathe for himself. A few hours later his condition deteriorated due to lung damage and pulmonary oedema caused by water inhalation (the case report refers to 'secondary drowning'). Despite mechanical ventilation on a respirator, his lungs lost their ability to capture oxygen, and he died from drowning.

◆ ◆ ◆

Lifesaving organisations consider that the removal of inhaled water is appropriate in some circumstances. Lifeguards and paramedics use suction apparatus at the accident scene.[63] In the Accident and Emergency Department, hospital staff may need repeated suction to remove fluid from the airway before the victim can be connected to a respirator.[64]

A mother who has just pulled her toddler from the bottom of a hotel swimming pool has no access to suction apparatus. But by applying Hunsucker's version of the Heimlich manoeuvre, she could clear some of the water in the child's lungs – and all of the water in the child's airways – before she starts mouth-to-mouth resuscitation. Hunsucker pointed out that the technique for giving in-water Heimlich manoeuvres to expel water from the airways is extremely easy to teach, to learn, and to carry out.[65]

One 'anecdotal' report by a trained water safety instructor describes just such an abdominal compression, applied unintentionally.[66] The instructor was standing in shallow water as he lifted a drowning child up to a lifeguard at the edge of a swimming pool. The lifeguard put his arms round the child's body with his hands just below the ribcage, took the weight of the child and stood up. His action pulled the child's upper abdomen up and in, whereupon: 'The child explosively released fluid from his mouth (could not tell you where it came from) and almost IMMEDIATELY regained consciousness.' [Emphasis in the original.] No further resuscitation was needed. The lifeguard returned the child to his parents, and advised them to take him to the doctor.

Dr Heimlich was responding to similar 'anecdotal' witness reports in 1981 when he published his paper 'Subdiaphragmatic Pressure to Expel Water From the Lungs of Drowning Persons' in the *Annals of Emergency Medicine*.[67] As a thoracic surgeon nearing retirement, he was not in a position to collect controlled experimental data or organise human trials to prove or disprove his theory.

In 2009, an editorial in *Anesthesiology* with the title 'Drowning: A Cry for Help' called for 'a new age of drowning research.' The writers ask, 'What interventions, achievable in the field, can be implemented to improve clinical outcome?'[68] At the World Conference on Drowning Prevention in Vietnam in 2011, Professor Joost Bierens, editor of the excellent *Handbook of Drowning*, reminded his audience of the many differences between cardiac arrest due to heart disease and cardiac arrest during drowning. He suggested that there are 'many reasons to believe that the standard resuscitation guidelines are suboptimal for drowning'. He called for 'small scaled research projects' that might improve resuscitation after drowning accidents.[69]

The introduction to the 2005 edition of the Guidelines admits that experts are often confronted with the need to make recommendations on cardiopulmonary resuscitation 'on the basis of results from human trials that reported only intermediate outcomes, non-randomised or retrospective observational studies, animal models, or extrapolations.'[70] Quoting this comment, John Hunsucker suggested that the documented resuscitations of 28 out of 32 unconscious non-breathing drowning victims reported in 2011 in the *International Journal of Injury Control and Safety Promotion* might be considered a 'retrospective observational study'.[71] He gave his strong support to 'any clinical study of drowning that would increase survival rates'.

The preferred methods of resuscitation of drowning victims have been subject to many changes over the previous century. Under careful scrutiny, it should be possible, and ethical, to organise a trial over the course of a summer season on a selected holiday beach, in which trained lifeguards use the Hunsucker technique before they start mouth-to-mouth resuscitation or CPR. The trial could be quickly abandoned if there appeared to be any harmful effects (if subdiaphragmatic compression caused injury, if it resulted in more frequent aspiration of vomit, if the start of mouth-to-mouth resuscitation or CPR was delayed for more than a few seconds, or if paramedics had to deal with an increased proportion of seriously affected drowning victims). But the trial could be continued if there appeared to be better outcomes after drowning accidents (if consciousness returned more quickly, if vomiting was more easily controlled, if time in hospital was shorter, if brain damage in survivors was less frequent and less severe, or if more drowning victims survived.)

◆ ◆ ◆

Since mouth-to-mouth resuscitation and CPR were introduced, very many lives have been saved. Their effectiveness is not in doubt. Yet despite fast and expert resuscitation, a distressing number of drowning victims die or suffer permanent brain damage. It seems curious that medical scientists have been so vehemently opposed to a simple first-aid measure designed to expel inhaled water from the airways of a drowning victim at the scene of the accident – a measure which might increase the effectiveness of mouth-to-mouth resuscitation and CPR. Perhaps it is time to question whether water in the airways is quite as innocuous as we have been led to believe. After all, our lungs evolved to capture oxygen from air, not from water.

CHAPTER THIRTY-THREE

INQUEST VERDICTS AND DROWNING STATISTICS

In England, Wales and Northern Ireland, every drowning death and all bodies found in water must be reported to the Coroner.[1] In Scotland, they must be reported to the Procurator Fiscal.[2] Police and forensic pathologists work together to investigate every incident. If the victim's identity is known, the medical records are examined for details of relevant physical or mental problems. After preliminary enquiries and post-mortem examination, an Inquest is opened to register the death, to record when and where the body was found, and to allow the Coroner to authorise burial. Then the Inquest is adjourned until further investigations have been completed.[3]

Inquests are also held on those who survived the immediate effects of drowning only to die in hospital after treatment failed to save their lives. Inquests are not held in Scotland, but the Procurator Fiscal may decide to hold a Fatal Injury Inquiry 'if it is in the public interest'.[4]

Police and pathologists search for factors that increased the risk of drowning or prevented escape from the water. Blood analysis may show a high level of alcohol, a low level of glucose or the presence of illegal drugs. Anticonvulsant medication is reviewed in a victim with epilepsy. If the victim dived into shallow water, X-rays of the cervical spine are taken. If the victim is unknown, dental X-rays and specimens for DNA analysis are preserved to aid future identification. Some time later, the Inquest is re-opened, the fatal incident is described, and a fuller hearing of evidence takes place to decide 'how the deceased came by his/her death'.[5] Sometimes a jury is appointed, but often the Coroner alone decides the Inquest verdict.

In the majority of cases, the verdict appears clear-cut: Accidental Death, Death by Natural Causes or Suicide. In other cases, the verdict suggests a degree of uncertainty about what happened: Open Verdict or Narrative Verdict. In a few cases, where a crime is suspected, the verdict in the Coroner's Court must await the verdict reached in a criminal trial.

ACCIDENTAL DEATH

If there is a clear explanation of why the victim entered the water and how the victim met his/her death, and if the pathologist finds evidence of drowning, the verdict is likely to be Accidental Death due to drowning.[6]

+ A child who could not swim fell into the swimming pool unobserved.[7]
+ A teenager jumped from the edge of a quarry and never surfaced.[8]
+ A canoeist capsized in white-water rapids and failed to reach the riverbank.[9]
+ An elderly tourist was snorkelling when he was found floating face down.[10]

An expert yachtsman was racing in the Bristol Channel in February 2015. He decided to wear his favourite trainers instead of the more appropriate deck shoes. He slipped on deck, skidded under the safety barrier, and fell into the sea.[11] He had stopped breathing by the time his crew mates pulled him back on board, and he did not respond to resuscitation. At post-mortem, there was no evidence of drowning, but the pathologist found that he had died from a massive heart attack, brought on, he considered, by the 'cold shock' of falling into water at 6°C. The Inquest verdict was Accidental Death.

DEATH BY NATURAL CAUSES

After a sudden fatality in the water, the pathologist may fail to find evidence of drowning but discovers signs that death was due to serious illness.[12] In 2013, a scuba-diver disappeared during a 50 metre descent in Wastwater, England's deepest lake. After two days of intensive searching, his body was recovered from the bottom of the lake.[13] At post-mortem, the pathologist found a blood clot in a coronary artery and damaged heart muscle typical of a heart attack. The Inquest verdict was Death by Natural Causes.[14]

In November 2014, a windsurfer was found floating in the Solent after falling into the sea. He had been prescribed medication for irregular heartbeats, but he never took the tablets. At post-mortem, there was no evidence of drowning, but his coronary arteries were almost totally blocked. The Inquest verdict was Death by Natural Causes.[15]

SUICIDE

Before a verdict of suicide is given, the Coroner must be satisfied, 'on the balance of probabilities', that the victim intended to take his or her own life.[16] In April 2010, a passenger jumped into deep water from a steamer on a Cumbrian lake. The next day, his wife received his letter telling her that he would not be coming home because of a 'financial disaster'. When his body was recovered six days later, a rucksack hidden under his jacket contained two large rocks.[17] At the Inquest, the required standard of proof was considered to have been established, and the Coroner delivered a verdict of Suicide.

OPEN VERDICT

Coroners in England and Wales bring in an Open Verdict when a death by drowning does not fit into one of the typical accident scenarios, there are no witnesses to describe what happened, no one can explain exactly 'how the deceased came by his/her death', and they are not satisfied 'on the balance of probabilities' that the facts prove suicidal intent.[18] Even when someone was seen to jump from a bridge, unless the evidence at the Inquest establishes that the victim intended to commit suicide by drowning and succeeded in carrying out the intention, the Coroner would usually give an Open Verdict.[19] And undoubtedly, Coroners pronounce an Open Verdict on some victims of fatal drowning accidents.

In April 2011, a 70-year-old man was found drowned in Sharpness Canal near Gloucester. Fraudsters had left him with debts of £40,000, and in 2009 he was briefly admitted to a mental hospital after cutting his wrists. The Coroner raised the possibility that he had been relieving himself in the canal before falling in. He said, 'I do not think that there is enough evidence to give a verdict of suicide, but I cannot be sure that it was an accident either.' He gave an Open Verdict.[20]

In 2013, the body of a young man was spotted floating face down in a Huddersfield canal. He had been walking to work along a muddy towpath. His wallet and mobile phone were still in his pocket. There was no evidence of a struggle, and no suggestion of self-harm.[21] The pathologist confirmed that the victim had been alive when he went into the water, where he drowned. He also discovered a small patch of fatty tissue in the heart. Laboratory investigations were under way to confirm or exclude a diagnosis of cardiomyopathy. Yet, even if the tests were positive, there was no way of knowing whether the onset of an abnormal heart rhythm caused the young man to lose consciousness and fall into the canal. His death may well have been an accident, but doubt remained. The Coroner said, 'There is a degree of unsolvable mystery as to how [he] came into the water.' He gave an Open Verdict.

NARRATIVE VERDICT

Instead of a standard 'short form' verdict, the Coroner sometimes chooses to give a Narrative Verdict, setting out the succession of events that led to the drowning death, giving additional details or a longer explanation of why things went wrong.[22] In 2010, fishermen found a man lying dead in a car parked on a tidal causeway in a nature reserve on the Kent coast.[23] At the Inquest, the police and the pathologist were able to explain how he came by his death. He had recently begun treatment for diabetes. He consulted his doctor two days before he died and was due to have a blood test on the day his body was found. While he was birdwatching, he lost consciousness due to the worsening state of his diabetes. Then the tide came in, submerging his car and drowning him. The Coroner delivered a Narrative Verdict confirming that he 'died by drowning having fallen into a coma as a result of severe ketosis caused by diabetes'.

The dead man had not told his father that he had diabetes, and after the verdict was delivered, his father said, 'I always knew he wouldn't have killed himself. There was never any doubt. But I couldn't work out why he wouldn't have woken up when the tide came in. That's the question I wanted an answer to, and I was given it today.'

In 2013, an elderly passenger became ill on a cruise to Norway. A lifeboat was sent to take her to hospital. As she was being transferred from the cruise ship to the lifeboat, she fell from a stretcher into the freezing water of the Arctic Ocean. She was wearing only a thin nightdress, without a life jacket. It was 8 minutes before she was pulled from the sea, and as she struggled to stay at the surface, she inhaled seawater. After spending two weeks in intensive care in Norway, she was flown back to hospital in England, where she died a week later.[24] The Coroner issued a Narrative Verdict, ruling that, while her death was as a consequence of her underlying health condition, the immersion in cold sea water was a significant factor and accelerated her death.

The Coroner sent an official 'Preventing Future Deaths Report' to the Norwegian rescue service and the cruise operators.[25] This formal document asked them to review the procedures, practices, risk assessments and equipment relating to ship-to-ship medical evacuations, and informed them that they were under a duty to respond within 56 days, giving details of the action they had taken or proposed to take to avoid similar tragedies in the future. (By demanding a written response to an official 'Preventing Future Deaths Report', the Coroner and the Crown Prosecution Service can apply pressure to improve safety standards.)

MURDER AND MANSLAUGHTER

A small number of drownings are the result of homicide. The Director of Public Prosecutions is informed if the evidence presented at the Inquest suggests that death was the result of unlawful killing. Since criminal liability is a matter for the Police and the Law Courts, the Inquest will be adjourned until the Court hearings are concluded. The evidence presented in the Law Court must prove homicide 'beyond reasonable doubt'.[26]

In Dorset, a group of schoolchildren were celebrating the end of a GCSE exam in June 2004 when two boys lifted a terrified classmate over the parapet of a bridge. Then, 'for a laugh', they dropped him into the River Stour 12 feet below, although they knew that he could not swim.[27] He resurfaced briefly, and another boy managed to hold on to him for a moment before he sank and disappeared. Police pulled his body from the water three hours later. At a trial at Bournemouth Crown Court, both schoolboys were found guilty of Manslaughter and jailed.

In Bedford two days after Christmas in 2006, two young men marched a seventeen-year-old student to a cash machine, forced him to enter his PIN number, punched him repeatedly, then dragged him to a bridge over the Great Ouse, and threw him into the swollen, icy river.[28] The student's body was found by police divers eight days later. The incident had been recorded on CCTV. Both attackers were found guilty of Murder. They were given long prison sentences.

Most childhood bathtub drownings are tragic accidents involving babies aged 8 to15 months who were left alone for a minute or two during their bath.[29] A different scenario warrants intense scrutiny. Police and pathologists look closely for signs of non-accidental injury. A full post-mortem X-ray survey may reveal old, healed fractures, signs of previous episodes of child abuse.[30]

In 2012, a day after celebrating her baby's first birthday, a young woman left him and his two-year-old brother alone in the bath for more than half an hour while she talked to neighbours in the garden. When she returned to her flat, she found the baby lying underwater, blue and unconscious.[31] Neighbours tried to revive the baby before the ambulance arrived. The mother told paramedics that her two-year-old son was jealous of the baby and pushed him into the water when her back was turned for a moment. The baby died in hospital four days later. At Guildford Crown Court, his mother admitted to a charge of Manslaughter. She was jailed for four years.

As well as Murder and Manslaughter, the Chief Coroner's advice on unlawful killing covers Infanticide.[32] These incidents are rare, and investigation is particularly difficult in cases of concealed pregnancy followed by the sudden onset of labour while the mother is taking a bath.[33] Pathologists can usually tell if a baby was stillborn or was born alive, attempted to breathe, and inhaled water instead of air. Nevertheless, the verdict at Inquest will depend on subjective interpretations of the circumstances, especially of the intention of the mother.

PROBLEMS IN DECIDING 'HOW THE DECEASED CAME BY HIS/HER DEATH'

The process of decision-making sounds fairly straightforward, but problems may arise at any stage in the investigation of a drowning death, and nuances may influence the choice of verdict. The diagnosis of death by drowning relies to some extent on circumstantial evidence, witness accounts, and exclusion of other causes.[34] At post-mortem, watery foam in the airways is suggestive of drowning, but similar foam may sometimes appear in severe pulmonary oedema due to congestive heart failure.[35] Death by drowning is probable if diatoms identified in the water are also founded in the victim's bone marrow, but a negative diatom test does not rule out drowning.[36] While the lungs are often bulky, heavy, and waterlogged, the signs of drowning may be limited to small areas of collapse or over-distension of the alveoli, or the presence of a small amount of oedema fluid.[37] Even these minimal signs may be absent when an infant or child drowns after a brief submersion.[38]

When someone falls or jumps into water below 15°C, the sudden 'cold shock' on entry may trigger an abnormal cardiac rhythm or stop the heartbeat completely due to the inhibition of the vagus nerve.[39] Such sudden deaths at the moment of immersion are known to occur in fit young adults, leaving no evidence of water inhalation.[40] In Cambridge in 2006, a 21-year-old graduate disappeared after attending a formal dinner at Magdalene College. A rower found his body in the River Cam,[41] and its positioning 'led the police to suspect he had been answering a call of nature when he fell into the water'. His blood alcohol level was twice the drink-drive limit. At the Inquest, the pathologist reported that he had died of 'acute heart failure consistent with sudden immersion in cold water'. The Coroner gave a verdict of Accidental Death.

Drowning deaths are uncommon in public swimming pools with lifeguards in attendance. But as Professor John Pearn pointed out, a fatal accident may be followed by 'major medicolegal, civil, insurance and regulatory consequences' which affect the careers of those involved in the rescue, the attempted resuscitation and the management of the pool.[42] He advised pathologists to carry out 'the widest array of appropriate post-mortem investigations' in order to discover and explain the likely cause of death.

Even when the pathologist finds clear evidence of death by drowning, with inhaled water in the victim's lung, it may not be possible to distinguish accidental death from death brought on by sudden incapacitating illness, or from suicide or homicide. Injuries to the skin and internal organs caused by buffeting after death must be distinguished from injuries suffered during life, whether caused by a fall (with impact against the water surface or projecting rocks), by self-inflicted wounding, or by a homicidal attack at the place where the body was found or elsewhere.[43]

It is possible that some drowning deaths are unrecognised murders. Without the reliable testimony of witnesses or recordings from surveillance cameras, it may be impossible to judge whether a drowning victim entered the water by choice or was pushed into the water by someone else (even by a member of the family). When a body is found in woodland, the police seek for signs of a crime. In contrast, a body found in a lake is likely to be considered the victim of a tragic accident rather than the victim of a homicidal attack.[44] Without a weapon or signs of a struggle, a crime is easily overlooked. Indeed, homicidal drownings may be medically undetectable. As American investigators point out, 'The body does not need to be disposed of, and the perpetrator often receives much sympathetic attention and possibly accidental death life insurance money.'

In the United Kingdom, a score of homicidal drownings are recognised every year. It is likely that a few are missed, especially those with no superficial injuries. In the notorious 'Brides in the Bath' murders, George Joseph Smith drowned three young women between July 1912 and December 1914. Each Inquest concluded that the victim had died of accidental drowning while taking a bath. Foul play was not suspected until police were sent newspaper clippings that described almost identical fatalities in Blackpool and London, suggesting the work of a serial killer.[45]

If a submerged body is discovered after weeks or months in the water, a decision on the cause of death may be impossible because of tissue decomposition.[46] In rare cases, the victim's body is never found, although the Coroner may be able to reach a verdict if the circumstantial evidence is strong enough. In 2005, a mother and her two children were overwhelmed by waves and washed from the promenade at Scarborough. Although the body of one of the children was not recovered, the Coroner pronounced a verdict of Accidental Death on all three victims.[47]

Conspiracy theories sometimes arise when the missing victim is a prominent public figure. In 1967, Harold Holt, the Australian Prime Minister disappeared, presumed drowned, while swimming in strong surf on a peninsula south of Melbourne. His death was announced two days later, after an intensive search failed to find his body, but it was not until 2005 that the state Coroner officially confirmed that he had drowned.[48] (Until the law was changed in 1985, Australian Coroners were unable to investigate a death in the absence of a body.)

In 1974, a British politician left a suicide note and a pile of clothes on an American beach, with a trail of footsteps leading out to the edge of the ocean. Although no body was found, he was thought to have drowned – until he was located in Australia three months later, alive and well.[49] (The Australian police thought at first that they had arrested Lord Lucan.)

PROBLEMS IN CODING AND PRESENTING STATISTICS ON DROWNING DEATHS

Once the Inquest verdict is reached, it is coded for statistical analysis. Researchers meet with unexpected difficulties when they try to compare the death rate from drowning in one country with another, or the pattern of drowning in one decade with another.[50] They discover that the collection of data on cause of death differs from country to country. Indeed, many countries have little or no organised records unless death takes place in hospital, so that rural drowning deaths are excluded from the statistics. Even in countries where the records are more complete and the statistical analysis is linked to the International Classification of Disease, researchers are confronted by categories and coding choices that have changed over the years. (The current version is the Tenth Revision – ICD-10).[51]

In 1996, researchers linked to the International Collaborative Effort (ICE) on Injury Statistics published a study on 15 years of official electronic mortality data files.[52] Over that time, 1977–1994, drowning deaths were recorded under six codes:

+ E910: Accidental drowning and submersion

- E954: Suicide and self-inflicted injury by submersion (drowning)
- E964: Assault by submersion (drowning)
- E984: Submersion (drowning) undetermined whether accidentally or purposely inflicted
- E830: Accident to watercraft causing submersion
- E832: Other accidental submersion or drowning in water transport accident

The researchers discovered that more than 20% of drowning deaths in New Zealand were not recorded under these official codes for drowning, but under a variety of other codes (including E816: Motor vehicle traffic accidents due to loss of control, without collision on a highway, or E957: Suicide and self-inflicted injuries by jumping from a high place).[53] The consequence of these 'hidden drownings' was to underestimate the frequency of drowning. Furthermore, the need for preventive measures remained unrecognised, delaying the provision of sturdy railings to stop cars from swerving off unprotected roads into the sea or the installation of safety barriers on bridges to deter people from jumping to their deaths.

◆ ◆ ◆

In 2011, an editorial in the *British Medical Journal* stated, 'Most deaths given open verdicts are likely to be suicides'.[54] The authors suggested that some Coroners may give open or accidental verdicts in the belief that this avoids adding to a family's distress. They commented, 'Suicides may be especially difficult to identify when methods such as drowning and overdose are used and the intent is unclear.' (A Coroner cannot give a verdict of Suicide without evidence that the victim intended to die.)

The Office of National Statistics (ONS) deals with the possibility of underestimating the number of suicides by the use of a simple but rather dubious expedient. In the official statistics for England and Wales, all deaths with an Inquest verdict of either Suicide or Open Verdict are counted as suicide.[55] In 2001, three distinguished psychiatrists queried this statistical manipulation. They wrote, 'The assumption that all open verdicts are suicides could substantially distort the 'true' suicide rate.'[56] They assessed all cases during the decade 1985–1994 in which the Coroner in Newcastle-upon-Tyne recorded a verdict of Suicide (188 cases) or Open Verdict (185 cases, which included 'babies found dead'). They considered that suicide was impossible or unlikely in 14% of the cases recorded as Open Verdict. They commented, 'Suicide rates based on suicide verdicts alone are clearly an underestimation of the true rate, yet including all open verdicts may equally be an approximation.'

An increasing number of Inquests end in a Narrative Verdict, often as an alternative to an Open Verdict in cases where the decision is difficult, or when the Coroner wishes to give more detail.[57] If a Coroner gives a Narrative Verdict with no indication of intention of the victim, a drowning death must be coded as an Accidental Death.[58] As an added complication, Narrative Verdicts are more frequent in some parts of the country than in others. The psychiatrists who drew attention to this geographical variation commented, 'Coroners who record a high proportion of narrative verdicts record fewer suicide verdicts,'[59] and warned that the frequent use of narrative verdicts may lead to the underestimation of the incidence of suicide and to the misinterpretation of the results of local attempts at suicide prevention.

There were no doubts about the verdicts of Suicide at the Inquest of two identical twin sisters, elderly and recently bereaved, who walked into the sea together in May 2004, with suicide notes in their pockets.[60] However, when someone drowns in a car driven at speed into water after an angry argument with a disaffected lover, the Inquest verdict often depends on guesses about circumstances and intent.

+ At the Inquest of a woman who was seen smoking a cigarette as her car slowly sank in Poole Harbour, the verdict was Suicide.[61]
+ At the Inquest of a mother and her two children who drowned when her car veered off a rutted track into a river, the three verdicts were Accidental Death.[62]
+ At the Inquest of a man who smashed his car through a barrier and drove over a cliff into the sea, the Coroner recorded an Open Verdict.[63]

+ + +

In 1996, another study linked to the ICE on Injury Statistics highlighted the problem of comparing the computerised mortality statistics of drowning deaths in different countries. (In these coded records, deaths due to accident or natural causes are classified as 'unintentional', deaths by suicide or homicide are classified as 'intentional', whereas deaths given an open verdict are classified as having 'undetermined intent').[64]

The researchers commented, 'The most surprising finding was the wide variation in the proportion of drowning deaths classified as of undetermined origin (E984)'. They noted that the total mortality from drowning was very similar in England and Wales (19.5 per million) and in USA (19.3 per million). But they pointed out that the rate for accidental drowning, coded E910, was much lower in England and Wales (8.0 per million) than in USA (13.8 per million), whereas the rate for drowning due to undetermined intent, coded E984, was much higher in England and Wales (7.7 per million) than in USA (1.0 per million). Discussing their findings, the researchers wrote, 'In England and Wales almost 40% of all drowning deaths are coded as of undetermined intent.' They had doubts about the low rate of accidental drowning deaths, and commented: 'A valid question to be answered is "Is England and Wales really so safe?", or is it just an artefact of difference in injury coding practices.'[65] In 2006, a review by British statisticians at the Office of National Statistics noted the very low rate of accidental drownings in England and Wales. They confirmed: 'This is largely an artefact due to a much larger proportion of our drowning deaths being coded as suicide and particularly "undetermined intent". The latter reflects coroners' "open" verdicts where there is no evidence of intent. In many countries such deaths are simply assumed to be accidental.'[66]

◆ ◆ ◆

The Royal Society for the Prevention of Accidents (RoSPA) is, of course, mainly involved in preventing accidents. On its website, a graph of the annual number of 'Accidental drowning deaths' from 1983 to 2013 shows a gratifying downward trend, from 637 in 1983 to 318 in 2013.[67] But the graph includes only those drowning deaths given an Inquest verdict of 'Accidental Death' or 'Death by Natural Causes' – less than 60% of the annual number of drowning deaths.[68]

More than 40% of annual drowning deaths – those given an 'Open Verdict' or a 'Narrative Verdict' at Inquest, and those where suicide or criminal intent is suspected or confirmed – appear on RoSPA's website only as brief notations on spreadsheets showing 'All activities and locations.'[69] Yet reporters often assume that RoSPA's annual announcements of 'Accidental drowning deaths' include all drowning deaths. As a result, the size of the drowning problem is seriously understated.

For instance, in June 2015 the headline on a government news wire announced: 'RoSPA – Number of UK drownings at lowest since records began.' The press release continued: 'There were a total of 338 water-related deaths from accidents or natural causes across the UK in 2014.'[70] Unfortunately, words had been cut and the meaning changed in the press release from the Royal Life Saving Society: 'Statistics revealed this week that 338 people lost their lives to drowning in 2014.'[71] Within days, articles in national and local newspapers repeated the phrase: '338 people lost their lives to drowning in 2014.'[72] In fact, the drowning statistics recorded on RoSPA's website show that 633 people lost their lives to drowning in 2014[73] – almost twice the number of 'Accidental drowning deaths' quoted in RoSPA's press release.

Over ten years, from 2009 to 2018, the number of 'Accidental drowning deaths' fell by 35% (from 405 fatal drowning accidents in 2009 down to 263 in 2018). However, the total number of drownings fell by only 15% (from 692 in 2009 down to 585 in 2018). Each drowning death is a tragedy. And the number of victims is startling. In the decade 2009–2018, RoSPA's drowning statistics show that 6,520 people died from drowning.[74]

CHAPTER THIRTY-FOUR

THE TRAGIC AFTERMATH OF A DROWNING ACCIDENT

When a life is cut short by drowning, the victim's family must cope with a most cruel shock. Most people who drown are young, fit and active until moments before their death. Their parents are totally unprepared for the sudden loss of their baby daughter, their busy toddler, their fun-loving teenager, their full-grown son who died with his adult life unlived.

Parents who witnessed the accident, took part in the rescue, or helped with resuscitation at the scene, are likely to suffer vivid flashbacks. If their child is admitted to hospital, they can only wait beside the bedside while purposeful doctors and nurses do their best to keep their loved one alive. If the struggle is lost, and the respirator is switched off, parents continue to suffer as they remember the days and nights of fading hope.

Such unexpected bereavement releases raw emotions quite different from the sadness felt after the predicted death of an elderly grandparent, where the rituals of the funeral, even the interment itself, may help the grieving family in their time of mourning. In contrast, when a child dies from drowning, the funeral signals the abrupt end of all their shared hopes and plans.[1] Moreover, as they try to come to terms with their loss, the family have to cope with a succession of additional stresses – investigations by the police and the coroner's officer, the post-mortem, the Inquest, journalists' inquiries and newspaper reports.[2]

For parents, the sudden death of their child may be overwhelming. They may blame a relative who failed to keep their daughter safe, blame friends who were swimming with their son but failed to save him, blame rescuers, lifeguards, even the victim for being so careless with his life. And parents often blame themselves.

Wrapped in their own sadness, some berate each other for showing too much emotion – or not enough. Nightmares, alcohol dependence and long-lasting depression are common, with the date of the accident and the victim's birthday acting as bitter reminders of their loss. Some parents never recover from the tragedy. An Australian study reported that a quarter of the couples whose children died in a drowning accident had separated, whereas the couples whose children survived a drowning accident were still together.[3]

Brothers and sisters suffer too. In the days that follow the fatal accident, parents may be so traumatised that they fail to respond to the pain of the children who remain.[4] Some parents find it impossible to talk about the child they lost, or they become over-protective or over-demanding.[5]

When children lose their father or mother in a drowning accident, their family unit is disrupted. Children are doubly distressed and bewildered if the surviving parent is too distraught to comfort them or explain what happened.[6] If the parent who drowned was the main breadwinner, the family has to cope with financial hardship as well as emotional turmoil. And if a parent drowned while trying to rescue a child who survived, the child may carry a lifelong burden of guilt.[7]

◆ ◆ ◆

Some families find a degree of consolation by granting immediate permission for organ donation so that other lives can be saved.[8] In 2008, a toddler died in a Southampton hospital three days after he was found unconscious in a garden pond. His mother said, 'At first I hadn't really wanted to donate [his] organs because he was still my little baby. But then I thought he was such a loving little boy and it was what he would have wanted.' His small kidneys were a perfect match for a young mother in Oxfordshire who had been on renal dialysis for seven years.[9] The two women exchanged letters. A year after the transplant, they were able to meet and begin a lasting friendship.[10]

A woman who drowned in a canoeing accident on New Year's Day 2012 had registered as an organ donor. Although her partner and their three-year-old daughter were devastated by her death, a letter helped them through their first Christmas without her. A young woman wrote to tell them that she was within weeks of dying from liver failure and the transplant had saved her life.[11]

Most families find ways to honour and commemorate the one who drowned. They display a favourite photograph. They place treasured items in a memory box – a lock of hair, a favourite toy, school reports, certificates, even swimming trophies. As a tribute to their daughter who drowned in a sailing accident, her parents commissioned the artist Marc Chagall to beautify their parish church in Kent by designing a huge east window.[12] When Chagall saw the window in place, he exclaimed, 'It's magnificent. I will do them all.' He designed eleven more windows for the church over the next fifteen years. (Seven were installed in 1974 and four in 1985, the year Chagall died aged 98.)

Some bereaved parents try to protect the lives of other people's children by running campaigns to increase water safety,[13] or by organising swimming lessons for babies and toddlers.[14] In 2006, a fifteen-year-old boy lost consciousness and drowned while swimming underwater at the end of a vigorous training session in his local pool in Gloucestershire. His death was due to shallow water blackout. His parents formed the Luke Jeffrey Memorial Trust and launched an awareness campaign to warn teenagers, parents and swimming coaches about the danger. Sharon Davies, the Olympic silver medallist, took part in an excellent video '*Shallow Water Blackout – This video could save your life!*'.[15]

In the same year, another fifteen-year-old boy drowned after sliding down a waterfall in Snowdonia. He was a strongly built rugby player and a capable swimmer. By the time the coroner held the Inquest and pronounced the verdict of accidental death, the boy's mother had raised £5,000 for a river safety campaign and had donated a lifesaving raft to the Ogwen Valley Mountain Rescue Team.[16] After giving a number of talks in schools, she set up River and Sea Sense to educate young people about the danger of drowning.[17]

In South Carolina in 2009, Jonathan Chase Lee was about to begin his degree course at Francis Marion University when he drowned in a nearby river. His parents established a scholarship in his memory, open to students who could not afford the university fees.[18] 'We see the benefit of doing it: to let his name live on,' his mother said. 'It's like he's not forgotten. To know that someone is going to school in his name, it's just really powerful.' For several years they raised funds by organising local turkey shoots and raccoon hunts. Then in 2015, an anonymous gift allowed the permanent endowment of the scholarship.[19]

Official memorials, too, can comfort and inspire. In 1900, the artist George Frederick Watts created a Memorial to Heroic Sacrifice in Postman's Park, a small green space close to St Paul's Cathedral.[20] Many of the plaques record drowning tragedies. Hand-painted tiles commemorate 'Harry Sisley of Kilburn. Aged 10. Drowned In Attempting To Save His Brother After He Himself Had Just Been Rescued. May. 24. 1878'. The artist's widow Mary Seton Watts continued the project, but after she died in 1938, no plaques were added for 70 years. Then in 2007, a young angler lost his life after jumping into a canal to rescue a nine-year-old boy. His fiancée and his colleagues approached the Diocese of London, and a plaque honouring his memory was unveiled at a ceremony in Postman's Park in June 2009:[21]

> 'Leigh Pitt, Reprographic Operator. Aged 30. Saved A Drowning Boy From The Canal At Thamesmead, But Sadly Was Unable To Save Himself . . June . 7 . 2007'.

◆ ◆ ◆

Parents also suffer a tragic loss if their child survives a drowning accident disabled by brain damage due to oxygen deprivation. Even though their child did not die, the parents no longer have the child they cherished, and the child has been robbed of much of his future.[22] An energetic toddler is reduced, within moments, to a permanent state of helpless babyhood. A happy, active teenager becomes a lifelong invalid, crippled in mind and body. The lives of other children in the family are profoundly affected. And the financial burden on the family and on society is high and prolonged.

Children who suffer brain damage in a drowning accident may lose control of movement, speech, sight, memory, reasoning, balance and spatial orientation. Many suffer recurrent seizures.[23] The pull of spastic muscles leads to permanent deformities of the spine, elbows, hands and feet. Walking may be impossible because of painful muscle spasms.[24] Some drowning victims are brought to hospital in a coma and never recover consciousness. Instead, they enter a vegetative state and remain unresponsive to human contact, without signs of awareness or emotion.[25]

Medical terminology can scarcely hint at the anguish of the victim's parents when they first realise how seriously their child has been injured.[26] Later, when the acute phase of hospital treatment is completed, parents face months or years of emotional and physical struggle as they care for a youngster suddenly deprived of the ability to talk, walk or play with friends, unable to wash, dress or eat unaided, with no control over bowel or bladder.[27] The director of a paediatric intensive care unit in California described the ongoing tragedy of brain damage in a drowning accident as 'one of the most disruptive and destructive events that can happen to a family.'[28]

After drowning accidents left their children with permanent brain damage, parents in Britain, Australia and America set up foundations to help others, to share information and equipment, to link charities and support groups, to foster research, to support water safety campaigns, and to promote and sponsor drowning prevention laws.[29] An American mother established the Drowning Prevention Foundation after a double tragedy. In 1978, both of her children fell into the family swimming pool. Her two-year-old daughter drowned, while her baby son survived with profound brain damage.[30] In 1997, her son reached his twentieth birthday, still completely paralysed, needing constant care, breathing through a tube in his windpipe, fed through a tube entering his stomach directly through his abdominal wall. 'The death is an awful experience, but life does go on eventually,' she told a reporter from the *Los Angeles Times*. 'With a child who is left with massive brain damage, you have a horrendous nightmare that never goes away.'

◆ ◆ ◆

Lifeguards, paramedics and doctors involved in the life-and-death drama of rescue and resuscitation at the scene of drowning accidents are trained to focus on the technical aspects of their task. Yet while they are doing their utmost to save the victim's life and comfort the victim's family, they also have to deal with their own high level of psychological stress.[31] If the victim survives, they feel intense satisfaction, even joy. If they are unable to save the victim, they inevitably feel despondent and distressed, even when they recognised at the outset that the child had virtually no chance of recovery.

An American textbook for first responders comments, 'In some situations, you cannot do anything, and the patient dies. In other situations, despite everyone's best effort, the patient still dies.'[32] Lifesavers have to prepare themselves for the possibility of failure, without becoming callous. Group debriefing meetings and advice from skilled, supportive colleagues help lifeguards and emergency teams to improve their morale by sharing their feelings about what went according to plan and what could be improved.[33]

In the *Daily Telegraph* in 2009, Dr Steve Walker, an Australian emergency physician, gave a vivid description of a helicopter flight in response to the summons: 'Toddler found unconscious in a pond. Not breathing. Frothing from mouth. Father doing CPR.'[34] The expert resuscitation team felt powerful emotions of hope and trepidation during their well-rehearsed actions. They slipped a tube into the boy's windpipe to get oxygen into his lungs. After 10 minutes of cardiac massage and several adrenaline injections, they succeeded in restarting his heart. As they transported him to hospital, they faced the possibility that he might die, and imagined his family plunged in grief. After their handover to hospital staff, they noticed the parents waiting to know if their son would survive, mother in tears, father pacing back and forth. The boy lived, but he suffered brain damage.

The team packed their gear and met at the helicopter base to discuss what for some was their first drowning. As Dr Walker explained, 'After such a case it helps if staff have an opportunity to wind down and discuss what they saw, did and felt. They need to understand the experience will live with them for a long while to come and that it is quite normal to be affected by it.' His account of this all-too-common family tragedy ended with the words: 'And to anyone who thinks this can't happen to them, please understand that the parents of every drowned child once believed that too.'

◆ ◆ ◆

In February 2015, the *British Medical Journal* published a study of 160 children who were admitted to university hospitals in Holland between 1993 and 2012 with hypothermia and cardiac arrest after drowning accidents in open water:[35] 116 children died, and only 17 of the 44 survivors escaped without serious, life-changing brain damage. Rescuers continued to give cardiopulmonary resuscitation to 98 of the children for longer than 30 minutes, but all had a poor outcome (87 died and 11 survived in a vegetative state or with severe brain damage). With such uniformly grim results, which failed to improve over the years, the researchers questioned the value of continuing cardiopulmonary resuscitation on chilled, pulseless children for more than 30 minutes.

Commenting on the Dutch study, an editorial in the same issue of the British Medical Journal concluded that, even if survival is unlikely, it is still appropriate to continue attempts at rescue and resuscitation for 60 minutes: 'It is important to remember that the parents are also victims. Parents need the reassurance that every effort has been made to resuscitate their child, particularly if a few more minutes of resuscitation might allow the parents to sit with their child before he or she is declared dead.' The editorial had the title: 'Resuscitating drowned children', but its subtitle warned: 'Outcomes are poor; we must focus on prevention.'[36]

PART FOUR

THE IMPORTANCE OF PREVENTION

Benjamin Franklin wrote, 'An Ounce of Prevention is worth a Pound of Cure'.[1] He was referring to house fires in eighteenth-century Pennsylvania, but experts involved with drowning agree: 'The vast majority of drownings can be prevented and prevention (rather than rescue or resuscitation) is the most important method by which to reduce the number of drownings.'[2]

Most of us underestimate how quickly and easily we could drown. Children don't recognise the danger, teenagers court the danger, adults dismiss the danger. Non-swimmers, poor swimmers, and skilled swimmers alike, we push our luck, and sometimes our luck runs out.

The drowning risk is strikingly higher than most people believe. Swimming appears to be less risky than travelling in a car, given that the mortality rate due to road traffic accidents is four times higher than the mortality rate due to drowning accidents.[3] But mortality rates give the number of victims per 100,000 of the entire population, including people who never travel in a car and people who never swim. When Australian researchers recalculated mortality rates taking into account the length of time that car occupants and swimmers are exposed to potential hazard (exposure-adjusted person-time estimates), they showed that the death rate from drowning was 200 times higher than the death rate in car accidents.[4]

Of course, drowning victims have a greater chance of survival if they are rescued quickly and if they receive early resuscitation followed by skilled hospital treatment. Yet many die at the scene of the accident despite the best efforts of relatives, bystanders and trained emergency crews. And the most expert treatment fails to preserve the lives of a quarter of those who survive long enough to reach hospital. Drowning accidents have a higher case fatality rate than yellow fever.[5]

In 2009, an American academic journal published a forceful editorial on the need for drowning prevention. Its author was Frederick P. Rivara, the professor of pediatrics and epidemiology at the University of Washington, Seattle. He wrote: 'Imagine a disease that kills nearly 200,000 children and adolescents around the world each year and for which even the most advanced high tech medical care is nearly ineffective in preventing mortality. Imagine this disease having a 50% case-fatality ratio. Imagine this disease affecting children in high-income countries as well as in the developing world.'[6] He warned that rescue and treatment often come too late: 'Many studies have shown that by the time the child with submersion arrives in the emergency department at even our most advanced children's hospitals, the die is cast.' His message was as clear as the title of his editorial – 'Prevention of Drowning: The Time Is Now.'

CHAPTER THIRTY-FIVE

A DOZEN WAYS TO PREVENT DROWNING ACCIDENTS

No simple set of precautions will protect all potential drowning victims (from babies to pensioners), in all locations (from bathtubs to holiday beaches), whatever the circumstances (from falling into a pond to falling off a trawler).[1] Yet the more we understand about drowning accidents, the earlier we can recognise potential drowning risks, and the better we can avoid them.

ENSURE THAT PARENTS UNDERSTAND THE NEED FOR CONSTANT CLOSE SUPERVISION OF YOUNG CHILDREN IN THE BATH, NEAR PONDS, BESIDE SWIMMING POOLS, AND ON THE BEACH

Babies, toddlers and preschool children are frequent victims of death by drowning. With their small immature lungs, they are particularly vulnerable to submersion injury, and it may be impossible to revive them after only one or two minutes underwater. Many drowning victims are dead or mortally injured within five minutes of sinking below the water surface. The great majority of these avoidable tragedies happen during brief lapses of parental supervision. The surest way to prevent drowning accidents in young children is constant friendly vigilance by a responsible adult.[2]

A baby must never be left alone while sitting in the bath, even for the short time it takes to fetch a towel, even when supported in a bath seat.[3] If the baby's mother leaves the bathroom, for whatever reason, she must take the baby with her. (The same rule applies when the father or any other responsible person is taking care of the baby at bath time.) Health visitors are well placed to advise new parents of the vital importance of constant supervision.

Toddlers and preschool children also die in bathtub drownings. Until the age of five, children in the bath need an adult watching over them. Parents should not leave older children in charge of a younger brother or sister.[4]

Young children are often fascinated by the water. Their growing ability to roam puts them at risk of drowning in ornamental ponds and swimming pools in their own garden or in the gardens of friends, relations or neighbours. Parents must keep their children within sight, watch what they are doing moment by moment, and stay close enough to grab them. Babysitters need clear instructions to do the same. And because minutes count if the child has fallen into the water, any search for a missing toddler should begin at the pond or the swimming pool.

When a party is held at a house with a swimming pool in the garden, the hosts are too busy to protect their own children adequately, let alone other people's children.[5] They need the help of a willing pool watcher to ensure the safety of the children who can swim and to prevent non-swimmers from getting into the water unobserved.

Swimming pools in holiday hotels or on caravan sites rarely have dedicated lifeguards, so parents must act as lifeguards themselves. Parents who rent a holiday villa with a swimming pool should realise that they have taken on a full-time job.[6] From early in the morning until late at night, they have to prevent their children from falling into the pool or jumping in unobserved. The risk of a drowning accident is greatest on the first day and the last day of the holiday while parents are occupied unpacking the luggage and settling in or collecting their belongings and loading the car.

On seaside holidays, parents should keep young children under close observation when they are playing ball games or building sandcastles on the beach. Toddlers easily lose their balance when they are paddling in shallow water, so they need a grown-up nearby to help them right themselves if they fall. Rock pools, waves, the incoming tide and outgoing rip currents provide a succession of instant dramas. When their children are swimming in the sea, parents must watch them constantly and be prepared to remove them from the water quickly if conditions deteriorate. Although parents provide the most important protection for their children, families are safer on beaches patrolled by trained lifeguards.[7]

Inflatable armbands have a role when a toddler is first learning to swim, but they should never be relied on as flotation aids.[8] Lilos and ride-on inflatable toys are best confined to a swimming pool, under close adult supervision. They are unsafe at the seaside because currents and offshore winds easily propel them away from the shore, taking terrified children into deep water.[9]

PLACE CHILDPROOF BARRIERS AROUND SWIMMING POOLS

All children in or near the water are at risk. A child who lives in a house with an unfenced swimming pool in the garden is four times more likely to drown than a child whose access to the family swimming pool is barred by a childproof fence.[10] However, the presence of a fence round the swimming pool does not release parents from their obligation to give close supervision to their children.[11] Drowning accidents happen even in fenced pools, usually when an open gate allows a child to enter the pool unobserved.[12]

In 1960, after an epidemic of drowning deaths, Mulgrave Shire in northern Queensland introduced the legal requirement that every swimming pool had to be surrounded by a childproof fence.[13] Although the measure was an outstanding success, the rest of Australia was slow to bring in similar legislation, and even slower to enforce it. New Zealand introduced mandatory swimming pool fencing in 1987, but many pool owners still fail to comply with the law, and many local authorities still fail to carry out inspections.[14]

In California, drowning is the leading cause of accidental death in children under the age of five, and the majority of drowning accidents happen in family swimming pools.[15] A senior firefighter in Orange County commented, 'There are 180,000 pools and spas that are not covered by present day codes and ordinances that require barrier fencing and other protective devices.' He described these unprotected pools as 'ticking time bombs that will contribute to the needless deaths of children in the future.'[16] Even where pool fences have been installed, they often allow access to the pool from the house.[17] And swimming pool fencing is still optional in Britain.[18]

A rigid swimming pool cover forms a protective barrier between a child and the water, but only as long as the cover is completely closed.[19] Children should be told, rather firmly, not to walk across a closed pool cover, even if the cover is designed to take the weight of a child.

Solar blankets are designed to trap the heat of the sun, cut heat loss due to evaporation, and keep grit and leaves out of the pool. They provide no protective barrier whatsoever. Instead, they become deadly traps, sinking under the weight of any child who steps onto the layer of bubble-wrap polythene floating on the surface of the water.

TEACH CHILDREN TO SWIM, THE YOUNGER THE BETTER

Teaching preschool children to swim.

Children are at the greatest risk of drowning between their second and fifth birthdays.[20] Unless they can swim, young children who fall into a pond or a swimming pool sink straight to the bottom. Yet until recently, experts ruled that young children were 'not developmentally ready for swimming lessons'.[21] Indeed, early swimming lessons were thought to increase the risk of drowning by encouraging young children to spend more time in the water, and by reducing their fear of water, thereby making them more likely to enter the water unobserved. Experts also decided that parents would become more casual about supervision once their children had learned to swim.

Then, in 2009, an American study confirmed the safety benefits of swimming lessons both for preschool children and for older children and teenagers.[22] Among children aged one to four years, only 3% of drowning victims had ever taken swimming lessons compared to 26% of matched controls who lived close by. Among children and teenagers aged five to nineteen years, only 27% of drowning victims had ever taken swimming lessons compared to 53% of matched controls. Within the year, the experts withdrew their advice against early swimming lessons.[23]

The main aim of early swimming lessons is to provide children with survival skills, not to teach them breaststroke or backstroke.[24] (Well-formed strokes can wait until later.) Toddlers practise getting their faces wet, blowing bubbles into the water, and opening their eyes beneath the surface.[25] As they stand in the shallow end of the pool, they learn to counter the pressure of the water on chest and tummy by working harder to fill their lungs with air.[26] They are shown how to press their arms and legs against the water, becoming aware of the resistance of the water and the destabilising lift of buoyancy as they start to step away from the side of the pool. They learn how to glide and how to replant their feet on the bottom of the swimming pool.[27]

They learn the beginnings of breath control (taking a deep breath before submerging their faces, holding their breath underwater, then lifting their mouths above the surface for their next intake of air), the beginnings of buoyancy control (floating face up, their lungs full of air, their muscles relaxed rather than clenched), and the beginnings of swimming proper (moving through the water using simple arm pulls and leg kicks). They learn how to enter and exit the water safely, and how to signal for help.

Even babies can be taught how to swim, but their puny neck muscles are not strong enough to lift their heads above the water surface. They need their own dedicated adult to support them in the water and to bring them up for air. Sometimes babies swallow excessive amounts of water from the pool,[28] a tendency that is easily controlled by holding them higher, with their mouths clear of the water. The water in a babies' training pool is usually warmer (30°C) than the water in a standard swimming pool (27°C to 28°C). Nevertheless, swimming lessons must be short because babies quickly become chilled.

Early swimming lessons introduce young children to elementary self-rescue techniques. They learn how to surface, roll onto their backs, fill their lungs with air and float face upwards. They practise treading water. They are shown how to look around them before heading for the nearest handhold. And importantly, they learn not to panic if they suddenly find themselves in the water.[29] In 2008, early swimming lessons helped a two-year-old girl to survive when she fell into the family swimming pool. Her mother found her clinging to the side of the pool, very cold but otherwise unhurt.[30] 'She was wearing a layered summer dress which became very heavy,' her mother reported. 'I had trouble pulling her out of the pool.'

Young children feel more confident and make faster progress during swimming lessons if their parents go into the water with them. Both children and parents benefit from their shared enjoyment. As an added bonus, swimming teachers can take the opportunity to speak to parents, emphasising that children will still need close parental supervision after their early swimming lessons come to an end.[31] Both children and their parents must be reminded that nobody can be 'drown-proofed', no matter how well they swim.

Teaching school-age children to swim

Most children are ready to be taught formal swimming strokes by the age of five.[32] Some make rapid progress. Others are slower to combine breathing, balance and forward motion. All have a justified sense of achievement when they succeed in swimming the length of the pool. Yet their lessons should not end there. To become competent in the water, children must be able to jump into deep water and regain the surface, roll and turn, swim comfortably on front or back, surface dive, swim underwater, and float face up without effort.[33]

Children should also be introduced to survival backstroke (sculling with the hands and an inverted whip kick with the legs).[34] Survival backstroke is safer than crawl or breaststroke if children become tired or cold. It allows swimmers to lift the mouth well clear of the water, to fill their lungs with air, and to take short rests while floating on their back. During a session of survival swimming in their familiar swimming pool, children can wear tracksuits over their swimsuits, which allows them to feel the increase in drag as they move through the water.[35]

Children usually learn to swim in the warmed calm water of the local swimming pool, but many drowning accidents follow a fall, fully clothed, into open water, often in the dark.[36] Sudden immersion in cold water triggers a gasp reflex followed by several minutes of excessively rapid breathing (hyperventilation).[37] Research at the University of Portsmouth has confirmed that people who fall into cold water increase their chances of survival if they 'Float First' for a minute or two, giving them time to get their breathing under control before they start to swim.[38] Some survival swimming courses add an extra level of protection by allowing closely supervised children to 'fall' fully clothed into the swimming pool.[39]

Children need further training if they intend to swim in the sea, where they have to cope with waves and currents as well as cold water. Norwegian researchers placed a wave-making machine in a swimming pool and compared the swimming, floating and diving ability of eleven-year-old children in calm water and waves. The children performed less well in the waves.[40] Children who take up water sports such as canoeing or sailing should have extra coaching to allow them to surface, float and swim to safety if they suddenly find themselves in the water.

School swimming lessons – 'Could do better'

Too many children have to wait until they are halfway through primary school, aged 7, 8 or 9 years old, before they have their first swimming lesson.[41] Inevitably, children who depend on school swimming lessons lag behind those who already know how to swim. Timid beginners may become increasingly self-conscious about their awkwardness in the water, and less eager to take part in swimming sessions with the school. They make slow progress unless their self-confidence is boosted by encouragement – and by extra swimming lessons.

By the time children leave their primary schools, they are expected, under the National Curriculum, to be able to swim the length of the swimming pool (25 metres). In March 2000, the Ofsted Report *Swimming in Key Stage 2* reported that almost one in five children left primary school unable to swim 25 metres. In some inner-city schools, one in three failed this modest test, and half of the schools inspected by Ofsted made no special provision for children with poor swimming ability.[42] In December 2000, Parliament discussed the Ofsted Report in an adjournment debate. Valerie Davey, MP for Bristol West, said, 'There is a danger that the opportunity to learn to swim will become a privilege for those in higher income groups, while others will be put further at risk as a result of not having the opportunity to develop this essential life skill.'[43]

Seven years later, in 2007, a further Ofsted Report *Reaching the Key Stage 2 Standard in Swimming* described inspections of 30 schools, all of which had received additional funding in a strategy to increase the proportion of pupils who could swim 25 metres by the time they left primary school. Despite three years of top-up swimming sessions for 'pupils who were unable to swim the expected distance', a quarter of these children were still unable to swim 25 metres.[44] Among the key findings was the comment, 'One in five of the lessons seen gave insufficient attention to the needs of the pupils least able to swim or those who had the least experience of swimming.' The needs of disabled children were often poorly served.[45]

By 2011, school swimming lessons were achieving even worse results. A third of English children were leaving primary school unable to swim 25 metres. Children were at a particular disadvantage if their parents could not swim. The Amateur Swimming Association noted in their report *Save School Swimming. Save Lives*: 'Parents who lack confidence in the water, or who cannot swim themselves are understandably reluctant to take their children to the swimming pool.'[46] The report warned, 'If these non-swimming parents don't have the financial means to pay for private lessons, and can't rely on school swimming provision, there's a good chance that their children, and even their children's children, will never learn to swim.'

In Scotland, swimming lessons are not considered part of the compulsory curriculum in primary schools, and in some areas, children have no access to swimming lessons at all. As a result, 40% of children leave primary school unable to swim.[47] Unfortunately, the right to roam allows them free access to the chilly water of countless lochs, reservoirs, rivers and beaches. Scotland has an annual drowning rate twice as high as England and Wales.[48]

Teaching swimming to children of immigrant and minority groups.

The 2007 Ofsted Report *Reaching the Key Stage 2 Standard in Swimming* noted: 'Evidence from this small sample of schools suggested that pupils from some black ethnic minorities often started with less experience of swimming and made the least progress.'[49] Providers did not always recognise or acknowledge this in their planning. Instances of good planning included consultation with a local mosque on how to improve the participation of Muslim pupils, and family swimming programmes to involve parents from black minority ethnic groups.' Alice Dearing, one of Britain's best open-water swimmers, often found herself the only black swimmer in her event at international competitions.[50] With the support of Swim England, she helped to found the Black Swimming Association in Britain in March 2020.

The crucial role of targeted swimming instruction is all too apparent in America. A majority of African American children either cannot swim or have poor swimming skills.[51] Among preschoolers, the death rate from drowning is lower for African American boys than for white boys, possibly because of differing access to both municipal and backyard swimming pools. After their fifth birthday, however, the death rate from drowning for African American children is three times higher than white children, while the death rate due to swimming pool drownings in black teenage boys is more than 12 times the death rate in white teenage boys.[52]

In August 2010, a family barbecue on a sandy bank beside the Red River in Louisiana ended in tragedy when seven teenagers decided to cool off by wading in the river. None of them could swim. Within minutes, six of the youngsters drowned, watched by their horrified, helpless parents who were also unable to swim.[53]

In Toledo, Ohio, Wanda Butts' father saw two children drown at a church picnic, so he would not let her go near the water, and she never learned to swim. In 2006, her sixteen-year-old son drowned while rafting on a lake with friends. He too was a non-swimmer. When Wanda Butts discovered how many African American children cannot swim, and how many drowned, she decided to cope with her personal tragedy by starting the Josh Project in his memory, organising low-cost swimming lessons that have already taught several thousand African American children to swim.[54] 'All children are at risk of drowning,' she said, 'But the majority of the children that the Josh Project serves are minority children, who we have found are more at risk.'

Keep local swimming pools open.

Most children have their first swimming lessons in their local public swimming pool – the safest place to learn to swim. At every level of skill, from beginners to experts, swimmers are much less likely to drown in public pools than in open water or at the seaside.

In the eighteenth century, a London jeweller transformed a pond in Finsbury known as 'Perrillous Pond, because diverse youths swimming therein have been drowned' into Peerless Pool, 'where Gentlemen may without danger learn to swim.'[55] Since then, Victorian philanthropists, Edwardian councillors, holiday camp developers and National Lottery punters have financed many splendid swimming pools.[56] However, maintenance and running costs are high. A few historic pools have been rescued from the wrecker's ball by determined campaigners and community fundraisers,[57] but a number have been abandoned or demolished.[58] Scores of schools have closed their on-site swimming pools. This means that pupils have to be ferried to a distant venue and back again, and a half-hour lesson may take three hours from the school day.[59] In London, despite local protests, a fitness company recently closed several swimming pools,[60] with plans to convert the premises into gym-only venues providing running machines and muscle-building equipment 24 hours-a-day.

According to Sport England, the recent proliferation of leisure clubs means that there are more pools than ever, but they 'tend to be small in size and have shallower water, being aimed primarily at the fitness/aerobic recreation market' rather than providing protective settings for swimming lessons, stroke development and racing.' Moreover, the high entrance costs of these 'private' pools puts them beyond the means of many families with young children.[61]

Rebecca Adlington, winner of two Gold Medals at the 2012 London Olympics, asked, 'Why are we closing swimming pools?' Concerned that so many children cannot swim, she pointed out, 'Swimming is not just a sport – you are cutting something that could save somebody's life. For me, that's much bigger than any Olympic medal.'[62]

TRAIN AND EMPLOY LIFEGUARDS TO PROTECT SWIMMERS AT SWIMMING POOLS AND ON BEACHES

All public swimming pools in Britain have lifeguards on patrol throughout opening hours, protecting young and old at every level of swimming skill. Public swimming pools are busy places. In England alone, over 2.6 million adults take part in swimming sessions every week, three million children swim regularly in their free time, and two million children attend school swimming classes.[63] Yet despite the enormous number of people entering the water, the presence of well-trained lifeguards ensures that drowning deaths are rare in public swimming pools.[64]

Lifeguards prevent countless accidents because they are trained to spot the early signs of an impending submersion, so they take action before anyone is harmed. They keep non-swimmers in the shallow end. They keep non-divers out of the diving pool. They steer beginners back from the deep end if their only stroke is dog paddle.[65] They calm over-excited youngsters, put a limit to horseplay, and intervene when teenagers endanger themselves or others. A blast on a lifeguard's whistle controls reckless behaviour on the poolside and in the water.

Lifeguards constantly scan the surface of the pool, ready to spring into action as soon as they see someone in difficulty in the water.[66] They also look intently beneath the water surface, checking that nobody is lying at the bottom of the pool, a task made surprisingly difficult by surface reflections, distortions and glare.[67] When the swimming pool is crowded, it is easy to miss a beginner who jumps into the pool and fails to come up for air, a weak swimmer who tires and sinks, or a teenager who suffers a shallow water blackout while swimming underwater. The lifeguard may only catch a glimpse of an indistinct, ill-defined shadow. The victim's friends or nearby swimmers may be the first to recognise that someone is in danger, and lifeguards must respond at once.[68]

Lifeguards are trained to rescue people from the water without putting themselves in danger. If a small child is panicking and sinking, the lifeguard may use a shepherd's hook to pull the victim to safety[69] or leap directly into the water to lift the victim to the surface. If an older, stronger victim is struggling in the water, the lifeguard approaches holding out a buoyant rescue buoy for the victim to grab, then tows the victim to safety.[70] If an unconscious victim is lying at the bottom of the pool, the lifeguard places the rescue buoy beneath the victim, using its buoyancy to help with lifting the inert heavy body to the surface and with towing it to the edge of the pool. Meanwhile, another member of staff calls 999 to summon an ambulance.

During their training, lifeguards practise lifting children and adults from the water onto the tiled pool surround. During a rescue, a lifeguard may need help from a colleague or a bystander, especially if the victim is unconscious and incapacitated. A backboard can be used as an aid for quick extraction from the pool or to prevent movement of a victim's head if neck injury is suspected after an ill-advised dive.[71]

If the victim has stopped breathing, the lifeguard will start mouth-to-mouth resuscitation as soon as the victim is lifted from the water. If the victim's heart has stopped beating, the lifeguard will start CPR at once, while a colleague readies the defibrillator for possible use at the poolside. Resuscitation will continue until the ambulance arrives and the paramedics take over. Lifeguards maintain their resuscitation skills, with the result that most children admitted to hospital after loss of consciousness in a public swimming pool 'had been effectively resuscitated at the poolside'.[72]

◆ ◆ ◆

At the beach, drowning accidents happen fast. Rip currents form wherever waves roll towards the shore. The higher the surf, the stronger the rip currents. Flatwater beaches may appear safer than surfing beaches, but even when the sea appears calm, the pattern of the waves changes from moment to moment, the water varies in depth because of undulations in the seabed, and the tide flows in and out with surprising speed and force.

Holidaymakers have more chance of successful rescue and resuscitation if they 'swim between the flags' on beaches that are patrolled by lifeguards.[73] Lifeguards save many lives by directing swimmers away from rip currents and by rescuing swimmers when rip currents drag them out to sea.[74] When someone is in danger, a beach lifeguard wades out, swims out carrying a buoyant rescue float, paddles out on an adapted surfboard, or drives out on a jet ski. Many beach lifeguards are able to give mouth-to-mouth resuscitation in deep water before towing the casualty back to the beach.[75] All beach lifeguards are trained to give immediate necessary mouth-to-mouth resuscitation or CPR as soon as they bring a drowning victim to the beach.[76]

Australian lifeguards rescue hundreds of tourists from drowning in the rip currents on Bondi Beach every year. Yet the lifeguards find that the surf is less of a problem than the ignorant behaviour of the tourists who 'think that the ocean is just another swimming pool' and ignore safety advice and warning signs.[77] In four weeks after Christmas in 2017, 12 swimmers had to be rescued twice because they failed to listen to the lifeguards or did not understand them.

In 2016, seven young men drowned while visiting Camber Sands in East Sussex.[78] In July, an offshore wind carried a teenager out to sea, and a man who tried to rescue him died in hospital four days later.[79] In August, five friends were trapped on a sandbank half a mile from the shore as the incoming tide filled gulleys around them, cutting off their retreat.[80] In 2017, four lifeguards were appointed to patrol Camber Sands from May to October.[81]

TEACH WATER SAFETY AWARENESS TO CHILDREN – AND THEIR PARENTS

It is not enough to teach children how to swim. From the start, swimming lessons should include training in water safety.[82] Toddlers must be drilled never to enter the water without the presence, attention and permission of an adult,[83] while parents must remember that young children have little understanding of danger, and if they are not watched all the time, the attraction of a nearby pond or swimming pool may be too great to resist.[84]

Many parents who bring their children to swimming lessons at the age of 2 to 3 years have an over-optimistic view of the protection that preschool swimming lessons can give.[85] Rather than claiming that early swimming lessons will protect children from drowning, swimming teachers should take the opportunity to remind parents that their children will continue to need constant supervision to keep them safe in the water.[86] Indeed, in New Zealand, parents who took part in a poolside water safety programme recognised that 'their toddler required more, not less, adult supervision after swimming lessons.'[87]

At American water parks, some parents fail to watch over their children, expecting the lifeguards to keep them safe. In 2008, responding to the drowning of a four-year-old boy in a huge, crowded wave pool, the Californian Senate passed the Wave Pool Safety Act which required wave pool operators to provide life jackets to all non-swimmers and all children less than 4 feet tall.[88] Since then, a number of swimming pools and water parks in America have introduced a 'Note and Float' drowning prevention programme.[89] To minimise the danger of submersion, non-swimmers are identified when they arrive at the pool, given coloured wrist bands, and fitted with child-sized life jackets so that they cannot sink below the surface. Parents are then expected to supervise their non-swimming children, and they are encouraged to organise swimming lessons for them. After an aquatics centre in Pennsylvania adopted the 'Note and Float' method for children attending birthday parties at the pool, lifeguards reported that far fewer children had to be rescued from the water.[90]

◆ ◆ ◆

As children grow older, their safety awareness should be sharpened by helpful explanations from parents and swimming teachers instead of being influenced by the risk-taking actions of their schoolmates or by their own experience of a frightening near-miss.[91] Children must be warned about the particular dangers of swimming in open water,[92] which is always much colder than the water in their local swimming pool. Parents should explain that they could be knocked off their feet when crossing a stream, swept away by river currents, tumbled over by waves and pulled out to sea by rip currents. Every winter, children must be reminded never to walk onto a frozen pond.[93]

The Ofsted Report *Swimming in Key Stage 2* noted that some schools gave their pupils no advice about water safety and survival.[94] Norwegian researchers found a similar failure to prepare children to deal with drowning risk: 'In too many cases, children are not taught what is necessary for them to cope with an unexpected submersion that could lead to drowning.'[95] In 2007, a nine-year-old boy drowned in a Warwickshire lake while trying to retrieve his fishing float.[96] After the accident, a spokesman for the Royal Society for the Prevention of Accidents visited the lake. He told a local reporter, 'Too many young people are no longer being equipped with either the swimming ability, water safety knowledge or survival and rescue skills to make them safe in the water.'[97] In a telling comment, he added, 'We believe that 25 metres is enough to get into danger, but not enough to save you.'

◆ ◆ ◆

At any age, walkers, joggers and runners are wise to stay back from the water margin. The Royal Life Saving Society (RLSS) reported that 54 people drowned in 2012 while walking or running beside a waterway or along the beach.[98] Some trip, slip or fall from a muddy canal towpath or an unstable riverbank and tumble into the water. Others are trapped by the incoming tide, or engulfed by unexpected waves. The RLSS advises joggers to take notice of any official warning signs, to take their mobile phone with them, to run with a friend – and to learn swimming and lifesaving skills. The Royal National Lifeboat Institute noted that more than half of the victims of drowning accidents at the coast had 'ended up in the water unexpectedly'.[99] They strongly advise people who find themselves in the sea to float on their backs until their breathing is under control.

EXPLAIN DROWNING RISKS TO TEENAGE BOYS

The risk of drowning is at its greatest among teenage boys, who are ten times as likely to drown as teenage girls.[100] They may be able to swim, but they frequently overestimate their swimming ability, and they almost always underestimate the danger of the water.[101] They set daring challenges for themselves and their friends[102] ('Race you to the island!'). They ignore official warning signs, assuming that an official notice telling them 'NO SWIMMING. DEEP WATER' was put up to spoil their fun, not to protect them.[103] They jump into water-filled quarries where their muscles fail in the unaccustomed cold.[104] They launch themselves from riverbanks, only to be swept away by the current.[105] They brave white-water rapids in flimsy canoes[106] and inflatable rafts. They risk paralysis by diving into shallow water,[107] or tombstoning from piers and breakwaters.[108] On surfing beaches, teenage boys stay longer in the surf, and swim out further from the shore.[109] They hire jet skis but fail to control them.[110]

Parents and teachers need to set protective boundaries. On a school rugby trip to Canada, a seventeen-year-old youngster drowned in a frigid lake when his high-spirited teammates began pushing each other off a jetty.[111] After recording the incident as an Accidental Death, the Coroner wrote a 'Regulation 28 Report to Prevent Future Deaths' to the Secretary of State for Education, noting that drowning accidents are particularly frequent in the teenage years because of 'ignorance of dangers and bravado and peer pressure'.[112] He criticised the risk assessments made while planning the trip and at the accident scene. He considered that children are not taught to swim to a sufficiently high standard – they had 'no skills to save themselves in deep, cold open water'. He advised that testing and recording of the swimming ability of every participant should be routine before every school trip involving water.

Explanations probably stick better than bald prohibitions, although 'sensible' advice is frequently disregarded during the teenage years. Weak swimmers are at the greatest risk of drowning in open water, but skilled swimmers drown too, overcome by cold shock or disabled by swimming failure, sometimes within a few feet of safety.[113] Teenagers who swim well are even at risk in their familiar local swimming pool if they over-breathe before trying to outswim each other underwater. Every young swimmer should understand that over-breathing cannot increase the amount of oxygen in his bloodstream. Instead, over-breathing washes carbon dioxide from his body, delaying the urge to take the next breath, sometimes with fatal consequences – the swimmer suddenly loses consciousness, and drowns due to 'shallow water blackout'.[114]

To protect themselves, teenage boys need to be made more risk-aware, not to make them timid, but to help them develop a prudent respect for the water, so that they avoid accidents by staying within their limits.[115] Peer pressure can be dangerous. Youngsters are warned: 'Never swim alone.' But swimming with others gives no guarantee of safety for teenage boys who are trying to keep up with their exuberant pals in open water. Most teenage drowning fatalities are witnessed.[116] Youngsters may lose their strength and sink beneath the surface so suddenly that their friends are powerless to save them. And non-swimmers often place themselves in jeopardy because they fail to recognise how vulnerable they are. When their friends go swimming, they decide to jump in too, only to panic and drown.[117]

NEVER DRINK ALCOHOL WHEN SWIMMING OR BOATING, AND NEVER WALK HOME BESIDE A RIVER OR CANAL AFTER AN EVENING DRINKING WITH FRIENDS

Teenage boys must be advised, emphatically and repeatedly, not to swim after drinking alcohol.[118] With even a small amount of alcohol in his bloodstream, a swimmer becomes more likely to make risky misjudgements and less able to extricate himself from danger. Alcohol disturbs his coordination, hampers his strokes, interferes with his breath control, and speeds the development of hypothermia.

In Durham between October 2013 and January 2015, three university students fell into the River Wear and drowned while walking home alone after prolonged sessions in city-centre pubs.[119] The police considered introducing on-the-spot fines for drunken behaviour to discourage undergraduates from putting themselves at risk.[120] Donations by the family and friends of one of the victims enabled the Royal Life Saving Society to provide water safety education packs to every school in County Durham, to organise demonstrations of rescue equipment by the Fire Service, and to teach lifesaving skills to more than one hundred pupils.[121]

The Royal Life Saving Society began its campaign 'Don't Drink and Drown' at the University of York in 2014 after three young people drowned within a few weeks of each other following evenings drinking with friends. A student and a musician fell from the riverbank and were swept away by the current. The third victim was a young soldier who set off to swim across the river. He was encouraged by his intoxicated companions, who failed to recognise the danger until they saw him disappear midstream.[122] The 'Don't Drink and Drown' campaign distributed leaflets that carried a photograph of a narrow canal-side footpath at night, with the caption: 'From happy hour to nightmare in just one slip'.[123] In Shropshire, publicans and club owners joined firefighters, police and Shrewsbury Town Council in a 'Don't Drink and Drown' campaign over Christmas and New Year, warning partygoers against walking home along the River Severn.[124]

The law demands that car drivers 'Don't drink and drive'. For anyone at the controls of a motorboat, 'Don't drink and drown' is equally relevant.[125] The helmsman is not the only one who should abstain. It is safer if everyone on board remains sober until the boat returns to its moorings.[126] However, when researchers in New Zealand asked teenage boys about their attitudes to water safety, more than half considered it acceptable to drink beer when fishing from a small boat on a calm day.[127] But if the boat pitches in the waves, young men who share a six-pack of beer increase their risk of falling into the sea, and decrease their survival chances should they become 'man overboard'.

ENCOURAGE PEOPLE TO PUT ON LIFE JACKETS BEFORE THEY ARE NEEDED

Life jackets are too often left in a locker when young people are boating or sailing close to shore. Many teenage boys consider that life jackets are unnecessary for good swimmers.[128] Every youngster ought to be aware that, however skilled his swimming, he will suffer exhaustion and cold-related swimming failure within twenty minutes of falling into the sea, whether he falls from a motorboat, a yacht or a jet ski.[129] By increasing his buoyancy, a life jacket will keep his head above the water surface after he becomes exhausted and incapacitated by the cold.

In 1999, a study of jet ski crashes on Arkansas waterways gave credit to the almost universal use of life jackets for saving the lives of 38 jet skiers over four years, including many non-swimmers who were thrown violently into the water.[130]

In Australia in 2005, the state of Victoria introduced regulations requiring recreational boaters to wear life jackets on small boats. Within two years, the percentage of boaters wearing life jackets rose from 22% to 63% – and fewer boaters lost their lives.[131] In the six years before the regulations came into force, 59 boaters drowned; in the five years after life jackets became compulsory, 16 boaters drowned.[132] In 2011, a coroner in New Zealand called for stricter regulations on the use of life jackets after two jet skis collided and both riders drowned.[133]

The choice of life jacket is crucial. Simple floatation aids are adequate for waterskiing in calm water close to the shore. A more buoyant life jacket is needed for everyone who ventures into turbulent water further out to sea. A crotch strap prevents the life jacket from riding up, so improving airway protection during accidental immersion and prolonged wave exposure.[134] In 2012, two men took four young children canoeing in Loch Gairloch off the north-west coast of the Scottish Highlands. When the canoe capsized, they fell into the sea. Neither of the men wore life jackets, and the children were dressed in floatation jackets that failed to hold their faces above the water surface. One man was able to reach his eight-year-old daughter and support her as they swam 500 yards to the shore. His friend and three of the children sank and drowned.[135]

People who fish from seaside rocks are at risk of being swept into the sea by sudden waves, but few choose to wear a life jacket. After a series of drowning fatalities on a 50km stretch of New Zealand's rugged west coast, researchers interviewed 250 rock fishermen at high-risk fishing locations during the summer months of 2005–2006.[136] A third of the rock fishermen admitted that they could not swim at all, or could swim 25 metres or less. The majority reported that they had poor or non-existent rescue skills and did not know how to carry out resuscitation. And although 70% of those interviewed agreed that 'getting swept off the rocks is likely to result in their drowning', 72% said that they 'never' wore a life jacket and less than 5% wore a life jacket 'often' or 'always'. A safety campaign was launched, promoting the use of life jackets, providing survival and water safety classes at community pools, and placing life rings at all high-risk fishing locations. Surveys five years and ten years later noted a rise in life jacket use from 4% to 40% between 2006 and 2015.[137]

Commercial fishermen, too, are often reluctant to put on a life jacket when working on deck, even though they recognise their risk of falling overboard. In 2005, the RNLI ran the *'Trawling-Which lifejacket for you?'* campaign:[138] 120 fishermen volunteered to wear a range of life jackets and buoyancy aids while working at sea and report on their comfort and durability. In 2011, the American National Institute for Occupational Safety and Health (NIOSH) published the results of a month-long evaluation of six modern life jackets and buoyancy aids (Personal Flotation Devices or PFDs) by 200 Alaskan fishermen:[139] crabbers, gillnetters, longliners and trawlermen each had different preferences. Both the RNLI and NIOSH noticed a moderate increase in life jacket/PFD use.

In 2013, the Fishing Industry Safety Group (FISG) introduced an intensive educational campaign to persuade British commercial fishermen to wear life jackets while working on deck, backed by the offer of hundreds of free or heavily subsidised life jackets.[140] Supporting the campaign, the Scottish Fishermen's Federation (SFF) publicised the survival of Dougie Brown, a lobster fisherman saved by his PFD when he fell overboard and was unable to pull himself back onto the boat. The SFF pointed out, 'If a fisherman's body is not found, it means families are unable to hold a funeral, close bank accounts or claim insurance for up to seven years.'[141] However, the campaign had little effect. Indeed, after a fisherman fell from a trawler and drowned in 2015, three life jackets were found on board – supplied by the SFF to the trawler's crew free of charge, never used and still in their original packaging.[142]

In April 2016, the Marine Accident Investigation Branch (MAIB) published an important report: *'Lifejackets: a review'*, which noted that the campaign had not increased the use of life jackets. Between 2000 and 2015, the MAIB listed '380 cases of persons in the water from UK fishing vessels', including 139 fatal drowning accidents, but only 17 of the drowned fishermen were wearing a life jacket.[143] MAIB statistics show that the chance of surviving a 'man overboard' incident is five times greater for someone wearing a life jacket than for someone without a life jacket, but the report concluded: 'Research has demonstrated that campaigns succeed in changing entrenched behaviours only when backed by mandatory regulations.'

Now that acceptable designs have become available, increasing numbers of commercial fishermen have the opportunity of protecting themselves by choosing a 'personal flotation device' that suits their method of working and by wearing it every time they go on deck. But fishermen are still drowning for want of a life jacket. In November 2016, after the deaths of several fishermen who might have survived had they been wearing life jackets, Steve Clinch, the Chief Inspector of Marine Accidents said, 'In the cold waters around the UK survival time can be measured in minutes unless a life jacket is worn. However, this message is not getting through despite a three year campaign that has seen almost every commercial fisherman in the UK receive a free life jacket. In order to prevent further unnecessary loss of commercial fishermen's lives, I am therefore recommending today that the Marine and Coastguard Agency moves quickly to introduce legislation making it compulsory for fishermen to wear personal flotation devices on the working decks of commercial fishing vessels while they are at sea.'[144] By 2018, against considerable opposition, the requested regulations were in place.[145]

PROTECT PEOPLE WITH MEDICAL PROBLEMS THAT INCREASE DROWNING RISK

People with epilepsy are liable to sudden loss of consciousness, which greatly increases their risk of drowning, both in the bath and while swimming.[146] However, they are not always made sufficiently aware of the risk by neurologists or family doctors.[147] A brief written 'official' warning notice of the danger of taking unsupervised baths should be handed to newly diagnosed patients and their families as part of the advice on coping strategies and medication, otherwise the drowning risk may be overlooked or forgotten.

When someone with epilepsy is taking a bath or a shower, the bathroom door must be left unlocked and a family member should stay within earshot. Showers are safer than baths, and the flow of water is best controlled by using a handheld shower head which releases water only while a button is pressed down.

The British Epilepsy Association encourages swimming, but parents of children with epilepsy must be warned about the high drowning risk and must ensure that their children always swim beside an adult companion, with the lifeguard informed. If people with epilepsy go boating, a life jacket is essential. They should never scuba-dive, even if their seizures are well controlled.

Swimming is often advised as a helpful form of exercise for people with asthma,[148] preferably in a supervised swimming pool. Swimming in the sea is unwise because the effort of swimming through waves may bring on an exercise-induced asthma attack, narrowing the airways at a time when the swimmer's life depends on increased capture of oxygen.[149] Most authorities advise people with asthma not to scuba-dive.

Several genetic abnormalities of the heart increase the risk of drowning. Swimming has been identified as a specific trigger for the sudden onset of excessively fast heartbeats in Long QT Syndrome.[150] Loss of consciousness may occur within moments, leading to submersion and drowning. Those who carry this inherited susceptibility to sudden death can be protected by medication or by the implantation of a small defibrillator under the skin.[151] They should be warned never to swim. Family members should be offered genetic screening.

Abnormalities of cardiac muscle have been found to be responsible for sudden deaths in competitive athletes, including swimmers.[152] Early investigations are warranted in any competitive swimmer who suffers warning symptoms such as fainting, palpitations or chest pain while swimming. Cardiac screening should also be considered for all competitive swimmers who survive a drowning accident or lose a family member younger than 50 years old from sudden cardiac death.

Many older people have narrowed coronary arteries. Others take medications which limit their exercise tolerance. They often underestimate the muscular effort involved if they take an unaccustomed swim in the sea or decide to try snorkelling.[153] Pensioners should protect themselves by making allowance for their reduced fitness and stamina. They should stand well back from the water's edge to avoid an unplanned immersion.[154]

DON'T DRIVE CARS THROUGH FLOODWATER OR INTO RIVERS

Car drivers are wise to ease their feet off the accelerator pedal when they are driving in the rain.[155] The spray from a shallow puddle can stall the engine, while less than a quarter of an inch of water may result in aquaplaning, with loss of control over steering or braking.[156] Drivers should pay attention to signs warning of flooded roads ahead. Two feet of water will lift the car off the road, leaving the driver powerless to direct the car to safety. Moving water may sweep the car off the road altogether.[157]

People often underestimate the risk of walking or driving through floodwater, especially as they near their destination.[158] Many flood victims place themselves in danger by voluntarily entering floodwater to continue a journey, to protect their possessions, livestock or pets, to help with evacuations, or to rescue others.[159]

The Royal Life Saving Society urges people to 'follow simple, common sense steps during periods of flooding to help ensure that they, and their families, stay safe':[160]

+ Never try to walk or drive through floodwater – six inches of fast-flowing water can knock an adult over and two feet of water will float a car.
+ Never try to swim through fast-flowing water or floodwater – you may get swept away or be struck or trapped by an object in the water.
+ Never allow children or pets to go near or play in floodwater. It is hazardous and may be contaminated with chemicals.
+ Keep an eye on weather reports for flooding in your area. Do not travel in heavy rainstorms unless absolutely necessary.
+ Prepare a flood kit in case your home floods or you are trapped in a vehicle.

◆ ◆ ◆

Whatever the weather, special care is needed when driving along winding roads beside rivers or lakes.[161] A car that skids and careers into the water will float briefly, but it soon sinks to the bottom, engine first, its doors held shut by the pressure of the water. The driver has only a couple of minutes to unclip the seat belt, wind down – or break – the window on the driver's side, and climb out through the window.[162] Passengers must also release themselves quickly, making their escape through the nearest window. If there are children in the back of the car, they will be trapped inside the flooded car unless adults in the front manage to release them from their seat belts and push them through an open window, before escaping themselves. Drivers whose daily route runs beside a waterway might consider carrying a centre punch or a rescue hammer for use in an emergency to break the glass.[163]

◆ ◆ ◆

On holidays abroad, travellers hope for sunny skies, but they are often surprised by dramatic changes in the weather. Tourists should choose their picnic places and campsites with care, avoiding ravines, arroyos or dry riverbeds.[164] A sudden thunderstorm in the mountains, perhaps miles away, often releases such a heavy downpour that the rain cannot soak into the ground. Instead, a flash flood rushes down its channelled route. The water, loaded with branches and boulders, carries away anyone and any car in its path. In Spain in 2011, a flash flood crashed through the open-air market in the picturesque town of Finestrat on the Costa Blanca, drowning a British couple as they sat drinking coffee at a market stall.[165] In America, more people die in flash floods than in river and coastal flooding combined.[166] The majority of victims drowned in their cars.

TAKE A TRAINING COURSE IN RESCUE AND RESUSCITATION

Between 1980 and 2012, 81 New Zealanders drowned while trying to rescue someone from drowning.[167] Nevertheless, the danger of in-water rescue remains unrecognised. In a survey in Auckland in 2013, the majority of young people admitted that they could swim less than a hundred metres, yet almost half of those questioned said that they would jump into the water to rescue a drowning victim.[168] Clearly, more caution is called for, with better training in non-contact rescue methods – especially Reach and Throw.[169]

In Britain, the Royal Life Saving Society (RLSS) runs courses which introduce youngsters to water safety, self-rescue, non-contact rescues and lifesaving skills. Children aged 8 to 12 years can join the Rookie Lifeguard programme.[170] Teenagers can develop their lifesaving skills in the Survive and Save programme, marking progress with awards at Bronze, Silver and Gold levels.[171]

The RLSS runs one-day Survive and Save Instructor Courses for school teachers, who are then able to pass on water safety and lifesaving skills to their pupils during school swimming sessions.[172] In 2014, as part of Drowning Prevention Week, an eight-year-old girl took part in a RLSS demonstration at her swimming pool. A few days later, she was playing beside a stream when she saw her friend fall off a rock into deep water.[173] She knew that her friend could not swim, but she managed to save her by reaching out and pulling her into the shallows. 'I knew what to do straight away,' she said, 'because I had learned basic lifesaving skills at my swimming lesson.' The RLSS presented her with a bravery award.[174]

The RLSS also runs lifesaving courses for adults, and advanced courses for lifeguards.[175] Three weeks after he qualified as a lifeguard, a 20-year-old university student put his training to the test on holiday in Majorca when a hired motorboat capsized in a sudden squall far from shore, pitching him, his parents and his girlfriend into the sea. Thanks to his training, all four survived.[176]

The Royal National Lifeboat Institute (RNLI) provides the thorough training needed by the volunteers who man the lifeboats around our coasts. The RNLI also trains increasing numbers of beach lifeguards who work with coastguards and lifeboat crews to protect people paddling, swimming and surfing around the coast of Britain.[177] As well as providing schools with free downloads to use in lessons on water safety, the RNLI works together with Swim England to provide Swim Safe sessions for children on many holiday beaches throughout the summer.[178]

The RNLI and the Fire Service train and equip swift water rescue teams, giving them the skills to cope with swollen rivers, swirling floodwaters and deadly recirculating currents in the stilling basins under weirs.[179]

◆ ◆ ◆

In 1961, the pioneers of resuscitation reported success in training Norwegian schoolchildren to give mouth-to-mouth resuscitation, using the recently developed Resusci Anne manikins.[180] A more recent study in British primary schools confirmed that pupils are quick to learn resuscitation skills, although some of the younger pupils were unable to compress the manikin's chest to the recommended depth.[181] During a family party in 2015, a seven-year-old American girl spotted her three-year-old nephew floating unconscious in the family swimming pool. She pulled him from the water and gave him mouth-to-mouth resuscitation, followed by several chest compressions.[182] He vomited a large volume of water. At this point, an adult continued the resuscitation. The little boy began to breathe again and made a full recovery. The girl's only training in CPR came from half-watching a video as she sat on a sofa beside her mother, a nurse.

The RLSS, the Ambulance Service, the St John Ambulance Service, the Red Cross and other organisations run a growing number of resuscitation courses in secondary schools, factories, offices, clubs, swimming pools and doctors' surgeries. Participants listen to explanations of what to do and why. Then they practise the techniques of mouth-to-mouth resuscitation and CPR on manikins. A drowning victim has a better chance of recovery if the rescuer has had training in resuscitation[183] and is prepared to continue CPR for an hour or more.[184]

At public swimming pools, lifeguards' skill in resuscitation ensures that 94% of children who lose consciousness in the water are successfully revived. In contrast, only 50% of children survive after losing consciousness in open water.[185] Ambulance staff are proficient in mouth-to-mouth and CPR, defibrillation and life support, but victims of drowning accidents are more likely to die if resuscitation is delayed until the ambulance arrives.[186]

To avoid delay, the rescuer should begin resuscitation at once. Even someone with no knowledge of resuscitation can be guided by telephone advice from the 999 call centre while the ambulance is on its way. On Boxing Day 2004, a mother found her two-year-old son face down in the garden pond, apparently dead. The boy's father phoned 999, and thanks to the operator's clear instructions, his mother began resuscitation immediately.[187] She continued to give the toddler rescue breaths and heart massage until the paramedics arrived and took over. Vigorous CPR continued in hospital for more than seven hours before his heart began to beat again. He made a complete recovery.

More than twenty years ago, Professor John Pearn made a persuasive argument that 'every parent should be a first aider' capable of giving their child pre-hospital resuscitation.[188] Parents with swimming pools in their garden should certainly heed his advice and take part in a training course on CPR,[189] although they hope that they will never need to put their training into practice.

SUPPORT LOCAL INITIATIVES AND NATIONAL DROWNING PREVENTION ORGANISATIONS

Many powerful drowning prevention campaigns grow from the efforts of bereaved parents to prevent a repetition of the fatal accident that lead to the death of their loved one. In 2005, a university student left a Kingston nightclub after spending an evening with friends. In the dark, he missed his footing, fell into the Thames and drowned. His mother was determined that railings should be placed along the stretch of riverbank where her son stumbled into the water.[190] On advice from the Royal Society for the Prevention of Accidents (RoSPA), Kingston Council agreed to a number of safety measures, and three years after her son's death, a protective barrier was installed along the steep riverbank near Richmond Bridge. The RNLI presented her with an award. As part of its 'Respect the Water' campaign, the RNLI enlisted the support of Thames-side pubs, explained how to promote water safety to their customers, supplied the pubs with rescue 'throw bags' and trained the staff how to use them in an emergency.[191]

Bereaved families have released powerful videos to warn others of drowning risks. In July 2010, two-year-old Oliver wore armbands while playing with his brothers and cousins in their grandfather's swimming pool, watched over by his parents. Then the children settled down to watch a DVD in the pool house, and the adults relaxed in the sun. Nobody noticed the little boy re-entering the pool. By the time he was spotted floating face down and unconscious, it was too late to save him.[192] In September 2011, during a visit to his grandmother, three-year-old Conwy wandered away from his parents and climbed a fence at the bottom of the garden. After a frantic search, his grandmother found him submerged in the pond beyond the fence.[193] The two families acted as advisers to the producer of *The Danger Age*, a short film that focuses on the drowning risk of preschool children by showing typical incidents in a swimming pool, a garden pond and a bathtub.[194]

In a heatwave in June 2005, two fifteen-year-old boys got into difficulties while swimming in Waterloo Lake near Leeds. Divers recovered their bodies from the depth of the lake.[195] On a sunny Sunday morning in March 2014, three fifteen-year-old school friends decided on an impromptu swim in a river pool in a Worcestershire park, but within minutes one of boys sank beneath the surface.[196] His body was recovered two hours later. In June 2014, the parents who lost their sons in these fatal accidents worked with the RLSS during Drowning Prevention Week to warn teenage boys about the danger of swimming in open water during hot weather.[197] The RLSS produced a striking video – *'Filling up'* – which has been shown in many schools and can be seen on YouTube.[198]

A year later, during Drowning Prevention Week in June 2015, the RLSS released a video – *'Beneath the Surface: the Families' Stories'* – that focused on the lasting grief of parents mourning for their drowned children.[199] Two of the victims were in their twenties, three were teenagers and two were toddlers. In seven brief interviews, mothers, fathers and sisters speak movingly about their feelings of regret and loss.[200] The RLSS thanked 'all the brave families who took part in the film'.

In 2015, a speedboat with a newly installed propeller skimmed past the breakwater of Brixham harbour, struck a large wave and capsized. Three people swam clear, but a fourteen-year-old girl drowned, trapped under the boat when her over-large buoyancy aid caught on a projecting cleat.[201] Her parents realised that recreational boaters are often too casual about safety, so they set out to provide clear, simple guidelines. After discussions with Richard Graham MP, the Royal Yacht Association, Her Majesty's Coastguard and the RNLI, they developed Emily's Code to improve the safety of recreational boaters – a ten-point checklist before they leave their mooring.[202]

Local councils have an important role in accident prevention. In 2009, a fifteen-year-old boy drowned in the Blue Lagoon, a disused limestone quarry near Buxton in Derbyshire.[203] After photographs of its turquoise water appeared on the internet, the quarry became an increasingly popular destination for young families despite signs warning that the water was intensely alkaline and contained 'car wrecks, dead animals, excrement and rubbish'. Draining the quarry was ruled out because the water was too toxic to pump into the local water supply. In 2013, High Peak Borough Council decided on a radical step and dyed the water black.[204]

Vigorous local campaigns are needed to change the mindset of youngsters caught up in the dangerous craze of tombstoning from cliffs and harbour walls. Between 2005 and 2015,

20 tombstoners were killed, while 83 suffered permanent paralysis or other serious injuries.[205] Plymouth City Council put up large signs on clifftops warning 'No Diving or Tombstoning. Danger of Death'. In spite of the signs, youngsters continued to hurl themselves into the sea, and in 2016, a father-of-three lost his life when he leaped from the cliffs of Plymouth Hoe.[206] The Maritime and Coastguard Agency, the RNLI and RoSPA joined forces to persuade young people to turn back from the brink. Eye-catching posters warned: 'Don't Jump into the Unknown'.[207]

In 2011, after a series of fatal drownings in the River Avon,[208] Bath Council asked the Royal Society for the Prevention of Accidents (RoSPA) for advice on improving safety for the increasing numbers of residents, students and tourists who were walking or cycling along the banks of the river.[209] RoSPA inspectors met with representatives from Avon Fire and Rescue, Avon and Somerset Police, the Environment Agency, British Waterways, the University of Bath and Bath Spa University and local businesses. They studied reports of the drowning accidents, including statements by the bereaved families and reports in local newspapers. They noted that, in the previous two years, the Fire Service had saved 16 people from drowning in the river. Then the inspectors walked along the riverside pathways in daylight and after dark, making notes and taking photographs.

RoSPA's Water Safety Review identified 'sections of the river where a simple stumble, trip or swerve while on a bicycle could result in a significant fall into the river. Once in the river, opportunities for self-rescue or for being rescued are few.' The edge of the river bank was interrupted by ill-defined cut-outs at ladder access points – danger points that were virtually invisible in the dark. Yet there were no signs along the route to indicate that special care should be taken. The Review recommended early action to increase the visibility of the ladder cut-outs with paint or 'cat's eyes', to cut back hedges and weeds to maintain the width of the pathway, to provide warning signs 'to dissuade people from using the route at night or if intoxicated', to repair the grab lines and ladders to help self-rescue, and to promote water safety awareness to residents, tourists and students, especially university 'Freshers'.

Between 2011 and 2016, Bath Council spent more than £500,000 on river safety, including erection of fences along several stretches of the riverbank.[210] Nevertheless, there have been 7 further fatal drowning accidents since the RoSPA Water Safety Review.[211] The father of a student who drowned in November 2016 described the River Avon as 'Bath's serial killer'.[212]

In June 2014, an unseen current trapped seven teenagers on a raft in the middle of Foulridge Reservoir near Colne in Lancashire. Two boys decided to swim to the shore to bring help, but they both suffered swimming failure. One youngster barely managed to return to the raft and had to be hauled out of the water, too chilled and exhausted to climb aboard himself. His friend drowned 30 yards from the shore.[213] The Lancashire Fire and Rescue Service immediately set to work on a water safety education package – *Dying for a Dip* – aimed at teenagers and young men, explaining that open water remains dangerously cold even in summer.[214] Firemen held demonstrations for Police Cadets and gave talks in schools throughout the county. Working with James's family and the youngster who survived, the Fire Service produced the video 'James Drowned, I nearly did'.[215]

Four days after this fatal accident in Foulridge Reservoir, during Prime Minister's Questions, the local Member of Parliament, Andrew Stevenson, told the packed House of Commons about his constituent: 'His death has left his family and friends, and the local community, in shock. As this week is Drowning Prevention Week, what can the Prime Minister do to raise awareness of the dangers of open water and to improve water safety, particularly during this warm summer?'[216] David Cameron answered, 'My heart goes out to the family that my honourable friend has mentioned. For anyone to lose a son in such a tragic way is absolutely heartbreaking. We need to spread better information about the dangers of swimming in open water. We also need to do more to teach swimming and lifesaving skills in schools.'

Within weeks, the All-Party Parliamentary Group (APPG) on Water Safety and Drowning Prevention held its first meeting, chaired by Nadhim Zahawi MP, Minister for Children and Families, with Andrew Stephenson MP as vice-chair.[217]

◆ ◆ ◆

In February 2016, Robert Goodwill, the Minister of State for Transport, announced the Government's support for the UK Drowning Prevention Strategy,[218] a 'call to action' developed by the National Water Safety Forum (NWSF) with contributions by the Royal Society for the Prevention of Accidents (RoSPA), the Amateur Swimming Association (since re-named Swim England), the Royal Life Saving Society (RLSS), the Royal National Lifeboat Institute (RNLI), the Maritime and Coastguard Agency (MCA), the Canal and Rivers Trust, the British Sub-Aqua Club (BSAC) and the National Fire Chiefs Council (NFCC) – each with its own 'wealth of knowledge, expertise and insight'. Their shared aim was to halve the number of fatal drownings by 2026.

The UK Drowning Prevention Strategy identified five targets:

+ provision of swimming lessons and water safety education for all schoolchildren
+ assessment of local water risks by every community, with water safety plans
+ better understanding of water-related self-harm
+ increased awareness of everyday risks in, on and around water
+ risk assessment by every recreational and water-sport organisation.

Although swimming lessons have been part of the curriculum for all primary schools since 1994, many children fail to swim 25 yards, a single length of the local swimming pool. In May

2016, the Government established the Curriculum Swimming and Water Safety Group, with representatives from education, sport and leisure. By March 2017, the Group published its report – *Recommendations to ensure all children leave primary school able to swim.*[219]

In June 2017, a new All-Party Parliamentary Group (APPG) on Swimming focussed on swimming lessons in primary schools.[220] Swim England and the Swimming Teachers Association (STA) delivered well-planned programmes for swimming lessons in and out of school hours.[221] The RNLI commissioned a review of research on swimming skills and aquatic safety training as drowning prevention for children and young people.[222] Together with Swim England, the RNLI expanded holiday Swim Safe programmes at chosen inland and coastal venues.[223]

To help local communities cope with the complexities of assessing water risks and planning water safety, the Local Government Association produced a 'Water safety toolkit' for councils,[224] and the RLSS published *A practical implementation guide to setting up a Water Safety Action Group and designing a local Water Safety Plan.*[225]

During the course of 2015, twelve people ended their lives by jumping from London Bridge.[226] The City of London Corporation reported that 'the most common method of suicide in the City is drowning in the Thames'.[227] The RNLI increased its involvement in suicide prevention, both by fast, skilled response to emergencies and by support for the crews who have to deal with people struggling or drowning in the river.[228] In July 2015, RNLI lifeguards also rescued a teenage prankster who had persuaded a friend to film him as he vaulted over the railing on Tower Bridge and plunged 40 feet into the river.[229] He screamed for help as the current carried him downstream. He was on the point of drowning when the lifeboat reached him near Butler's Wharf. An ambulance took him to St Thomas's Hospital.

To gain more understanding of 'water-related self-harm', the RNLI became a member of the National Suicide Prevention Alliance.[230] In April 2016, the RNLI joined *The Bridge Pilot*, a multi-agency project working to reduce the number of people jumping from the bridges over the Thames.[231] Members include the City of London Corporation, the Metropolitan Police, London Fire Brigade, the Ambulance Service, the Samaritans, the Port of London Authority, the London Coastguard and the Marine and Coastguard Agency (MCA).

The Samaritans placed six signs on London Bridge, showing their free telephone number and encouraging troubled people to seek help.[232] More signs followed on Tower Bridge, Waterloo Bridge, Blackfriars Bridge, Southwark Bridge, the Millennium Bridge and other danger spots. Transport for London installed life rings on bridges, river walls and piers, each placement bearing prominent advice to phone 999 and ask for the Coastguard if someone is spotted in the water, so ensuring that a lifeboat can be dispatched without delay.[233] Volunteers patrolled Waterloo Bridge, trained to intercept people who looked as if they were about to jump.[234] In May 2016, Prince William paid an official visit to the RNLI Tower Lifeboat Station near Waterloo Bridge, where he met 'volunteer and full-time members of the lifeboat crew and representatives of other front-line organisations who are often the first on scene in dealing with self-harm'.[235]

The organisations involved with *The Bridge Pilot* became members of the Tidal Thames Water Safety Forum, working together to reduce the number of accidents, fatal drownings and suicides along the entire tidal Thames as it winds for 95 miles from Teddington to the Thames Estuary.[236] In May 2019, Prince William showed his continued support when he travelled by boat to *HMS President*, the Royal Navy establishment close to Tower Bridge, for the opening of the Safer Thames Campaign.[237]

By 2017, as many as 120 organisations were involved with the UK Drowning Prevention Strategy, often forming progressive new partnerships to promote increased awareness of everyday risks in, on and around water.[238] The RLSS, the Canal and Rivers Trust and the Fire and Rescue Service joined forces to warn runners and walkers about the risk of falling into the water from riverside paths.[239] The Fire and Rescue Service ran the annual campaign 'Be Water Aware', alerting people to the danger of cold shock after a fall into open water and endorsing the advice from the RNLI that it is safer to float for a couple of minutes until breathing is under control instead of trying to swim to safety at once.[240] The RLSS continued the annual campaign 'Don't Drink and Drown'.[241] The Police and Fire Services worked with the Mineral Products Association (MPA) to prevent drownings in quarries.[242] The RLSS introduced the National Water Safety Management Programme, which provides training modules on risk management and accident prevention for employers and employees working in, on or near water.[243] In 2019, RoSPA published *Managing Safety at Inland Waters*, which offers advice to managers, land owners and local authorities in charge of bodies of water such as canals, reservoirs and lakes accessible to the public.[244]

The National Trust looks after 800 miles of coastline in England, Wales and Northern Ireland.[245] To improve beach safety, the National Trust works increasingly closely with the RNLI and the MCA.[246] In 2018, to inform its prevention strategies and campaigns, the RNLI commissioned the report *Personal narratives of serious incidents at sea and on the coast*, based on structured interviews with people who had witnessed a water-related fatality or survived a life-threatening incident.[247] In 2019, the MCA published *Managing Beach Safety*, which sets out 'to assist coastal local authorities and others who are in effective control of the beach, to help keep people safe by assessing the risks and preventing incidents'.[248]

As well as providing a 24-hour search and rescue service around the coast and at sea, the MCA is responsible for 'implementing British and international law and safety policy',[249] protecting all on board vessels in UK waters and all 'seafarers on UK flagged vessels'. The government publishes advice on health and safety on ships, on weather, tides and navigation, and on the choice of emergency and life-saving equipment on ships.[250] The BBC has produced a series of advisory leaflets for film-makers, listing what can go wrong when working in or near bodies of water, while filming from boats, when scuba diving, or while filming adventure activities and high risk sports.[251]

Scotland has a higher death rate from drowning than the rest of Britain: the public has legal access to the extensive coastline, many islands, tumbling mountain streams, fast-flowing rivers, numerous reservoirs and thousands of lochs filled with extremely chilly water. *Scotland's Drowning Prevention Strategy 2018–2026* was published in January 2018.[252]

◆ ◆ ◆

The Royal Society for the Prevention of Accidents (RoSPA) warned that a third of accidental drowning deaths in inland waters occur when the victims are taking part in a specific water sport, such as swimming in open water, canoeing, kayaking, motorboating and angling. RoSPA noted a 'very high' death rate for motorboating and scuba-diving, a 'high' death rate for canoeing, kayaking, sailing, jet skiing and angling, a 'moderate' to 'high' death rate for outdoor swimming, a 'moderate' death rate for wind/kite surfing, but a 'very low' death rate for swimming indoors in the warm, clear water of lifeguarded swimming pools.[253]

For each water sport, a national governing body defines the rules, encourages safe practice and identifies the risks. British Canoeing governs canoeing, kayaking and white-water rafting.[254] The Royal Yachting Association governs sailing, motorboating, jet skiing and windsurfing.[255] The Angling Trust governs inland and sea angling.[256] The Sub-Aqua Club governs recreational diving.[257] UK Canyon Guides govern canyoning and gorge walking.[258] The British Caving Association governs caving and potholing.[259] British Rowing governs indoor and on-water rowing.[260] Swim England, Scottish Swimming, Swim Wales and Swim Ulster govern swimming, high diving, synchronised swimming, water polo and open-water swimming,[261] while the British Triathlon Federation governs triathlon and associated multisport competitions.[262] Recently, British Triathlon and the Royal Life Saving Society (RLSS) developed SH2OUT, a project which promotes safer open-water swimming.[263]

In April 2017, the government published *Duty of Care in Sport: Independent Report to Government*, which dealt with 'the safety, wellbeing and welfare' of athletes and participants involved in organised, competitive sport, including 'both elite and grassroots sports'.[264] The working group included Olympic, Paralympic and international athletes with impressive careers in wheelchair racing, basketball, football and rugby union, but only one member, an Olympic rower, was involved in any water sports. The review referred to first aid, cardiac screening, CPR and the provision of defibrillators, and asked the Government to consider the feasibility of a national register for serious sporting injuries and fatalities. However, the emphasis was on terrestrial sports, rather that water sports. The phrase 'catastrophic sporting injuries' was defined as 'life-changing injuries received in the field of play'. And although the word 'concussion' appeared thirty times, there was no mention of 'drowning'.

◆ ◆ ◆

The Adventure Activities Licensing Authority (AALA) was established in 1996, following the drowning of four young kayakers in Lyme Bay. The providers of commercial adventure activities have to satisfy AALA inspectors that their risk management procedures are in good order before they are granted a licence to take young people caving, climbing, trekking, kayaking, canoeing, white-water rafting, waterskiing, sailing and windsurfing.[265] In 2007, after a series of 'incidents and near-misses', coasteering was added to the list of adventure activities that had to be inspected and licensed by AALA.[266]

In 2008, the Council for Learning Outside the Classroom (CLOtC) introduced a national accreditation scheme which awards the LOtC Quality Badge to 'organisations that provide good quality educational experiences and manage risk effectively'.[267] Teachers planning school trips are encouraged to make use of providers awarded the LOtC Quality Badge.

The Outdoors Education Advisers' Panel (OEAP) organises training courses for leaders of outdoor learning and educational visits,[268] and publishes helpful guidance for schools including group management and supervision, risk management, adventure activities, safe use of swimming pools during off-site visits, swimming and paddling in natural water (river, canal, sea or lake) and an improved version of the 2003 leaflet 'Group Safety at Water Margins'.[269] The OEAP also discusses what the law requires, how to comply with it, and what may happen following an accident and incident.[270] In November 2018, the Department of Education published 'Health and safety on educational visits', praising the Adventure Activities Licensing Authority, the Council for Learning Outside the Classroom, the LOtC Quality Badge and the Outdoors Educator Advisers' Panel.[271]

◆ ◆ ◆

The RNLI held its first drowning prevention campaign 'Respect the Water' in Brighton in 2013, with an award-winning road show aimed at those who are at the greatest risk of drowning at the coast – adult men.[272] Specially designed pint glasses and beer mats warned about the danger of drinking before swimming, while a custom-built transparent cube holding a 'tonne of water' allowed people who planned to swim in the sea to picture the impact of water on the move. (A cubic metre of water weighs one thousand kilograms.)

Over half of the people who drown at the coast each year never meant to be in the water. During the annual 'Respect the Water' campaigns, the RNLI reminds people that they can reduce risk by changing their behaviour.[273] They can keep away from the edge of a cliff when they follow a coastal path. They can check on the tides before walking along the beach. They can stay out of reach of crashing waves. They can wear a life jacket when they fish from seaside rocks. They should know about rip currents and how to avoid them. They must remember the power of the sea and the coldness of the water.

Since 2017, the '*Respect the Water*' campaigns have emphasised the year-round danger of 'cold water shock', advising people who fall into open water or the sea to float on their backs for a minute or two, until their breathing is under control.[274] Posters and adverts in cinemas and on TV show young men and women floating on their backs in the water, with compelling captions such as:

'FIGHT YOUR INSTINCT, NOT THE WATER';
'EXTEND YOUR ARMS, LEGS AND LIFE EXPECTANCY';
'GIVE YOURSELF A FLOATING CHANCE';
'FLOAT FOR YOUR LIFE'.

A news release in May 2019 noted that 11 people credited the *Respect the Water* float advice with saving them from drowning at the coast.[275] In July 2019, a man in his twenties was about to drown in the Thames when he recalled seeing an RNLI 'Float to Live' poster – and was able to keep himself afloat until officers on a Metropolitan Police launch grabbed him and RNLI lifeboatmen pulled him from the water.[276] In August 2020, the Scarborough lifeboat rescued a ten-year-old boy who had floated on his back for nearly an hour as the tide and wind swept him across South Bay. He told the crew that he had been watching the TV programme '*Saving Lives at Sea*' and had followed the advice given on the show.[277] The coxswain said, 'We're very much in awe of this incredible lad, who managed to remain calm and follow safety advice to the letter in terrifying and stressful circumstances. Had he not, the outcome might have been very different.'

❖ ❖ ❖

Since the UK Drowning Prevention Strategy began in 2016, the published number of 'Accidental drowning deaths' has fallen year by year, from 300 in 2016 to 223 in 2019.[278] However, it is important to point out that the total number of drowning deaths annually is far higher. In 2017, for instance, there were 255 'Accidental drowning deaths'. These deaths, however, were only 43% of the **total** of 592 drowning deaths that year.[279] Although the experts and enthusiasts involved in the UK Drowning Prevention Strategy 'are making good progress throughout the country to keep people informed of the risks',[280] too many people ignore their warnings. The RNLI '*Respect the Water*' campaigns remind us all that it is our own responsibility to think twice before putting our children and ourselves in danger:

'TREAT THE WATER WITH RESPECT: NOT EVERYONE CAN BE SAVED.'[281]

REFERENCES

INTRODUCTION: WHY SO MANY PEOPLE DROWN.

1   Drowning Statistics, The Royal Society for the Prevention of Accidents, UK

2   Drowning Statistics, Centers for Disease Control and Prevention, Atlanta, GA

3   Malcolm Read with Paul Wade (1997), *Sports Injuries*, 2nd Edn. Butterworth-Heinemann, Oxford, p.178

4   American Academy of Pediatrics, Committee on Sports Medicine and Fitness, (2001) 'Medical conditions affecting sports participation', *Pediatrics*, 107: 1205-1209

5   Nigel Warburton (1998), 'Freedom to Box', *Journal of Medical Ethics*, 24: 56-60

6   RoSPA Drowning Statistics 1998 to 1992, Royal Society for the Prevention of Accidents, http://www.rospa.com/leisure-safety/statistics

7   B. J. Gabbe, C. F. Finch, P. A. Cameron and O. D. Williamson (2005), 'Incidence of serious injury and death during sport and recreational activities in Victoria, Australia', *British Journal of Sports Medicine*, 39: 573-577

8   Kevin M. Harris et al (2010) 'Sudden death during the triathlon.' *Journal of the American Medical Association*, 303: 1255-1257

9   William C. McMaster, Andrew Roberts and Terry Stoddard (1998), 'A correlation between shoulder laxity and interfering pain in competitive swimmers', *American Journal of Sports Medicine*, 26: 83-86

10   Alison Kemp and J. R. Sibert (1992), 'Drowning and near drowning in children in the United Kingdom: lessons for prevention', *British Medical Journal*, 304: 1143-1146

11   T. Noguchi (1994) 'A survey of spinal cord injuries resulting from sport', *Paraplegia*, 32: 170-173

12   Malcolm Read with Paul Wade, *Sports Injuries*, p.166

13   Angus Watson, 'Deep and meaningful', *Financial Times*, 30/31 July 2005, W4

14   Virginia Hunt Newman (1968), *Teaching an Infant to Swim*, Angus and Robertson, London, p.5

15   The Australian Council for the Teaching of Swimming and Water Safety: Austswim (2001), *Teaching Infant and Preschool Aquatics: Water experiences the Australian Way*, Human Kinetics, Champaign, IL.

16   The Australian Council for the Teaching of Swimming and Water Safety: Austswim (2001),

Teaching Swimming and Water Safety: Learning Aquatics the Australian Way, Human Kinetics, Champaign, IL.

17   Frank Pia (1974), 'Observations on the Drowning of Nonswimmers', *Journal of Physical Education*, YMCA Society of North America, July 1974, http://www.pia-enterprises.com/observation.rtf

18   Geoffrey Clarkson, MBE, international fly fishing champion: personal communication.

19   'Drowning Facts and Figures', International Life Saving Federation, http://ilfs.org/content/drowning-facts-and-figures

20   'Drowning Facts and Figures', International Life Saving Federation, http://ilfs.org/content/drowning-facts-and-figures

21   Office of National Statistics (2006), *Mortality Statistics: Deaths Registered in 2006*, Her Majesty's Stationary Office, London

22   Greg R. McLatchie (1986), *Essentials of Sports Medicine*, Churchill Livingstone, Edinburgh, p.119

23   Royal Society for the Prevention of Accidents (2005) 'Accidental Drowning 2005', http://www.rospa.co.uk/leisuresafety/water/statistics/2005statistics.htm

PART 1: WATER IS A HOSTILE ENVIRONMENT

CHAPTER 1: WEIGHED DOWN AND BUOYED UP

1   Malcolm Jobling (1995), *Environmental Biology of Fishes*, Chapman and Hall, London, p.541

2   R. McN. Alexander (1959), 'The densities of the Cyprinidae', *Journal of Experimental Biology*, 36: 333-340

3   Alwyne Wheeler (1983), *Freshwater Fishes if Britain and Europe*, Kingfisher Books London, p.20

4   Peter Whitehead (1975), *How Fishes Live*, Elsevier-Phaidon, Oxford, p.151

5   Pamela Bristow (1987), *The Illustrated Book of Fishes*, Octopus Books, London, p.12

6   Pamela Bristow, *The Illustrated Book of Fishes*, p.12

7   Pamela Bristow, *The Illustrated Book of Fishes*, p.283

8   R. M. Forbes, A. R. Cooper and H. H. Mitchell (1953), 'The Composition of the adult human body as determined by chemical analysis', *Journal of Biological Chemistry*, 203: 359-366

9   Lauralee Sherwood (1994), *Fundamentals of Physiology: a human perspective*, West Publishing, St Paul, MN., p.336

10   Cyril A. Keele and Eric Neil (1961), *Sampson Wright's Applied Physiology*, 10th Edn, Oxford University Press, Oxford, p.166

11   E. R. Donoghue and S. C. Minnigerode (1977), 'Human Body Buoyancy: a study of 98 Men', *Journal of Forensic Sciences*, 22:573-579

12   John Verrier (2001), *Swimming: The Skills of the Game*, Crowood Press, Marlborough, p.10-12

13   Lauralee Sherwood, *Fundamentals of Physiology: a human perspective*, p.338

14   Cyril A. Keele and Eric Neil, *Sampson Wright's Applied Physiology*, p.166

15   James A. Allardice (1972), *The Medical Aspects of Competitive Swimming*, Pelham Books, London, p.56

16 David Wilkie and Kelvin Juba (1996), *The Handbook of Swimming*, Pelham Books, London, p.4

17 'History of Lifejackets' (2005), http://comingbackalive.com/lifejackets.html

18 Richard J Fantus, 'Flotation devices – Mae West style', *The Bulletin of the American College of Surgeons*, 1 June 2014

19 'Choose it. Wear it', *The RNLI Guide to lifejackets and buoyancy aids* , https://rnli.org

20 Stan Bradshaw (2004), *River safety: a floater's guide*, Lyons Press, Connecticut, Chapter 1 quoted in Topkayaker.net http://www.sit-on-topkayaking.com/Articles/instruction/PDF.html

21 John Langley, 'Personal Flotation Devices' and Chris Brooks, Gunther Cornelissen and Rolf Popp 'Lifejackets', in *Handbook on Drowning* (2006), Joost Bierens (ed.), Springer Verlag, Berlin, p.71-72 and 226-229

22 'Personal flotation devices for the ones you love' (1997), http://www.boatingsafety.com/lifejack.htm

CHAPTER 2: BALANCE OF FORCES

1 Austswim, *Teaching Swimming and Water Safety*, p.30

2 Austswim, *Teaching infant and preschool Aquatics*, p.117-121

3 Austswim, *Teaching Swimming and Water Safety*, p.38-39

4 David Wilkie and Kelvin Juba, *The Handbook of Swimming*, p.33

5 Austswim, *Teaching Swimming and Water Safety*, p.74-76

6 Austswim, *Teaching Swimming and Water Safety*, p.74

7 A. G. Need, J. M. Wishart, F. Scopacasa, M. Horiwitz, H. A. Morris and B. E. C. Nordin (1995), 'Effect of physical activity on femoral bone density in men', *British Medical Journal*, 310:1501-1502

8 Sir James Gray (1959), *How Animals Move*, Penguin Books, London, p.53-55

9 Pamela Bristow, *The Illustrated Book of Fishes*, p.278

10 Ruth Brenner, Kevin Moran, Robert Stallman, Julie Gilchrist and John McVan, 'Swimming Abilities, Water Safety Education and Drowning Prevention' in *Handbook on Drowning*, p.112-117

CHAPTER 3: SHAPES IN THE WATER

1 Peter Whitehead, *How Fishes Live*, p.18

2 'Sticky Water', http://www.exploratorium.edu/ronh/bubbles/sticky_water.html

3 Mark M. Denny (1993), *Air and Water: the Biology and Physics of Life's Media*, Princeton University Press, Princeton, NJ, p.59

4 Peter Whitehead, *How Fishes Live*, p.30

5 Terry Laughlin (2002), 'Swimming. The smarter way to speed: get slippery'. http://www.adksportsfitness.com/october2002/articles/swimming.html

6   E. J. Anderson, W. R. McGillis and M. A. Grosenbaugh (2001), 'The boundary layer of swimming fish', *The Journal of Experimental Biology*, 204: 81-102

7   Amy Lang, Maria L. Habegger and Philip Motta (2012), 'Shark Skin Drag Reduction', *Encyclopedia of Nanotechnology*, Part 19, B. Bushan (ed.) Springer, Berlin, p.2394-2400

8   Huub Toussaint and Martin Truuens (2005), 'Biomechanical aspects of peak performance in human swimming', *Animal Biology*, 55: 17-40

9   Ross Vennell, Dave Pease and Barry Wilson (2006) 'Wave drag on human swimmers', *Journal of Biomechanics*, 39:664-671

10   Pamela Bristow, *The Illustrated Book of Fishes*, p.149

11   N.F. Hughes (2004), 'The wave-drag hypothesis: an explanation for size-based lateral segregation during the upstream migration of salmonids', *Canadian Journal of Fisheries and Aquatic Sciences*, 61:103-109

12   Andrew D. Lyttle, Brian A. Blanksby, Bruce C. Elliott and David G. Lloyd (1998), 'The effect of depth and velocity on drag during streamlined glide', *Journal of Swimming Research*, 13:15-22

13   Ross Vennell, Dave Pease and Barry Wilson (2006), 'Wave drag on human swimmers', *Journal of Biomechanics*, 39:664-667

14   Ross Vennell, Dave Pease and Barry Wilson (2006), 'Wave drag on human swimmers', *Journal of Biomechanics*, 39:664-667

15   David Wilkie and Kelvin Juba, *The Handbook of Swimming*, p.67-68

16   Terry Laughlin with John Delves (1996), *Total Immersion*, Simon and Schuster, New York, p. 34-44

17   R. L. Sharp and D. L.Costill (1989) 'Influence of body hair removal on physiological responses during breaststroke swimming.' *Medicine and Science in Sports and Exercise* 21: 576-580

18   Jean-Claude Chatard and BarryWilson (2008), 'Effect of fastskin suits on performance, drag, and energy cost of swimming', *Medicine and Science in Sports and Exercise*, 40:1149-1154

19   'World Records at Sydney Olympic Park Aquatic Centre', http://www.aquaticcentre.com.au/attractions/path_of_champions/world_records

20   'Ian Thorpe', Wikipedia, https://en.wikipedia.org/wiki/Ian_Thorpe

21   Steve Conner, 'Sharkskin swimsuits lead to hi-tech bid for Olympic gold', *The Independent*, 17 March 2000

22   Anna Quarrel (2003), 'Adidas presents new bodysuit: the JETCONCEPT', http://www.eurekalert.org/pub_releases/2003-07/aa-apn071803.php

23   'Ian Thorpe', Wikipedia https://en.wikipedia.org/wiki/Ian_Thorpe

24   'Record Breaking Benefits: A Speedo-NASA partnership after the 2004 Olympics resulted in a swimsuit worthy of world records', http://www.nasa.gov/offices/oct/home/tech_record_breaking.html

25   'Michael Phelps', Wikipedia, https://en.wikipedia.org/wiki/Michael_Phelps

26   Tom Scocca, 'Pool Hustlers: Swimming's polyurethane-assisted glide off the deep end', *Slate*, 31 July 2009, http://www.slate.com/articles/sports/sports_nut/2009/07/pool_hustler

27   Leon Foster, David James and Steve Haake (2012), 'Influence of full body swimsuits on competitive performance', *Procedia Engineering*, 34:712-717

28   'Swimming world records in Rome', BBC Sport http://newsvote.bbc.co.uk/mpapps/pagetools/print/news.bbc.co.uk/sp

29   Brent S. Rushall (2001), 'Why floatation bodysuits are unfair', http://coachsci.sdsu.edu/swim/bodysuit/unfair.htm

30   N. Benjanuvatra, G. Dawson, B. A. Blanksby and B. C. Eliott (2002), 'Comparison of buoyancy, passive and net active drag forces between FastskinTM and standard swimsuits', *Journal of Science and Medicine in Sport*, 5:115-123

31   'The Bodysuit Problem: What the scientists report', http://coachsci.sdsu.edu/swim/bodysuit/science.htm

32   Lachlan Thompson, 'How do polyurethane suits work?' http://www.crikey.com.au/2009/07/30/crikey-clarifier-how-do-full-polyurethane

33   Jim Morrison, 'How Speedo Created a Record-Breaking Swimsuit', *Scientific American*, 27 July 2012

34   'Hi-tech suits banned from January', BBC Sport, http://news.bbc.co.uk/sport1/hi/other_sports/swimming/8161867.stm

35   Kevin Moran (2014), 'Can You Swim in Clothes? An Exploratory Investigation of the Effect of Clothing on Water Competency', *International Journal of Aquatic Research and Education*, 8:338-350

36   Martin James Barwood, Victoria Bates, Geoffrey Long and Michael J. Tipton (2010), '"Float First": Trapped Air Between Clothing Layers significantly Improves Buoyancy After Immersion', *International Journal of Aquatic Research and Education*, 5:147-163

CHAPTER 4: SAFE ENTRY

1   Austswim, *Teaching infant and preschool aquatics*, p.109-116

2   J. D. Blitvich, G. K. McElvoy and B. A Blanksby (2000), 'Risk reduction in diving spinal cord injury: Teaching safe diving skills', *Journal of Science and Medicine in Sport*, 3: 120-131

3   Barth A. Green, Alexander Gabrielsen, Wiley J. Hall and James O'Heir (1980), 'Analysis of swimming pool accidents resulting in spinal cord injury' *Paraplegia*, 18: 94-100

4   P. Barss, H. Djerrari, B. E. Leduc, Y. Lepage and C. E. Dionne (2008), 'Risk factors and prevention for spinal cord injury from diving in swimming pools and natural sites in Quebec, Canada: a 44-year study', *Accident Analysis and Prevention*, 40: 787-797

5   Craig Ferrell (1999), 'The spine in swimming', *Clinics in Sports Medicine*, 18: 389-393

6   Gale M. Gehlsen and John Wingfield (1998), 'Biomechanical analysis of competitive swimming starts and spinal cord injuries', *Journal of Swimming Research*, 13 :23-30

7   'Fast waters run deep for Olympic swimmers', *New Scientist*, 20 August 2008

8   P.T. Perry (1965), *Diving into shallow water*, Amateur Swimming Association, Loughborough.

9   *Guidance Notes: Safe Operation of Public Diving Facilities* (2004), Great Britain Diving Federation, www.diving-gbdf.com

10   H. L. Frankel, F. A. Montero and P. T. Perry (1980), 'Spinal cord injuries due to diving', *Paraplegia*, 18: 118-122

11   *Guidance Notes: Safe Operation of Public Diving Facilities* (2004), Great Britain Diving Federation, www.diving-gbdf.com

12   'How to Do an Underwater Save in Diving', (2012), iSport.com

13   Michael Hutson and Cathy Speed (2011), *Sports Injuries*, Oxford University Press, Oxford. p. 465

14   Vadimir M. Zatsiorsky (2000), *Biomechanics in Sport: performance enhancement and injury prevention*, Vol 9, John Wiley and Son, New York, p.346-347

15   Benjamin D. Rubin (1999), 'The basics of competitive diving and its injuries', *Clinics in Sports Medicine*, 18: 293-303

16   'Diving Queen Threatened by detached retina', September 14, 2008, http://www. womenofchina.cn/womenofchina/html

17   Matt Somerford, 'Diving: Team GB's Chris Mears completes remarkable journey from near death to Olympic finalist', *The Independent*, 7 August 2012

18   Ian Prior, 'Chris Mears and Jack Laugher win gold in synchronised springboard diving', *The Guardian*, 10 August 2016

CHAPTER 5: DIVING UNDER PRESSURE

1   Frances Ashcroft (2001), *Life at the Extremes: the science of survival*, Flamingo, London, p.68

2   David J. Steedman (1994), *Environmental Medical Emergencies*, Oxford University Press, Oxford, p.55

3   Frances Ashcroft, *Life at the Extremes: the science of survival*, p.7

4   Sharon Krum, 'Takes your breath away', *The Times*, 11 October 2003

5   Peter Lindholm (2006), 'Physiological mechanisms involved in the risk of loss of consciousness during breath-hold diving', Breath-Hold Diving, Proceedings of the Undersea and Hyperbaric Medical Society/Divers Alert Network 2006 June 20-21 Workshop, Durham, NC.

6   Dale Sheckter 'Snorkels – How They Work; How They Work Best', *California Diving News*, 22 June 2002

7   Jacques Y. Cousteau with Frédéric Dumas (1953), *The Silent World*, Hamish Hamilton, London. p.12

8   Jacques Y. Cousteau with Frédéric Dumas, *The Silent World*, p.1-13

9   'Diving Regulator', Wikipedia, https://en.wikipedia.org/Diving_regulator

10   Jacques Y. Cousteau with Frédéric Dumas, *The Silent World*, p.20-22 and 80-83

11   Peter Wilmshurst (1998), 'Diving and oxygen', *British Medical Journal*, 317: 996-999

12   Jacques Y. Cousteau with Frédéric Dumas, *The Silent World*, p.10-11

13   Peter Wilmshurst (1998) 'Diving and oxygen', *British Medical Journal*, 317: 996-999

14   F. K. Butler Jr and N. Gurney (2001), 'Orbital haemorrhage following face-mask barotrauma', Undersea and Hyperbaric Medicine, 28:31-34. See Minerva, *British Medical Journal* (28 October 2006) 333:926

15   Joseph C. Farmer, 'Otologic and paranasal sinus problems in diving', Chapter in *The Physiology and Medicine of Diving*, 3rd Edn. (1982), Peter B. Bennett and David H. Elliott (ed.) Balliere Tindall, London, p.532-533

16   Frances Ashcroft, *Life at the Extremes: the science of survival*, p. 70-71

17   Joseph C. Farmer (1997), 'Ear and Sinus Problems', Chapter in *Diving Medicine*, 3rd Edn. Edited by Alfred A. Bove, W. B. Saunders, Philadelphia, Ep. 240-243

18   Bernard Empleton, Robert W. Hill (1980), *The New Science of Skin and Scuba Diving*, 5th Edn. Council for National Cooperation in Aquatics, Association Press New Century Publishers, New Jersey, p.143-145

19   Joseph C. Farmer (1997), 'Ear and Sinus Problems', Chapter in D*iving Medicine*, p.238

20   Joseph C. Farmer, 'Ear and Sinus Problems', Chapter in *Diving Medicine*, p.515-521

21   Peter Whitehead, *How Fishes Live*, p.112-117

22   Peter Whitehead, *How Fishes Live*, p.24

23   Peter Moyle and Joseph Cech (1996), *Fishes: An introduction to ichthyology*, 3rd Edn. Prentice Hall, New Jersey, p.69

24   William N. McFarland, F. Harvey Pough, Tom J. Cade and John B. Heiser (1979), *Vertebrate Life*, Macmillan, New York, p.210-212

25   Quentin Bone, Norman B. Marshall and J. H. S Baxter (1995), *The Biology of Fishes*, 2nd Edn. Blackie, London. p.79

26   Rich Novak (2000), 'Release Techniques for Marine Fish', University of Florida, http://edis. ifas.ufl.edu/sg047

27   James Francis (1997), 'Pulmonary Barotrauma: A new look at mechanisms.' *SPUMS Journal*, 27: 205-218 (South Pacific Underwater Medical Society)

28   E.W. Russi (1998), 'Diving and the risk of barotrauma', *Thorax*, 53: S20-S24

29   British Sub Aqua Club (1978), *BSAC Diving Manual* 10th Edn. The British Sub Aqua Club, London, p.78-84

30   Bernard Empleton, Robert W. Hill and Edward Lanphier, *The New Science of Skin and Scuba Diving*, p.150-155

31   Bernard Empleton, Robert W. Hill and Edward Lanphier, *The New Science of Skin and Scuba Diving*, Bernard Empleton, p.150

32   Larry D. Weiss and Keith W. Van Meter (1995), 'Cerebral Air Embolism in Asthmatic Scuba Divers in a Swimming Pool', *Chest*, 107: 1653-1654

33   D. Steedman (1997), 'Near drowning and diving injuries', Chapter in *Cambridge Textbook of Accident and Emergency Medicine*, D. V. Skinner (ed.) Cambridge University Press, Cambridge, p.758

34   The British Sub-Aqua Club, *Sport Diving*, p.104

35   S. Gribben (2006) 'Depth Limits', British Sub-Aqua Club, www.bsac.com/page. asp?section=2674&sectionTitle=Depth+Limits

36   David J. Steedman, *Environmental Medical Emergencies*, p. 56

37   Beekman Wines and Liquors (1999), 'Champagne – How Many Bubbles? http://www. beckmanwine.com/prevtopam.htm

38   Frances Ashcroft, *Life at the Extremes: the science of survival*, p.58

39   D. H. Elliot, and E. P. Kindwall, 'Manifestations of the Decompression Disorders', (1982) Chapter in *The Physiology and Medicine of Diving*, 3rd Edn. Peter B. Bennett and David H. Elliott (ed.) Balliere Tindall, London, p.460-468

40   The British Sub-Aqua Club, *Sport Diving*, p.104

41 D. H. Elliot and E. P. Kindwall, 'Manifestations of the Decompression Disorders' (1982), Chapter in *The Physiology and Medicine of Diving*, p.460-46

42 B. D. Butler and B. A. Hills, 'The lung as a filter for microbubbles', *Journal of Applied Physiology: respiratory, environmental and exercise physiology*, 47: 537-543

43 J. D. King (2006), 'Skin Bends', *London Diving Chamber*, http://www.londondivingchamber. co.uk/php?id=dci&page=12

44 The British Sub-Aqua Club, *Sport Diving*, p.105

45 The British Sub Aqua Club (1978), *Diving Manual*, 10th Edn. The British Sub Aqua Club, London, p.109-112

46 Professor Sir Brian Smith. Personal communication.

47 John H. Lienhard (1997), 'Engines of Our Ingenuity: Under Pressure', http://www.uh.edu/ engines/epi1093.htm

48 *Encyclopaedia Britannica* (1911), 'Caisson Disease', http://en.wikisource.org/wiki/1911_ Encyclop%C3%A6dia_Britanni

49 Frances Ashcroft, *Life at the Extremes: the science of survival*. P.59

50 Ernest S. Campbell, 'Flying after Diving', http://www.scuba-doc.com/flyngaft.htm

51 Peter Wilmshurst (1997), 'Brain damage in divers', Editorial in *British Medical Journal*, 313: 689-690

52 Michael Knauth et al (1997), 'Cohort study of multiple brain lesions in sport divers: role of a patent foramen ovale', *British Medical Journal*, 314: 701-705

53 P. Gremonpré, P. Dendale, P. Unger and C. Balestra (1998), 'Patent foramen ovale and decompression sickness in sports divers', *Journal of Applied Physiology*, 84: 1622-1626

54 Arthur C. Guyton and John E. Hall (1996), *Textbook of Medical Physiology*, 9th Edn. W. B. Saunders Company, Philadelphia, PA. p.1051

55 Sanjay Sastry and Charles McCollum (2001), 'Patent Foramen Ovale and Stroke', *Circulation*, 103: 46-47

56 D. B. Butler and B. A. Hills (1979), 'The lung as a filter for microbubbles', *Journal of Applied Physiology*, 47: 537-543

57 P. Gremonpré, P. Dendale, P. Unger and C. Balestra (1998), 'Patent foramen ovale and decompression sickness in sports divers', *Journal of Applied Physiology*, 84: 1622-1626

58 Peter Glanvill (1994), 'Deep-sea GP under pressure', *Monitor Weekly*, 22 June 1994, p.46.

59 Yehuda Melamed, Avi Shupar and Haim Betterman (1992), 'Medical problems associated with underwater diving', *New England Journal of Medicine*, 326:30-35

60 King, J. D. (2006), 'Skin Bends', London Diving Chamber, http://www.londondivingchamber. co.uk/php?id=dci&page=12

61 Nigel Hewitt, 'Decompression theory for goats', http://www.nigelhewitt.co.uk/diving/maths. deco.html

62 A. E. Boycott, G. C. C. Damant and J. S. Haldane (1908), 'The Prevention of Compressed-air Illness', *The Journal of Hygiene*, 8: 342-443

63 J. S. Haldane quoted by Frances Ashcroft, *Life at the Extremes: the science of survival*, p.60-61

64 Carl Edmonds and Douglas Walker (1989), 'Scuba diving fatalities in Australia and New Zealand. 1. The human factor', *SPUMS Journal*, 19: 94-104. Carl Edmonds and Douglas

Walker (1990) 'Scuba diving fatalities in Australia and New Zealand. 2. The environmental factor', *SPUMS Journal*, 20: 2-4. Carl Edmonds and Douglas Walker (1991) 'Scuba diving fatalities in Australia and New Zealand. 3. The equipment factor', *SPUMS Journal*, 21: 2-4. Peter J. Denoble, Alessandro Marroni and Richard D. Vann (2011), 'Annual Fatality Rates and Associated Risk Factors for Recreational Scuba Diving', Recreational Diving Fatalities Workshop, Divers Alert Network, Durham, NC. Ben Davison, 'Scuba Drowning Deaths And Those Who Survive', *Undercurrent*, January 1999, www.undercurrent.org/UCnow/dive_magazine/1999/ScubaDrowningDeaths199901.htr

65   'Drowning statistics – RoSPA', http://www.rospa.com/leisure-safety/statistics/drowning/

CHAPTER 6: LOOKING OUT FOR TROUBLE

1   'Refractive index', https://en.wikipedia.org/wiki/Refractive_index (Note: the refractive index of a transparent substance is the ratio of the speed of light in a vacuum to the speed of light in the substance. The refractive index of air is 1.0003: the refractive index of water is 1.333.)

2   Keith Johnson (1980), *Physics for You*, Hutchinson and Co., London, p.184

3   Keith Johnson, *Physics for You*, p.185-186

4   'Blink', http://wilipedia.org/wiki/Blinking

5   Austswim, *Teaching Swimming and Water Safety*, p.31 and 35-37

6   Virginia Hunt Newman, *Teaching an Infant to Swim*, p.45-48

7   R. Colin Black, Swimming Pool Superintendent. Personal communication.

8   Brian T. C. Moore (1977), *Introduction to the Psychology of Hearing*, The Macmillan Press, London, p.169-170

9   R. Colin Black, Swimming Pool Superintendent. Personal communication.

10   'Swimming', 19 July 2012, http://www.eyeway.org/?q=swimming

11   Bill Briggs, 'From darkness to gold: Blinded Navy swimmer set to race in Paralympics', NBC News, 29 August 2012

12   'List of refractive indices', https://en.wikipedia.org/wiki/List_of_refractive_indices (Note: The refractive indices for water (1.333) and cornea (1.37) are very similar.)

13   Peter B. Moyle and Joseph J Cech (1996), *Fishes: An Introduction to Ichthyology*, 3rd Edn. Prentice Hall, New Jersey, p.152 (Note: The refractive index of the central zone of the fish's lens may be as high as 1.65.)

14   Keith Johnson (1980), *Physics for you*, Hutchinson and Co., London, p.200 (Note: The refractive indices of air (1.0003) and cornea (1.37) differ.)

15   Jane F. Koretz and George H. Handelman, 'How the Human Eye Focuses', *Scientific American*, July 1988

16   'The Physics of Diving: Light and Vision', http://www,dtic.mil/dtic/tr/fulltext/u2/693472,pdf

17   'The Physics of Diving: Light and Vision', http://library.thinkquest.org/28170/35.html

18   'Snellen chart', http://en.wikipedia.org/wiki/Snellen_chart

19   S. M. Luria and Jo Ann S. Kinney (1969), 'Visual Acuity under water without a face mask', Submarine Medical Research Laboratory, Groton, CT. http://www.dtic.mil/dtic/tr/fulltext/u2/693472.pdf

20   Michael Shapinker, 'I can swim clearly now', *Financial Times: Weekend*, W7 30/31 July 2005

21   'The Physics of Diving: Light and Vision', http://library.thinkquest.org/28170/35.html

22   British Sub-Aqua Club Diving Manual, *Sport Diving*, p. 3023

23   'Problems in Using Common Diving Equipment', http://www.scuba.net.hk/medicine/problem1.htm

24   Jan Kranhouse (2007), 'HydroOptix Double-Dome Dive Mask', http://www.scubaherald.com/hydrooptix-double-dome-dive-mask/

25   Mark W. Denny, *Air and Water: the biology and physics of life's media*, p. 251

26   Melanie Reid, 'In the watery darkness men work by feel alone', *The Times*, 1 December 2008

27   Max Garth, 'Colour in the Fishes Eye – The Light (part 2), www.sexyloops.com/articles/light.shtml

28   'Biology of fishes. Fish 311', http://www.washington.edu/classes/fish311/lecture-22.pdf

29   'Snell's window', https://en.wikipedia.org/wiki/Snell's_window

30   Keith Johnson, *Physics for You*, p.190 (Note: Light arriving from the scene above the water, 180° from horizon to horizon, is refracted into a 97.2° cone of light.)

31   Laura Nott, H. G. Reza and Molly Hennessy-Fiske, 'A strike from below, and a triathlete is gone', *Los Angeles Times*, 26 April 2008

32   David Knight, 'Refraction & Snell's Law', http://www.camerasunderwater.co.uk/articles/optics/refraction (Note: Light that strikes the water/air boundary at an angle greater than the 'critical angle' of 48.6 is totally reflected – not refracted.)

33   Max Garth, 'Colour in the Fishes Eye – The Light' (part 2), www.sexyloops.com/articles/light.shtml

34   Frances Ashcroft, *Life at the Extremes*, p.86

35   Max Garth, 'Colour in the Fishes Eye – The Light (part 2), www.sexyloops.com/articles/light.shtml

36   C. Guyton and John E. Hall, *Textbook of Medical Physiology*, p.637-639

37   Carl Bianco, 'How Vision Works: Perceiving Light', http://health.howstuffworks.com/eye2.htm

38   'Fovea centralis', http://en.wikipedia.org/wiki/Fovea_centralis

39   Marc Green, 'Night Vision', http://visualexpert.com/Resources/nightvision.html

40   Malcolm Jobling, *Environmental biology of fishes*, p.19-21

41   Malcolm Jobling, *Environmental biology of fishes*, p.18-19 (Note: 'Eyeshine' in nocturnal land animals, including cats, is also caused by a layer of guanine crystals.)

42   Francis G. Carey (1982), 'A Brain Heater in the Swordfish', *Science* 216:1327-1329

43   Kerstin A. Fritches, Richard W. Brill and Eric J Warrant (2005), 'Warm Eyes Provide Superior Vision in Swordfishes', *Current Biology*, 15: 55-58

44   Marc Green, 'Night Vision', http://visualexpert.com/Resources/nightvision.html

45   S. Plainis, I. J. Murray, and I. G. Pallikaris (2006), 'Road traffic casualties: understanding the night-time death toll', *Injury Prevention*, 12: 125-138

46   The Physics of Diving: Light and Vision 'Insufficient Light', Marc Green, 'Night Vision', http://visualexpert.com/Resources/nightvision.html

47   Robert Rossier, 'More than darkness: Human factors and the Night Diver,' *Dive Training*. https://www.dtmag.com/thelibrary/more-than-darkness…

48   J. Floor Anthoni (2005), 'Water and light in underwater photography', http://www.seafriends. org.nz/phgraph/water.html

CHAPTER 7: FEELING THE VIBRATIONS

1   Arthur C. Guyton and John E. Hall, *Textbook of Medical Physiology*, p.583-597

2   Peter B. Moyle and Joseph J. Cech, *Fishes: An Introduction to Ichthyology*, p.149

3   Sven Dijkgraaf (2012), 'Lateral line organs', *Encyclopedia Brittanica*, http://www.britannica. com/EBchecked/topic/371976/mechanoreception

4   F. Harvey Pough, John B. Heiser, and William N. McFarland (1989), *Vertebrate Life*, 3rd Edn. Macmillan Publishing Company, New York, p.284-285

5   Andrew Sherwood Romer and Thomas S. Parsons (1986), *The Vertebrate Body*, 6th Edn. Saunders College Publishing, Philadelphia, p.526

6   Brian C. J. Moore (1977), *Introduction to the Psychology of Hearing*, The Macmillan Press, London, p.15-17

7   *The Encyclopedia of Man and Medicine* (1973), Marshall Cavendish, London. p.164

8   'Ossicles and their function', http://hyperphysics.phy-ast.gsu.edu/hbase/sound/oss.html (Note: The ossicles have Latin names referring to their shape: maleus (hammer), incus (anvil) and stapes (stirrup).)

9   James W. Kalat (1998), *Biological Psychology*, 6th Edn. Brooks/Cole Publishing, Pacific Grove, p.182

10   *Gray's Anatomy*, 15th Edn. Bounty Books, New York, p.863-868

11   James W. Kalat, *Biological Psychology*, p.182-184.

12   Mark M. Denny, *Air and Water: the biology and physics of life's media*, p.215

13   'Sound Absorption', *Discovery of Sound in the Sea*, http://www.dosits.org/science/ soundmovement/soundweaker/absorption/

14   W. Kalat, *Biological Psychology*, p.186-187

15   'Speed of sound', https://en.wikipedia.org/wiki/Speed_of_sound

16   Brian C.J. Moore, *Introduction to the Psychology of Hearing*, p.169-190

17   'Speed of sound', https://en.wikipedia.org/wiki/Speed_of_sound

18   'The Physics of Diving: Sound and Hearing', http://library.thinkquest.org/28170/36.html

19   'How does sound travel long distances? The SOFAR Channel', *Discovery of Sound in the Sea*, http://www.dosits.org/science/soundmovement/sofar

20   Arthur N. Popper, John Ramcharitan and Steven E. Campana (2005), 'Why otoliths? Insights from inner ear physiology and fisheries biology', *Marine and Freshwater Research*, 56: 497-504

21   Jennifer M. Allen, J. H. S Blaxter and E. J. Denton (1976), 'The Functional Anatomy and Development of the Swimbladder-Inner Ear-Lateral Line System in Herring and Sprats', *Journal of the Marine Biological Association of the United Kingdom*, 56: 471-486.

22   Walter Lechner and Friedrich Ladich (2008), 'Size Matters: Diversity in swimbladders and Weberian ossicles affects hearing in catfishes', *Journal of Experimental Biology*, 211: 1681-1689

23   Arthur N. Popper, Dennis T. T. Plachta, David A. Mann and Dennis Higgs (2004), 'Response of clupeid fish to ultrasound: a review', *ICES Journal of Marine Science*, 61: 1057-1061

24   Richard R. Fay and Peggy L. Edds-Walton (2000), 'Directional encoding by fish auditory

systems', *Philosophical Transactions of the Royal Society*, London, B. 355:1281-1284

25   Malcolm Jobling, *Environmental biology of fishes*, p.25-28

26   Sir James Gray, *How Animals Move*, p.54-55

27   *Gray's Anatomy*, p.859-868

28   Lauralee Sherwood, *Fundamentals of Physiology*, p.145-147

29   Lauralee Sherwood, *Fundamentals of Physiology*, p.145-148

30   Arthur C. Guyton and John E. Hall, *Textbook of Medical Physiology*, p.707-712

31   'Vestibulo-ocular reflex', http://wikipedia.org/wiki/Vestibulo-ocular_reflex

32   Arthur C. Guyton and John E. Hall, *Textbook of Medical Physiology*, p.709.

33   Bayard Tarpley (2011), 'The Importance of Spotting in Ballet', http://ehow.com/info_8782804_importance-spotting-ballet.html

34   Rachel (2006), 'Diary of an Amateur Triathlete: Seasickness Advice?' http://amateurtrigirl.blogspot.com/2006/06/seasickness-advice.html

35   Penny Lee Dean, *Open Water Swimming: A Complete Guide for Distance Swimmers and Triathletes*, p.177 and 181.

36   'Sea Sickness', Channel Swimming Association, www.channelswimmingassociation.ccom/swim-advice/sea-sickness

37   G. Yancey Mebane (1995), 'Motion Sickness', Divers Alert Network: Divers Helping Divers. http://www.diversalertnetwork.org.medical/articles/Motion_Sickness

38   Carl Edmonds (1971), 'Vertigo in diving', *Royal Australian Navy School of Underwater Medicine*, Report No.1/71

39   David J. Doolette and Simon J. Mitchell (2003), 'Biophysical basis for inner ear decompression sickness', *Journal of Applied Physiology*, 94: 2145-2150

40   Simon J. Mitchell and David J. Doolette (2009), 'Selective vulnerability of the inner ear to decompression sickness in divers with right-to-left shunt: the role of tissue gas supersaturation', *Journal of Applied Physiology*, 106: 298-301

41   R. Douglas Fields (July 2007), 'The shark's electric sense', *Scientific American*, p.59-65

42   A. J. Kalmijn (1971), 'The electric sense of sharks and rays', *Journal of Experimental Biology*, 55: 371-383

43   Shaun P. Collin and Darryl Whitehead (2004), 'The functional roles of passive electroreception in non-electric fishes', *Animal Biology*, 54: 1-25

44   Ito Kazuaki (2005), 'Catfish and Earthquakes in Folklore and Fact', *Nipponia, Discovering Japan*, No. 3

45   Theodore H. Bullock, 'Electric fish, electric organ discharges, and electroreception', *Electrochemistry Encyclopedia*, http://electrochem.cwru.edu/ed/encycl/

46   Mark M. Denny, *Air and Water: The Biology and Physics of Life's Media*, p. 185

47   'Electric catfish.' Wikipedia, https://en.wikipedia.org/wiki/Electric_catfish

48   'Electric eel', Wikipedia, http://en.wikipedia.org/wiki/Electric_eel

49   R. Aidan Martin, 'Electric rays: a shocking use of muscle power,' Biology of Sharks and Rays, http://elasmo-research.org/education/topics/p_electric_rays.htm

50   'Torpedo ray attacks diver's camera', *Jonathan Bird's Blue World*, BlueWorldTV, Short video on YouTube,

51   Jim Grier (2006), Further note by Jay Sissons, Woods Hole Oceanographic Institute, 'Torpedo Ray Injury', http://tenfootstop.blogspot.com/2006/07/torpedo-ray-injury.html

52   Mark M. Denny, *Air and Water: The Biology and Physics of Life's Media*, p.174 and 183

CHAPTER 8: FLEXING THE MUSCLES

1   Barbara Tyldesley and June Grieve (2002), *Muscles, Nerves and Movement in Human Occupation*, 3rd Edn. Blackwell Publishing, Oxford, p.30, 131-152

2   Sir James Gray, *How Animals Move*, p.43

3   John D. Altringham and David J. Ellerby (1999), 'Fish swimming: patterns in muscle function', *The Journal of Experimental Biology*, 202: 3397-3403

4   'Fish Muscles', http://www.earthlife.net/fish/muscles.html

5   John D. Altringham and David J. Ellerby (1999), 'Fish swimming: patterns in muscle function', *The Journal of Experimental Biology*, 202: 3397-3403

6   Malcolm Jobling, *The Environmental Biology of Fishes*, p.263

7   D. G. Mackean (1978), *Introduction to Biology*, John Murray, London, p.160

8   Sir James Gray, *How Animals Move*, p.49

9   Malcolm Jobling, *The Environmental Biology of Fishes*, p.242

10   R. Aidan Martin, 'What's the Speediest Marine Creature?' http://www.elasma-research.org/education/topics/r_haulin'_bass.htm

11   John H. Hebrank, Mary R. Hebrank, John H. Long Jr, Barbara A. Block and Stephen A. Wainwright (1990), 'Backbone mechanics of the blue marlin makaira nigricans (pisces, istiophoridae)', *The Journal of Experimental Biology*, 148: 449-459

12   'Billfish', http://en.wikipedia/wiki/Billfish

13   Peter B. Moyle and Joseph J. Cech, *Fishes: An Introduction to Ichthyology*, p.28

14   Pamela Bristow, *The Illustrated Book of Fishes*, p.249

15   Peter B. Moyle and Joseph J. Cech, *Fishes: An Introduction to Ichthyology*, p.28

16   Richard Henderson (1992), *Singlehanded Sailing: The Experiences and Techniques of Lone Sailors*, 2nd Edn. McGraw Hill Professional, New York, p.259

17   Simon de Bruxelles, 'Fisherman impaled on marlin's bill as it leaps across boat', *The Times*, 4 August 2006

18   Terry Laughlin with John Delves, *Total Immersion: The Revolutionary Way to Swim Better, Faster, and Easier*, p.48-53

19   David Wilkie and Kelvin Juba, *The Handbook of Swimming*, p.71-73

20   'Rio 2016 100m freestyle men – Olympic Swimming', https://www.olympic.org/rio-2016/swimming

21   'Rio 2016 100m men – Olympic athletics', https://www.olympic.org/rio-2016/athletics

22   M. P. Schwellnus (2008), 'Cause of Exercise Associated Muscle Cramps (EAMC) – altered neuromuscular control, dehydration or electrolyte depletion?' *British Journal of Sports Medicine*, 43: 401-408

23   *Gray's Anatomy*, p.436-438

24   Montague A. Holbein (1914), *Swimming*, First published in 1914 by Arthur Pearson, reissued in 2005 by Bloomsbury Publishing, London, p.47-48

25   Webbed glove – Manuscript B, folio 81 v. Produced between 1487 and 1490, 'Leonardo: Water and Land Machines', http://digilander.libero.it/debibliotheca/Arte/Leonardowater_file/pa

26   David Wilkie and Kelvin Juba, *The Handbook of Swimming*, p.11

27   David Wilkie and Kelvin Juba, *The Handbook of Swimming*, p.11

28   'Swimfin.' https://en.wikipedia.org/wiki/Swimfin

29   The British Sub-Aqua Club, *Sport Diving*, p.150

30   'Tidal Thames – A Guide for Users of Recreational Craft' (2005), The Port of London Authority, London

CHAPTER 9: GOING THE DISTANCE

1   Terry Laughlin with John Delves, *Total Immersion: The Revolutionary Way to Swim Better, Faster, and Easier*, p.13 and 20

2   Arthur C. Guyton and John E. Hall, *Textbook of Medical Physiology*, p.1064

3   Arthur C. Guyton and John E. Hall, *Textbook of Medical Physiology*, p.857

4   Lauralee Sherwood, *Fundamentals of Physiology: a human perspective*, p.23-29

5   Arthur C. Guyton and John E. Hall, *Textbook of Medical Physiology*, p.905

6   Lauralee Sherwood, *Fundamentals of Physiology: a human perspective*, p.26

7   Arthur C. Guyton and John E. Hall, *Textbook of Medical Physiology*, p.81-82

8   Lauralee Sherwood, *Fundamentals of Physiology: a human perspective*, p.22-23

9   Arthur C. Guyton and John E. Hall, *Textbook of Medical Physiology*, p.81-82

10   Mike Stroud (2004), *Survival of the Fittest*, Yellow Jersey Press, London, p.42

11   Lauralee Sherwood, *Fundamentals of Physiology: a human perspective*, p.186

12   Arthur C. Guyton and John E. Hall, *Textbook of Medical Physiology*, p.1064

13   Arthur C. Guyton and John E. Hall, *Textbook of Medical Physiology*, p.1065

14   Arthur C. Guyton and John E. Hall, *Textbook of Medical Physiology*, p.1065

15   Arthur C. Guyton and John E. Hall, *Textbook of Medical Physiology*, p.1065

16   Arthur C. Guyton and John E. Hall, *Textbook of Medical Physiology*, p.1065

17   Peter B. Moyle and Joseph J. Cech, *Fishes: An Introduction to Ichthyology*, p.24

18   'Oxygen Solubility in Fresh and Sea Water', http://www.engineeringtoolbox.com/oxygen-solubility-water-d_841.html

19   Q. Bone, N. B. Marshall and J. H. S. Blaxter, *The Biology of Fishes*, p.49 and 53

20   Peter B. Moyle and Joseph J. Cech, *Fishes: An Introduction to Ichthyology*, p.24

21   Nicholas C. Wegner, Chugey A. Sepulveda, Kristina B. Bull and Jeffrey B. Graham (2010), 'Gill Morphometrics in Relation to Gas Transfer and Ram Ventilation in High-Energy Demand Teleosts: Scombrids and Billfishes', *Journal of Morphology*, 271: 36-49

22   J. J. Videler and D. Weihs (1982), 'Energetic advantages of burst-and-coast swimming of fish at high speeds', *The Journal of Experimental Biology*, 97: 169-178

23   Q. Bone, N. B. Marshall and J. H. S. Blaxter, *The Biology of Fishes*, p.53

24   Q. Bone, N.B. Marshall and J.H.S. Blaxter, *The Biology of Fishes*, p.53

25   Chris M. Wood (1991), 'Acid-base and ion balance, metabolism, and their interactions, after exhaustive exercise in fish', *The Journal of Experimental Biology*, 160: 285-308

26   James D. Keifer (2000), 'Limits to exhaustive exercise in fish', *Comparative Biochemistry and Physiology*, Part A 126: 161-179

27   C. G. Lee, A. P. Farrell, A. Lotto, S. G. Hinch and M. C. Healey (2003), 'Excess post-exercise oxygen consumption in adult sockeye (Oncorhynchus nerka) and coho (O. Kisutch) salmon following critical speed swimming', *The Journal of Experimental Biology*, 206: 3253-3260

28   C. M. Wood, J. D. Turner and M. S. Graham (2006), 'Why do fish die of severe exercise?' *Journal of Fish Biology*, 22: 189-201

29   Michael Doherty and Lygeri Dimitriou (1997), 'Comparison of lung volume in Greek swimmers, land based athletes, and sedentary controls using allometric scaling', *British Journal of Sports Medicine*, 31: 337-341

30   Arthur C. Guyton and John E. Hall, *Textbook of Medical Physiology*, p.1066-1068

31   'Lactic Acid Not Athlete's Poison, But An Energy Source If You Know How To Use It', *Science Daily*, 21 April 2006

32   George A. Brooks (1986), 'The lactate shuttle during exercise and recovery', *Medicine and Science in Sports and Exercise,* 18: 360-368

33   'VO2 max', http://en.wikipedia.org/wiki/VO2_max

34   'Lactate threshold', http://en.wikipedia.org/wiki/Lactate_threshold

35   D. L. Costill (1992), 'Lactate metabolism for swimming,' in *Biomechanics and Medicine in Swimming: Swimming Science V1*, D. Maclaren, T. Reilly and A. Rees (ed.) Chapman and Hall, London

36   Ingvar Holmér (1972), 'Oxygen uptake during swimming in man', *Journal of Applied Physiology*, 33: 502-509

37   R. K. Stallman, J. Major, S. Hemmer, G. Haarvaag, 'Movement Economy in Breaststroke Swimming: A Survival Perspective', XIth International Symposium for Biomechanics and Medicine in Swimming, Oslo, June, 2010

38   Arthur C. Guyton and John E. Hall, *Textbook of Medical Physiology*, p.1062

39   H. M. Toussaint (1992), 'Performance determining factors in front crawl swimming', Chapter in *Biomechanics and Medicine in Swimming: Swimming Science VI*, D. MacLaren, T. Reilly and A. Lees. E. and F. N. Spon (ed.) London, p.13-32

40   Lauralee Sherwood, *Fundamentals of Physiology: a human perspective*, p.186

41   Penny Lee Dean, *Open Water Swimming*, p.163-164

42   R. K. Stallman, J. Major, S. Hemmer, G. Haarvaag, 'Movement Economy in Breaststroke Swimming: A Survival Perspective', XIth International Symposium for Biomechanics and Medicine in Swimming, Oslo, June, 2010

43   Arthur C. Guyton and John E. Hall, *Textbook of Medical Physiology*, p.1062 (Note: He must replace the two litres of oxygen extracted from his lungs, bloodstream, muscles and tissues, and in addition, more than nine litres of oxygen are needed to deal with the excessive amounts of lactic acid produced by anaerobic muscle contraction.)

44   Lauralee Sherwood, *Fundamentals of Physiology: a human perspective*, p.185

45   Lauralee Sherwood, *Fundamentals of Physiology: a human perspective*, p.514-515

46   D. L. Costill (1992), 'Lactate metabolism for swimming', Keynote Address in *Biomechanics and Medicine in Swimming: Swimming Science V1*, D. Maclaren, T. Reilly and A. Rees (ed.) Chapman and Hall, London, p.3-11. K. Tolfrey and N. Armstrong (1995), 'Child-adult differences in whole blood lactate responses to incremental treadmill exercises', *British Journal of Sports Medicine*, 29: 196-199

CHAPTER 10: FRESH WATER AND SALT WATER

1   Arthur C. Guyton and John E. Hall, *Textbook of Medical Physiology*, p.297-298

2   Mike Stroud, *Survival of the Fittest*, p.113

3   C. Guyton and John E. Arthur Hall, *Textbook of Medical Physiology*, p.298

4   Louise Burke (2007), 'Refuelling and rehydration workouts during workouts', Chapter in *Practical Sports Nutrition*, Human Kinetics, Champaign, IL. p.150-152

5   G. R. Cox, E. M. Broad, M. D. Riley and L. M. Burke (2002), 'Body mass changes and voluntary fluid intakes of elite level water polo players and swimmers', *Journal of Science and Medicine in Sport*, 5: 183-193

6   Matt Schudel, 'J. Robert Cade, 80; Gatorade Inventor', Obituary, *The Washington Post*, 28 November 2007 (Note: Professor Cade was a renal physician at the University of Florida. When the coach of the university football team – the Gators – asked him, 'Doctor, why don't football players wee-wee after a game?' he realised that their problem was dehydration and invented a drink that replaced the water and salt lost in sweat.)

7   B. Murray (2007), 'Hydration and physical performance', *Journal of the American College of Nutrition*, 26 (5 Suppl):543S-548S

8   George M. Dallam, Steven Jonas and Thomas K. Miller (2005), 'Medical Considerations in Triathlon Competitions: Recommendations for Triathlon Organisers, Competitors and Coaches', *Sports Medicine*, 35: 143-161

9   Arthur C. Guyton and John E. Hall, *Textbook of Medical Physiology*, p.916-917

10   William Wurts, 'Why can some fish live in freshwater, some in salt water, and some in both?' http://www.ca.uky.edu/wkrec/VertebrateFishEvolution.htm

11   Lauralee Sherwood, *Fundamentals of Physiology*, p.46-49

12   William N. McFarland, F. Harvey Pough, Tom J. Cade and John B. Heiser (1979), *Vertebrate Life*. Macmillan Publishing, New York, p.251-252

13   David H. Evans, Peter M. Piermarini, and W. T. W. Potts (1999), 'Ionic Transport in the Fish Gill Epithelium', *Journal of Experimental Zoology*, 283: 641-652

14   'Seawater', Wikipedia, https://en.wikipedia.org/wiki/Seawater

15   William N. McFarland, F. Harvey Pough, Tom J. Cade and John B. Heiser, *Vertebrate Life*, p.250

16   Pamela Bristow, *The Illustrated Book of Fishes*, p.77

17   David H. Evans, Peter M. Piermarini, and W. T. W. Potts (1999), 'Ionic Transport in the Fish Gill Epithelium', *Journal of Experimental Zoology*, 283: 641-652

18   Pamela Bristow, *The Illustrated Book of Fishes*, p.14

19   I. Homer, W. Smith (1936), 'The Retention and Physiological Role of Urea in the Elasmobranchii', *Biological Reviews*, 11:49-82, http://onlinelibrary.wiley.com/doi/10.1111/j.1469-185X.1936.TB0049

20   William N. McFarland, F. Harvey Pough, Tom J. Cade and John B. Heiser, *Vertebrate Life*, p.253

21   William N. McFarland, F. Harvey Pough, Tom J. Cade and John B. Heiser, *Vertebrate Life*, p.257-258

22   'How Bull Sharks Can Live in Both Ocean and Fresh Water', https://www.sharksavers.org/en/education/biology/how-bull-sharks-s

23   J. P. O'Hare, A. Heywood, C. Summerhayes, G. Lunn, J. M. Evans, G. Walters, R. J. M. Corrall and P. A. Dieppe (1985), 'Observations on the effects of immersion in Bath spa water', *British Medical Journal*, 291: 1747-1751

24   Audrey Heywood, H. A. Waldron, P. O'Hare and P. A. Dieppe (1986), 'Effect of immersion on urinary lead excretion', *British Journal of Industrial Medicine*, 43: 713-715

25   Robert J. Ellis (1997), 'Severe Hypernatremia From Sea Water Ingestion During Near-Drowning in a Hurricane', *Western Journal of Medicine*, 167:430-433

26   Frank Golden and Michael Tipton (2002), *Essentials of Sea Survival*, p.149-155

27   Frank Golden and Michael Tipton, *Essentials of Sea Survival*, p.52-53

28   'Immersion Diuresis (Urge to Urinate)', DAN Divers Alert Network, http://www.diversalertnetwork.org/medical/faq/faq.aspx?faqid=165. Frank Golden and Michael Tipton, *Essentials of Sea Survival*, p.52-53

29   Donal Buckley (2013), 'What is cold immersion diuresis in swimmers? (aka why do you pee after swimming?)' https://loneswimmer.com/2013/02/12/what-is-cold-immersion-diuresis

30   Penny Lee Dean, *Open Water Swimming*, p.17-18

31   Freda Streeter, 'Thoughts on Equipment for swims', http://www.thechannelswimmer.com/Resources/Swimmer

32   'Captain Webb', http://www.dover-kent.co.uk/people/capt_webb.htm

33   Ernest Campbell, 'To pee or not to pee?' *Scuba Diving Magazine*, November 2001, http://www.scubadiving.com/pee-or-not-pee

34   Mark Brill and Laura Harris, 'The Art of Drinking and Diving. The importance of being well-hydrated while diving', https://www.danap.org/DAN_diving_safety/DAN_Doc/pdfs/hydration.pdf

35   'More Water Less Bubbles', http://www.dansa.org/more-water-less-bubbles.htm

36   E. Gempp, J. E. Blateau, J-M Pontier, C. Balestra and P. Louge (2009), 'Preventive effect of pre-dive hydration on bubble formation in divers', *British Journal of Sports Medicine*, 43:224-228

37   Duane E. Graveline, and Michael McCally (1962), 'Body Fluid Distribution: Implications For Zero Gravity', *Aerospace Medicine*, 33:1281-1290 (Note: This paper was republished in 2009, as one of twenty-three 'Classics in Space Medicine'. See *Aviation, Space and Environmental Medicine*, 80:993)

38   P. A Nyquist, J. Schrot, J. R. Thomas, D. Hyde and W. R. Taylor (2005), 'Desmopressin prevents immersion diuresis and improves physical performance after long duration dives', Naval Medical Research Institute, MD.

39   Frank Golden and Michael Tipton, *Essentials of Sea Survival*, p.245-264

40   F. St C. Golden, G. C. David and M. J. Tipton (1997), 'Review of Rescue and Immediate Post-Immersion Problems: A Medical/Ergonomic Viewpoint', *Health and Safety Executive – Offshore Technology Report*, HMSO, Norwich

CHAPTER 11: CHILLED TO THE BONE

1   Lauralee Sherwood, *Fundamentals of Physiology: A Human Perspective*, p.471-474

2   Frank Golden and Michael Tipton, *Essentials of Sea Survival*, p.32-38

3   Frank Golden and Michael Tipton, *Essentials of Sea Survival*, p.29

4   Frank Golden and Michael Tipton, *Essentials of Sea Survival*, p.28

5   Frank Golden and Michael Tipton, *Essentials of Sea Survival*, p.29

6   'Swimming Pools', Sport England 2013, https://sportengland.org/media/187196/swimming-pool

7   R. E. G. Sloan and W. R. Keatinge (1973), 'Cooling rates of young people swimming in cold water', *Journal of Applied Physiology*, 35:371-375

8   W. R. Keatinge, C. Prys-Roberts, K. E. Cooper, A. J. Honour and J. Haight (1969), 'Sudden Failure of Swimming in Cold Water', *British Medical Journal*, 1:480-483

9   W. R. Keatinge C. Prys-Roberts, K. E. Cooper, A. J. Honour and J. Haight (1969,) 'Sudden Failure of Swimming in Cold Water', *British Medical Journal*, 1:480-483

10   David McFarland (1985), *Animal Behaviour*, Pitman Publishing, London, p.190

11   'Are Fish Cold Blooded?: Thermoregulation in Fish', http://www.earthlife.net/fish/tregulate.html

12   William N. McFarland, F. Harvey Pough, Tom J. Cade and John B. Heiser, *Vertebrate Life*, p.275

13   P. B. Moyle, and J. J. Cech, *Fishes: An introduction to ichthyology*, p.26

14   'Are Fish Cold Blooded?: Thermoregulation in Fish', http://www.earthlife.net/fish/tregulate.html

15   R. Aidan Martin, 'Fire in the belly of the beast', http://www.elasmo-research.org/education/topics

16   Francis G. Carey and John M. Teal (1969), 'Regulation of body temperature by the bluefin tuna', *Comparative Biochemistry and Physiology*, 28:205-213

17   R. Aidan Martin, 'Circulation and a wonderful net', http://www.elasmo-research.org/education/white

18   Ben S. Roesch, 'White Shark Physiology: Warm Bodied and Ready to Go', http://web.ncf.ca/bz050/wsphysio.html 1997

19   Francis G. Carey (1984), 'Bluefin Tuna Warm Their Viscera During Digestion', *Journal of Experimental Biology*, 109:1-20

20   D. Scott Linthicum and Francis G. Carey (1972), 'Regulation of brain and eye temperatures by the bluefin tuna', *Comparative Biochemistry and Physiology Part A, Physiology* 43: 425-433

21   Frank Golden and Michael Tipton, *Essentials of Sea Survival*, p.59-64

22   M. J. Tipton, P. C. Kelleher and F. S. Golden (1994), 'Supraventricular arrhythmias following breath-hold submersions in cold water', *Undersea and Hyperbaric Medicine*, 21: 305-313

23   M. J. Shattock and M. J. Tipton (2012), '"Autonomic conflict": a different way to die during cold water immersion?' *Journal of Physiology*, 590:3219-3230

24   Frank Golden and Michael Tipton, *Essentials of Sea Survival*, p.62-64

25   Kevin H. Monahan, 'The Chilling Truth About Cold Water', *Pacific Yachting Magazine*, February 2006

26   Martin J. Barwood, Victoria Bates, Geoffrey M. Long and Michael J. Tipton (2011), '"Float First": Trapped Air Between Clothing Layers Significantly Improves Buoyancy after Immersion', *International Journal of Aquatic Research and Education* 5:147-163

27   Frank Golden and Michael Tipton, *Essentials of Sea Survival*, p.66-67

28   Donal Buckley, 'Understanding The Claw as a hypothermia indicator', http://loneswimmer.com/2012.11/22/understanding-the-claw

29   Frank Golden and Michael Tipton, *Essentials of Sea Survival*, p.134-135

30   Frank Golden and Michael Tipton, *Essentials of Sea Survival*, p.71-75 and 103

31   'Density of Water (g/mL) vs. Temperature (°C)', http://www2.volstate.edu/CHEM/Density_of_Water.html

32   'Water – Absolute or Dynamic Viscosity', http://www.engineeringtoolbox.com/absolute-dynamic-viscosity-wat

33   'Surface Tension of Water in contact with Air', http://engineeringtoolbox.com/water-surface-tension-d_597.html

34   Lee A. Fuiman and Robert S. Batty (1997), 'What a drag it is getting cold: partitioning the physical and physiological effects of temperature on fish swimming', *The Journal of Experimental Biology*, 200: 1745-1755

35   'Report of the working party on water safety' (1977), Home Office Report, HMSO, London. Quoted by: Frank Golden and Michael Tipton, *Essentials of Sea Survival*, p.57

36   Matthew Parris (2002), *Chance Witness*, Viking, London, p.197 (Long quotation by permission of the author.)

37   Keith C. Heidorn (2005), 'Fall/Spring Lake Turnover', http://www.islandnet.com/~see/weather/elements

38   Robert K.Lane (2009), 'Lake', *Encyclopedia Britannica*, Inc. http://www.history.com/topics/lake

39   Steve Tatlock, Lake District National Park, http://www.lakedistrict.gov.uk/index/visiting/outdoors/on_the_water

40   'Swimmer dies in Windermere', *News and Star*, Cumberland, 29 June 2015

41   Russell Jenkins, 'Three drown where still waters run deep', *The Times*, 25 September 2006

42   Marc Martin (2008), 'The Thermocline's Effect on Fishing', http://www.washingtonlakes.com/ReadArticle.aspx?id=300

43   'Harris' Taylor (2004), 'Practical Buoyancy Control', www.personal.umich.edu/lpt/practical.htm

44   The British Sub-Aqua Club, *Sport Diving*, p.85

45   Carolyn Rainey (1998), 'Wet Suit Pursuit: Hugh Bradner's Development of the First Wet Suit', *Archives of Scripps Institute of Oceanography*, SIO Reference Number 98-16

46   'Woman diver died after rapid ascent', *Leicester Mercury*, 21 July 2011

47   L. G. C. Pugh and O. G. Edholm (1955), 'The physiology of Channel swimmers', *Lancet* 269: 761-768

48   Wildswimmer Pete's Cold Water Acclimatisation Tips, http://www.swimclub.co.uk/forum/showthread.php?t=14855

49   Janet Smith (2005), *Liquid Assets: the lidos and open air swimming pools of Britain*, English Heritage, Swindon. p.52-57

50   Frances Klemperer and Emily Simon-Thomas, 'Captain Webb's legacy: the perils of swimming the English Channel', *British Medical Journal*, 349:5-7

51   Danny Boyle, 'Student Sophie Mills rescued in Channel swim drama', http://kentonline.co.uk/kent/student. Maik Grossekathöfer, 'From England to France: The Attraction of Swimming the Channel', Speigel Online International 6 September 2013. David Brown, 'Dying while trying to swim the English Channel', Washington Post, 14 July 2014

52   'Channel swimming fatalities', https:www.dover.uk.com/channel-swimming/fatalities . Maxine Frith and Peter Allen 'British woman, 34, dies as she swims the Channel to help charities', *London Evening Standard*, 15 July 2013. Emma Glanfield and Alexander Ward 'Channel swimmer and father-of-two, 45, dies after getting into difficulties just ONE MILE from Calais having battled tides for 16 hours in a bid to complete the crossing for the second time', *Daily Mail*, 28 August 2016

53   Mario Vittone (2010), 'The Truth About Cold Water', http://gcaptain.com/cold_water/?11198

54   J. S. Hayward, J. D. Eckerson, M. L. Collis (1975), 'Effect of behavioural variables in cooling rate of man in cold water', *Journal of Applied Physiology*, 38: 1073-1077

55   Alan Steinman and Gordon Giesbrecht, 'The Four Stages of Cold-Water Immersion', *On Scene: The Journal of US Coast Guard Search and Rescue*, Fall 2006

56   Frank Golden and Michael Tipton, *Essentials of Sea Survival*, p.114

57   Frank Golden and Michael Tipton, *Essentials of Sea Survival*, p.95-107

58   Arthur C. Guyton and John E. Hall, *Textbook of Medical Physiology*, p.917

59   Francis Ashcroft, *Life at the Extremes: the science of survival*, p.160-162

60   Lorenz E. Witters and Margaret V. Savage (2001), 'Cold Water Immersion', Chapter 17 in *Medical Aspects of Harsh Climates*, Vol 1, Kent B. Pandolf and Robert E. Burr (ed.) Office of the Surgeon General, Borden Institute, TX.

61   Peter Stark, 'As Freezing Persons Recollect the Snow – First Chill – Then Stupor – Then Letting Go', *Outside Magazine*, January 1997, also published in 2000 in *Last Breath: Cautionary Tales from the Limits of Human Endurance*, Macmillan, London, p.11-24 (Note: The title of the article is a quotation from a poem by Emily Dickinson 'After great pain, a formal feeling comes.')

62   'Density of Water (g/mL) vs. Temperature (°C), http://www2.volstate.edu/CHEM/Density_of_Water.html

63   Pamela Bristow (1987), *The Illustrated Book of Fishes*, Octopus Books, London, p.22

64   'About Sea Ice', The National Snow and Ice Data Center, University of Colorado, Boulder, USA

65   Peter B. Moyle and Joseph J. Cech, *Fishes: An Introduction to Ichthyology*, p. 91

66   Ferris Jabr, 'How the Antarctic Icefish Lost its Red Cells But Survived Anyway', *Scientific American*, 3 August 2012

67   'Reveller drowns after attempting polar bear swim', *The Globe and Mail*, Ottawa, 3 January 2000

68   'Dicing with death', *Oxford Journal*, February 27/28 1986 (In a 'Picture Exclusive', a girl of about twelve crosses the thin, cracking ice on the frozen Oxford Canal.). Chris Brooke 'Tragedy of Six-year-old stuck beneath the ice: The desperate attempts to save boy who died in frozen pond', *Daily Mail*, 9 February 2008 . 'Big Freeze Brings Ice Danger', 2 January 2010, http://www.tameside.gov.uk/pressreleases/bigfreeze

69   Neil Millard, 'Matt the cocker spaniel saved from a watery grave by 17 firefighters and a very long ladder', *Daily Mail*, 31 December 2009

70   Royal Society for the Prevention of Accidents, 'Ice Safety', https://www.rospa.com

71   Dominick J. DiMaio and Vincent J. M. DiMaio (1989), *Forensic Pathology*, Elsevier, New York, p.359

72   A. M. Kemp and J. R. Sibert (1991), 'Outcome in children who nearly drown: a British Isles study', *British Medical Journal*, 302:931-933

73   Wilfred G. Bigelow (1984), *Cold Hearts: The Story of Hypothermia and the Pacemaker in Heart Surgery*, McLelland and Stewart, Toronto (Note: During his surgical training in Toronto, Wilfred Bigelow often had to amputate frost-bitten fingers and toes. As a cardiac surgeon, his research on hypothermia was crucial for the development of open heart surgery. Ivan Oransky 'Wilfred Gordon Bigelow', Obituary, *Lancet*, 7 May 2005.)

74   Jerome H. Modell, Ahamed H. Idris, Jose A. Pineda and Janet H. Silverstein (2004), 'Survival After Prolonged Submersion in Freshwater in Florida', *Chest* 125: 1948-1961

75   Robert G. Bolte, Philip G. Black, Robert S. Bowers, J. Kent Thorne and Howard M. Cornell (1988), 'The use of Extracorporeal Rewarming in a Child Submerged for 66 Minutes', *The Journal of the American Medical Association*, 260: 377-379

76   Mike Stroud (2004) *Survival of the Fittest: Understanding health and peak physical performance*, Yellow Jersey Press, London, p.147

CHAPTER 12: CAPTURING OXYGEN

1   Barbara J. Becker (2006) 'Scientific Societies', Lecture 12, Department of History, University of California, Irvine. http://eee.uci.edu/clients/bjbecker/NatureandArtifice/lecture12.html

2   Joseph Wright 'of Derby', An Experiment on a Bird in the Air Pump, NG725, https://www.nationalgallery.org.uk/paintings/joseph-wright-of-derby

3   'An Experiment on a Bird in the Air Pump', Wikipedia, https://en.wikipedia.org/wiki/An_Experiment_on_a_Bird_in_the_Air_Pump

4   Joseph Priestley (1790), *Experiments and Observations on different kinds of air and other branches of natural philosophy connected with the subject*, Thomas Pearson, Birmingham, http://www.lakesidepress.com/pulmonary/papers/ox-hist/ox-hist1.htm (Note: Priestley called the gas 'dephlogisticated air'.)

5   Stephen Jay Gould (1991), 'The Passion of Antoine Lavoisier', Chapter in: *Bully for Brontosaurus: Reflexions in Natural History*, Hutchinson Radius, London, p.359

6   Jere H. Mitchell and Bengt Saltin (2003), 'The oxygen transport system and maximal oxygen uptake', Chapter in *Exercise Physiology: People and Ideas*, Charles M. Tipton (ed.) Oxford University Press, Oxford

7   Stephen Jay Gould, *Bully for Brontosaurus: Reflexions in Natural History*, p.362-363

8   Lauralee Sherwood, *Fundamentals of Physiology: a human perspective*, p.279 and 345-346

9   Arthur G. Guyton and John E. Hall, *Textbook of Medical Physiology*, p.513

10   Pamela Bristow (1987), *The Illustrated Book of Fishes*, Octopus Books, London, p.22

11   William N. McFarland, F. Harvey Pough, Tom J. Cade and John B. Heiser, *Vertebrate Life*, p.278

12   Peter B. Moyle and Joseph J. Cech, *Fishes: An Introduction to Ichthyology*, p.34-37

13   William N. McFarland, F. Harvey Pough, Tom J. Cade and John B. Heiser, *Vertebrate Life*, p.277-279

14   'Fish Respiration', http://bio-isu.tripod.com/id3.html

15   Malcolm Jobling, *The Environmental Biology of Fishes*, p.93

16   William N. McFarland, F. Harvey Pough, Tom J. Cade and John B. Heiser, *Vertebrate Life*, p.278

17   Pamela Bristow, *The Illustrated Book of Fishes*, p.18

18   Malcolm Jobling, *The Environmental Biology of Fishes*, p.270

19   Peter B. Moyle and Joseph J. Cech, *Fishes: An Introduction to Ichthyology*, p.35

20   William N. McFarland, F. Harvey Pough, Tom J. Cade and John B. Heiser, *Vertebrate Life*, p.265

21   Peter B. Moyle and Joseph J. Cech, *Fishes: An Introduction to Ichthyology*, p.24

22   Lewis Smith, 'Even the fish are gasping for air in the heat wave', *The Times*, 17 June 2006, p.13

23   Pamela Bristow, *The Illustrated Book of Fishes*, p.12

24   William N. McFarland, F. Harvey Pough, Tom J. Cade and John B. Heiser, *Vertebrate Life*, p.279

25   Pamela Bristow, *The Illustrated Book of Fishes*, p.12

26   Pamela Bristow, *The Illustrated Book of Fishes*, p.138

27   Lexi Krock (2003), 'Other Fish in the Sea', http://www.pbs.org/wgbh/nova/nature/other-fish-sea.html

28   J.N. Maina (2002), *Functional Morphology of the Vertebrate Systems*. Science Publishers, Enfield, New Hampshire, p.31

29   'Mudskipper', https://en.wikipedia.org/wiki/Mudskipper

30   Neil Santaniello, 'Walking Catfish Let Off The Hook', *Sun-Sentinel*, 9 January 2005, http://articles.sun-sentinel.com/2005-01-09/news/0501080450_i_pau

31   Robert H. Robins, 'Walking Catfish', Ichthyology at the Florida Museum of Natural History, Biological Profiles, http://www.flmnh.ufl.edu/fish/gallery/descript/walkingcatfish/walkin

32   William N. McFarland, F. Harvey Pough, Tom J. Cade and John B. Heiser, *Vertebrate Life*, p.181-184

33   'Lungfish', Wikipedia, https://en.wikipedia.org/wiki/Lungfish

34   William N. McFarland, F. Harvey Pough, Tom J. Cade and John B. Heiser, *Vertebrate Life*, p.280

35   Ewald R. Weibel (1963), *Morphometry of the Human Lung*, Springer-Verlag, Berlin

36   Robert M. Berne and Matthew N. Levy (1996), *Principles of Physiology*, 2nd Edn. Mosby, St.Louis, p.372

37   Robert M., Berne and Matthew N. Levy, *Principles of Physiology*, p.381 (Note: Only 360ml of the inspired air reach the alveoli, while 140ml get no further than the 'dead space' of the trachea and bronchi.)

38   P. T. Marshall and G. M. Hughes, *Physiology of mammals and other vertebrates*, p.72-73

39   Frank Golden and Michael Tipton, *Essentials of Sea Survival*, p.82

40   Arthur C. Guyton and John E. Hall, *Textbook of Medical Physiology*, p.479-481

41   P.T. Marshall and G.M. Hughes, *Physiology of mammals and other vertebrates*, p.69

42   Lauralee Sherwood, *Fundamentals of Physiology: a human perspective*, p.204-207

43   Arthur C. Guyton and John E. Hall, *Textbook of Medical Physiology*, p.503 and 516

44   Lauralee Sherwood, *Fundamentals of Physiology: a human perspective*, p. 207

45   Robert M. Berne and Matthew N. Levy, *Principles of Physiology*, p.219

46   Arthur C. Guyton and John E. Hall, *Textbook of Medical Physiology*, p.162

47   Lauralee Sherwood, *Fundamentals of Physiology: a human perspective*, p.207

48   Arthur C. Guyton and John E. Hall, *Textbook of Medical Physiology*, p.223

49   Arthur C. Guyton and John E. Hall, *Textbook of Medical Physiology*, p.223

50   Arthur C. Guyton and John E. Hall, *Textbook of Medical Physiology*, p.484

51   Arthur C. Guyton and John E. Hall, *Textbook of Medical Physiology*, p.1065

52   Arthur C. Guyton and John E. Hall, *Textbook of Medical Physiology*, p.496

53   Arthur C. Guyton and John E. Hall, *Textbook of Medical Physiology*, p.1067

54   Arthur C. Guyton and John E. Hall, *Textbook of Medical Physiology*, p.241-242 and 1067

55   Lauralee Sherwood, *Fundamentals of Physiology: a human perspective*, p.343

56   Arthur C. Guyton and John E. Hall, *Textbook of Medical Physiology*, p.1063

57   Michael Doherty and Lygeri Dimitriou, 'Comparison of lung volume in Greek swimmers, land based athletes, and sedentary controls using allomeric scaling', *British Journal of Sports Medicine*, 31:337-341

58   Lauralee Sherwood, *Fundamentals of Physiology: a human perspective*, p.330-331

59   E. M. Mostyn, S. Helle, J. B. L. Gee, L. G. Bentivoglio, and D. V. Bates, 'Pulmonary diffusing capacity of athletes', *Journal of Applied Physiology*, 18:687-695

60   George A. Brooks (1986), 'The lactate shuttle during exercise and recovery', *Medicine and Science in Sports and Exercise*, 18:360-368

PART 2: A CLOSER LOOK AT DROWNING ACCIDENTS

1   John Fletemeyer, 'Light! Camera! Action! Hollywood's take on drowning is a distorted view of a quiet killer', *Aquatics International*, May 2013

2   Frank Pia (1974), 'Observations on the Drowning of Nonswimmers' *Journal of Physical Education*,

YMCA Society of North America, July 1974, http://www.pia-enterprises.com/observation.rtf

3   Juan Forero, 'Orchard Beach Journal: Slice of the Riviera, With a Familiar Bronx Twist', *New York Times*, 9 July 2000

4   Jonathan Howland, Ralph Higson, Thomas W. Mangione, Nicole Bell and Sharon Bak (1996), 'Why Are Most Drowning Victims Men? Sex Differences in Aquatic Skills and Behaviours', *American Journal of Public Health*, 86:93-96

5   L. Quan and P. Cummings (2003), Characteristics of drowning by different age groups', *Injury Prevention*, 9:163-168

6   Royal Society for the Prevention of Accidents 'Drowning Statistics – RoSPA', www.rospa.com/leisure-safety/statistics/drowning

7   A. C. Queiroga and A. Peden (2013), 'Drowning Deaths in Older People: A 10 year analysis of drowning in people aged 50 years and over in Australia', Royal Life Saving – Australia, Sydney

CHAPTER 13: DANGER IN THE GARDEN

1   Alison Kemp (2002), 'Preventing deaths from drowning in children in the United Kingdom: have we made progress in 10 years? Population based incidence study', *British Medical Journal*, 342:1070-1071

2   BBC News, 'Dover toddler's pond death ruled accidental', 10 March 2010

3   'Pond & Garden Water Safety' (2008), Royal Society for the Prevention of Accidents, http://www.rospa.com/leisuresafety/adviceandinformation/watersafety

4   Elizabeth Walker, 'Potential perils of a pond', *RoSPA Staying Alive Magazine*, August 2007

5   Dean Kirby, 'Tot drowned after dog opened door', *Manchester Evening News*, 2 July 2003

6   'Pond & Garden Water Safety' (2008), Royal Society for the Prevention of Accidents, http://www.rospa.com/leisuresafety/adviceandinformation/watersafety

7   Monty Don, 'Just add water', *The Observer*, 7 September 2003

8   Austswim, *Teaching Infant and Preschool Aquatics*, p.14

9   'Devastated mother reveals how her 16-month-old son drowned in garden pond', *Daily Mail*, 29 August 2008

10   Britton, 'Father pays tribute to drowned three-year-old boy', *Daily Telegraph*, 20 June 2008

11   John Pearson and Peter Davies (2000), 'Drowning accidents in the garden involving children under five', Consumer Affairs Directorate, Department of Trade and Industry, London

12   Rob Cooper, 'Toddler drowned in garden pond after wandering off from parents in grandmother's garden', *Daily Mail*, 25 February 2012

13   Nick Britten, 'Girl of three drowns in pond near nursery', *The Telegraph*, 29 November 2002

14   Frank Golden and Michael Tipton, *Essentials of Sea Survival*, p.62

15   'Toddler found in pond died accidentally, inquest finds', *Southern Daily Echo*, Southampton, 11 September 2008

16   'Daughter drowned in garden pond; toddler tragedy', *Daily Post*, Liverpool, 31 May 2002

17   'Grandmother of boy who drowned in pond "thought he was in bed"', *Evening Standard*, 10 March 2010

18   Lucy Thornton and Martin Charlesworth, '"Livewire" toddler drowned in fish pond while grandad was distracted looking after his sister', *The Daily Mirror*, 27 November 2015

19   Ron Quenby, 'Father found his daughter drowning in garden pond', *Daily Post*, Liverpool, 31 May 2002

20   James McClean (2008), 'Experts warn of garden pond danger', Press Release, University Hospitals of Leicester, http://www.uhl-tr-nhs.uk/formedia/press-releases/copy-of-jan-dec-08

21   John Pearson and Peter Davies (2000), 'Drowning accidents in the garden involving children under five', Consumer Affairs Directorate, Department of Trade and Industry, London

22   'Toddler drowned in old bin', *The Daily Mirror*, 5 July 2007

23   Severin Carrell, 'Police contact safety watchdog over twin's fish tank deaths', *The Guardian*, 14 March 2016

24   Mary I. Jumbelic and Michael Chambliss (1990), 'Accidental Toddler Drowning in 5-Gallon Buckets', *Journal of the American Medical Association*, 263:1952-1953

25   News from CPSC (1989), 'Large Buckets Are Drowning Hazards For Young Children' U.S Consumer Product Safety Commission, Washington

26   Sambrook Research International (1996), 'Fatal Drowning Accidents: 5-Gallon Buckets', Department of Trade and Industry, London

27   'Swimming Pool Safety', http://cityofpasadena.net/Fire/Swimming_Pool_Safety. Kiran Randhawa, 'US Olympic skier "devastated" after daughter drowns in pool', *Evening Standard*, 12 June 2018

28   Ken LaMance (2011), 'Landowner Liability For A Child Drowning In A Swimming Pool', https://www.legalmatch.com/law-library/article/landowner-liability-for-a-child-drowning-in-a-swimming-pool.html

29   Steve Levitt, 'Pools more dangerous than guns', *Chicago Sun-Times*, 28 July 2001, Steven D. Levitt and Stephen J. Dubner (2005), *Freakonomics*, William Morrow, Harper Collins, New York, p.149

30   W. Barry, T. M. Little and J. R. Sibert (1982), 'Childhood drownings in private swimming pools: an avoidable cause of death', *British Medical Journal*, 285:542-543

31   G. J. Wintemute, C. Drake and M. Wright (1991), 'Immersion events in residential swimming pools. Evidence for an experience effect', *American Journal of Diseases of Children*, 145:1200-1203

32   Ruth Brenner, Gitanjali Saluja Taneja, Denise L. Haynie, Ann C. Trumble, Cong Qian, Ron M. Klinger and Mark A. Klebanoff (2009), 'Association Between Swimming Lessons and Drowning in Childhood' *Archives of Pediatrics and Adolescent Medicine*, 163:203-210.'Boy, 2, drowns day before his first swimming lesson', *Evening Standard*, 28 September 2010

33   Virginia Hunt Newman, *Teaching an Infant to Swim*, p.3-6

34   'Parents' vain bid to save toddler', *Abingdon Herald*, 2 December 1993. John Stevens 'Toddler, 2, dies after being found at the bottom of grandparent's swimming pool moments after he was seen playing happily with his brother', *The Daily Mail*, 31 July 2013. 'CPSC Warns Backyard Pool Drownings Happen "Quickly and Silently"', US Consumer Product Safety Commission, 1 July 2003, http://www.cpsc.gov/cpscpub/prerel/prhtml03/03151.html

35   CPSC, 'How to plan for the unexpected: Preventing Child Drownings', http://www.cpsc.gov/cpscpub/pubs/359.pdf

36   Alex Benady, 'Troubled waters', *Daily Telegraph*, 22 August 2002

37   Joe Torg, 'Children can drown in a flash', *The Philadelphia Enquirer*, 24 June 2008

38   Anna Bradley, personal communication.

39   John H. Pearn and James Nixon (1997), 'Swimming pool immersion accidents: an analysis from the Brisbane Drowning Study', *Injury Prevention* 3:307-309

40   Anjali Athavaley, 'When Pet Doors Cause Child Deaths', *The Wall Street Journal*, 12 August 2009

41   Kate Connolly, 'Child drownings in Germany linked to parents' phone 'fixation', *The Guardian*, 15 August 2018

42   Virginia Hunt Newman, *Teaching an Infant to Swim*, p.90

43   A. B. Craig (1976), 'Summary of 58 cases of loss of consciousness during underwater swimming and diving', *Medicine and Science in Sports*, 8:171-175

44   Barth A. Green, M. Alexander Gabrielsen, Wiley J. Hall and James O'Heir (1980), 'Analysis of swimming pool accidents resulting in spinal cord injury', *Paraplegia*, 18:94-100 (Note: Young children weigh so little that the resistance of the water slows their descent, but when they grow older and heavier, their impetus carries them deeper into the water. Teenagers risk concussion, neck fractures and life-threatening paralysis if their heads strike the tiles at the bottom of the pool.)

45   Virginia Hunt Newman, *Teaching an Infant to Swim*, p.84

46   Tom Griffiths, 'Killer parties', *Aquatics International*, 1 March 2008

47   Richard Savill, 'I saw son lying in friend's party pool', *The Daily Telegraph*, 8 January 2004

48   Ian Scott (2003), 'Prevention of drowning in home pools – lessons from Australia', *International Journal of Injury Control and Safety Promotion*, 10:227-236

49   N. Milliner, J. Pearn and R. Guard (1980), 'Will fenced pools save lives? A 10-year study from Mulgrave Shire, Queensland', *The Medical Journal of Australia*, 2:510-511

50   J. Thompson (1977), 'Drowning and near-drowning in the Australian Capital Territory: a five-year total population study of immersion accidents', *The Medical Journal of Australia*, 1:130-133

51   Ian Scott (2003), 'Prevention of drowning in home pools – lessons from Australia', *International Journal of Injury Control and Safety Promotion*, 10:227-236

52   Victor F. Carey (1993), 'Childhood drownings: who is responsible?' *British Medical Journal*, 307:1086-1087

53   Diane C. Thompson and Fred Rivara, 'Pool fencing for preventing drowning of children', *Cochrane Database of Systematic Reviews* 1998

54   Mark R. Stevenson, Miroslava Rimajovic, Dean Edgecombe and Ken Vickery (2003), 'Childhood Drowning: Barriers Surrounding Private Swimming Pools', *Pediatrics*, 111:115-119. 'Fencing Regulations in Australia', https://www.tradeglassdepot.com.au/pool_fencing_regulations. Luke Morrison, David J. Chalmers, John D. Langley, Jonathan C. Alsop and Catriona McBean (1999), 'Achieving compliance with pool fencing legislation in New Zealand: a survey of regulatory Authorities', *Injury Prevention*, 5:114-118.

55   A. A. Ellis and R. B. Trent (1997), 'Swimming pool drownings and near-drownings among Californian preschoolers', *Public Health Reports*, 112:73-77

56   Anna Mauremootoo, 'Coroner's barrier call after tragic pool drowning of three-year-old J… R…' *Wiltshire Gazette & Herald*, 15 April 2015

57   CPSC 'Safety Barriers Guidelines for Home Pools.' US Consumer Product Safety Commission, Washington, http://www.cpsc.gov/cpscpub/pubs/pool.pdf

58   'Toddler drowned after foster father left gate unlatched', *The Daily Telegraph*, 8 December 2009

59   Kim Briscoe, 'Warning after toddler drowns in pool', *Norwich Evening News*, 22 July 2009

60   M. V. Ridenour (2001), 'Climbing performance of children: is the above-ground pool wall a climbing barrier?' *Perceptual and Motor Skills*, 92:1255-1282

61   'Coroner urges parents to supervise toddlers after two-year-old drowned in paddling pool', *Skegness Standard*, 28 October 2008

62   'Southampton toddler drowned in parents' swimming pool', BBC News, 15 November 2011

63   Brenda J. Shields, Carol Pollack-Nelson and Gary A. Smith (2011), 'Pediatric Submersion Events in Portable Above-Ground Pools in the United States, 2001-2009', *Pediatrics*, 128:45-52

64   Beverley Norris (1994), 'The Design and Safety of swimming pool covers', *International Journal of Injury Control and Safety Promotion*, 1:163-174

65   'CPSC Warns Consumers Of Potential Drowning Hazard Posed By Solar Pool Covers Used On Swimming Pools', US Consumer Product Safety Commission, Washington, DC

66   Troy W. Whitfield (2000), 'An Evaluation of Swimming Pool Alarms', US Consumer Product Safety Commission, http://www.cpsc.gov/PartFiles/116873/alarms.pdf

67   Jo R. Sibert, Ronan A. Lyons, Beverley A. Smith, Peter Cornall, Valerie Sumner, Maxine A. Craven and Alison Kemp (2002), 'Preventing deaths from drowning in children in the United Kingdom: have we made progress in 10 years? Population based incidence study', *British Medical Journal*, 342:1070-1071

68   'Child Accidental Drownings 2005', Royal Society for the Prevention of Accidents, http://www.rospa.com/leisuresafety/statistics/child-accidental-drownings-2005

69   'Layers of Protection Around Aquatic Environments to Prevent Child Drowning' (2011), National Drowning Prevention Alliance, Fort Lauderdale, Florida, http://ndpa.org/home/resources/safety-tips/layers-of-protection/

CHAPTER 14: OPEN WATER

1   Arthur Ransome (1930), *Swallows and Amazons*, Jonathan Cape, London (paperback edition by Random House, London, 2001) p.2

2   Royal Society for the Prevention of Accidents (RoSPA), *Drowning Statistics – All activities and locations*, 2009-2013

3   Sarah Harris, 'Pond swing game ends in tragedy for two playmates', *Daily Mail*, 25 August 2000

4   '12-year-old boy drowns in pond while playing with friends', *Daily Mirror*, 8 May 2008

5   'Drowning teenager's last cry as pals battled to save him', *The Star*, Sheffield, 5 May 2011

6   'Teenage boys drown in reservoir after reeds tangled around feet as they cooled off with friends', *Daily Mail*, 23 July 2008

7   'Teen saw pals drown', *Manchester Evening News*, 18 April 2010

8   'Ice Safety', Royal Society for the Prevention of Accidents (RoSPA), www.rospa.com/leisure-safety/water/advice/ice

9   Tim Bugler, 'David said to the ducks: "We'll not hurt you"... then he slipped in: Five-year-old tells of frozen pond tragedy', *Daily Record*, Glasgow, 6 November 2001

10   C. E. Lewis and M. A. Lewis (1984), 'Peer pressure and risk-taking behaviours in children', *American Journal of Public Health*, 74:580-584

11   Vicki Kellaway, 'We thought John was on his computer... but he was on that lake; Distraught parents tell of family heartbreak', *Liverpool Echo*, 15 February 2008

12   Dean I. Manheimer, and Glen D. Mellinger (1967), 'Personality characteristics of the child accident repeater', *Child Development* 38: 491-513 (reprinted thirty years later in *Injury Prevention* (1997) 3:135-145)

13   'Peter, 15, drowns fishing for golf balls', *Daily Record*, Glasgow, 10 January 1997

14   Tom Morgan, 'Police search for missing "skinny dipping" best man', *Daily Express*, 9 August 2010

15   Peter Dominiczak, 'Student drowned in the Serpentine after end of term drinking game', *Evening Standard*, 21 June 2011

16   Louie Smith and Adam Aspinall, 'UK weather: Heatwave drowning death toll reaches 13 after four more swimmers lose their lives', *Daily Mirror*, 17 July 2013

17   Daniel Johnson, 'Heatwave: keen rugby player drowns after jumping into reservoir to cool off', *Daily Telegraph*, 17 July 2013

18   'A week in the sun turned to tragedy with the death of talented artist M. J., who drowned in the Lake District on Sunday', *Romford Recorder*, 12 July 2013

19   'Father's drowning in Ullswater was 'misadventure' – coroner', *The Westmorland Gazette*, 23 February 2005 (Note: A verdict of 'misadventure' is given when the death is due to the unexpected consequences of a lawful action. It is similar to, but distinct from, a verdict of 'accidental death'.)

20   Charlotte Wareing, 'Kayak teen died in weekend trip', *Leyland Guardian*, 19 January 2011

21   'Teenager who drowned in lake was not wearing life jacket', *Cumberland and Westmorland Herald*, 21 January 2011

22   'Businessman died from "secondary drowning" two hours after capsizing canoe', *The Daily Telegraph*, 4 June 2015

23   Frank Golden and Michael Tipton, *Essentials of Sea Survival*, p.80 and 265

24   David O'Leary, 'Reservoir death: Drowned boy just 20ft from safety', *The Scotsman*, 7 July 2014

25   'Don't dice with death at our Harrogate reservoirs says Yorkshire Water', *Harrogate News*, 30 March 2012

26   Helen McArdle and Rebecca Gray, 'Community in mourning after teenager drowns in loch'

The Herald Scotland, 21 August 2010 (Note: Man-made Banton Loch in Lanarkshire supplies water to the Forth and Clyde Canal through sluice gates on its southern bank.)

27   Anthony Bond, 'Mum watched in horror as teenager drowned in reservoir – unaware it was her own son' *Daily Mirror*, 24 September 2014

28   'Retired Banker drowns in sluice on Scottish estate', *The Scotsman*, 2 March 2009

29   'United Utilities', https://en.wikipedia.org/wiki/United_Utilities

30   'Man's body recovered from Thirlmere', *Cumberland and Westmorland Herald*, 21 July 2001

31   'Man's body recovered from Thirlmere', *Cumberland and Westmorland Herald*, 21 July 2001

32   'North West reservoir swim peril campaign', *The Lancashire Telegraph*, 3 August 2011

33   '*It's Not a Game*', YouTube, https://youtube.com/watch?v=i27iGYgYBqc

34   'Teens warned – 'It's not a game', *ITV News*, 18 June 2012

35   'Warning – Reservoir dip can be deadly' http://corporate.unitedutilities.com/3795.aspx

36   Paul Crute, 'Lakes warning after man drowns in Thirlmere', *News and Star*, Carlisle, 23 December 2013

37   Kelly Williams, 'Hard-hitting video shows final moments of teens drowned in reservoir', *Daily Post North Wales*, 20 June 2016

38   'Merthyr Tydfil mother backs reservoir danger campaign', BBC, 19 July 2014

39   'Public access to Scotland's outdoors', Scottish Outdoor Access Code Approved by the Scottish Parliament on 1 July 2004

40   'Paper for National Access Forum from Scottish Water on Reservoir Signage', Scottish National Heritage, Inverness, www.snh.gov.uk/docs/A1176113.pdf

41   Alison Campsie (2013), 'Call for new safety measures to reduce drowning fatalities', *The Herald*, Scotland, 21 August 2013

42   Stuart Woledge, 'Angler, 20, drowns while trying to save life of his friend, 17, after he got into difficulty swimming in a reservoir', *Daily Mail*, 6 August 2014

43   Stuart Jeffries, 'Cyclists beware: don't stray from the towpath', *The Guardian*, 31 July 2009

44   'Boy, four, drowns trying to get his football out of canal', *Evening Standard*, 1 June 2004

45   'Father drowned in canal after a night out', *This is Staffordshire*, 12 June 2009.

46   'Rochdale Canal', Wikipedia. https://en.wikipedia.org/wiki/Rochdale_Canal

47   'The Caen Hill Flight', http://www.ukcanals.net/caen.html

48   Michael E. Ware (2003), *Canals and Waterways*, Shire Publications, Princes Risborough, p.18

49   'Tring Reservoirs – Grand Union Canal', http://www.wow4water.net

50   Michael E. Ware, *Canals and Waterways*, p.15-16

51   'Deepest Canal Locks in England', http://www.penninewaterways.co.uk/locks.htm

52   'Boy, 13, drowns after falling 20ft into canal in front of horrified onlookers', *The Daily Mail*, 3 June 2011

53   Katriona Ormiston, 'Anglers save elderly woman from drowning', *Oxford Mail*, 9 October 2013

54   John Gagg (1977), *Canals in a Nutshell*, John Gagg, Princess Risborough, p.16

55   Lynn David, lock-keeper. Personal communication – the year was 1987.

56   Yakub Qureshi and Blaise Tapp, 'Boy dies in canal tragedy', *Manchester Evening News*, 25 April 2005

57   Dan McMullan, 'Canal-surfing kids dice with death', *Manchester Evening News*, 27 June 2006

58   'Make sure your child knows canal dangers', Wigan Council 6 July 2007, http://www.wigan.gov.uk/News/Archive/July2007/CanalDangers.htmp.9

59   'Safety on your canal boat holiday', http://www.canalboatholiday.net/safety.php

60   'Boy aged 2 drowns in boatyard', *The Daily Mirror*, 6 April 1999

61   Andrew Saunders, Post in forum 'Safe operation of locks', The Inland Waterways Association of Ireland, January 2001, http://iwai.ie/archive/forum/read.php?1,5133

62   'How to work Canal Locks', Canal Junction, http://www.canaljunction.com/canal/lock_skills.htm

63   'Holiday horror: Mother fell to her death under the propellers of narrowboat as her children looked on', *Daily Mail*, 20 January 2010

64   'Investigations Into Canal Boat Tragedy Which Claimed Four Lives', *Thurrock Gazette*, 1 January 2000

65   'Report on the investigation of the foundering of the narrow boat *Drum Major* with the loss of four lives at Steg Neck Lock near Gargrave, North Yorkshire on 19 August 1998', Marine Accident Investigation Branch, Southampton

66   Mike Laycock, 'York man saves wife from drowning in canal', *The York Press*, 14 October 2013

67   Royal Society for the Prevention of Accidents (RoSPA), Drowning Statistics – All activities and locations, 2009-2013

68   'General Risks Associated With Angling', Keswick Angling Association, Cumbria.

69   Gemma Anderson, 'Fisherman drowned after falling into Kennet and Avon canal', 22 October 2014, https://www.getreading.co.uk/news/localnews/fisherman

70   'Body of boy, nine, found in lake', BBC News, 13 June 2007

71   'Family pay tribute to a drowning tragedy angler', *Peterborough Telegraph*, 19 March 2009

72   'Small Boat Safety', http://watercraft.ohiodnr.gov/education-safety/safety-tips-

73   Geoffrey Clarkson, fly fishing champion. Personal communication.

74   Debbie White 'Boy, 6, missing after falling into the River Stour while fishing with his dad who dived in to try and save him', *The Sun*, 18 August 2019

75   Peter Barss, Shelley Dalke, Jane Hamilton and Myke Dwyer (2011), '3000 lives, 6 billion dollars lost, 18 years of surveillance – What have we learned about safe boating in Canada?' World Conference on Drowning Prevention, DaNang

76   Auslan Cramb, 'Divers find body of man who drowned trying to save grandson', *The Daily Telegraph*, 27 August 2002

77   'Schoolboy, 15, drowns on fishing trip after inflatable dinghy springs leak and sinks', *The Daily Mail*, 23 March 2011

78   Jim McBeth and Catriona Stewart, 'We could only listen to their cries, say rescuers stuck on shore as four men drown in loch', *The Daily Mail*, 22 March 2009

79   Mike Wade, 'Wildlife expert dies with son, 7, as canoe overturns', *The Times*, 25 August 2009

CHAPTER 15: 'WILD SWIMMING' AND OPEN-WATER RACES

1   Roger Deakin (1999), *Waterlog: a swimmer's journey through Britain*, Chatto and Windus, London.

2   John Cheever (1982), *The Stories of John Cheever*, Penguin Books, Harmondsworth, p.603-612

3   Roger Deakin, *Waterlog: a swimmer's journey through Britain*, p.7-8

4   Roger Deakin, *Waterlog: a swimmer's journey through Britain*, p.50 and 251 (Note: in 2003, five endurance swimmers swam across the Corryvreckan from Jura to Scarba at slack tide. 'Sink or swim for whirlpool daredevils', *The Scotsman*, 24 August 2003.)

5   Suzy Bennett, 'Isles of Scilly: In a gaggle of flippers and goggles', *The Daily Telegraph*, 29 May 2004

6   Kate Rew (2008), *Wild Swim*, Guardian Books, London. Daniel Start (2008), *Wild Swimming: 150 hidden dips in the rivers, lakes and waterfalls of Britain*. Punk Publishing, London. Daniel Start (2012), *Wild Swimming Coast*, Wild Things Publishing, Bath Daniel Start (2013), *Wild Swimming: 300 hidden dips in the rivers, lakes and waterfalls of Britain*, Wild Things Publishing, Bath

7   Chris Coffey, 'Thanks for all your memories', *St Helens Star*, 12 November 2009

8   T. Horne, 'Man drowned in lake after midnight swim', *North West Evening Mail*, 3 April 2007. http://test.cnmedia.co.uk/man-drowned-in-lake

9   Emma Lidiard, 'Tributes to "wonderful" student in Coniston Water drowning', *The Westmorland Gazette*, 25 June 2010

10   Piers Meyler, 'Warning after man's Coniston Water lake death', *Brentwood Gazette*, 10 July 2013

11   'Adventure hols leader is swept to death saving boy from river', *The Daily Mirror*, 26 June 1998

12   'Investigation into river drowning', BBC News, 3 August 2001

13   'Tragic drowning at Lady Falls', http://www.walesonline.co.uk/news/wales-news/tragic-drowning

14   'Llyn Gwynant', http://www.wildswimming.co.uk/map/llyn-gwynant

15   'Mountain swimming inquest warning', BBC News, 21 November 2006

16   'Drowning Memorial – Linn of Dee', http://www.geograph.org.uk/photo/2983065

17   'Poignant Dee Tragedy', *Aberdeen Journal*, 17 October 1927

18   'Girl slipped and drowned at beauty spot while having her picture taken', *The Scotsman*, 9 May 2006

19   Janice Burns, 'Mum's warning as tragic young fisherman dragged to his death in Loch Lomond', *Daily Record*, 7 June 2008

20   Craig McQueen, 'The water rangers fighting to prevent tragedies in Loch Lomond', *Daily Record*, 2 July 2008

21   Donald Fullarton, 'Loch Lomond's Rescue Boat', *Helensburgh Heritage*, 26 April 2010

22   Daniel Start, *Wild Swimming: 150 hidden dips in the rivers, lakes and waterfalls of Britain*, p.97.

23   Robert Hale, 'Draining ruled out, but Gullet Quarry, near Castlemorton in the Malvern Hills, should never be the same again', *Malvern Gazette*, 22 January 2014

24   'Police and Fire Warn Taking A Dip Could Be Deadly', Hereford and Worcester Fire and Rescue Services, 14 June 2010

25   Harriet Arkell, 'Teenager who disappeared while swimming in abandoned quarry on hottest day of the year', *The Daily Mail*, 8 July 2013. Melanie Hall, 'Body of drowned man pulled from quarry', *The Daily Telegraph*, 12 July 2013

26 'Coroner's praise for those who tried to save young men who drowned in Herefordshire quarry', *Worcester News*, 14 November 2013

27 'Access closed to Gullet Quarry', https://www.facebook.com/malvernhillsconservators

28 'Gullet Quarry, Ledbury', http://wildswim.com/gullet-quarry-ledbury

29 James Connell and Ian Craig, 'Swimmers ignore safety warnings at Gullett Quarry', *Malvern Gazette*, 15 July 2013. 'Why are so many young men drowning?' BBC News, 28 July 2014. James Forrest, 'Swimmers flock to death quarry despite safety concerns', *Malvern Gazette*, 2 July 2015

30 Daniel Start, 'Wild swimming isn't dangerous, but our behaviour around water can be', *The Guardian*, 18 July 2013

31 Michael Tipton and Carl Bradford (2014), 'Moving in extreme conditions: open water swimming in cold and warm water', *Extreme Physiology and Medicine*, 3:12 http://extremephysiolmed.com/contents/3/1/12

32 'Teen drowns trying to swim across quarry to meet girls', *The Daily Telegraph*, 21 January 2015

33 'Teenager drowned in quarry leap', *Birmingham Post*, 25 October 2002

34 Liz Roberts, 'Lake District police urge wild swimming caution after two deaths', http://www.grough.co.uk/magazine/2013/07/27/lake-district-police

35 'UK Drowning Statistics 2002. The Dangers in the Water – The Lies and the Statistics', http://www.river-swimming.co.uk/stats.htm

36 Jonathan Knott, 'The truth about reservoir swimming', July 2010, http://www.outdoorswimmingsociety.com/index.php?p=news&start=70

37 Nichola Davies, 'Teenager drowned in Appleton Reservoir in Warrington is named', This Is Cheshire, 3 June 2009, http://www.thisischeshire.co.uk/news/4416127.Tragic_Appleton_Reservoir

38 Vicki Stockman, 'Seeing your friend die is the worst thing imaginable', This Is Cheshire, 3 December 2009, http://www.thisischeshire.co.uk/news/4774457._Seeing_your_friend_die

39 'Praise for the success of Ironman's Pennington Flash swimming event', *The Bolton News*, 1 August 2010

40 Frank Golden and Michael Tipton, *Essentials of Sea Survival*, p.59-64

41 Donal Buckley, 'Ten Common Myths of Cold Water Swimming', https://loneswimmer.com/2014/10/21/ten-common-myths

42 J. S. Hayward and C. D. French (1989), 'Hyperventilation response to cold water immersion reduced by staged entry', *Aviation, Space and Environmental Medicine*, 60:1163-1165

43 F. S. Golden and M. J. Tipton (1988), Human adaptation to repeated cold immersions', *The Journal of Physiology* 396:349-363. Michael Tipton, Clare M. Eglin and Frank St C. Golden (1998), 'Habituation of the initial responses to cold water immersion in humans: a central or peripheral mechanism?' *Journal of Physiology*, 512:621-628. Donal Buckley, 'Cold Water Acclimatization', http://loneswimmer.com/2014/02/10/acclimatization

44 Flora L. Bird (2011), 'A study of the physiological and subjective responses to repeated cold water immersion in a group of 10-12 year olds', PhD Thesis, University of Portsmouth.

45 R. E. G. Sloan and W. R. Keatinge (1973), 'Cooling rates of young people swimming in cold water', *Journal of Applied Physiology*, 35:371-375

46   Frank Golden and Michael Tipton, *Essentials of Sea Survival*, p.61 and 66-68

47   Donal Buckley, 'Understanding The Claw as a hypothermic indicator', http://loneswimmer.com/2012/11/22/understanding-the-claw

48   Frank Golden and Michael Tipton, *Essentials of Sea Survival*, p.71-74

49   W. R. Keatinge, C. Prys-Roberts, K. E. Cooper, A. J. Honour and J. Haight (1969), 'Sudden Failure of Swimming in Cold Water', *British Medical Journal* 1(5642):480-483

50   Frank Golden and Michael Tipton, *Essentials of Sea Survival*, p.101-107

51   Donal Buckley, 'How To: Understanding hypothermia in swimmers – Mild Hypothermia', http://loneswimmer.com/2012/05/28/understanding-hypothermia. Armin Schubert (1995), 'Side Effects of Mild Hypothermia', *Journal of Neurosurgical Anaesthesiology*, 7:139-147

52   Donal Buckley, 'The relevance of shivering in open water swimming', http://loneswimmer.com/2013/03/1/the-relevance-of-shivering

53   D. Brannigan, I. R. Rogers, I. Jacobs, A. Montgomery, A. Williams and N. Khangure (2009), 'Hypothermia is a significant medical risk of mass participation long-distance open water swimming', *Wilderness & Environmental Medicine*, 20:14-18

54   Frank Golden and Michael Tipton, *Essentials of Sea Survival*, p.255

55   Donal Buckley, 'Why would you swim in cold water?' http://loneswimmer.com/2012/12/10/why-would-you-swim

56   Michael Tipton and Carl Bradford (2014), 'Moving in extreme environments: open water swimming in cold and warm water', *Extreme Physiology & Medicine*, doi: 10.1186/2046-7648-3-12

57   D. Brannigan, I. R. Rogers, I. Jacobs, A. Montgomery, A. Williams and N. Khangure (2009), 'Hypothermia is a significant medical risk of mass participation long-distance open water swimming', *Wilderness & Environmental Medicine*, 20:14-18

58   Penny Lee Dean, *Open Water Swimming*, p.31-35

59   Chrissie Wellington 'Chrissie's Guide to the Open Water. Part 1' 220 *Triathlon Magazine*, June 2012 p.32-42, www.22triathlon.com

60   'Lake District swimmers urged to don coloured caps', BBC News, 8 April 2014

61   'The Great Tow Float For & Against', http://www.outdoorswimmingsociety.com/news/514-the=great-tow-float

62   'Great Swim', Wikipedia, https://en.wikipedia.org/wiki/Great-Swim

63   'Great North Swim hailed a success', BBC News, 14 September 2009

64   'Great North Swim called off due to poisonous algae in lake', *The Journal*, Newcastle, 3 September 2010

65   'Woman, 35, in Great East Swim race in Suffolk dies', BBC News, 19 June 2010

66   Steven Bell, 'Call for medical checks after Great North Swim death', *Lancashire Telegraph*, 23 June 2011

67   Adam Sherwin, 'Safety concerns after man dies during Great North Swim', *The Independent*, 20 June 2011

68   Steven Bell, 'Call for medical checks after Great North Swim death', *Lancashire Telegraph*, 23 June 2011

69   'Woman airlifted from Great North Swim', http://www.itv.com/news/border/

update/2013-06-16

70   'Great North Swim: Man dies after taking part in event on Windermere in the Lake District', *The Daily Mirror*, 13 June 2014

71   Kevin M. Harris, J. T. Henry, E. Rohman, T. S. Haas and B. J. Maron (2010), 'Sudden death during the triathlon', *Journal of the American Medical Association*, 303:1255-1257 (Note: Thirteen competitors died during the swimming phase of the race and one competitor fell during the cycling phase and died from neck injuries.)

72   Ulla Ilnytzky, 'Second death in Nautica New York City Triathlon', *Huffington Post*, 8 August 2011

73   Larry Greenemeir, 'Why Is Swimming the Most Deadly Leg of a Triathlon?' *Scientific American*, 9 August 2011

74   Richard E. Moon, Stephanie D. Martina, Dionne F. Peacher and William E. Kraus (2016), 'Deaths in triathletes: immersion oedema as a possible cause', *British Medical Journal Open Sport & Exercise Medicine*, 2:e000146 (Note: 42 competitors died while swimming, 11 while cycling, and 5 while running.)

75   Jeff Z. Klein, 'Preparing Triathletes for the Chaos of Open Water', *New York Times*, 14 August 2011

76   Matt Majendie, 'When it becomes brutal, I just kick harder', *Evening Standard*, 19 May 2011

77   David Brown, 'Deaths in triathlons may not be so mysterious; panic attacks may be to blame', *The Washington Post*, 14 November 2011

78   Frank Golden and Michael Tipton, *Essentials of Cold Water Survival*, p.61. Michael J. Tipton (2014), 'Sudden cardiac death during open water swimming', *British Journal of Sports Medicine*, 48:1134-1135

79   Kevin M. Harris, J. T. Henry, E. Rohman, T. S. Haas and B. J. Maron (2010), 'Sudden death during the triathlon', *Journal of the American Medical Association*, 303:1255-1257

80   'Wolff-Parkinson-White syndrome', Wikipedia, https://en.wikipedia.org/wiki/Wolff-Parkinson-White

81   H. J. H. Colebatch and D. F. J. Halmagyi (1962), 'Reflex airway reaction to fluid aspiration', *Journal of Applied Physiology*, 17:787-794

82   H. J. H. Colebatch and D. F. J. Halmagyi (1963), 'Reflex pulmonary hypertension of fresh-water aspiration', *Journal of Applied Physiology*, 18:179-185

83   Charles C. Miller, Katherine Calder-Becker and Francois Modave (2010), 'Swimming-induced pulmonary edema in triathletes', *The American Journal of Emergency Medicine*, 28:941-946. Eric A. Carter and Michael S. Koehle (2011) 'Immersion Pulmonary Edema in Female Triathletes', *Pulmonary Medicine*, 2011, Article ID 261404, 4 pages. Richard E. Moon, Stephanie D. Martina, Dionne F. Peacher and William E. Kraus (2016) 'Deaths in triathletes: immersion oedema as a possible cause', *British Medical Journal Open Sport & Exercise Medicine*, 2:e000146

84   P. T. Wilmshurst, M. Nuri, A. Crowther and M. M. Webb-Peploe (1989), 'Cold-induced pulmonary oedema in scuba divers and swimmers and subsequent development of hypertension', *The Lancet*, 1(8629):62-65. Carl Edmonds. John Lippmann, Sarah Lockley and Darren Wolfers (2012), 'Scuba divers' pulmonary oedema: recurrences and fatalities', *Diving and Hyperbaric Medicine*, 42:40-44

85   D. Weiler-Ravell, A. Shupak, I Goldenberg, P. Halpern, O. Shoshani, G. Hirschhorn and A Margulis (1995), 'Pulmonary oedema and haemoptysis induced by strenuous swimming', *British Medical Journal*, 311: 361-362

86   D. F. Peacher S. D. Martina, C. E. Otteni, T. E. Wester, J. F. Potter and R. E. Moon (2015), 'Immersion pulmonary edema and comorbidites: Case series and updated review', *Medicine and Science in Sports and Exercise*, 47:1128-1134

87   Kevin Becker and Katherine Calder-Becker, 'Swimming Induced Pulmonary Edema (SIPE)', http://www.endurancetriathletes.com/sipe.html

88   R. Biswas, P. K. Shibu and C. M. James (2004), 'Pulmonary oedema precipitated by cold water swimming', *British Journal of Sports Medicine*, 38:e36

89   Eric A. Carter and Michael S. Koehle (2011), 'Immersion Pulmonary Edema in Female Triathletes', *Pulmonary Medicine*, 2011/261404, 4 pages. Andrea Himmel, 'Why I Almost Died During the 2012 New York City Ironman Triathlon', 25 August 2012, https://mic.com/articles/13194/why

90   Maya Rao and Chris Melchiorre, 'Philadelphia Triathlon drowning victim a first-time competitor', 28 June 2010, http://articles.philly.com/2010-06-28/24966911_1_bevan-doc

91   'Open-Water Death Brings Scrutiny to a Sport', *The New York Times*, 30 October 2010

92   'FINA Report. FINA Appointed Task Force. Part 7 – Francis Crippen', http://www.fina.org/H2O/docs/report/FCrippen/Report_FC_O.pdf

93   D'Darcy Doran, 'Swimming tragedy sparks hi-tech safety drive', *Herald Sun*, Melbourne, 20 June 2011

94   'Shunyi Olympic Rowing-Canoeing Park', https://en.wikipedia.org/wiki/Shunyi

95   'Olympic swimming leads to swans moving from Serpentine', BBC News, 9 August 2012

96   Renato R. T. Castro, Fernanda S. N. S. Mendes and Antonio Caudio L. Nobrega (2009), 'Risk of hypothermia in a new Olympic event: the 10-km marathon swim', *Clinics* (São Paulo), 64:351-356

97   Coral Barry, 'Absolute chaos at the start of the Women's Marathon swimming in Rio Olympics', *Metro*, 15 August 2016

98   Daniel Schofield, 'Rio Olympics: Controversy in the women's marathon swimming with competitor disqualified for dunking', *The Daily Telegraph*, 15 August 2016. Mike Walters, 'Team GB Olympic swimmer Jack Burnell slams judges who disqualified him after controversial end to 10km open-water race', *Daily Mirror*, 16 August 2016

CHAPTER 16: HARD LANDINGS

1   Barth A. Green, Alexander Gabrielsen, Wiley J. Hall and James O'Heir (1980), 'Analysis of swimming pool accidents resulting in spinal cord injury', *Paraplegia*, 18: 94-100. P. Barss, H. Djerrari, B. E. Leduc, Y. Lepage and C. E. Dionne (2008), 'Risk factors and prevention for spinal cord injury from diving in swimming pools and natural sites in Quebec, Canada: a 44-year study', *Accident Analysis and Prevention*, 40:787-797

2   Howard B. Cloward (1980), *Acute Cervical Spine Injuries, Clinical Symposia*, Vol 32. Ciba

Pharmaceutical Company. p.9-15 and 29-31. S. Aito, M. D'Andrea and L. Werhagen (2005), 'Spinal cord injuries due to diving accidents', *Nature*, 43-109-116

3   Peter Wernicki, Peter Fenner and David Szpilman (2006), 'Immobilisation and Extraction of Spinal Injuries', Chapter in *Handbook on Drowning*, p.291-297

4   L. S. Kewalramani, and Jesse f. Kraus (1977), 'Acute Spinal-Cord Lesions from Diving – Epidemiological and Clinical Features', *Western Journal of Medicine*, 126:353-36

5   Barth A. Green, Alexander Gabrielsen, Wiley J. Hall and James O'Heir (1980), 'Analysis of swimming pool accidents resulting in spinal cord injury', *Paraplegia*, 18: 94-100

6   D. Pang and I. F. Pollack (1989), 'Spinal cord injury without radiographic abnormality in children – the SCIWORA syndrome', *The Journal of Trauma*, 29: 654-664

7   David Grundy, Philip Penny and Lucy Graham (1991), 'Diving into the unknown', Editorial, *British Medical Journal*, 302:670-671

8   Howard B. Cloward, *Acute Cervical Spine Injuries*, p.2

9   Milton Gabrielsen, James McElhaney and Ronald F. O'Brien (2001), *Diving injuries: Research Findings and Recommendations for Reducing Catastrophic Injuries*, Informa Healthcare, St Helier, p.105

10   David Grundy, Philip Penny and Lucy Graham (1991), 'Diving into the unknown', Editorial, *British Medical Journal*, 302: 670-671

11   'Spinal injuries', http://www.usoceansafety.com/safety/popup/spinal.asp

12   S. K. Chang et al (2006), 'Risk factors for water sports-related cervical spine injuries', *The Journal of Trauma and Acute Care Surgery*, 60:1041-1046

13   Patrick Foster, 'Bridge-leaping ban is thwarted as the local powers dither', *The Times*, 12 November 2005

14   Roger Deakin, *Waterlog*, p.222-224

15   Steven Bell, 'Coroner's "think twice" plea to people who jump from Kirby Lonsdale's Devil's Bridge', *The Westmorland Gazette*, 22 November 2012

16   Emily Allen '"Tombstoning" boy, 15, dies after leaping off 20ft bridge into river in bid to cool off during heatwave', *The Daily Mail*, 25 July 2012

17   'Rescuers tell of 'impossible' battle to save boy from flooded quarry', *Yorkshire Post*, 14 May 2001

18   'Teen drowns trying to swim across quarry to meet girls', *Daily Telegraph*, 21 January 2015

19   'Mother of drowned boy in Hansen quarry threatens legal action', 2 November 2003, https://www.aggregateresearch.com/news/mother

20   'Army recruit, 19, killed in "tombstoning" tragedy at notorious water-filled quarry', *Daily Mail*, 15 June 2010

21   'Chesterfield boy drowned in quarry, inquest hears', BBC News, 29 January 2010

22   Andy Dolan, 'The poison Blue Lagoon: It might look inviting, but the water is almost as toxic as bleach', *Daily Mail*, 12 August 2012

23   Anthony Bond, 'Derbyshire's Blue Lagoon dyed black to deter people taking a dip in water that is almost as toxic as bleach', *Daily Mail*, 11 June 2013

24   Elliot Furniss, 'Pair plunged to deaths despite warnings', *Eastern Daily Press*, 25 April 2008

25   '"Tombstoner" paralysed from the waist down', *The Plymouth News*, 14 May 2008

26   'Tombstoning victim Sonny Wells in Dover safety DVD', BBC News, 16 February 2011

27   *The Dangers of Tombstoning*, YouTube, Uploaded by Kent Police, 24 February 2011

28   Laura Fennimore, 'Tombstoning campaigner to receive award from RNLI', http://rnli.org/NewsCentre/Pages/Tombstoning 4 May 2011

29   'Definition of Tombstoning', BuzzWord, www.macmillandictionary.com. Keith Rossiter 'New laws could put the leash on dog owners', *Plymouth Herald*, 9 August 2013, credited the journalist Jane Slavin with the first use of the word 'tombstoning' in 1995.

30   'Teenagers filmed tombstoning at cliff edge where man plunged to death', ITV Report, 10 April 2017

31   David Leafe, 'Tombstoning: The terrifying new craze in Britain in which children as young as 11 leap from clifftops into the sea ... and some children are paying a terrible price', *Daily Mail*, 4 August 2012

32   'Warning over harbour tombstoning', BBC News Scotland, 5 August 2008

33   'Harbour "tombstoner" swept away.' BBC News, 22 June 2008

34   'Tombstoning man rescued after 100ft jump goes wrong', *Daily Mail*, 13 August 2007

35   Ian McDonald, '"Stupid" Durdle Door leaper saved', *Dorset Echo*, 13 August 2007

36   Neil Shaw, 'Naked teenagers risking their lives', *Plymouth Herald*, 9 July 2013

37   Gareth Roberts, '"Deadly" flying belly-flop craze goes viral as youngster dives 65ft and lands on stomach', *Daily Mirror*, 4 September 2014

38   Mike Smallcombe, 'Youngsters are risking death by tombstoning in Newquay hardour', http://www.cornishguardian.co.uk/people/CGMikeS/profile.html

39   RoSPA – Drowning Statistics, www.rospa.com/leisure-safety/statistics/drowning

40   'Tombstoning – "Don't jump into the unknown"', http://www.rospa.com/leisure-safety/water/advice

41   Max Cannon, 'How I got hooked on tombstoning – just like a glorified Tom Daley', *Plymouth Herald*, 2 September 2015

42   Michael Savage, 'Tombstoning craze claims its first victim of the summer', *The Independent*, 1 August 200

43   'Youth charged over "tombstoning"', BBC News Scotland, 1 June 2009

44   Sarah Ann Henley, http://en.wikipedia.org/wiki/Sarah_Ann_Henley

45   Mike Nowers and David Gunnell (1996), 'Suicide from the Clifton Suspension Bridge in England', *Journal of Epidemiology and Community Health*, 50:30-32

46   Olive Bennewith, Mike Nowers and David Gunnell (2007), 'Effect of barriers on the Clifton Suspension Bridge, England, on local patterns of suicide: implications for prevention', *The British Journal of Psychiatry*, 190:266-267

47   Edward Guthman, 'Lethal Beauty/The Allure: Beauty and an easy route to death have long made the Golden Gate Bridge a magnet for suicides', *San Francisco Chronicle*, 30 October 2005

48   Tad Friend 'Jumpers. The fatal grandeur of the Golden Gate Bridge', *The New Yorker*, 13 October 2003

49   John Bateson (2012), *The Final Leap*, University of California Press, Berkeley, CA., http://thefinalleap.com/about-the-book.html

50   John Bateson, *The Final Leap*, http://thefinalleap.com/about-the-book.html

51   John Bateson, *The Final Leap*, http://thefinalleap.com/about-the-book.html

52   Richard Seiden (1978), 'Where are they now? A Follow-up Study of Suicide Attempters from the Golden Gate Bridge', *Suicide and Life Threatening Behavior*, 8:203-216

53   G. M. Lukas, J. E. Hutton Jr, R. C. Lim and C. Mathewson Jr., 'Injuries sustained from high velocity impact with water: an experience from the Golden Gate Bridge', *Journal of Trauma and Acute Care Surgery*, 21:612-618

54   Tad Friend, 'Jumpers. The fatal grandeur of the Golden Gate Bridge', *The New Yorker*, 13 October 2003

55   Joe Fitzgerald Rodriguez, 'Decline in suicides at Golden Gate Bridge in 2017 attributed to new patrols', *San Francisco Examiner*, 22 January 2018

56   Katy Steinmetz, 'The Golden Gate Bridge Is a "Suicide Magnet". So Officials Are Adding a Net', *Time*, 12 April, 2017

CHAPTER 17: SWEPT DOWNSTREAM

1   Drowning Statistics – All activities and locations. Royal Society for the Prevention of Accidents, http://rospa.com/leisuresafety/statistics/.

2   'Army platoon "hit by water surge" in flooded beck', *The Yorkshire Post*, 21 November 2008

3   John E. Saunders (1981), *Principles of Physical Geology*, John Wiley, New York, p.289-290

4   Alastair Gowans (2007), 'Fly fishing instruction wading safely – keep one step ahead of trouble', http://letsflyfish.com/fly_fishing_wading.htm. Keith McElroy, Jennifer Blitvich, Lauren Petrass and Andrew McKinley (2011), 'Drowning Prevention strategies for fishers wearing waders', World Conference on Drowning Prevention, DaNang. Tony Miles, Martin Ford, and Peter Gathercole (2001), *The Practical Fishing Encyclopedia*, Select Editions, London p.206-208

5   'Renowned biochemist in drowning tragedy', *Fife Herald*, 26 August

6   'River Safety – Crossing', http://riversafe.org.nz/facts/rivercrossing.shtml

7   Will Pavia, 'Divers in desperate search as father and girl, 3, fall to death at beauty spot waterfall', *The Times*, 20 July 2009

8   'Man swept to his death over one of England's highest waterfalls', *Daily Mail*, 11 October 2010

9   Daniel Start, *Wild Swimming: 300 hidden dips in the rivers, lakes and waterfalls of Britain*, p.22 and 265

10   'Waterfall death: Cold water shock killed Low Force swimmer', BBC News, 15 September 2015

11   Mike Hardisty, 'Ceunant Mawr Waterfall: Say It With A Camera', https://mikehardisty. wordpress.com/2011/07/24'. Ceunant Mawr Waterfall', http://wildswimming.co.uk/map/ ceunant-maur

12   Luke Traynor, 'Two men who died in 80ft Snowdonia waterfall could have been pulled under by deadly whirlpool', *The Daily Mirror*, 8 June 2015. 'Drowned men "unaware of danger" at Llanberis waterfall', BBC News, 21 October 2015

13   John McHale, 'Mountain rescuers risk their lives to save swimmer as two die in Llanberis river',

https://www.grough.co.uk>magazine>2015/06/07

14   Lewis Pennock, 'Tragic teen drowned saving friend from rapid current near Scottish waterfall', *Daily Record*, Glasgow, 14 December 2017 (Note: Lower Falls of Bruar now joins other listed 'wild swimming' sites where people have drowned.)

15   'Galloway rock pool death sparks regulation review call', BBC News South Scotland, 28 June 2010. (Inquiry under the Fatal Accidents and Inquiries (Scotland) Act 1976 into the sudden death of Laura McDairmant, 24 June 2010, Sheriff J. Johnson. http://scotcourts.gov.uk/opinions/2010fai2) (Note: An expert witness concluded that the jump site 'was not suitable for the activity with a client group of young people… if a participant slipped, tripped or went weak at the knees or changed their mind mid-stride, they would almost certainly die.')

16   'Canyoning', Wikipedia, http://en.wikipedia.org/wiki/Canyoning

17   'Lawyer Dies in Canyon Horror; Scot Plunges to Death on Adventure Holiday', *Daily Record* Glasgow, 15 June 2002

18   Ian Sparks, 'Father watches helpless as girl, 13, drowns in rock pool', *The Daily Mail*, 1 August 2006

19   Paul Lashmar and Imre Karacs 'Swiss river disaster: Black wall of water swept down the gorge, crushing everyone in its path.' *The Independent* 29 July 1999

20   Martha McKenzie-Minifie, 'Survivor's Story: Taken by the river', *The New Zealand Herald*, 17 April 2008. Alexander Brunt (2011), 'Responding to the Mangatepopo canyoning tragedy that claimed seven lives in 2008', World Conference on Drowning Prevention 2011, DaNang, Vietnam.

21   'Corsica flash floods kills five in French canyoning group', BBC, 2 August 2018 Kiran Randhawa, '11 hikers swept to death in Italian floods', *Evening Standard*, 21 August 2018

22   Mark Hilton-Jones, 'Porth yr Ogof', http://wwwukcaves/cave-porthyrogof

23   Simon O'Hagan 'Eyewitness: Mysterious, idyllic, alluring and lethal', *The Independent* 28 July 2002

24   Ray Kershaw, 'What lies beneath: Mossdale caving disaster', *The Independent*, 15 March 2008

25   Paul Cook, 'Escape bid sealed fate of tragic cave pair', *The Northern Echo*, 16 May 2008

26   Thair Shaikh, 'Students rescued after four-hour ordeal in flooded cave system', *The Guardian*, 19 January 2008

27   'Lessons Learned from two managed incidents in Long Churn', LLG Case 5. Lessons Learned Group. http://www.lessonslearned.org.uk/docs/case%205%20Long%Churn.pdf. British Cave Rescue Council. Incident Report for 2008, http://www.caverescue.org.uk. 'Pothole rescue mission saves 25', *Daily Post*, Liverpool, 6 October 2008

28   Ric Halliwell, 'A Century of British Caving', Craven Pothole Club, http://www.sat.dundee.ac.uk/arb/cpc/century.html. British Cave Rescue Council. Incident Reports for years 2008 to 2013, http://www.caverescue.org.uk

29   Tamia Nelson, 'Eddies, Up Close And Personal', http://www.paddling.net/sameboat/archives/sameboat202.html. William McGinnis, 'Whitewater Glossary for Guides', http://www.highdesertriver.com/terms2.htm

30   Martin Williams, 'Are risks too great for river kayakers?' *The Herald*, Scotland 9 April 2010

31   Peter Cornall, until recently head of leisure safety at RoSPA. Quoted by Martin Williams in

'Are risks too great for river kayakers?' *The Herald*, Scotland 9 April 2010

32   'Tragic canoeist drowned "doing what she loved"', *Yorkshire Evening Post*, 1 May 2012

33   Eryl Crump 'Student's tragic death as his kayak capsized and his foot got stuck in fast-flowing river', *The Daily Mirror*, 15 March 2017

34   Katie Timms 'Heroic strangers battled to save man's life in Dartmoor kayaking tragedy, Plymouth inquest hears', *Plymouth Herald*, 2 June 2018

35   'Canoeist dies on River Dart "mad mile"', BBC News, 22 November 2009

36   'How to Read a Whitewater River: A Few Tips for Kayak and Rafting Trips', http://www.wildasia.org/main.cfm/Travel/Reading_the_Whitewater

37   Amar Singh and Frank Thorne, 'Pupils see teacher die on rapids rafting trip', *Evening Standard*, 1 August 2007

38   Samantha Healy, 'Inquest into deaths of five tourists while white water rafting in north Queensland recommends rapids be risk assessed', *The Sunday Mail*, Queensland 30 June 2012

39   Robin Pagnamenta, 'Briton dead, second is missing after "crocodile attack" on raft', *The Times*, 26 May 2012, G. R. Istre, R. E. Fontaine, J. Tarr and R. S. Hopkins (1984), 'Acute schistosomiasis among Americans rafting the Omo River, Ethiopia', *Journal of the American Medical Association*, 251:508-510

40   'Rafting deaths in 2006', CNN, 5 September 2006, http://cnn.com/2006/US/09/05/whitewater.deaths/index.html

41   David Brown, 'Guided Rafting Accident Statistics, http://www.americanwhitewater.org/content/Article/view/articleid/2

42   'Review of Commercial Whitewater Rafting Safety Standards. Final Advisory Report', Maritime Safety Authority, Wellington, N.Z. (1995), quoted in Ralph Buckley (2006) *Adventure Tourism*, CABI Publishing, Wallingford, England, p.439

43   Carl L. Cater (2005), 'Playing with risk? Participant perceptions of risk and management implications in adventure tourism', *Tourism Management*, 27:317-325

44   Peter Allen, 'Minutes after this photograph was taken, four of them were dead. Survivor of the whitewater rafting tragedy tells how she saw her boyfriend swept away', Report from Salzburg of fatal white-water rafting incident at Zell am See, Austria, *Daily Mail*, 9 June 1999

45   Luna B. Leopold and W.B. Langbein (1966), 'River Meanders', *Scientific American*, 1966: 60-70. John Connelly (2014), 'Drowning: The Exit Problem', *International Journal of Aquatic Research and Education*, 8:73-97

46   Tony Miles, Martin Ford and Peter Gathercole (2001), *The Practical Fishing Encyclopedia*, Select Editions, London, p.116-118

47   'Parents of Bridgnorth river death man call for warning signs', *The Shropshire Star*, 1 October 2013

48   'Barney Davies "Forever loved"; 17-year-old boy who drowned in Thames on hottest day', *Evening Standard*, 26 June 2017

49   Felix Allen, 'Friend's desperate fight to save student who drowned in the Thames', *Evening Standard*, 24 September 2009

50   'Pupils mourn friend who drowned in river tragedy', *The Abingdon Herald*, 9 September 1999

51   Jamie Micklethwaite, 'Teenage boy feared drowned after taking a dip in Thames in hot weather',

Evening Standard, 9 May 2016

52   Chris Musson, 'Teenage hero drowns after rescuing best friend from stretch of river known as "Dead Man's Pool"', *Daily Record*, Glasgow, 21 June 2010

53   Steve Robson, 'Schoolboy drowned in river playing "piggy-in-the-middle" – after pals thought cries for help were part of a game', *Daily Mirror*, 3 September 2015

54   Kevin Moran (2014), 'Getting Out of the Water: How Hard Can That Be?' *International Journal of Aquatic Research and Education*, 8:321-337

55   J. W. Keeling, J. Golding and H. K. G. R. Millier (1985), 'Non-natural deaths in two health districts', *Archives of Disease in Childhood* 60:525-529

56   'Holidaymaker fell into river in Grasmere and drowned – inquest', *The Westmorland Gazette*, 14 October 2011. Tristan Kirk, 'Highgate woman's body found in lake in Lake District', *The Tottenham and Wood Green Independent*, 14 February 2011

57   Ben Ellery and Amanda Perthen, 'We did it for a joke. I had no idea he couldn't swim: Sobbing schoolgirl reveals how her friend, 15, drowned after they jumped from river bridge to cool down', *Daily Mail*, 28 May 2012

58   Rashid Razaq, 'Law student died trying to swim the Thames after a drinking spree', *Evening Standard*, 7 September 2011

59   'Student played drinking games before drowning in York river, inquest hears', *The Yorkshire Post*, 1 October 2014. 'River searches for man missing after York night out', BBC News, 4 March 2014. Victoria Prest, 'York river tragedy inquest: Friend told Tyler to get in the water then failed to help police – Parents tell of heartache', *The York Press*, 5 March 2015

60   Dan Bean, 'Royal Life Saving Society launch Don't Drink And Drown campaign at University of York', *The York Press*, 3 October 2014

61   John Connelly (2014), 'Drowning: The Exit Problem', *International Journal of Aquatic Research and Education*, 8:73-97

62   'Water safety Review for: City of York Council', 4 September 2014

63   Jeremy Culley, 'Tragic soldier drowned after "drinking 16 pints" and going for a late night swim', *Daily Star*, 20 July 2016. Mike Laycock 'Police find body of missing York student in river', *The York Press*, 10 October 2016. 'Family of York student who drowned calls for greater safety measures', ITV News, 21 December 2016. Victoria Prest, 'York river safety: More improvements still needed, say experts', *The York Press*, 24 June 2016

64   'Missing Student: Police Find Body In River', Sky News, 23 January 2015. 'Death of Durham University student S… P… was accidental inquest hears', *Evening Chronicle*, April 2014. 'Student "drowned after night out"', *Daily Mail*, 1September 2014. 'Durham river deaths: Safety work follows student drownings', BBC News, 24 June 2015. Will Metcalfe, 'Generation of Durham school children given water safety training', *Chronicle Live*, 23 July 2015 . Sean O'Neill, 'Threat of drinking Fines after student drownings', *The Times*, 13 February 2015

65   RLLS leaflet 'From happy hour to nightmare in just one slip'

66   Laura Churchill, 'Some of the most tragic deaths in and around Bristol's waterways', *The Bristol Post*, 13 June 2017

67   Bronwen Weatherby 'Bristol student freshers warned about dangers of drinking near docks after drowning deaths', *The Bristol Post*, 26 September 2017

68   'Locks and weirs on the River Thames', Wikipedia, http://en.wikipedia.org/wiki/Locks_and_weirs_on_the_River_Thames

69   'Weirs – nice to play in? Or deadly – can you tell the difference?', http:www.internationalrafting.com/2013/01/weirs-nice-play-in-deadly. Edward William Kern (2014), 'Public Safety at Low-Head Dams: Fatality Database and Physical Model of Staggered Deflector Retrofit Alternative', M.Sc. Thesis. Department of Civil and Environmental Engineering, Brigham Young University, Utah. Ed Kern 'Low Head Dams – A dangerous current (presentation), YouTube, August 2013. Environment Agency/Rescue 3(UK) Weir Assessment System, Environment Agency Wales, September 2009

70   'Guyzance Tragedy, 17th January 1945', http://www.fusilier.co.uk/brainshaugh_acklington/guyzance_river_co

71   Sophie Doughty, 'Young man's death leaves friends aching; Swimmer pulled under by strong currents near weir', *The Newcastle Journal*, 3 August 2009

72   Nick Britten and Martin Evans, 'Father and son killed on first outing in new boat just yards from their home', *Daily Telegraph*, 14 May 2012

73   '300 Troop, 131 Independent Parachute Squadron. Royal Engineers Territorial Army. Cromwell Lock … River Trent in Nottinghamshire September 28th 1975', http://www.parachuteregimenthsf.org/131Independent'. 'Cromwell Weir tragedy: Service remembers the dead soldiers', BBC News, 27 September 2015

74   Mike Smith (1995), 'Inland and estuarial navigation safety system for small craft based on SGNI/CEVNI standards', British Waterways North East Region, Leeds

75   Thomas Petch, 'Tragedy in the "weir of death" – update', *Angler's Mail*, 27 January 2012

76   'Dramatic rescue over treacherous weir', *Newark Advertiser*, 1 November 2013

77   Adam Poole, 'Old London Bridge', Engineering Timelines, http://www.engineeringtimelines..com/why/bridgesOfLondon.

78   Christopher Hibbert (1969), *London: the biography of a city*, Longmans, Green and Co, London, p.108

79   'Memories of Tower Beach', Historic Royal Palaces, Tower of London, www.hrp.org.uk

80   *2012 London Bridge and Thames Estuary Tide Tables*. Tidal Press, Steeple Ashton

81   'H. M. Coastguard London: Now on guard on the Thames', Leaflet published by the Maritime and Coastguard Agency, Southampton

82   Marine Accident Investigation Branch (1990), 'Report of the Chief Inspector of Marine Accidents into the collision between the passenger launch Marchioness and MV Bowbelle with loss of life on the River Thames on 20 August 1989', http://www.maib.gov.uk/publications/investigation_reports/popular

83   Glenda Cooper, 'Pleasure boat disaster on River Thames was "a predictable event that was utterly preventable"', *The Independent*, 8 April 1995

84   Louise Butcher (2010), 'Shipping: safety on the River Thames and the Marchioness disaster', Commons Briefing Paper SN00769, Library of the House of Commons

85   David van Vlymen (2004), 'The History and Re-Organization of London's Thames Police: The RNLI Takes Over', http://www.policespecials.com/thames.html. 'Tower Lifeboat Station', Wikipedia https://en.wikipedia.org/wiki/Tower_Lifeboat_Station

86   'Lifeboat station on London's River Thames was the busiest in Britain in 2013', *Evening Standard*, 28 January 2014

87   Rosa Silverman, 'Mother jumps into freezing river to save toddler', *The Daily Telegraph*, 28 January 2013. 'Mother and child, two, rescued from Thames after little one slips through railings', *Evening Standard*, 28 January 2013. 'Chiswick RNLI rescues mother and baby girl', http://chiswicklifeboat.org.uk/latestnews.htm

88   'Swimming in the Thames – the new arrangements', Port of London Authority 2012

89   'River Thames fatalities: RNLI says "fight with your instincts, not with the water" to help', *Lifeboat News*. Release 26 May 2017, https://rnli.com/news-and-media/2017/may/26/river-thames-fatalities

CHAPTER 18: FLOODS AND STORM SURGES

1   J. Hume Brown (1972), *Weather and Climate*, Blackie and Sons, Glasgow, p.56-57

2   Charlie Furniss, 'Saturation Point', *Geographical, Magazine of the Royal Geographical Society*, June 2006

3   'Floods and Coastal Surges.' BBC safety advice 18 September 2014, https://www.bbc.c0.uk/safety/resources/aztopics/floods-inland.htm

4   Sir Michael Pitt (2008), *Learning Lessons from the 2007 Floods*, Foreword. Cabinet Office, London.

5   'Two die in Sheffield flood chaos', BBC News, 25 June 2007

6   'Flood victim boy had "no chance"', BBC News, 20 September 2007

7   'Bravery nomination for flood heroes', *The Star*, Sheffield, 20 September 2007

8   Nicholas Walliman, 'Evaluating flood damage. Case Study: Gloucestershire, GB flood 2007', www. floodprobe.eu. Oxford Brookes University, Oxford

9   Nicholas Walliman, 'Evaluating flood damage. Case Study: Gloucestershire, GB flood 2007', www. floodprobe.eu. Oxford Brookes University, Oxford

10   'Teenager who drowned in floods had "consumed 13 pints", inquest hears', *The Daily Telegraph*, 17 November 2008

11   '2007 Summer floods: Tackling surface water flooding in Hull', The Environment Agency, www.environment-agency.gov.uk

12   Professor Tom Coulthard et al, 'The June 2007 floods in Hull', Final Report by the Independent Review Body, 21 November 2007, www.coulthard.org.uk/downloads/floodsinhull3.pdf

13   '"Emergency services not trained for underwater rescue," says judge at flood-death inquest', *Daily Mail*, 14 December 2007

14   'Flood death – rescue workers speak', *BBC, Inside Out* – Yorkshire & Lincolnshire, 39 April 2008

15   Sir Michael Pitt (2008), *Learning Lessons from the 2007 Floods*, Cabinet Office, London. 'Key points: Pitt report on floods', BBC News, 25 June 2008

16   Jamie Grierson, 'Cumbria deluge breaks historic rainfall record', *The Independent*, 20 November 2009

17   Sam Knight, 'The extreme floods in Cumbria', *Financial Times*, 5 February 2010. 'Cockermouth Floods: How community coped with 2009 devastation', BBC News, 11 June 2013

18   'The flood in Workington, Cumbria, 19th-20th November 2009', http://visitcumbria.com/workington-floods.htm

19   Kiran Randhawa, 'Swept away: Hero PC lost as bridge collapses – hundreds trapped in homes', *Evening Standard*, 20 November 2009

20   'Keswick flood defence gates and barriers completed', BBC News, 27 September 2012

21   'Keswick. Flood Investigation Report. Flood Event 5-6th December 2015', Environmental Agency and Cumbria County Council, 2016. 'Last winter's floods "most extreme on record in UK", says study', *The Guardian*, 5 December 2016

22   Geoffrey Lean, 'UK flooding: How a Yorkshire town worked with nature to stay dry', *The Independent*, 2 January 2016

23   'Sustainable drainage system', Wikipedia, https://wikipedia.org/wiki/Sustainable_drainage

24   Ben McAlinden, 'Slowing the Flow at Pickering', Institution of Civil Engineers, https://www.ice.org.uk/disciplines-and-resources/

25   Jeremy Biggs, '"Working with nature" didn't save Pickering from the floods – it just didn't rain much', *The Guardian*, 7 January 2016. Geoffrey Lean, 'Despite the critics, Pickering's natural flood defences do seem to have saved the town (it rained plenty!)', http://Geoffreylean.tumblr.com/post/13716709588

26   'Rescuers lose race against time as boy, 17, dies in flooded drain', *Daily Mail*, 5 June 2008

27   'St Asaph flood inquest: 91-year-old woman refused to leave', BBC News, 14 October 2014

28   'Don't Underestimate The Power Of Floodwater', Royal Life Saving Society, 22 October 2014

29   S. N. Jonkman and E. Penning-Rowsell, 'Human Instability in Flood Flows' *Journal of the American Water Resources Association*, 8 October 2008

30   'Lynmouth Flood', Wikipedia, https://en.wikipedia.org/wiki/Lynmouth_Flood. David Huxtable, 'The 1952 Flood Disaster in Context: Exmoor National Park Authority', www.exmoor-nationalpark.gov.uk

31   'Dozens rescued from flash floods', BBC News Channel, 17 August 2004

32   'Tales of Heroism At Flood-Hit Boscastle', Sky News, 18 August 2004

33   Lech Mintowt-Czyz, '£15m flood clear-up: Sixty cars and six buildings washed away', *Evening Standard*, 18 August 2004

34   'Driving through heavy rain and floods. Stay safe on the road when it's wet, wet, wet', The Automobile Association, https://www.theaa.com/driving-advice/seasonal/driving-through-flood-water

35   'Turn Around: Don't Drown', National Oceanic and Atmospheric Administration (NOAA), www.nws.noaa.gov/os/water/tadd/recources/TADD_6_Ariel.pfd

36   Nigel Bunyan and Nick Britten, 'Judge drowns after car is swamped in floods', *Daily Telegraph*, 28 June 2007

37   'Sent to drown by his satnav? Motorist dies after getting trapped in notorious ford', *Daily Mirror*, 1 April 2012

38   'Family's vain fight to save girl, 5, as car sank in river', *The Guardian*, 4 February 2000

39   Henry Holland, 'Tot drowned as mum's car rolled into river after being left alone for just two

minutes', *Daily Star*, 22 November 2018

40   G. G. Giesbrecht and G. K. McDonald (2010), 'My car is sinking: automobile submersion, lessons in vehicle escape', *Aviation, Space, and Environmental Medicine*, 81:779-84. John Lienhard, 'A Sinking Automobile', University of Houston's College of Engineering, http://www.uh.edu/engines/epi1173

41   Jaap Molenaar and John Stoop, 'Submerged Vehicle Rescue', Chapter in *Handbook on Drowning* (2006) p.239-241

42   Jim Lawson, 'What a braveheart; Ross, 14, braves icy river to save drowning sisters', *Sunday Mail*, Glasgow 14 June 1998

43   Helen Studd, 'Mother's grief after children drown trapped in sunken car', *The Times*, 30 July 2001

44   Sean McCarthaigh, Stephen Maguire and Nadeem Badshah, 'Father helped his baby to escape as family died in sinking car', *The Times*, 22 March 2016

45   'Medal for Buncrana pier tragedy hero', *Donegal News*, 14 October 2016

46   Elizabeth Walker 'Cars and water... avoiding a tragedy', *Staying Alive*, February 2008, http://www.RoSPA.com/leisuresafety/info/SA_cars_water.pdf

47   Alan Middleton, 'Draining the Fens', *Lincolnshire Life*, March 2014

48   'Inquest: Coroner backs safety calls after father and son die', *Peterborough Telegraph*, 27 June 2006

49   'Inquests into fens river deaths', BBC News, 8 February 2006

50   'Pregnant mother's agony as she saved her daughter from a sinking car... but was forced to abandon her 16-year-old son', *Daily Mail*, 20 April 2011

51   G. J. Wintemute, J. F. Kraus, S. P. Teret and M. A. Wright (1990), 'Death resulting from motor vehicle immersions: the nature of the injuries, personal and environmental contributing factors, and potential interventions', *American Journal of Public Health*, 80:1068-1070

52   'Atmospheric pressure', Wikipedia, http://en.wikipedia.org/wiki/Atmospheric_pressure

53   'Floods', Royal Meteorological Society, http://www.metlink.org/weather-climate-resources-teenagers/what-i

54   'Storm Surge Overview', http://www.nhc.noaa.gov/surge/

55   Erik Larson (1999), *Isaac's Storm: The Drowning of Galveston: 8 September 1900*, Fourth Estate, London

56   '1970 cyclone changes the course of history', http://thisinnocentcorner.wordpress.com/2010/10/07/1970-cyclone

57   Peggy Mihelich, 'Storm surge the fatal blow for New Orleans', CNN, 7 September 2005. 'Hurricane Katrina', Wikipedia, https://en.wikipedia.org/wiki/Hurricane_Katrina

58   'January 1953 lunar eclipse', Wikipedia, http://wikipedia.org/wiki/January_1953_Lunar_eclipse

59   '1953 U.K. Floods: 50-year Retrospective' (2003), Risk Management Solutions, https://www.rms.com/Publications/1953_Floods_Retrospective.pdf

60   Peter Cockroft, 'Canvey Island – floods of '53', BBC London, 4 September 2008

61   Kathy Taylor 'The Great Flood of Canning Town.' http://www.docklandsmemories.org.uk/Floods1953.pdf

62   '1953 U.K. Floods: 50-year Retrospective' (2003), Risk Management Solutions, https://www.rms.com/Publications/1953_Floods_Retrospective.pdf

63   Herman Gerritsen (2005), 'What happened in 1953? The Big Flood in the Netherlands in retrospect', *Philosophical Transactions of the Royal Society*, A 363:1263-1270

64   Terence Vickress (1986), *The Thames Barrier*, Wilson Publications, p.24. C-H. C. Bae and H. W. Richardson, 'Not Katrina: The Thames Barrier Decision', http://create.usc.edu/assets/pdf/51951.pdf

65   Rob Williams, 'Near miss as HMS Ocean squeezes through Thames Barrier', *The Independent*, 4 May 2012

66   Michael Hanlon, 'The Thames Barrier has saved London – but is it time for TB2', *The Daily Telegraph*, 18 February 2014

67   Mark Rowe, 'Come hell or high water', *Geographical*, April 2011, p.34-41

68   Sarah Lavery and Bill Donovan (2005), 'Flood risk management in the Thames Estuary looking ahead 100 years', *Philosophical Transactions of the Royal Society*, A 363: 1455-1475

CHAPTER 19: TIDES, WAVES AND RIP CURRENTS

1   'Coastline of the United Kingdom', https://en.wikipedia.org/wiki/Coastline

2   'Beach Safety: Facts and Figures', http://www.nationalbeachsafety.org.uk/facts.htm

3   David Crawley A Guide to Morecambe Bay, http://www.travelintelligence.com/travel-writing/a-guide-to-morecambe-bay

4   Roy Shepherd, 'Discovering Fossils: Lyme Regis (Dorset), http://www.discoveringfossils.co.uk/lyme_regis_fossils.htm

5   'Tides and Water Levels: What Causes Tides?' http://oceanservice.noaa.gov/education/tutorial_tides/tides02_cause.html

6   'Tides and Water Levels: Gravity, Inertia, and the Two Bulges', http://oceanservice.noaa.gov/education/kits/tides/tides03_gravity.html

7   'Bay of Fundy', http://en.wikipedia.org/wiki/Bay_of_Fundy

8   The world's Highest Tides – here in Wales, http://www.newportunlimited.co.uk/newsandevents/newsarchive.htm

9   'Bendrick Rock', http://geoconservationlive.org/site/bendrick-rock

10   'Teenager drowned after getting into difficulty off Barry Island', *Barry & District News*, 18 November 2013

11   'Sully Island', Wikipedia, http://en.wikipedia.org/wiki/Sully_Island

12   Robert James Owen, 'Man "lucky to be alive" after plunging into icy water,' *The Argus*, South Wales, 9 December 2010

13   'Father and son-in-law risk their lives to save couple caught by the tide', *Walesonline*, 29 May 2013

14   'Tidal traffic lights! Sully Island installs warning to stop daytrippers getting stranded in Bristol Channel… half a mile from land', *Daily Mail*, 2 July 2014

15   Jessica Walford, 'RNLI Sully Island rescues at all-time high this year after people get caught

out when tide comes in', 16 September 2017, www.walesonline.co.uk

16   'Devon beach death: Mother drowned after playing in waves', BBC News, 9 October 2014. 'Bristol mum drowned after being swept out to sea by huge rip current', *The Bristol Post*, 10 October 2014

17   'Coastguards' fears over "lethal" mudflats', BBC News, 5 August 2005

18   Richard Alleyne and Nicole Martin, 'Rescuers tell of desperate fight to save mud flats girl', *The Daily Telegraph*, 25 June 2002

19   'Burnham Area Rescue Boat', http://en.wikipedia.org/wiki/Burnham_Area_Rescue_Boat

20   'Hovercraft and Mud Rescue Teams Save Trapped Teens trapped at Brean', http://www. burnham-on-sea.com/barb/rescue2.html. Alastair Jamieson, 'Massive rescue operation to save herd of cows trapped in mud', *The Daily Telegraph*, 18 April 2009

21   'A brief explanation of tides and how they affect people', http://www.uktides.net/tides_ explanation.php

22   Bob Wilson 'Derivation of Spring and Neap Tides', *SW Soundings*, No. 50 (Feb 2001), The South West Maritime History Society

23   Rebecca Seales and Mark Duell, 'Rocky horror: Terrified woman rescued by helicopter after Cornish tides trap her just 1ft from rising waves', *Daily Mail*, 13 April 2012

24   Nick Anthony Florenza (2013), 'Lunar Perigee and Apogee', http://lunarplanner.com/ LunarPerigee/index.html. 'A perigean spring tide occurs when the moon is either new or full and closest to Earth', http://oceanservices.noaa.gov/facts/perigean-spring-tide.html

25   'Perigean spring tides. Predicting Potential Disasters. How Tidal Information May Save You From a Coastal Crisis', Woods Hole Oceanographic Institution, Massachusetts, http://www. whoi.edu/seagrant/education/buletins/tides.html.'

26   David Walker, 'Coasteering safely', *Staying Alive*, August 2011, Royal Society for the Prevention of Accidents (RoSPA)

27   'Brother tells of moment he left his sister to drown in tiny cave while he searched for help after they were swept into hole on "coasteering" trip', *The Daily Mail*, 18 March 2014

28   Steven Morris, 'Family of teacher who died in Dorset sea cave calls inquest "total sham"', *The Guardian*, 19 March 2014

29   'Tides. Tidal Variations', British Marine Life Study Society, http://glaucus.org.uk/Tides.htm

30   Carolyn Fry, 'Water's out, moon's up – go for a night hike on the seabed', *The Independent on Sunday*, 8 August 2004

31   'Sharing knowledge – All about tides', National Tidal and Sea Level Facility, http://www.ntslf. org/sharing-knowledge/tides

32   *London Bridge and Thames Estuary Tide Tables*, Tidal Press, Steeple Aston.

33   Simon Cable, 'Schoolboy dies as the tide sweeps him away', *The Daily Mail*, 9 July 2007

34   'Drowning fears at Hunstanton', *Lynn News*, 12 September 2008

35   Robin Turner, 'We drove down there in our patrol vehicle and shouted to them to come back off the bank and it was then the sand started to give way: Lifeguards at Tenby praised as heroes', *Western Mail*, Cardiff, 27 July 2009

36   'Camber Sands deaths: Victims "could have underestimated the tide"', BBC News, 25 August 2016

37   'Blackpool. Historic Town Assessment Report', April 2005, Lancashire County Council with English Heritage, p.28-29

38   'Two Scots drown in seaside accident on stag party weekend BLACKPOOL: TRAGEDY Coastguard warns of hazards as young men die', *The Herald*, Glasgow, 28 July 2006

39   Alex Ross, 'Drowning tragedy as man is swept off sea wall' *The Gazette*, Blackpool, 5 April 2011

40   'Crew life on the ocean wave', BBC Local: Lancashire 28 September 2009, http://news.bbc. co.uk/local/lancashire/low/people_and_places/nature

41   'Sea warning after Wrexham man drowns in Blackpool', BBC News, 22 April 2010, http:// news.bbc.co.uk/1/hi/wales/north_east/8635624.stm

42   Jon Ungoed-Thomas and John Elliott, 'Tragedy of cockle pickers ruled by ruthless gangmasters', *Sunday Times*, 8 February 2004

43   David Fickling, 'Cockler deaths jury shown film of survivor's rescue' *The Guardian*, 20 September 2005

44   Nigel Bunyan, 'We heard cries, but couldn't reach them', *The Daily Telegraph*, 7 January 2002

45   Sinead McIntyre, 'Father and son perish on sandbank', *The Daily Mail*, 8 January 2002.

46   'Wind wave', Wikipedia, http://en.wikipedia.org/wiki/Wind_wave

47   'High Surf Safety', Pacific Disaster Centre, http://www.pdc.org/iweb/high_surf_safety. jsp?subg=1

48   Stephen Leatherman (2005), *Dr Beach's Survival Guide*, Yale University Press, New Haven, p.85-87

49   Richard Savill, 'Boy tells of night friends were swept to their deaths', *The Daily Telegraph*, 22 December 2005

50   Martha Busby, 'Freak wave on Dorset walk carried girl, 5, out to sea, inquest told', *The Guardian*, 3 October 2018

51   'Isle of fright! Giant wave sends man and boy flying', *Metro*, 5 November 2019

52   Stephen Maguire, 'Freak Wave Swept Away Climber; Daredevil Michael was washed out to sea as he chatted to photographer', *The Sunday Mirror*, 15 July 2007

53   Alana Fearon, 'Primark Founder's son drowns in doomed bid to save his son and girlfriend from freak wave', *The Daily Mirror*, 2 July 2015

54   'Single wave on loch brings tragedy. Warnings over deceptive spring conditions of bright sunshine on freezing water that make boating treacherous', *The Herald*, Scotland, 14 March 2005

55   John Simpson, 'Remains found at loch from teenager drowned eight years ago', *The Times*, 8 February 2013

56   Ron Brackett, 'Lake Michigan's Deadly Waves Prompt Warnings for Swimmers and Boaters in Illinois and Michigan', *The Weather Channel*, 27 August 2019, https://weather.com/news/ news/2019-08-27-lake-michigan-beach

57   'Tragedy of young sailor who held on to drowning brother for over half an hour but had to let him go', *Daily Express*, 15 November 2012

58   'Girl swept into sea at Newquay dies days after her father', BBC News, 24 August 2016

59   Kevin Moran (2008), 'Rock fisher's practice and perception of water safety', *International*

Journal of Aquatic Research and Water Safety 2: 128-139. Kevin Moran (2011), 'Rock-based fishing safety promotion: Five years on', *International Journal of Aquatic Research and Water Safety* 5:164-173

60   'Investigation into the coronial files of rock fishing fatalities that have occurred in NSW between 1992 and 2000', Safe Waters. A New Government Safety Initiative. September 2003. Matthew Thomson, Anthony Bradstreet and Shauna Sherker (2011) 'Don't put your life on the line – A strategy to reduce rock fishing deaths', World Conference on Drowning Prevention, DaNang.

61   Tracy V. Wilson 'Breaking Waves. How Surfing Works', http://adventure.howstuffworks.com/outdoor-activities/water-sports/s

62   'Breaking Waves', NOAA Ocean Explorer. National Oceanic and Atmospheric Administration, US Department of Commerce, http://oceanexplorer.noaa.gov/edu/learning/9_ocean_waves/activities

63   'Swash', Wikipedia, http://en.wikipedia.org/wiki/Swash

64   'So You're Swimming… And The Water Is Moving Too?', http://tripadvisor.com/ShowTopic-g150792-1257-k6358871-S

65   S. Crawford, 'Wayne drowned after tragic game of dare', *News and Star*, Cumberland, 6 September 2005

66   'Breaking Waves', NOAA Ocean Explorer. National Oceanic and Atmospheric Administration,. US Department of Commerce. http://oceanexplorer.noaa.gov/edu/learning/9_ocean_waves/activities

67   'Nine Dangers at the Beach' (2013), http://oceanservice.noaa.gov/news/features/july13/beachdangers.html

68   Caren Chesler, 'Do serious beach injuries come in waves?' *Scientific American*, 7 June 2013

69   Tom Avril, 'Research aims to predict dangerous surf', philly.com 14 June 2014 http://articles.philly.com

70   Jessica McDonald, 'Delaware beaches serve as living laboratories for understanding surf zone injuries', *The Pulse*, 31 July 2014, http://www.newsworks.org

71   'Shipwrecks and Those in Peril on the Sea', http://www.weymouth-dorset.co.uk/shipwrecks.html

72   Ian West (2008), 'Safety on Geological Field Trips. Geology of the Dorset Coast', Southampton University, http://www.southampton.ac.uk/imw/safety.htm

73   'Breaking Waves', NOAA Ocean Explorer. National Oceanic and Atmospheric Administration, US Department of Commerce. http://oceanexplorer.noaa.gov/edu/learning/9_ocean_waves/activities

74   'Rip Current Safety', National Weather Service, http://www.ripcurrents.noaa.gov/overview.shtml

75   'Rip Current Research', Tim Scott, http://www.rospa.com/events/pastevents/water2012/info/tim-scott.pdf

76   R. Brander, D. Dominey-Howes, C. Champion, O. Del Vecchio and B. Brighton (2013), 'Brief Communication: A new perspective on the Australian rip current hazard', *Natural Hazards and Earth System Sciences*, 13:1687-1690

77   Stephen P. Leatherman, 'Five Types of Rip Currents. Don't get ripped', California Sea Grant College, 26 October 2012, http://www.csgc.ucsd.edu/NEWSROOM/NEWSRELEASES/2012/d

78   James B. Lushine (1991), 'A study of rip current drownings and related weather conditions', *National Weather Digest*, 16:13-19

79   Stephen P. Leatherman, 'Five Types of Rip Currents. Don't get ripped', California Sea Grant College, 26 October 2012

80   'Rip Basics', Dynamics of Rip Currents and Implications for Beach Safety (DRIBS), http://www.research.plymouth.ac.uk/coastal

81   Dave Benjamin, 'Drowning experience turns into community project', The Great Lakes Surf Rescue Project, 9 January 2012, http://david-benjamin.blogspot.com/2012/01/matteson-man-drown and, http://glsrp.org/flip-float-follow (Note: Prevailing west winds blow across the Great Lakes in Canada and America, raising waves and driving strong longshore and rip currents.)

82   'Break the grip of the rip', http://www.ripcurrents.noaa.gov

83   Jamie MacMahan, Ad Reniers, Jenna Brown, Rob Brander, Ed Thornton, Tin Stanton, Jeff Brown and Wendy Carey (2011), 'An Introduction to Rip Currents Based on Field Observations', *Journal of Coastal Research*, 27:iiii-vi. R. Brander, D. Dominey-Howes, C. Champion, O. Del Vecchio and B. Brighton (2013), 'Brief Communication: A new perspective on the Australian rip current hazard', *Natural Hazards and Earth System Sciences* 13:1687-1690

84   Eleanor Woodward, Emily Beaumont, Paul Russell, Adam Wooler and Ross Macleod (2103), 'Analysis of Rip Current Incidents and Victim Demographics in the UK', *Journal of Coastal Research*, Special Issue 65, p.850-855 (Proceedings of the 12th International Coastal Symposium).

85   'Lifeguards rescue 30 people from dangerous rip current in Croyde', *North Devon Journal*, 15 August 2012. Beach Rips', Dynamics of Rip Currents and Implications for Beach Safety (DRIBS), http://ripcurrents.co.uk

86   'The Myth of the Collapsing Sandbar', Dynamics of Rip Currents and Implications for Beach Safety (DRIBS), http://www.ripcurrents.co.uk

87   T. Scott, P. Russell, G. Masselink. A. Wooler and A. Short (2007), 'Beach Rescue Statistics and their Relation to Nearshore Morphology and Hazards: A Case Study for Southwest England', *Journal of Coastal Research*, Special Issue 50, p.1-6. (Proceedings of the 9th International Coastal Symposium.)

88   'Beach Rips: Tides', Dynamics of Rip Currents and Implications for Beach Safety (DRIBS), http://www.ripcurrents.co.uk

89   'Lifeguards rescue 30 people from dangerous rip current in Croyde', *North Devon Journal*, 15 August 2012

90   Richard Savill, 'Surfing safety alert as father and son die', *Daily Telegraph*, 8 September 2001

91   Dominic Kennedy, 'Half-term horror as three die in Cornwall surfing tragedy', *The Times*, 27 October 2014. 'Mawgan Porth drowning inquests: Father died trying to save sons', BBC News, 17 August 2015

92   'Topographical Rips', Dynamics of Rip Currents and Implications for Beach Safety (DRIBS),

      http://www.ripcurrents.co.uk/site/303/default.aspx

93   'Thomas Hardy and Dorset', http://thewordtravels.com/thomas-hardy-and-dorset.html

94   Thomas Hardy (1874), *Far from the Madding Crowd*, Penguin Books, Harmondsworth, p.383

95   Sam Webb, 'Company director, 41, who promoted "wild swimming" dies in hospital after being found unconscious in English Channel', *The Daily Mail*, 18 June 2013

96   'Beach children feared drowned. Members of the public join authorities in search as police praise parents' courage', *The Herald*, Scotland, 20 August 1996

97   'Safety plea by father of beach death children', *The Herald*, Scotland, 8 October 1996

98   'Jake disappeared in "blink of an eye"', *Lynn News*, 6 October 2000

99   'Mother's plea for beach safety', BBC News, 20 July 2001

100   Simon de Bruxelles, 'Trust loses battle with eroding seas: Rising tides have forced authorities to abandon beach huts and coastal homes', *The Times*, 12 January 2004.

101   Catherine Zandonella (2001), 'Pressure waves crack open sea defences', *New Scientist*, 171(2308):17

102   'Perigean Spring Tides. Predicting Potential Disasters: How Tidal Information May Save You From a Coastal Crisis', Woods Hole Oceanographic Institution, Massachusetts, August 1998, http://www.whoi.edu/seagrant/education/bulletins/tides.html

103   'Cornish bodyboarders dice with death in huge storm surge waves.' *The Sun*, 21 August 2008

104   Emma Hartley, 'Wave-watchers dice with deadly seas', *The Times*, 4 February 2002

105   Anthony France, '"Wave dodger" feared drowned', *Evening Standard*, 27 February 2002

106   Emma Hartley, 'Wave-watchers dice with deadly seas', *The Times*, 4 February 2002

107   Tom Michael, 'Risking their lives. Three teenagers dice with death as they dodge huge waves crashing over Cornish pier', *The Sun*, 19 August 2017

108   Mark Branagan, 'How horseplay in the waves led to tragedy', *The Yorkshire Post*, 23 August 2005

CHAPTER 20: PERIL ON THE SEA

1   Keith C. Heidorn (1998), 'The Weather Legacy of Admiral Sir Francis Beaufort', http://www.islandnet.com/~see/weather/history/beaufort.html

2   'Beaufort scale', Wikipedia, http://en.wikipedia.org/wiki/Beaufort_scale

3   'Saffir-Simpson hurricane wind scale', Wikipedia, http://en.wikipedia.org/wiki/Saffir-Simpson_hurricane_wind_scale

4   Nadeem Badshah and Dominic Kennedy, 'Sailors get safety warning after dinghy disaster that killed two', *The Times*, 18 March 2006

5   Joseph Sienkiewicz (2000), 'Forecasting extreme ocean waves', Proceedings of the International Fishing Industry Safety and Health Conference, Woods Hole, Massachusetts, October 2000. 'Freak waves spotted from space', BBC News, 22 July 2004

6   Rose George, 'Worse things still happen at sea: the shipping disasters we never hear about', *The Guardian*, 10 January 2015

7   Tony Paterson, 'Hell and high water: The Fastnet disaster', *The Independent*, 18 July 2009

8   Sir Hugh Forbes, Sir Maurice Laing and Lieutenant-Colonel James Myatt (1979), '1979 Fastnet Race Inquiry: Report', http://www.blur.se/images/fastnet-race=inquiry.pdf

9   Jack Hunter, *The loss of the Princess Victoria* (1998), Stranraer and District Local History Trust, Stranraer

10   'Princess Victoria: The disaster that sank from memory', *The Belfast Telegraph*, 4 January 2013

11   'MV Herald of Free Enterprise. Report of Court No. 8074 Formal Investigation', Department of Transport, September 1987

12   'Squat effect', Wikipedia, http://en.wikipedia.org/wiki/Squat_effect

13   'MV Herald of Free Enterprise. Report of Court No. 8074 Formal Investigation', Department of Transport, September 1987.

14   '1987: Zeebrugge heroes honoured', BBC Home: On this day, http://news.bbc.co.uk/onthisday/hi/dates/stories

15   'Carley float', Wikipedia, http://en.wikipedia.org/wiki/Carley_floats

16   Robert C. Fisher (1997), 'Within Sight of Shore: The Sinking of HMCS Esquimalt, 16 April 1945', http://www.familyheritage.ca/Articles/esquimalt

17   W. R. Keatinge (1969), *Survival in Cold Water*, Blackwell Scientific Publications, Oxford, p.ix

18   W. R. Keatinge and J. A. Nadel (1965), 'Immediate respiratory response to sudden cooling of the skin', *Journal of Applied Physiology*, 20:65-69. W. R. Keatinge, M. B. McIlroy and A. Goldfien (1964), 'Cardiovascular responses to ice-cold showers', *Journal of Applied Physiology*, 19:1145-1150

19   W. R. Keatinge (1961), 'The effect of work and clothing on the maintenance of the body temperature in water', *Quarterly Journal of Experimental Physiology*, 46:69-82

20   W. R. Keatinge (1960), 'The effects of subcutaneous fat and of previous exposure to cold on the body temperature, peripheral blood flow and metabolic rate of men in cold water', *The Journal of Physiology*, 153:166-178

21   W. R. Keatinge, C. Prys-Roberts, K. E. Cooper, A. J. Honour and J. Haight (1969), 'Sudden Failure of Swimming in Cold Water', *British Medical Journal*, 1(5642):480-483. R. E. G. Sloan and W. R. Keatinge (1973), 'Cooling rates of young people swimming in cold water', *Journal of Applied Physiology*, 35:371-375

22   'TSMS Lakonia', Wikipedia, http://en.wikipedia.org/wiki/TSMS_Lakonia

23   W. R. Keatinge (1965), 'Death after Shipwreck', *British Medical Journal*, 2(5477):1537-1541

24   W. R. Keatinge (1965), 'Death after Shipwreck', *British Medical Journal*, 2(5477):1537-1541

25   C. J. Brooks (2003), 'Survival in Cold Water: Staying Alive', Transport Canada, http://www.tc.gc.ca/publications/EN/TP13822/HR/TP13822E.pdf. Kevin H. Monahan, 'The Chilling Truth About Cold Water', *Pacific Yachting Magazine*, February 2006, http://www.shipwrite.bc.ca/Chilling_truth.htm

26   Frank Golden and Michael Tipton, *Essentials of Sea Survival*, p.57-66

27   Frank Golden and Michael Tipton, *Essentials of Sea Survival*, p.67-75 and 109-113

28   'Triple Trawler Tragedy', BBC Humberside, 24 September 2014

29   L. G. C. Pugh and O. G. Edholm (1955), 'The Physiology of Channel Swimmers', *The Lancet* 2:761-768. L. G. C. Pugh (1964) 'Deaths from exposure on Four Inns Walking Competition, March 14-15, 1964: Report to Medical Commission on Accident Prevention', *The Lancet*,

283:1210-1212. L. G. C. E. Pugh (1966), 'Accidental Hypothermia in Walkers, Climbers, and Campers: Report to the Medical Commission on Accidental Prevention', *British Medical Journal* 1(5840):123-129. L. G. C. E. Pugh (1967), 'Cold Stress and Muscular Exercise, with Special Reference to Accidental Hypothermia', *British Medical Journal*, 2(5548):333-337

30   Harriet Tuckey (2013), *Everest: The First Ascent*, Rider Books, London

31   L. G. Pugh (1968), 'Isafjordur trawler disaster: medical aspects', *British Medical Journal*, 1(6595) 826-829

32   'So many lives lost to the cruel sea', *The Yorkshire Post*, 30 January 2008

33   L. G. Pugh (1968), 'Isafjordur trawler disaster: medical aspects', *British Medical Journal*, 1(6595) 826-829

34   Richard S. Schilling (1998), *A Challenging Life: Sixty Years in Occupational Medicine*, Canning Press, London. (Book reviewed in 2000 by Jean Spencer Felton in *Occupational Medicine* 50:147-150)

35   M. L. Newhouse (1966), 'Dogger Bank Itch: Survey of Trawlermen', *British Medical Journal*, 1:1142-1145

36   R. S. Schilling (1966), 'Trawler fishing: an extreme occupation', *Proceedings of the Royal Society of Medicine*, 59:405-410

37   R. S. F. Schilling (1971), 'Hazards of deep-sea fishing', *British Journal of Industrial Medicine*, 28:27-35

38   F. P. Ellis (1970), 'Medical aspects of trawler safety', *British Journal of Industrial Medicine*, 27:78-85

39   Sir Deric Holland-Martin (Chairman) (1969), '*Trawler Safety: Final Report of the Committee of Inquiry into Trawler Safety*', HMSO, London

40   D. Hunter (1978) *Diseases of Occupations*, 6th Edn. Hodder and Stoughton, London, p.1161-1166 cited by M. S. Reilly (1985), 'Mortality from occupational accidents to United Kingdom fishermen 1961-1980', *British Journal of Occupational Medicine*, 42:806-814

41   M. S. J. Reilly (1985), 'Mortality from occupational accidents to United Kingdom fishermen 1961-80', *British Journal of Industrial Medicine*, 42:806-814

42   M. S. Reilly (1987), 'Have "formal investigations" into fishing vessel losses ceased?' *British Journal of Industrial Medicine*, 44:7-13

43   'Marine Accident Investigation Branch', Wikipedia, http://en.wikipedia.org/wiki/Marine_Accident_investigation

44   Stephen Roberts (2002), 'Seafaring – Britain's most dangerous occupation', *The Lancet*, 360:543-544. S. E. Roberts (2010), 'Britain's most hazardous occupation: commercial fishing', *Accident Analysis and Prevention*, 42:44-49. Stephen E. Roberts and Judy C. Williams (2007), 'Update of mortality for workers in the UK merchant shipping and fishing sectors', Report for the Maritime and Coastguard Agency and Ministry of Transport. 'Analysis of UK Fishing Vessel Safety 1992-2006', Marine Accident Investigation Branch, Southampton

45   'Mevagissey fisherman death: Boat capsized due to weight of catch', *Cornish Guardian*, 5 November 2014. 'RNLI Commercial Fishing Stability Films – Free surface effect', https://vimeo.com/109138195

46   Stephen E. Roberts and Judy C. Williams (2007), 'Update of mortality for workers in the UK

merchant shipping and fishing sectors', Report for the Maritime and Coastguard Agency and Ministry of Transport

47   Stephen E. Roberts and Judy C. Williams (2007), 'Update of mortality for workers in the UK merchant shipping and fishing sectors', Report for the Maritime and Coastguard Agency and Ministry of Transport

48   Peter P. Abrahams (2000), 'International Comparison of Occupational Injuries Among Commercial Fishers of Selected Northern Countries and Regions', *Proceedings of the International Fishing Industry Safety and Health Conference*, Woods Hole, Massachusetts. Jennifer Lincoln and Devin Lucas (2011), 'Preventing commercial fishing deaths in the United States', World Conference on Drowning Prevention, DaNang. P. Hasselback and C. I. Neutel (1990), 'Risk for commercial fishing deaths in Canadian Atlantic provinces', *British Journal of Industrial Medicine*, 47:498-501. T. R. Driscoll, G. Ansari, J. E. Harrison, M. S. Frommer and A. E. Ruck (1994), 'Traumatic work related fatalities in commercial fishermen in Australia', *Occupational and Environmental Medicine*, 51:612-616. A. E. Norrish and P. C. Cryer (1990), 'Work related injury in New Zealand commercial fishermen', *British Journal of Industrial Medicine*, 47:726-732

49   Peter P. Abrahams (2000), 'International Comparison of Occupational Injuries Among Commercial Fishers of Selected Northern Countries and Regions', *Proceedings of the International Fishing Industry Safety and Health Conference*, Woods Hole, Massachusetts, October 2000

50   Rear Admiral John S. Lang (2000), 'Fishing Vessel Safety – A Marine Investigator's Perspective', *Proceedings of the International Fishing Industry Safety and Health Conference*, Woods Hole, Massachusetts

51   C. J. Brooks (2008), 'All You Need to Know About Life Jackets: A Tribute to Edgar Pask', Chapter in *Survival at Sea for Mariners, Aviators and Search and Rescue Personnel*, Research and Technology Organisation, NATO

52   R. R. Macintosh and E. A. Pask (1957), 'The Testing of Life-jackets', *British Journal of Industrial Medicine*, 14: 168-176

53   Gary Enever (2005), 'Resuscitation Great: Edgar Alexander Pask – a hero of resuscitation', *Resuscitation*, 67:7-11

54   E.A. Pask (1961), 'The Design of Life-jackets', *British Medical Journal*, 2(5260): 1140-1142

55   C. J. Brooks (2008), 'All You Need to Know About Life Jackets: A Tribute to Edgar Pask', Chapter in *Survival at Sea for Mariners, Aviators and Search and Rescue Personnel*, Research and Technology Organisation, NATO

56   E. A. Pask (1961), 'The Design of Life-jackets', *British Medical Journal*, 2(5260): 1140-1142

57   M. J. Tipton and C. J. Brooks (2008), 'The Dangers of Sudden Immersion in Cold Water', Chapter in *Survival at Sea for Mariners, Aviators and Search and Rescue Personnel*, Research and Technology Organisation, NATO

58   J. S. Weiner (1970), 'Recent developments in personal protective clothing and equipment', *Proceedings of the Royal Society of Medicine* 63:1003-1005 (Volume 63:1003-1028 presented brief communications dealing with problems in several industries.)

59   M. L. Newhouse (1970), 'Protective Clothing for Fishermen', *Proceedings of the Royal Society of Medicine*, 63:1005

60   C. R. Constable (1970), 'Protective Clothing for Fishermen: Design Factors', *Proceedings of the*

Royal Society of Medicine, 63:1006-1007. G. W. Crockford (1970), 'Protective Clothing for Fishermen: Assessment of Design', *Proceedings of the Royal Society of Medicine*, 63:1007-1008

61   C. R. Constable (1970), 'Protective Clothing for Fishermen: Design Factors', *Proceedings of the Royal Society of Medicine* 63:1006-1007

62   I. H. Geving, J. Reitan, M. Sandsund, H. Faerevik, R. Reinertsen and H. Aasjord (2006), 'Safer Work Clothing for Fishermen.' *International Maritime Health* 57:1-4

63   I. H. Geving, J. Reitan, M. Sandsund, H. Faerevik, R. Reinertsen and H. Aasjord (2006), 'Safer Work Clothing for Fishermen.' *International Maritime Health* 57:1-4

64   'Fishermen's Oilskins with buoyancy', http://www.don-mor.co.uk/FishermensOilskins.page10b.html

65   'Fishermen's Oilskins with buoyancy', http://www.don-mor.co.uk/FishermensOilskins.page10b.html

66   'Irish lobsterman saved from drowning by Regatta Fishermen trousers', http://www.fishupdate.com/irish-lobsterman-saved. 'Report of the investigation into the sinking of the *FV Carraig An Iasc* on 20th January 2011', The Marine Casualty Investigation Board, Dublin, 22 December 2011

67   'Life saving suits at sea', http://www.sintef.no/home/news/sintef-fisheries-and-aquaculture

68   P. G. Schnitzer, D. D. Landen and J. C. Russel (1993), 'Occupational injury deaths in Alaska's fishing industry, 1980 through 1988', *American Journal of Public Health*, 83:685-688

69   'Background on PFD Study in Alaska Fisheries', www.cdc.gov/niosh/topics/fishing

70   'Background on PFD Study in Alaska Fisheries', www.cdc.gov/niosh/topics/fishing

71   Devin Lucas, Jennifer Lincoln, Philip Somervell and Theodore Teske (2012), 'Worker satisfaction with personal flotation devices (PDFs) in the fishing industry: evaluations in actual use', *Applied Ergonomics*, 43:747-752

72   'PDFs That Work: Overview', Centers for Disease Control and Prevention, Atlanta, GA. http://www.cdc.gov/niosh/docs/2013-131/pdfs/2013-131.pdf

73   'Background on PFD Study in Alaska Fisheries', www.cdc.gov/niosh/topics/fishing

CHAPTER 21: HAZARDS ON HOLIDAYS ABROAD – SWIMMING POOLS AND WATER PARKS

1   Roger Milward (2011), 'Safer Facilities or Using Facilities Safely', Swimming Teachers' Association Presentation at World Conference on Drowning Prevention, DaNang. http://www.sta.co.uk/downloads.php?id=197

2   Natalie Norman and Joanne Vincentin, 'Child safety: Swimming pools on holiday properties', Section in 'Protecting children and youths in water recreation: Safety guidelines for service providers', European Child Safety Alliance, http://www.childsafetyeurope.org/.../watersafetyguidelines/swimming-pools

3   P. Cornall, S. Howie, A. Mughal, V. Sumner, F. Dunstan, A. Kemp and J. Silbert (2005), 'Drowning of British children abroad', *Child: Care, Health and Development*, 31:611-613

4   Donna Watson, 'Family tell of torment over boy's pool death: Gran flies home from sun isle where police question lifeguard's absence', *Daily Record*, Glasgow, 12 June 2002

5   Henry McDonald, 'Boy, five, drowns in Disneyland Paris hotel pool while on Christmas break', *The Guardian*, 23 December 2008

6   'Door lock plea after boy drowns', BBC News, 3 August 2006

7   John H. Pearn and James Nixon (1977), 'Swimming pool immersion accidents: an analysis from the Brisbane Drowning Study', *Medical Journal of Australia* 1:432-437 (Reprinted as an Injury Classic (1997) in *Injury Prevention* 3:307-309)

8   Natalie Norman and Joanne Vincentin, 'Child safety: Swimming pools on holiday properties', Section in 'Protecting children and youths in water recreation: Safety guidelines for service providers', European Child Safety Alliance, http://www.childsafetyeurope.org/.../ watersafetyguidelines/swimming-pools

9   Peter Cornall, head of water and leisure safety at the Royal Society for the Prevention of Accidents (RoSPA), quoted in: 'Holiday drownings on the rise', *The Daily Telegraph*, 17 July 2004

10   Bryn Littleton and Nick Woods, 'Schoolboy dies in holiday resort pool', *The Journal*, Newcastle, 18 July 2003

11   Ekin Karasin, 'Devastated father tells how his four-year-old daughter drowned in Spanish villa pool', *Daily Mail*, 14 June 2017

12   Clair Weaver, 'Two-year-old Girl Dies in Holiday Pool as Her Parents Pack the Bags', *Evening Standard*, 27 July 2003

13   Paul Gallagher, 'Boy aged 3 drowns in hols pool tragedy; Parents were loading car nearby', *The Daily Mirror*, 5 August 2003

14   Daniel O'Mahony, 'Boy aged 16 months drowns in pool after slipping out through door of villa,' *Evening Standard*, 12 June 2019

15   Tom Griffiths, 'Risk Management: Killer parties', *Aquatics International*, March 2008

16   Janice Burns ,'Family's horror as tot, 3, drowns in swimming pool while on holiday in Spain', *Daily Record*, 1 July 2012

17   Benedict Moore-Bridger, John Dunne and Tom Brooks-Pollock, 'Family "devastated" at pool death of boy, 4, at Costa del Sol villa', *Evening Standard*, 19 August 2015

18   'Shock increase in Britons drowning on holiday abroad', http://www.rospa.com/news/ releases/detail/default.aspx?id=387. 'Toddler, 3, drowned 'in the blink of an eye' in swimming pool on family holiday', *Daily Mail*, 4 August 2009

19   Judith Larner, 'France sounds the alarm on safety', *The Guardian*, 5 July 2005

20   'Law and safety for your swimming pool', *The Connection* (France's English-language newspaper), http://www.connectionfrance.com/law-and-safety-for-your-swimming-pool

21   'Law and safety for your swimming pool', *The Connection* (France's English-language newspaper), http://www.connectionfrance.com/law-and-safety-for-your-swimming-pool

22   V. Carey, S. Chapman and D. Gaffney (1994), 'Children's lives or garden aesthetics? A case study in public health advocacy', *Australian Journal of Public Health*, 18:3

23   D. C. Thompson and F. Rivera (1998), 'Pool fencing for preventing drowning in children', *Cochrane Database of Systematic Reviews*, John Wiley and Son, New Jersey

24   R. A. Brenner, G. S. Taneja, D. L. Haynie, A. C. Trumble, C. Qian, R. M. Klinger and M. A. Klebanott (2009), 'Association between swimming lessons and drowning in childhood: a case-control study', *Archives of Pediatrics and Adolescent Medicine*, 163:203-210

25   Malcolm Read with Paul Wade (1997), *Sports Injuries*, Butterworth-Heinemann, Oxford, p.47. Barth A. Green, Alexander Gabrielsen, Wiley J. Hall and James O'Heir (1980), 'Analysis of swimming pool accidents resulting in spinal cord injury', *Paraplegia*, 18:94-100

26   Jack Mathieson, 'Teenager drowns trying to save boot; Spanish pool tragedy', *Daily Record*, Glasgow 11 September 2002

27   Laurie Lawrence (2006), 'Community Campaign in Australia Targeted Towards Parents and Children', Chapter in *Handbook on Drowning*, p.121-122

28   Douglas Martin, 'Water Parks Elicit Fears In Aftermath Of a Death', *New York Times*, 31 July 1997

29   'Wave Pool Drownings – Waterpark Accidents, Injuries & Deaths', http://www.texastriallawyer. com/wavepool/Default.htm

30   'Tsunami Generator Will Help Protect Against Future Catastrophe', *Science Daily*, 24 March 2010

31   Tom Harris, 'How Wave Pools Work', http://tlc.howstuffworks.com/family/wave-pool.htm. (Note: Some wave pools can generate waves more than 9 feet high – e.g. Sian Park in Tenerife.)

32   Kendra Kozen, 'New Laws Promote Aquatic Safety', *Aquatics International*, 1 September 2008, http://aquaticsintl.com/2008/sept/0809_n_newlaws.html

33   'Wavepool Lifeguard Rescue' videos on YouTube

34   Frank Pia (1998), 'Reflections on lifeguarding surveillance programs', *Parks and Recreations* Ontario, Aquatics Branch Annual Conference

35   Douglas Martin, 'Water Parks Elicit Fears In Aftermath Of a Death', *New York Times*, 31 July 1997

36   Joshua Brener and Michael Oostman, 'Lifeguards watch, but they don't always see', *World Waterpark Magazine*, May 2002

37   'Wave-pool lifeguards failed to spot lifeless child, inquest hears', CBC News, 28 June 2006

38   Robert J. Viteretti, Karen L. Lupuloff and Robert Werner, 'An investigation into the drowning of D… M… during the I.S 166 Eight Grade trip to Dorney Park and Wildwater Kingdom', Special Commissioner of Investigation for the New York City School District, November 1994

39   Tom Harris, 'How Water Slides Work', http://tlc.howstuffworks.com/family/water-slide.htm

40   Tom Harris, 'Putting the Water into Water Slides', http://tlc.howstuffworks.com/family/ water-slide4.htm

41   'Water park closed as injured swimmer is trapped in high-speed flume', *The Daily Mail*, 30 August 2010

42   Natalie Norman and Joanne Vincentin, 'Child safety: waterslides' in 'Protecting children and youths in water recreation: Safety guidelines for service providers', European Child Safety Alliance, http://www.childsafetyeurope.org/…/watersafetyguidelines/swimming-pools

43   Milton Gabrielsen (2000), 'Sliding headfirst down a water slide', Chapter in *Diving Injuries: Research Findings and Recommendations for Reducing Catastrophic Injuries*, CRC Press, BocaRaton, Florida, p.135-136

44   P. D. O. Davies, (1990) Water slide aquaplaning injury', Letter, *British Medical Journal*, 300:1401

45   David J. Ball (1998), 'Some observations on waterslide injuries', *Injury Prevention*, 4:225-227

46   Mark Henderson, 'Warning: water slides are more risky than rugby', *The Times*, 8 September 1998

47   Jesse Guerra, 'Man Dies In Water Park Slide Drowning Accident', *Swimming Pool Safety News*, 11 July 2011

48   Robert Hanley, '8-Year-Old Boy Drowns in Pool At a Water Park in New Jersey', *The New York Times*, 30 July 1997

49   Olivia Katrandjian, 'Woman Dies in Pool After Boy Reports Her Drowning To Lifeguard Who Said She Was On a Break', ABC News, 2 July 2011

50   'Waterpark Safety: Accident Prevention Measures to Avoid Injuries', http://accidentandinjuries. blo.com/waterpark-safety-accident=preve

51   Kevin Gipson, '1999-2009 Reported Circulation/Suction Entrapments Associated with Pools, Spas, and Whirlpool Tubs, 2010 Memorandum', US Consumer Product Safety Commission, Bethesda, MD. Matthew Hnatov, '2009-2013 Reported Circulation/Suction Entrapment Incidents Associated with Pools, Spas, and Whirlpool Tubs, 2014 Report', US Consumer Product Safety Commission, Bethesda, MD

52   Tom Harris, 'Putting the Water in Water Slides', http://tlc.howstuffworks.com/family/water-slide4.htm

53   'Epidemiologic Notes and Reports. Fatality at a Waterslide Amusement Park – Utah', *Mortality and Morbidity Weekly Report* (MMWR) 4 July 1986, 35(26):429-430, Centers for Disease Control and Prevention, Atlanta

54   'Boy, 7, critical after water park accident', RideAccidents.com – 2006 Accident Reports and News, http://www.rideaccidents.com/2006.html

55   'Children killed at water park', *The Independent*, 31 July 1993

56   Allison Meier, 'The Water Gardens of Otherworldly Modernism', https://hyperallergenic. com/91947/the-water-gardens-of-otherworldly-modernism

57   Thomas Corosec, 'Fort Worth seeks answers after four drown', *Houston Chronicle*, 18 June 2004

58   Tom Harris, 'How Swimming Pools Work', http://home.howstuffworks.com/swimming-pool.htm

59   'Historical Note on Attractive Nuisances in Washington', Municipal Research and Services Center of Washington (Seattle, WA), http://www.mrsc.org/Subjects/Legal/nuisances/nu-attractHist.aspx

60   Virginia Hunt Newman, *Teaching an Infant to Swim*, p.90

61   T. Llewellyn and L. Manerios (1994), 'Bruising after swimming pool filtration', Short notation and photograph in 'Minerva', *British Medical Journal*, 1 October 1994, (309: 888)

62   'Holiday teenager drowned after being sucked underwater by powerful swimming pool filter', *The Daily Mail*, 5 March 2009

63   Laura Dixon, 'Boy, 14, drowns trying to reach goggles in holiday pool vent', *The Times*, 13 July 2009

64   Shabnam Mogharabi and Bob Dumas, 'Avoiding the trap: suction entrapment is an emotional issue fraught with frightening stories and costly litigation. But just how widespread is the problem and what should the aquatics industry do about it?' *Aquatics International*, 1 June 2004

65   N. A. Toosy and B. Brookes (2005), 'Limb entrapment in a swimming pool suction outlet: A multidisciplinary approach to in-hospital extrication', *Injury Extra*, 37:225-227

66   Erin Ansley, 'Drain entrapment. Florida Code May "Create Hazards"', *Pool and Spa News*, 10 August 2012

67   'Suction-Drain Injury in a Public Wading Pool – North Carolina, 1991', *Mortality and Morbidity Weekly Report (MMWR)*, 15 May 1992/41:333-335, Centers for Disease Control and Prevention, Atlanta. 'Pool Drain Eviscerates 6-year-old girl', *Journal of Emergency Medical Services*, 19 July 2007

68   W. S. Cain, C. G. Howell, M. M. Ziegler, A. J. Finley, M. J. Asch and J. P. Grant(1983), 'Rectosigmoid perforation and intestinal evisceration from transanal suction', *Journal of Pediatric Surgery*, 18:10-13 (5 cases). C. S. Hultman and R. Morgan (1994), 'Transanal intestinal evisceration following suction from an uncovered swimming pool drain: case report', *Journal of Trauma and Acute Care Surgery* 37:843-847 . J. Juem, D. Schmeling and B. Feltis (2010), 'Transanal wading pool suction-drain injury resulting in complete evisceration of the small intestine: case report and review of the literature', *Journal of Pediatric Surgery*, 45:E1-3 . Neil R. Price, S.V. Soundappan, Anthony L. Sparnon and Danny T. Cass (2010), 'Swimming pool filter-induced transrectal evisceration in children: Australian experience', *Medical Journal of Australia* 192:534-536(3 cases). P. Debeugny, M. Bonnevalle, R. Besson and T. Basset (1990) 'A recto-sigmoid lesion due to trans-anal aspiration. Apropos of a case', (in French), *Chirurgie Pediatrique*, 31:191-194. M. Gomez-Juarez et al (2001), 'Complete evisceration of the small intestine through a perianal wound as a result of suction at a wading pool', *Journal of Trauma and Acute Care Surgery*, 51:398-399. E. Valletta, N. Zampieri, M. Fomaro, P. Biban, S. Marzini and F. S. Camogli (2007), 'Transanal intestinal evisceration from swimming pool skimmer suction: a spur to prevention', *Acta Paediatrica*, 96:1376-1377

69   A. Davison and J. W. I. Puntis (2003), 'Awareness of swimming pool suction injury among tour operators', *Archives of Disease in Childhood* 88:584-586

70   Bob Dumas, 'Troubled Waters: Suction entrapment is an emotional issue fraught with frightening stories and costly litigation. But just how widespread is the problem and what should the pool and spa industry do about it?' *Pool & Spa News* October 2003, http://www.poolspanews.com/2003/102/102entrapment.html

71   'Guidelines for Entrapment Hazards: Making Pools and Spas Safer', US Consumer Product Safety Commission, Washington, March 2005

72   'Virginia Graeme Baker Pool and Spa Safety Act', http://www.cps.gov/PageFiles/105477'vgb.pdf

73   Rebecca Robledo, 'Drained! The VGB drain cover recalls and recants have left operators exhausted', *Aquatics International*, 1 May 2012

74   'Long hair cost Amanda her life', *The Sydney Morning Herald*, 12 September 2006

75   'Guidelines for Entrapment Hazards: Making Pools and Spas Safer', US Consumer Product Safety Commission, Washington, March 2005

76   Kevin Gipson (2013), '2008-2012 Reported Circulation/Suction Entrapments Associated with Pools, Spas, and Whirlpool Bathtubs, 2013', US Consumer Product Safety Commission, Bethesda MD. Qian Zhang (2019), '2014-2018 Reported Circulation/Suction Entrapments Associated with Pools, Spas, and Whirlpool Bathtubs, 2019 Report', US Consumer Product Safety Commission, Bethesda MD. Hannah Al-Othman, 'Mother's terror as boy, five, almost

drowns during horrific hot tub ordeal', *Evening Standard* 20 June 2016

77   Bob Dumas, 'Troubled Waters: Suction entrapment is an emotional issue fraught with frightening stories and costly litigation. But just how widespread is the problem and what should the pool and spa industry do about it?' *Pool & Spa News* October 2003, http://www.poolspanews.com/2003/102/102entrapment.htm

78   A. Davison and J. W. I. Puntis (2003), 'Awareness of swimming pool suction injury among tour operators', *Archives of Disease in Childhood*, 88:584-586

CHAPTER 22: HAZARDS ON HOLIDAYS ABROAD – BY THE SEA

1   Richard Dawood (1996) *Traveller's Health: How to stay healthy abroad.* Oxford University Press, Oxford. 3rd Edn. p.vii and 211-212. P. Cornall, S.Howie, A. Mughal, V. Sumner, F. Dunstan, A. Kemp and J. Sibert (2005) 'Drowning of British children abroad', *Child: Care, Health and Development* 31:611-613. Gary Stoller, 'The No.1 Killer of Americans Abroad', 3 August 2017, https://www.forbes.com/sites/garystoller/2017/08/03/the-no-1-

2   'Raft girl rescued', *The Daily Mirror*, 16 July 2007

3   Kevin Moran (2007), 'Water safety supervision of young children at beaches', Report to the Faculty of Education Research Committee, University of Auckland, Surf Life Saving New Zealand and Watersafe Auckland. Kevin Moran (2009) 'Parent/caregiver perceptions and practice of child water safety at the beach', *International Journal of Injury Control and Safety Promotion* 16:215-221. Kevin Moran (2010) 'Watching Parents, Watching Kids: Water Safety Supervision of Young Children at the Beach.' *International Journal of Aquatic Research and Education* 4:269-277

4   Kevin Moran (2007), 'Water safety supervision of young children at beaches' Report to the Faculty of Education Research Committee, University of Auckland, Surf Life Saving New Zealand and Watersafe Auckland

5   Robert Booth, Esther Addley and Brendan de Beer, 'Holiday turns to tragedy as parents drown trying to save children in current', *The Guardian* 24 October 2007

6   David Brown and Thomas Catán, 'Huge wave claims father and son', *The Times* 22 November 2007

7   Kevin Moran (2011), '(Young) Men behaving badly: Dangerous masculinities and the risk of drowning in aquatic leisure activities', *Annals of Leisure Research* 14:260-272

8   Donal Buckley, 'HOW TO: Understand Rough Water: Force Three', http://loneswimmer.com/2012/05/21/understanding-rough-water-force-three.

9   Dennis K. Graver (2004), *Aquatic Rescue and Safety*, Human Kinetics, Champaign, p.123-124

10   Lucy Crossley, 'Cambridge student drowned on family holiday after taking warning not to swim in underwater tunnel as a 'challenge''', *Daily Mail*, 26 November 2013

11   Peter Allen, 'Tributes for British rugby coach who died after being hit by giant wave in France', *The Daily Express*, 11 August 2015

12   Peter Allen, 'British schoolboy feared drowned off coast of France', *Daily Telegraph*, 26 July 2013

13   Peter Allen, 'Seven people drown in seas off France's Mediterranean coast after 80mph winds and unpredictable currents cause treacherous conditions', *The Daily Mail*, 29 July 2013

14   'Thailand, Bali and Other Beaches and Islands: Some Tips on Not Drowning', http://tezzasthaiinfo.blogspot.co.uk/2007/04/some. 'Warning for swimmers in Thailand after series of drownings in Phuket' *The Daily Mail*, 30 July 2012. 'Missing Cambodia backpacker drowned', BBC News, 1 November 2019

15   Jonathan Howland, Ralph Higson, Thomas W. Mangione, Nicole Bell and Sharon Bak (1996), 'Why Are Most Drowning Victims Men? Sex Differences in Aquatic Skills and Behaviours' *American Journal of Public Health* 86: 93-96

16   J. S. J. Haight and W. R. Keatinge (1973), 'Failure of thermoregulation in the cold during hypoglycaemia induced by exercise and ethanol' *The Journal of Physiology* 229:87-97. Frank Golden and Michael Tipton, *Essentials of Sea Survival*, p.105 and 135-137

17   Craig Mills and Kevin Moran (2008), 'Do Alcohol and Aquatics Mix? The Context of Youth Alcohol Consumption and Aquatic Recreation', Auckland. WaterSafe Auckland Inc. www.watersafe.org

18   Frank Pia (1974), 'Observations on the drowning of nonswimmers', *Journal of Physical Education*, YMCA Society of North America

19   A. Nathanson, P. Haynes and D. Galanis (2002), 'Surfing Injuries', *American Journal of Emergency Medicine* 20:155-160

20   Margaret Donlon (2013), 'Concussion in surfers', *Surfing Medicine*, Fall 2013, Issue #26.

21   Luis A. Robles (2006), 'Cervical Spine Injuries in Ocean Bathers: Wave-related incidents', *Neurosurgery* 58:920-923

22   Jonathan Peartman, 'British surfer killed on one of Sydney's most dangerous beaches', *Daily Telegraph*, 3 July 2014

23   Cornelia Dean, 'Stalking a Killer That Lurks a Few Feet Offshore', *The New York Times* 7 June 2005. D. Morgan, J. Ozanne-Smith and T. Triggs (2009), 'Direct observation measurement of drowning risk exposure for surf beach bathers', *Journal of Science and Medicine in Sport*, 12:457-462

24   Flora Drury, 'British father-of-two is swept out to sea in front of his fiancée during family holiday on Canary Islands', *The Daily Mail*, 26 August 2015

25   'Rip Currents: Break the Grip of the Rip', http://www.ripcurrents.noaa.gov/brochures/rip_brochures_final051309.pdf. Shauna Sherer, Matt Thompson, Peter Agnew, Norm Farmer, Anthony Bradstreet, Robert Brander and Danielle Drozdzewski (2011), 'Swim or Float? An evidence-based approach to reducing the risk of rip-related drowning in Australia', World Conference on Drowning Prevention, Da Nang, Vietnam. Kirk Mason of the Great Lakes Surf Rescue Project, Michigan, 'Flip, float and follow: surviving dangerous currents,' WWMT Television News Channel 3, 18 May 2018, https://wwmt.com/news/local/flip-float-follow

26   'Swim between the flags', http://safewaters.nsw.gov.au/flags.htm

27   P. J. Fenner, S. L. Harrison, J. A. Williamson and B. D. Williamson (1995), 'Success of surf lifesaving resuscitations in Queensland, 1973-1992', *Medical Journal of Australia* 163:580-583

28   Ann Williamson, Julie Hatfield, Shauna Sherker and Rob Brander (2011), 'Why were you swimming there? Analysis of risky swimming behaviour on Australian beaches', World Conference on Drowning Prevention, DaNang.

29   Rob Taylor, 'Tourists to Australia told not to fear spiders but surf', *Reuters*, 16 September 2010

30   Julie Hatfield, Ann Williamson, Shauna Sherker and Rob Brander (2011), 'Improving beach safety: The Science of the Surf (SOS) research project', World Conference on Drowning Prevention, DaNang

31   Damian Morgan, Joan Ozanne-Smith and Tom Triggs (2008), 'Descriptive epidemiology of drowning deaths in a surf beach swimmer and surfer population', *Injury Prevention* 14:62-65

32   I. J. Mackie (1999), 'Patterns of drowning in Australia, 1992-1997', *Medical Journal of Australia*, 171:587-590

33   Jeffrey Wilks and Michael Coory (2000), 'Overseas Visitors admitted to Queensland hospitals for water-related injuries', *The Medical Journal of Australia* 173:244-246

34   Jeffrey Wilks (2000), 'Scuba Diving and Snorkelling Safety on Australia's Great Barrier Reef', *Journal of Travel Medicine* 7: 283-289

35   Richard Dawood (1996), *Traveller's Health: How to stay healthy abroad*, 3rd Edn. Oxford University Press, Oxford, p.257-261. Ben Morgan 'Coroner condemns diver training after gap-year woman dies', *Evening Standard*, 31 May 2018

36   J. L. Caruso, J. A. Hopgood, D. M. Uguccioni and P. B. Bennett (1998), 'Inexperience kills: The relationship between lack of diving experience and fatal diving mistakes', Undersea and Hyperbaric Medical Society, http://archive.rubicon-foundation.org/665

37   Steve Warren 'Growing up too fast?', http://www.mavericksdiving.co.uk/education/kids.html. N. Norman and J. Vincenten (2008), 'Child safety: Scuba diving' in 'Protecting Children and Youths in Water Recreation: Safety Guidelines for Service Suppliers', European Child Safety Alliance, Eurosafe, Amsterdam

38   David F. Colvard, 'Fathoms of Fear: A Case Study of Panic in a Recreational Scuba Diver', *Alert Diver Magazine*, Asia Pacific Edition, May/June 2009

39   Steve Warren, 'Growing up too fast?' http://www.mavericksdiving.co.uk/education/kids.html

40   Amy Iggulden, 'Father and son drown on diving holiday of a lifetime', *The Daily Telegraph*, 4 August 2006

41   Carl Edmonds, and Douglas Walter (1989), 'Diving Fatalities in Australia and New Zealand: Part 1: The human factor', *Journal of the South Pacific Underwater Medicine Society (SPUMS)* 19:94-104

42   Yehuda Melamed, Avi Shupak and Haim Bitterman (1992), 'Medical Problems Associated with Underwater Diving', *The New England Journal of Medicine* 326:30-35. Yehuda Melamed, Avi Shupak, Haim Bitterman and Daniel Weiler-Ravell (1992) 'The medical problems of underwater diving.' Letter. *The New England Journal of Medicine* 326:1498

43   Jeffrey Wilks (2000), 'Scuba Diving and Snorkelling Safety on Australia's Great Barrier Reef', *Journal of Travel Medicine* 7: 283-289

44   '79 tourists die while snorkeling in Hawaiian waters over five year period', KITV, Honolulu, 26 August 2014

45   John Pearn and John Lippman (2011), 'Snorkelling fatalities in Australia: Pathophysiological considerations', World Conference on Drowning Prevention, DaNang

46   N. Norman and J. Vincenten (2008), 'Child safety: Snorkelling' Chapter in 'Protecting Children and Youths in Water Recreation: Safety Guidelines for Service Suppliers.' European Child Safety Alliance, Eurosafe, Amsterdam.

47   'Hypoxic blackout at recreational snorkelling workplaces', http://www.deir.qld.gov.au/workplace/publications/alerts/hypoxicbl

48   Barron H. Lerner (1992), 'The medical problems of underwater diving', Letter, *The New England Journal of Medicine* 326:1498

49   Laura Nott, H. G. Reza and Molly Hennessy-Fiske, 'A strike from beneath, and a triathlete is gone', *Los Angeles Times*, 26 April 2008. Andrew Malone 'Have WE turned sharks into maneaters: Baiting leads to rise in Great White attacks', *Daily Mail*, 15 January 2010. Jonathan Paige 'Missing British tourist eaten by shark', *The Times*, 7 November 2019

50   Richard Shears, 'Scuba diver mauled to death by saltwater crocodile off coast of Australia', *The Daily Mail*, 6 December 2011

51   Roy Caldwell, 'Death in a Pretty Package: The Blue-Ringed Octopus', The Cephalopod Page, http://thecephalopodpage.org/bluering1.php

52   Peter J. Fenner and John A. Williamson (1996), 'Worldwide deaths and severe envenomation from jellyfish stings' *The Medical Journal of Australia* 165:658- 661

53   'Another Box Jellyfish Sting – Timely Reminder', http://www.lonelyplanet.com/thorntree/thread.jspa?threadID=1601542

54   N. Norman and J. Vincenten (2008), 'Child safety: Snorkelling' in 'Protecting Children and Youths in Water Recreation: Safety Guidelines for Service Suppliers', European Child Safety Alliance, Eurosafe, Amsterdam

55   Katie Zezima, 'Death Does Not Deter Jellyfish Sting' *The New York Times* 22 July 2010

56   Luca Cegolon, William C. Heymann, John H. Lange and Guiseppe Mastrangelo (2013) 'Jellyfish Stings and Their Management: A Review', *Marine Drugs* 11:523-550

57   Kenneth D.Winkel, Gabrielle M. Hawdon, Peter J. Fenner, Lisa-ann Gershwin, Allen G. Collins and James Tibbals (2003), 'Jellyfish Antivenoms: Past Present, and Future', *Journal of Toxicology, Toxin Reviews* 22: 115-127

58   Peter J. Fenner and John C. Hadok (2002), 'Fatal envenomation by jellyfish causing Urukandji syndrome', *The Medical Journal of Australia*, 177:362-363

59   Bernard Lagan, 'Killer Irukandji jellyfish threaten tourists at Great Barrier Reef', *The Times*, 3 October 2018. Lydia Lynch '"The problem will get worse": Three suspected Irukandji stings in SEQ' [South East Queensland] *The Brisbane Times*, 20 December 2018

60   'Portuguese Man-of-War: A Dangerous Ocean Organism of Hawaii', http://aloha.com/~lifeguards/portugue.html

61   'Watersports Safety Abroad', (2015) Royal Society for the Prevention of Accidents. David Szpilman and James P. Orlowski (2016), 'Sports related to drowning', *European Respiratory Review* 25:348-359

62   'Windsurfing – preventing injury', (2016) Department of Health and Human Services, State Government of Victoria, Australia. http://www.betterhealth.vic.gov.au/health/healthy

63   Natasha Donn, 'Englishman airlifted to hospital after horrific kitesurf accident in Ria de Alvor', *Portugal Press* 8 June 2015

64   G. Hummel and B. J. Gainor (1982), 'Water-skiing-related injuries', *The American Journal of Sports Medicine* 10:215-218

65   'Avoiding Propeller Strike Injuries', http://www.boat-ed.com/nv/handbook/ski.htm

66   'Water Tubing-Related Injuries Up 250 Percent', *Science Daily*, 4 February 2013

67   'British man missing after banana boat accident in Portugal', BBC News, 22 August 2018

68   'Tourist killed in pedalo accident', BBC News, 31 August 2001

69   Auslan Cramb, 'Daughter sees her parents drown on holiday isle', *The Daily Telegraph*, 9 July 2003

70   Paul Rockwell, 'Why jet skis kill: Reckless Endangerment on the Water' *In Motion Magazine*, 16 August 2001

71   N. Norman and J. Vincenten (2008), 'Child safety: Personal watercraft', Section in 'Protecting Children and Youths in Water Recreation: Safety guidelines for service providers', European Child Safety Alliance, Eurosafe, Amsterdam

72   Paul Rockwell, 'Why jet skis kill: Reckless Endangerment on the Water' *In Motion Magazine* 16 August 2001. Clare Garner (1998) '"Swimmers only" zones proposed after jet-ski deaths', *The Independent*, 8 October 1998 (Note: Government concern after a series of deaths and injuries at home and abroad.). Mark Dowdney 'Holiday girl, 17, killed as jet-ski rams her pedalo; Police chief calls for hire ban', *The Daily Mirror*, 23 October 1998 (Note: The accident happened near Tarragona on the Mediterranean coast of Spain.). Thomas Penny (2001) 'Family sees girl die in jetski collision', *Daily Telegraph*, 23 May 2002 (Note: Teenager lost control of jet ski and died instantly when she crashed into yacht at Altinkum on Aegean coast of Turkey.)

73   Gary Polson(2013), 'UK History of the Boat Kill Cord & Propeller Safety Movement', http://www.propellersafety.com/7608/history-propeller/uk-history-boat-kill-cord

74   Anna Edwards, 'Mother who lost husband and daughter in Padstow speedboat tragedy has leg amputated as it emerges safety device which could have cut engine was not properly attached', *Daily Mail*, 16 May 2013

75   Richard Smith, 'Padstow speedboat crash: Brave water-ski instructor hailed for risking life to save four from out-of-control boat' *Daily Mirror*, 7 May 2013

76   Dennis K. Graver, Aquatic Rescue and Safety, p.173-177

77   '10 Lessons From the USCG's 2009 Accident Report', http://www.boat.test.com/resources/view_news.aspx?newsid=4264

78   Frank Golden (2006) 'Let's Have A Beer Before We Start' *On Scene, The Journal of US Coast Guard Search and Rescue.* p.25-26

79   Jonathan Howland, Gordon S. Smith, Thomas Mangione, Ralph Hingson, William DeJong and Nicole Bell (1993) 'Missing the Boat on Drinking and Boating', *Journal of the American Medical Association* 270:91-9. J. Howland, T. W. Mangione and S. Minsky (1996), 'Perceptions of risks of drinking and boating among Massachusetts boaters', *Public Health Reports* 111:372-377

80   Gordon S Smith, Penelope M. Keyl and Jeffrey A. Hadley (2001), 'Drinking and Recreational Boating Fatalities. A Population-Based Case-Control Study', *Journal of the American Medical Association* 286:2974-2980

81   'Man overboard', https://en.wikipedia.org/wiki/Man_overboard. Frank Woodford (2016), 'Man Over Board – What should you consider?' https://safetyatsea.blogspot.com. Frank Woodford (2016), 'Man Overboard – Making Contact', https://safetyatsea.blogspot.com. Frank Woodford (2016), 'Man overboard – Recovering Your Casualty', https://safetyatsea.blogspot.com

82   'How to call for help at sea', Royal National Lifeboat Institution, https://rnli.org/how-to-call-for-help

83   Mario Vittone (2010), 'Drowning Doesn't Look Like Drowning', http://mariovittone.com/2010/05/15/

84   'Failure to recover a man overboard', Adventure Activities Licensing Authority, Health and Safety Executive, May 1999, http://www.hse.hse.gov.uk/aala/articles/man-overboard.htm

85   Frank Golden (2006), 'Let's Have A Beer Before We Start' *On Scene. The Journal of US Coast Guard Search and Rescue.* p.25-26

86   '10 Lessons From the USCG's 2009 Accident Report' http://www.boat.test.com/resources/view_news.aspx?newsid=4264

87   D13 Recreational Boating Statistics 1999, United States Coast Guard, http://www.uscg.mil/d13/dep/boatsafstats_99.asp. 'Choose it: Wear it. The RNLI Guide to Lifejackets and buoyancy aids', https://rnli.org

88   Richard Dawood, *Traveller's Health: How to stay Healthy abroad*, p. 218. Benedict Moore-Bridger, 'Briton Drowns on Trip of a Lifetime; Tourists Trapped in Their Cabins as Boat Sinks off Vietnam', *The Evening Standard*, 17 February 2011

89   Abigail S. Golden and Roberta E. Weisbrod (2016), 'Trends, Causal Analysis, and Recommendations from 14 Years of Ferry Accidents', *Journal of Public Transport*, 19:17-27

90   David Langton, 'Briton among eight dead in Thai cave hit by flash flood', *The Independent*, 15 October 2007. Gerard Couzens, Hannah Strange and Patrick Sawyer, 'Majorca floods: Two Britons among eight dead amid 'biblical' scenes', *Daily Telegraph*, 10 October 2018

91   Nick Allen and Rob Crilly, 'British man dies as Royal Navy helicopter rescues wife and children from capsized ship near Puerto Rico after Hurricane Maria', *Daily Telegraph*, 22 September 2017

92   Mia de Graaf, 'British father, his partner and daughter, three, feared dead in the wake of Philippines typhoon which hit during family holiday', *Daily Mail*, 15 November 2013

93   '2004 Indian Ocean earthquake and tsunami', Wikipedia, https://en.wikipedia.org/wiki/2004_Indian_Ocean_earthquake_and_tsunami

CHAPTER 23: SHALLOW WATER BLACKOUT

1   J. S. Haldane and J. G. Priestley (1905), 'The Regulation of Lung-Ventilation', *Journal of Physiology*, 32:225-266

2   'Shallow Water Blackout', Fact Sheet 23, Royal Life Saving, Australia. https://www.royallifesaving.com.au/facts-and-figures/key-facts/medical/shallow-water-blackout

3   A. B. Craig (1976) 'Summary of 58 cases of loss of consciousness during underwater swimming and diving.' *Medicine and Science in Sports* 8:171-175

4   Walter Griffiths and Tom Griffiths, 'Dying for Air', *Aquatics International*, February 2005. http://www.aquaticsintl.com/2005/feb/0502_perspectives.html

5   Arthur C. Guyton and John E. Hall, *Textbook of Medical Physiology*, p.516

6   'Oxygen saturation (medicine)', http://en.wikipedia.org/wiki/Oxygen_saturation

7   Albert B. Craig (1961), 'Underwater Swimming and Loss of Consciousness', *The Journal of the American Medical Association*, 176:225-228

8   Amanda Levy et al, 'Fatal and Nonfatal Drowning Outcomes Related to Dangerous Underwater Breath-holding Behaviours – New York State, 1988-2011', Centers for Disease Control and Prevention (CDC), Atlanta, 22 May 2015

9   Terry Maas (1997), 'Physiology Chapter: Shallow Water Blackout' in his book *Blue Water Hunting and Freediving*, Blue Water Freedivers Publishing, Ventura, CA. http://www.freedive. net/chapters/SWB3.html

10   Angela Thompson, 'Game that took Jack's life occurs every day', *Illawarra Mercury*, 15 February 2013

11   Lucy Thackray, '"He was in the shallow end – he could have just stood up": The deadly phenomenon that tragically killed this 12-year-old boy… and strikes without warning in just TWO minutes', *Daily Mail*, 26 October 2014.

12   Bruce Wigo, 'The Dangers of Underwater swimming are real – Mexican Player, Omar Ortega, Drowns at Practice', Lifesaving Resources Inc. Reprinted from: *Water Polo Scoreboard*, February 1999 (Note: The oldest son, Wolf Wigo, was a member of the American water polo team in three Olympic Games – in 1996 in Atlanta, 2000 in Sydney and 2004 in Athens. 'Wolf Wigo', Wikipedia https://en.wikipedia.wiki/Wolf_Wigo)

13   Dr Kathleen M. Belk, 'Drowning in good, healthy swimmers', *Pulse*, 30 July 1988

14   Judith Sperling, 'Coming Up for Air', *Aquatics International*, 1 February 2008

15   Leanora Minae, 'Teenager drowns at college', *St Petersburg Times*, 20 July 1999

16   Arthur C. Guyton and John E. Hall, *Textbook of Medical Physiology*, p.516

17   C. Pancaro, E. Diaz, P. Lindholm and M. Ferrigno (2007), 'Cerebral Oxygenation and Neurological Problems during Prolonged Breath-holds', Abstract of paper presented at the Undersea and Hyperbaric Medical Society Annual Scientific Meeting, 14-16 June, 2007, Hawaii.

18   G.M.Woerlee (2009), 'Anesthesia & Hypoxia', Anesthesia problems and Answers, http:// anesthesiaweb.org/hypoxia.php (Note: The water of the swimming pool, warmed to 25°C or so, offers none of the protective slowing of brain metabolism that sometimes allows the survival of a child who falls into an icy pond.)

19   Letitia Rowlands, 'Shallow water blackout kills fit, healthy dad', *Canberra Times*, 20 February 2015

20   Laura Elder, 'Swimmer Discovers Dangers of Shallow Water Blackout', *The Daily News*, Lifesaving Resources Inc. 23 November 2003, http://www.lifesaving.com/news/news-articles/news1/N_04-20-17.html

21   Laura Elder, 'Swimmer Discovers Dangers of Shallow Water Blackout', *The Daily News*, Lifesaving Resources Inc. 23 November 2003, http://www.lifesaving.com/news/news-articles/news1/N_04-20-17.html

22   P. Lindholm (2007), 'Loss of motor control and/or loss of consciousness during breath-hold competitions', *International Journal of Sports Medicine*, 28:295-299

23   Sebastian Naslund, 'Seven Sambas: Static apnoea measurements', http://www.fridykning.se/ freediving/features/samba.html

24   B. N. Davies, G. C. Donaldson and N. Joels (1995), 'Do the competition rules of synchronised swimming encourage undesirable levels of hypoxia?' *British Journal of Sports Medicine* 29:16-19

25   Dan Moran, Ming Lueng, Phil Wyman and David Williams, 'Saint Edward State Park Carole Ann Wald Swimming Pool Injury Investigation Report, April 21, 2008 Incident', Public Health – Seattle & King County. Final Report: October 2008, https://www.kingcounty. gov. Linda Quan, Bruce N. Culver and Roy R. Fielding (2010), 'Hypoxia-Induced Loss of Consciousness in Multiple Synchronised Swimmers During a Workout', *International Journal of Aquatic Research and Education* 4:379-389 (Note: The girls involved were 13 to 15 years old. Their punishing training session was rather more than a workout. It began with 28 lengths of the 25 yard pool, including crawl and butterfly, followed by a progressive hypoxic drill of four rounds, each of 4 lengths, with a brief rest between each round. On each round, the swimmers increased the distance swum underwater without coming up for air – half a length on the first round, working up to 2 lengths by the fourth round. On the night of the drama, a fifth round had been added 'to prepare the team for up-coming competitions'. The girls were coming to the end of the fifth round when they developed symptoms of marked oxygen deprivation. Four girls lost consciousness and sank. Two needed mouth-to-mouth resuscitation. One needed cardiopulmonary resuscitation.)

26   S. S. Spanoudaki, M. D. Maradaki, P. M. Myrianthefs and P. J. Baltopoulos (2004), 'Exercise induced arterial hypoxaemia in swimmers', *The Journal of Sports Medicine and Physical Fitness* 44:342-348

27   Terry Laughlin, 'Taking the Hype Out of Hypoxic', http://www.alexandriamasters.com/ articles/hypoxic.htm. Judith Sperling (2011), 'Hyperventilation and hypoxic training: What's the difference and are they both dangerous?' *Risk Management, Training & Development*, UCLA Recreation, http://www.sportrisk.com/2011/04/hyperventilation-and-hypoxic-training

28   Craig Lord, 'Shallow Water Blackout Prevention: The NCAA experience that can save Lives', 27 March 2015, http://www.swimvortex.com/shallow-water-blackout (Note: The National Collegiate Athletic Association (NCAA) is involved with sports in USA and Canada.)

29   Associated Press and Snejana Farberov, 'Dartmouth college swimmer, 21, drowns as he tries to complete four laps underwater', *Daily Mail*, 28 December 2015

30   Ernest W. Maglisco (2003), *Swimming Fastest*, Human Kinetics, Champaign, IL. p.219

31   Craig Lord, 'This month in history: when breaststroke went underwater' *Swim Vortex* 1 October 2013, http://www.swimvortex.com

32   'Breaststroke', Wikipedia, https://en.wikipedia.org/wiki/Breaststroke (Note: Since September 2005, they are also allowed one dolphin kick at the start and after each turn.)

33   Chuck Warner, 'Lessons from legends, Berkoff blast-off', https://swimswam.com/lessons-from-legends-berkoff-blast-off

34   Howard Berkes, 'Dolphin Kick Gives Swimmers Edge', National Public Radio, http://www. npr.org/templates/story/story.php?storyId=93575235

35   Robin Kiefer, '15 meter resurfacing markers – the underwater swimming rule', https://www. kiefer.com/blog/15-meter-resurfacing-markers

36   Robin Kiefer, '15 meter resurfacing markers – the underwater swimming rule', https://www. kiefer.com/blog/15-meter-resurfacing-markers

37   Jon Meoli, 'Towson teen swimmer's death was accidental, autopsy shows', *The Baltimore Sun* 13 February 2013

38   W. C. McMaster, T. Stoddard and W. Duncan (1989), 'Enhancement of blood lactate clearance following maximal swimming', *The American Journal of Sports Medicine* 17:472-476

39   B. Albert and J.R. Craig (1961), 'Causes of loss of consciousness during underwater swimming', *Journal of Applied Physiology* 16:583-586

40   Adam Fresco, 'Teenage Olympic hopeful drowns while training', *The Times*, 22 April 2006

41   'Luke's Story – What is Shallow Water Blackout?' www.shallowwaterblackout.org.uk/luke-jeffrey-memorial-trust/lukes-story (Note: This YouTube site includes a video about shallow water blackout, including thoughtful comments by Luke's mother, Dr Rik Jones and Sharon Davies, the Olympic swimmer.). 'Breathing Techniques and Over-Breathing.' http://www.londonswimming.org>category>108-health-and-safetyy

42   Terry Maas (1997) 'Physiology Chapter: Shallow Water Blackout', Section in his book *Blue Water Hunting and Freediving*, Blue Water Freedivers Publishing, Ventura, CA. http://www.freedive.net/chapters/SWB3.html. B.M. Male (1961), 'Underwater Swimming', Letter:15 July 1961, *British Medical Journal*, 2(5245):174

43   'Shallow water blackout', http://en.wikipedia.org/wiki/Shallow_water_blackout

44   'Deep water blackout', Wikipedia, http://en.wikipedia.org/wiki/Deep_water_blackout (Note: People with a diving background often use the term 'shallow water blackout' for these incidents because the skin-divers affected usually black out as they near the surface or immediately after they break the surface.)

45   Arthur C. Guyton and John E. Hall, *Textbook of Medical Physiology*, p.514

46   P. Lindholm (2007), 'Loss of motor control and/or loss of consciousness during breath-hold competitions', *International Journal of Sports Medicine* 28:295-299

47   Peter Wilmshurst (1998), 'Diving and oxygen', *British Medical Journal* 317:996-999

48   Sebastian Naslund (2008), 'How to handle a freediver suffering from blackout due to hypoxia', www.freediving.biz/education/laryngospasm.html

49   Terry Maas (1997), Physiology Chapter: Shallow-Water Blackout. Section in his *Blue Water Hunting and Freediving*. Blue Water Freedivers Publishing, Ventura, CA. http://www.freedive.net/chapters/SWB3.html

50   Peter B. Bennett and David H. Elliott (1982), *The Physiology and Medicine of Diving*. 3rd Edn. Balliere Tindall, London, p.39

51   M. J. Parkes (2005), 'Breath-holding and its breakpoint', *Experimental Physiology* 91:1-15

CHAPTER 24: HEALTH WARNINGS

1   J. P. Orlowski, A. D. Rothner and H. Lueders (1982) 'Submersion accidents in children with epilepsy' *The American Journal of Diseases of Children* 136:777-780

2   G. S. Bell, A. Gaitatzis, C. L. Bell, A. L. Johnson and J. V, V, Sander (2008), 'Drowning in people with epilepsy: how great is the risk?' *Neurology*, 71:578-582

3   D. S. Diekema, L. Quan and V. L. Holt (1993), 'Epilepsy as a risk factor for submersion injury in children.' *Pediatrics* 91:612-616

4   C. Anthony Ryan and Graeme Dowling (1993), 'Drowning deaths in people with epilepsy', *Canadian Medical Association Journal* 148:781-784

5   Niall O'Donohoe (1983), 'What should the child with epilepsy be allowed to do?' *Archives of Diseases of Childhood* 58:934-937

6   'Flickering light may have led to accident', *The Southland Times*, New Zealand 1 January 2009

7   Niall O'Donohoe (1983), 'What should the child with epilepsy be allowed to do?' *Archives of Diseases of Childhood* 58:934-937

8   J. Pearn, R. Bart and R. Yamaoka (1978), 'Drowning risks to epileptic children: a study from Hawaii', *British Medical Journal*, 2 (6147):1284-1285

9   John H. Pearn (1977), 'Epilepsy and drowning in childhood', *British Medical Journal* 1(6075):1510-1511

10   Niall O'Donohoe (1983), 'What should the child with epilepsy be allowed to do?' *Archives of Disease in Childhood* 58:934-937

11   Alison M. Kemp and J. R. Sibert (1993), 'Epilepsy in children and the risk of drowning', *Archives of Disease in Childhood* 68:684-685

12   'Teaching children with epilepsy', The Epilepsy Association, https://www.epilepsy.org.uk/teaching-children-epilepsy#.V8WnStZBVUT

13   Frank M.C. Besag (2001), 'Tonic seizures are a particular risk factor for drowning in people with epilepsy.' (Lesson of the week.) *The British Medical Journal* 322:975-976

14   G. S. Bell, A. Gaitatzis, C. L. Bell, A. L. Johnson and J. V. V, Sander (2008), 'Drowning in people with epilepsy: how great is the risk?' *Neurology* 71:578-582

15   Frank M.C. Besag, 'Re: Swimming proficiency irrelevant if person is unconscious', Letter, *British Medical Journal* 25 June 2001

16   Stephanie Milne and Andrew Cohen (2006), 'Secondary drowning in a person with epilepsy', (Lesson of the week.) *British Medical Journal* 332:775-776

17   The Cornishman, 'Teacher who died surfing "wanted to live life to the full"', *The Western Morning News*, 2 February 2012

18   John M. Lippman and John H. Pearn (2012), 'Snorkelling-related deaths in Australia, 1994-2006', *The Medical Journal of Australia*, 197:230-232

19   C. Anthony Ryan and Graeme Dowling (1993), 'Drowning deaths in people with epilepsy', *Canadian Medical Association Journal* 148:781-784. 'Teenage canoeist who died after capsizing on the Thames "after having a fit during race"', *The Daily Mail*, 7 August 2011

20   *'Epilepsy in later life',* Joint Epilepsy Council (1998) Department of Health, Whitehall, London

21   Ernest Campbell (2008) 'Epilepsy and Diving', http://www.scuba-doc.com/epildv.htm

22   K. A. Corre and R. J. Rothstein (1985), 'Assessing severity of adult asthma and need for hospitalization' *Annals of Emergency Medicine* 14:45-52 (Note: the expected FEV1 varies with the subject's age and height. In adults, the FEV1 is normally between 3 to 4 litres, but the FEV1 may fall below 2 litres in an asthma attack.)

23   K. D. Fitch and A. R. Morton (1971), 'Specificity of Exercise in Exercise-induced Asthma', *British Medical Journal* 4:577-581

24   K. D. Fitch, A. R. Morton and B. A. Blanksby (1976), 'Effects of swimming training on children with asthma', *Archives of Disease in Childhood* 51:190-194

25   'Asthma Facts and FAQs', http://www.asthma.org.uk

26   Margo Mountjoy, Ken Fitch and Mohamed Koudri (2008), 'Asthma in Aquatic Athletes', FINA (Federation Internationale de Natation) Conference 2008, http://coachesinfo.com/ index.php?option=com_content&view=a (Note: Before 2010, all Olympic athletes with asthma had to submit a detailed medical dossier and were permitted to use their inhalers only after being granted official 'Therapeutic Use Exemption'. Since 2010, athletes with asthma may use inhalers after making a simplified declaration of use. Sophie Arie (2012), 'Asthma in elite athletes. What can we learn?' *British Medical Journal* 344: 20-22)

27   K. M. Thickett, J. S. McCoach, J. M. Gerber, S. Sadhra and P. S. Burge (2002), 'Occupational asthma caused by chloramines in indoor swimming-pool air', *European Respiratory Journal* 19:827-832. Andrew B. Lindstrom, Joachim D. Piell and David C. Berkoff (1997), 'Alveolar Breath Sampling and Analysis to Assess Trihalomethane Exposures during Competitive Swimming Training', *Environmental Health Perspectives* 105:636-642

28   Margo Mountjoy, Ken Fitch and Mohamed Koudri (2008), 'Asthma in Aquatic Athletes' FINA (Federation Internationale de Natation) Conference 2008, http://coachesinfo.com/ index.php?option=com_content&view=a

29   R. U. Lee, K. M. Woessner and D. A. Mathison (2009), 'Surfer's asthma', *Allergy and Asthma Proceedings* 30:202-205

30   James D. M. Douglas (1985), 'Medical problems of sport diving', *The British Medical Journal*, 291:1224-1226

31   Yehuda Melamed, Avi Shupak, Haim Bitterman and Daniel Weiler-Ravell (1992), 'The Medical Problems of Underwater Diving', Letter, *New England Journal of Medicine*, 326:1497-1499

32   Larry D. Weiss and Keith W. Van Meter (1995), 'Cerebral Air Embolosm in Asthmatic Scuba Divers in a Swimming Pool', *Chest* 107:1653-1654

33   Yehuda Melamed, Avi Shupak Haim Bitterman and Daniel Weiler-Ravell (1992) 'Medical Problems Associated with Underwater Diving', *New England Journal of Medicine*, 326:1497-1499

34   Lawrence Martin (1997) 'Should Asthmatics Not Scuba Dive?' Scuba Diving Explained, http://www.lakesidepress.com/pulmonary/books/scuba/asthma.htm

35   P. J. S. Farrell and P. Glanvill (1990), 'Diving Practices of scuba divers with asthma', *The British Medical Journal* 300:166

36   John Parker (1991), 'The relative importance of different parts of the diving medical in identifying fitness to dive and the detection of asthma', *The Journal of the South Pacific Underwater Medical Society* 21:145-153

37   Carl Edmonds and Douglas Walker (1989), 'Scuba diving fatalities in Australia and New Zealand. 1. The human factor', *The Journal of the South Pacific Underwater Medical Society* 19:94-104

38   David Clinton-Baker (1982), 'A case of pulmonary barotrauma in an asthmatic diver', *The Journal of the South Pacific Underwater Medical Society*, 12:17-18

39   'Long QT syndrome', NHS, https://www.nhs.uk/conditions/long-qt-syndrome

40   M. J. Ackerman, D. J. Tester and C. J. Porter (1999), 'Swimming, a gene-specific arrhythmogenic

trigger for inherited long QT syndrome', *Mayo Clinic Proceedings* 74:1088-1094. T.Bradley, J. Dixon and R. Easthope (1999), 'Unexplained fainting, near drowning and unusual seizures in childhood: screening for long QT syndrome in New Zealand families' *New Zealand Medical Journal*, 112:299-302. *Drug and Therapeutics Bulletin* (2016) 'QT interval and drug therapy', *British Medical Journal*, 353:538-540

41   Grace Choi et al (2004), 'Spectrum and Frequency of Cardiac Channel Defects in Swimming-Triggered Arrhythmia Syndromes', *Circulation* 110:2119-2124

42   M. J. Ackerman and C. J. Porter (1998), 'Identification of a Family With Inherited Long QT Syndrome After a Pediatric Near-drowning', *Pediatrics* 101:306-308

43   'Torsades de pointes', http://en.wikipedia.org/wik/Torsades_de_pointes

44   M. J. Ackerman and C. J. Porter (1998), 'Identification of a Family With Inherited Long QT Syndrome after a Pediatric Near-drowning', *Pediatrics* 101: 306-308

45   Michael J. Ackerman et al (1998), 'A Novel Mutation in KVLQT1 Is the Molecular Basis of Inherited Long QT Syndrome in a Near-Drowning Patient's Family', *Pediatric Research* 44:148-153

46   Michael Ackerman, David J. Tester, C. J. Porter and William D. Edwards (1999) 'Molecular Diagnosis of the Inherited Long-QT Syndrome in a Woman Who Died after Near-Drowning' *The New England Journal of Medicine* 341:1121-1125

47   M. J. Ackerman, D. J. Tester and C. J. Porter (1999) 'Swimming, a gene-specific arrhythmogenic trigger for inherited long QT syndrome', *Mayo Clinic Proceedings* 74:1088-1094

48   Minerva (1999), *British Medical Journal* 319:1446

49   Carlo Napolitano, Sylvia G, Priori and Raffaella Bloise (2004), 'Catecholaminergic Polymorphic Ventricular Tachycardia', Gene Reviews, University of Washington, Seattle. http://www.ncbi.nih.gov/books/NBK1289/

50   Grace Choi et al (2004), 'Spectrum and Frequency of Cardiac Channel Defects in Swimming-Triggered Arrhythmia Syndromes', *Circulation* 110:2119-2124

51   Daniel Davies, 'I left him at the school gates to go on a school trip and he never came back: Heart defect killed teen after canoe capsized', *Western Mail*, Cardiff, 2 Dec 2010

52   Claire A. Martin, Gareth D. K. Matthews and Christopher L-H Huang (2012), 'Sudden cardiac death and inherited channelopathy: the basic electrophysiology of the myocyte and myocardium in ion channel disease', *Heart* 98:536-543. Maura M. Zylla and Dierk Thomas (2016) 'Inherited Arrhythmias: Of Channels, Currents, and Swimming', *Biophysical Journal*, 110:1017-1022

53   E. M. Harris, J. F, Knapp and V. Sharma (2002), 'The Romano-Ward Syndrome: a case presenting as near drowning with a clinical review.' *Pediatric Emergency Care* 8:272-275. Wojciech Zareba et al (2003), 'Implantable Cardioverter Defibrillator in High-Risk Long QT Syndrome Patients', *Journal of Cardiovascular Electrophysiology* 14:337-341

54   Peter Ott, Frank I. Marcus and Arthur J. Moss (2002), 'Ventricular Fibrillation During Swimming in a Patient with Long-QT Syndrome', *Circulation* 106:521-522

55   Grace Choi et al (2004), 'Spectrum and Frequency of Cardiac Channel Defects in Swimming-Triggered Arrhythmia Syndromes', *Circulation* 110:2119-2124

56   Peter Ott, Frank I. Marcus and Arthur J. Moss (2002), 'Ventricular Fibrillation During Swimming in a Patient with Long-QT Syndrome', *Circulation* 106:521-522. Christopher J.

Acott (2004), 'Prolonged QT syndrome: a probable cause of a drowning death in a recreational scuba diver', *The Journal of the South Pacific Underwater Medical Society* 34:209-313

57   Robin Martin (2009), 'Swimming Pools and Cardiovascular Collapse in the Young', Bristol Congenital Heart Centre, http://www.heartrhythmcharity.org.uk.

58   D. Kenny and R. Martin (2010) (Leading article.) 'Drowning and sudden cardiac death' *Archives of Disease in Childhood* 96:5-8

59   Robin Martin (2009), 'Swimming Pools and Cardiovascular Collapse in the Young', Bristol Congenital Heart Centre, http://www.heartrhythmcharity.org.uk

60   Kevin Donald, 'Council is fined over pool death', *The Journal*, Newcastle, 16 December 2003

61   Barry Nelson, 'Just a heartbeat away from tragedy', *The Northern Echo*, 3 September 2004

62   Paul James, 'Heart blip theory in double tragedy', *The Journal*, Newcastle 1 September 2004

63   Chris Miles, Zephryn Fanton, Maite Tome and Elijah R. Behr (2019), 'Inherited cardiomyopathies', *British Medical Journal* 365:157-159

64   'Hypertrophic cardiomyopathy', http://en.wikipedia.org/wiki/Hypertrophic_cardiomyopathy

65   'Arrhythmogenic Right Ventricular Cardiomyopathy Pathology', http://emedicine.medscape.com/article/2017949-overviewf

66   Gerard Couzens, 'Second British holidaymaker drowns in Canary Islands in less than 48 hours', *The Daily Mail*, 19 October 2010

67   Imogen Blake, 'Drowned photographer in Hampstead Heath bathing pond told not to swim in open water', *Hampstead and Highgate Express*, 26 December 2013

68   G. Thiene, A.Nava, D. Corrado, L Rossi and N. Pennelli (1988), 'Right ventricular cardiomyopathy and sudden death in young people', *New England Journal of Medicine*, 318:129-133

69   Domenico Corrado, Cristina Basso, Maurizio Schiavo and Gaetano Thiene (1998), 'Screening for Hypertrophic Cardiomyopathy in Young Athletes', *New England Journal of Medicine*, 339:364-369

70   Domenico Corrado et al (2006), 'Trends in Sudden Cardiovascular Death in Young Competitive Athletes After Implementation of a Preparticipation Screening Program', *The Journal of the American Medical Association* 296:1593-1601 (Note: In the Veneto region, the annual death rate from sudden cardiovascular death in athletes in the years 1979-1980 [before screening began] was 3.6 per 100,000. By 2001-2004 [after twenty years of screening] the annual death rate from sudden cardiovascular death in athletes had fallen to 0.43/100,000. In the unscreened 'non-athlete' population of the same age, the annual death rate from sudden cardiovascular death was 0.79 per 100,000 in 1979-1980, and the rate remained unchanged in 2001-2004.)

71   J.F. Goodwin (1997), 'Sudden cardiac death in the young', Editorial. *British Medical Journal*, 314:843

72   'Policy statement: Cardiac screening for professional athletes', http://www.bhf.org.uk

73   Hansard 12 March 2004, *Cardiac Risk in the Young (Screening) Bill*, House of Commons Debate vol.418 cc1755-1815

74   R. M. Campbell, S. Berger and J. Drezner (2008), 'Sudden cardiac arrest in children and young athletes: the importance of a detailed personal and family history in the pre-participation

evaluation', *British Journal of Sports Medicine*, 43:336-341. J. A. Drezner, S. Sharma, A. Baggish et al (2017), 'International criteria for electrocardiographic interpretation in athletes: Consensus statement', *British Journal of Sports Medicine*, 51:704-731 (also published in the *Journal of the American College of Cardiology* in February 2017, and in the *European Heart Journal* in April 2018.). 'New international recommendations will pave the way for more routine cardiac screening in athletes', https://www.c-r-y.org.uk/new-international-recommendations

75   Lauralee Sherwood, *Fundamentals of Physiology*, p.229

76   Chris Henwood, 'Grandfather beat cancer twice but died of a heart attack as he swam in Spanish sea; Lifeguards spent an hour trying to resuscitate family man', *Birmingham Mail*, 23 April 2012

77   Carl Edmonds and Douglas G. Walker (1999), 'Snorkelling deaths in Australia, 1987-1996', *The Medical Journal of Australia* 171: 591-594

78   John M. Lippman and John H. Pearn (2012), 'Snorkelling-related deaths in Australia, 1994-2006', *The Medical Journal of Australia*, 197:230-232

79   Mat Luebbers 'Lap swimming Etiquette', About.com Swimming, http://swimming.about.com/od/swimworkouts/qt/Lap-Swimming-Etiquette

80   *'Beating Heart Disease'*, Booklet by Health Education Council and Scottish Health Education Group 1989

81   'OAP plucked from Purley pool by a lifeguard after heart attack', *Coulsdon and Purley Advertiser*, 14 September 2012

82   Emma Wilkinson, 'My heart stopped in the pool', BBC News, 18 July 2008

CHAPTER 25: DROWNING IN THE BATH

1   Peter Davies and John Pearson, *Drownings in the Home and Garden*, Consumer Affairs Directorate, Department of Trade and Industry, London, 2001

2   'Assessing Inland Accidental Drowning Risk', RoSPA/BNFL, 3 September 2013 (Note: This study reported that '20 people drown in baths per year... with people aged over 65 years having the highest rate.')

3   Alison Kemp and J.R. Sibert (1992), 'Drowning and near drowning in children in the United Kingdom: lessons for prevention', *British Medical Journal* 304:1143-1146

4   Matt Claridge and Alan Muir (2011), 'Water Safety NZ and Plunket bath mat campaign – Practice' World Conference on Drowning Prevention, DaNang, Vietnam, May 2011

5   'Case study: Towards zero fatalities for under fives', *Report on Drowning* 2012, Water Safety New Zealand, p.7

6   Renae Rauchschwalbe, Ruth A. Brenner and Gordon S. Smith (1997), 'The Role of Bathtub Seats and Rings in Infant Drowning Deaths' *Pediatrics*, 100:e1

7   J. Sibert, N. John, D. Jenkins, M. Mann, V. Sumner, A. Kemp and P. Cornall (2005), 'Drowning of babies in bath seats: do they provide false reassurance?' *Child: Care, Health & Development*, 31:255-259

8   Renae Rauchschwalbe, Ruth A. Brenner and Gordon S. Smith (1997), 'The Role of Bathtub

Seats and Rings in Infant Drowning Deaths', *Pediatrics*, 100; e1

9   Mary Sheila Gall (2001), 'CPSC Votes to Begin Rulemaking to Improve the Safety of Baby Bath Seats', US Consumer Product Safety Commission, Washington, http://www.cpsc.gov/CPSCPUB/PREPEL/prhtml01/01163.html

10   Mary Sheila Gall (2001) 'CPSC Votes to Begin Rulemaking to Improve the Safety of Baby Bath Seats', US Consumer Product Safety Commission, Washington. http://www.cpsc.gov/CPSCPUB/PREREL/prhtml101/01163.html

11   'CPSC Approves New Federal Safety Standard for Infant Bath Seats', US Consumer Product Safety Commission, Washington. http://www.cpsc.gov/cpscpub/prerel/prhtml10/10237.html

12   J. Sibert, N. John, D. Jenkins, M. Mann, V. Sumner, A. Kemp and P.Cornall (2005), 'Drowning of babies in bath seats: do they provide false reassurance?' *Child: Care, Health & Development*, 31:255-259

13   'Drowning in baths a risk for young children warns PHE', Public Health England. Press Release, 2 February 2015, https://www.gov.uk/government/news/drowning-in-baths

14   'Baby girl drowned in bath after mother left her for a moment', *The Daily Telegraph*, 26 August 2009

15   Chris Clements, 'Grieving parents whose baby died in a bathtime accident campaign to prevent future tragedies', *The Daily Record*, 25 July 2013

16   Renae Rauchschwalbe, Ruth A. Brenner and Gordon S. Smith (1997), 'The Role of Bathtub Seats and Rings in Infant Drowning Deaths', *Pediatrics* 100;e1

17   'Mother's safety warning after baby "almost drowns" in bath incident', *The Ulster Herald*, 14 September 2013

18   Peter Davies and John Pearson, *Drownings in the Home and Garden*

19   'Baby girl, 1, drowned after falling into parents' bath as her mother slept downstairs', *The Daily Mail*, 3 September 2009

20   Jane Tyler, 'Girl of 3 in bath tragedy; She drowns trying to retrieve dummy', *Birmingham Evening Mail*, 12 June 2000

21   Peter Davies and John Pearson, *Drownings in the Home and Garden*

22   Alison Kemp and J. R. Sibert (1992), 'Drowning and near drowning in children in the United Kingdom: lessons for prevention', *British Medical Journal* 304:1143-1146

23   H. K. Simon, T. Tamura and K. Colton (2003), 'Reported level of supervision of young children while in the bathtub', *Ambulatory Pediatrics* 3:103-108

24   Sue Castle, 'Baby drowns; Tot dies as he plays with twin sister in bath', *Daily Mirror*, 9 May 2002

25   Sandra Spencer, Brenda J. Shields and Gary A. Smith (2005), 'Childhood Bathtub-related Injuries: Slip and Fall Prevalence and Prevention', *Clinical Pediatrics* 44:311-318

26   Debra Sweet (2002), 'Fatality Information for the In-Home Drowning Prevention Campaign', (Memorandum to Mark Ross), US Consumer Product Safety Commission Release 02-169, https://www.cpsc.gov/PageFiles/18414/drwnstat.pdf

27   J. H. Pearn, J. Brown, R. Wong and R. Bart (1979), 'Bathtub drownings: report of seven cases', *Pediatrics*, 64:68-70

28   R. Byard, C. de Koning, B. Blackbourne, J. Nadeau and H. F. Krous (2001), 'Shared bathing and drowning in infants and young children', *Journal of Paediatrics and Child Health* 37:542-544

29   Lloyd R. Jensen, Scott D. Williams, David J. Thurman and Patricia A. Keller (1992), 'Submersion Injuries in Children Younger Than 5 Years in Urban Utah', *Western Journal of Medicine*, 157:641-644

30   J. M. Lavelle, K. N.Shaw, T. Seidl and S. Ludwig (1995), 'Ten-year review of pediatric bathtub near-drownings: evaluation for child abuse and neglect', *Annals of Emergency Medicine* 25:344-348. Deborah Ann Mulligan and Kim Burgess (2011), 'At risk family: Correlation between drowning and child abuse/neglect in Broward County Florida', World Conference on Drowning Prevention, DaNang, Vietnam

31   Beth Hale, 'Jailed for four years, cannabis-smoking mother who downed vodka and chatted as her baby drowned in the bath', *The Daily Mail*, 30 July 2012. Nigel Bunyan, 'Childminder too drunk to stop baby drowning', *The Daily Telegraph*, 23 March 2005

32   A. M. Kemp, A. M. Mott and J. R. Sibert (1994), 'Accidents and child abuse in bathtub submersions', *Archives of Diseases in Childhood*, 70:435-438

33   Philip Delves-Broughton, 'Mother drowns her five children in bath', *The Daily Telegraph*, 21 June 2001

34   James Nixon and John Pearn (1977), 'Non-accidental immersion in bath-water: another aspect of child abuse', *British Medical Journal*, i:271-272

35   Alison Kemp and J. R. Sibert (1992), 'Drowning and near drowning in children in the United Kingdom: lessons for prevention', *British Medical Journal*, 304:1143-1146

36   Alison Kemp and J. R. Sibert (1992), 'Drowning and near drowning in children in the United Kingdom: lessons for prevention', *British Medical Journal*, 304:1143-1146

37   Lawrence D. Budnick and David A. Ross (1985) 'Bathtub-Related Drownings in the United States, 1979-81', *American Journal of Public Health* 75:630-633

38   Lawrence D. Budnick and David A. Ross (1985) 'Bathtub-Related Drownings in the United States, 1979-81', *American Journal of Public Health* 75:630-633

39   Peter Davies and John Pearson, *Drownings in the Home and Garden*

40   D. S. Diekema, L. Quan and V. L. Holt (1993), 'Epilepsy as a risk factor for submersion injury in children', *Pediatrics*, 91:612-616

41   C. Anthony Ryan and Graeme Dowling (1993), 'Drowning deaths in people with epilepsy', *Canadian Medical Association Journal* 148:781-784

42   E. Cihan, D. C. Hesdorfer, M. Bradsoy, L. Li, D. R Fowler, J. K. Graham, E. J. Donner, O. Devinsky and D. Friedman (2018), 'Dead in the water: Epilepsy-related drowning or sudden unexpected death in epilepsy?' *Epilepsia* 59:1966-1972 (Note: This study reviewed all deaths of people with epilepsy in New York City, San Diego County and Maryland over a timescale of at least 8 years. As in the Canadian study, the majority of epilepsy-related drowning deaths occurred in the bath.)

43   Stephanie Todd 'Epileptic drowns in bath after a fit.' *The Scotsman*, 21 September 2002. Brent Orton, 'Tragedy as boy, 15, found dead in bath after epileptic fit', *Express and Star*, Wolverhampton 21 July 2010

44   G. S. Bell, A. Gaitatzis, C. L. Bell, A. L. Johnson and J. V. V, Sander (2008), 'Drowning in people with epilepsy: how great is the risk?' *Neurology*, 71:578-582. D. S. Diekema, L. Quan and V. L. Holt (1993), 'Epilepsy as a risk factor for submersion injury in children', *Pediatrics* 91:612-616. Thomas Burrows, 'Epileptic woman, 21, who was scared of sleeping in case she had a seizure drowned after suffering a fit in the bath but did not have her medication', *The Daily Mail* 26 May 2015

45   Helen Weathers 'Epilepsy... Here, in a painfully poignant account, the father of one beautiful young victim asks why so little is being done', *The Daily Mail* 7 August 2001. 'Nurse, 27, drowns in bath', *Evening Chronicle*, Newcastle, 5 March 2005

46   Stuart Arnold, 'Epileptic mother hails her little hero', *The Northern Echo* 16 July 2010

47   Kim Janssen, 'Girl found mum dead in bath', *Camden New Journal* 29 July 2004

48   Guiseppe Erba (2006), 'Photosensitivity and epilepsy', Epilepsy Foundation, http://www. epilepsyfoundation.org/about/photosensitivity/gerba.cfm

49   Graham F. R. Harding and Peter M. Jeavons (1994), *Photosensitive Epilepsy*, Cambridge University Press, Cambridge

50   John H. Pearn (1977), 'Epilepsy and drowning in childhood', *British Medical Journal* i:1501-1502

51   Lauren Turner (2006), 'Mother died while bathing her little boy' *South Wales Echo*, 12 January 2006.

52   P. Satishchandra, A. Shivaramakrishana, V. G. Kaliaperumal and B. S. Schoenberg (1988), 'Hot-water epilepsy: a variant of reflex epilepsy in southern India', *Epilepsia* 1988:52-56

53   H. S. Pall and A. C. Williams (1987), 'Hot-bath epilepsy', *Postgraduate Medical Journal*, 63:975-976

54   C. D. Binnie (1988), 'Self-induction of seizures: the ultimate non-compliance', *Epilepsy Research*, Supplement 1:153-158

55   P. Satishchandra (2003), 'Hot-Water Epilepsy', *Epilepsia*, 44 (Suppl.1):29-32

56   Samuel Livingston, Lydia L. Pauli and Irving Pruce 'Epilepsy and Drowning in childhood.' Letter, *British Medical Journal*, 2 August 1977. p. 515-516

57   Alexandra Klausner and Snejana Farberov ' Top food critic drowned in the shower while having a seizure at his hotel room.' *The Daily Mail*, 28 May 2015

58   Maev Kennedy 'Bank of America intern died from epileptic seizure in shower, inquest told', *The Guardian*, 22 November 2013

59   Lawrence D. Budnick and David A. Ross (1985), 'Bathtub-Related Drownings in the United States, 1979-81' *American Journal of Public Health*, 75:630-633. Peter Davies and John Pearson *Drownings in the Home and Garden*, Consumer Affairs Directorate, Department of Trade and Industry, London, 2001

60   Alexandra Fernandez-Morera, 'Someone drowns in a tub nearly every day in America', Scripps Howard News Service, 13 April 2006

61   Philippe Lunetta, Gordon S. Smith, Pirjo Lillsunde, Erkki Vuori, Kai Valonen and Ilkka Ojanpera (2011), 'Drowning under the influence of drugs and alcohol', World Conference on Drowning Prevention, DaNang, Vietnam

62   Edwin L. Alderman and D. John Coltart (1982), 'Alcohol and the heart', *British Medical Bulletin* 38:77-80

63 Eugene C. Rich, Constance Siebold and Brian Campion (1985), 'Alcohol-related Acute Atrial Fibrillation', *Archives of Internal Medicine*, 145:830-833

64 Philip O. Ettinger, Chia F. Wu, Catalino de la Cruz, Allen B. Weisse, S. Sultan Ahmad and Timothy J. Regan (1978) 'Arrhythmias and the "Holiday Heart": Alcohol-associated cardiac rhythm disorders', *American Heart Journal*, 95:555-562

65 Edward Press (1991) 'The Health Hazards of Saunas and Spas and How to Minimize Them', *American Journal of Public Health* 81:1034-1037

66 C. Miwa, T. Matsukawa, S. Iwase, Y. Sugiyama, T. Mano, J. Sugenoya, H. Yamaguchi and KA Kirsch (1994) 'Human cardiovascular responses to a 60-min bath at 40 degrees C', *Environmental Medicine* 38-77-80.

67 'U.S. Consumer Product Safety Commission Warns Of Hot Tub Temperatures', US CPFC, Washington, 31 December 1979, http://www.cpsc.gov/en/Newsroom/News-Releases/1979/CPSC-Wa

68 V. A. Tron, V. J. Baldwin and G. E. Pirie (1985), 'Hot Tub Drownings', *Pediatrics*, 75:789-790. Debra Sweet 'Fatality Information for the "In-Home Drowning Prevention Campaign"', Consumer Product Safety Commission, Washington, DC.

69 Christian S. Shinaberger, Craig l. Anderson and Jess F. Kraus (1990) 'Young Children Who Drown In Hot Tubs, Spas, and Whirlpools In California: A 26-Year Survey', *The American Journal of Public Health* 80:613-614

70 'Toddler dies after "hot tub fall" in Bathgate', BBC News, 22 March 2012

71 Lawrence D. Budnick and David A. Ross (1985), 'Bathtub-Related Drownings in the United States, 1979-81', *American Journal of Public Health* 75:630-633

72 Virginia Center on Ageing (2010), 'Medications and Falls', http://www.seniornavigator. org/vaprovide/consumer/snArticle.do?c. Abir Mullick (2005), 'Bathing For Older People With Disabilities', Center for Inclusive Design and Environmental Access, State University of New York at Buffalo, NY. http://www.ap.buffalo.edu/idea/Publications/Bathing_for_Older_People. S. Brannan, C. Dewar, J. Sen, D. Clarke, T. Marshall and P. I. Murray (2002), 'A prospective study of the rate of falls before and after cataract surgery', *British Journal of Ophthalmology*, 87:560-562

73 Lawrence D. Budnick and David A. Ross (1985), 'Bathtub-Related Drownings in the United States, 1979-81' *American Journal of Public Health* 75:630-633

74 B. J. Vellas, S. J. Wayne, L. Romero, R. N. Baungartner, L. Z. Rubenstein and P. J. Garry (1997), 'One-leg balance is an important predictor of injurious falls in older persons', *Journal of the American Geriatrics Society* 45:735-738

75 Abir Mullick (2005), 'Bathing For Older People With Disabilities', Center for Inclusive Design and Environmental Access, State University of New York at Buffalo, NY. http://www.ap.buffalo.edu/idea/Publications/Bathing_for_Older_People

76 Susan E. Carter, Elizabeth M. Campbell, Rob W. Sanson-Fisher, Selina Redman and William J. Gillespie (1997) 'Environmental hazards in the homes of older people', *Age and Ageing* 26:195-202

77 Thomas M. Gill, Christianna S. Williams, Julie T. Robinson and Mary E. Tinetti (1999), 'A Population-Based Study of Environmental Hazards in the Homes of Older Persons', *American*

Journal of Public Health 89:553-556

78   S. L. Murphy, L. V. Nyquist, D. M. Strasburg and N. B. Alexander (2006), 'Bath transfers in older adult congregate housing residents: assessing the person-environment interaction', *Journal of the American Geriatrics Society*, 54:1265-1270

79   Susan E. Carter, Elizabeth M. Campbell, Rob W. Sanson-Fisher, Selina Redman and William J. Gillespie (1997) 'Environmental hazards in the homes of older people', *Age and Ageing*, 26:195-202. Thomas M. Gill, Christianna S. Williams, Julie T. Robinson and Mary E. Tinetti (1999) 'A Population-Based Study of Environmental Hazards in the Homes of Older Persons', *American Journal of Public Health*, 89:553-556

80   Takahito Hayashi, Kazutoshi Ago, Mihoko Ago and Mamoru Ogata (2010), 'Bath-related deaths in Kagoshima, the southwest part of Japan', *Medicine, Science and the Law* 50:11-14

81   Takahito Hayashi Kazutoshi Ago, Mihoko Ago and Mamoru Ogata (2010), 'Bath-related deaths in Kagoshima, the southwest part of Japan', *Medicine, Science and the Law* 50:11-14

82   Ryotaro Takahashi (2003), 'How bathing accidents occur among old people?' Human Care Research Group, Tokyo Metropolitan Institute of Gerontology, http://www.tmig.or.jp/topics/topics_02_1.html

83   George Orwell (1937), *The Road to Wigan Pier*, first published by Victor Gollancz, London. Issued in 1981 by Penguin Books, Harmondsworth. p.114

84   Chris Brooke, 'Doctors save baby when mother-to-be drowns in bath after nurses ignore fainting warning', *The Daily Mail*, 28 February 2008

85   'Care unit death "contributed to by neglect"', BBC News, 18 October 2015. Sara Ryan (2018), *Justice for Laughing Boy*, Jessica Kingsley Publishers, London

86   'Oxfordshire Learning Disability Services move to Oxford Health', Oxford Health NHS Foundation Trust, https://www.oxfordhealth.nhs.uk/news/oxford-learning-disability-sevices

87   Sam Russell, 'Woman who drowned in Norwich hospital bath died accidentally, inquest concludes', *Lowestoft Journal* 5 March 2016

88   Christine Cunningham, 'Norfolk and Suffolk NHS Trust fined £366k after 78-year-old patient drowned in hospital bath', *Eastern Daily Press*, Norwich, 28 November 2016

89   Julia Breen, 'Paranoid schizophrenic mother was left alone to drown in mental hospital bath', *The Northern Echo*, Darlington, 5 February 2016

90   Elizabeth R. Cluett, Ethel Burns and Anna Cuthbert (2018) 'Immersion in water during labour and birth', Cochrane Database Systematic Review, https://doi.org/10.1002/14651858.CD000111.pub4

91   Liz Baxter (2006) 'What a difference a pool makes: Making choice a reality', *British Journal of Midwifery* 14:368-372

92   Ruth E. Gilbert and Pat A. Tookey (1999), 'Perinatal mortality and morbidity among babies delivered in water: surveillance study and postal survey', *British Medical Journal* 319:483-487. Z. G. Hodgson, L. R. Comfort and A. A. Y. Albert (2019), 'Water Birth and Perinatal Outcomes in British Columbia: A Retrospective Cohort Study', *Journal of Obstetrics and Gynaecology Canada* https://doi.org/10.1016/j.jogc.2019.07.007. [Epub ahead of print.]

93   'How to labour in water', National Childbirth Trust, https://www.nct.org.uk/labour-birth/different-types-birth/water-birth

94   Zainab Kassim, Maria Sellars, and Anne Greenough (2005) 'Underwater birth and neonatal distress', Lesson of the week. *British Medical Journal* 330:1071-1072 (Note: Chest X-rays showed 'widespread bilateral patchy, ill-defined air space shadowing consistent with water aspiration.'). Ioannis N. Mannas, and Prakash Thiagarajan (2008), 'Water aspiration syndrome at birth – report of two cases', *The Journal of Maternal-Fetal and Neonatal Medicine* 22:365-367. E. Sotiridou, S. Mukhopadhyay and P. Clarke (2010), 'Neonatal aspiration syndrome complicating a water birth' *Journal of Obstetrics and Gynaecology* 30:631-633. M. G. Pinette, J. Wax and E. Wilson (2004) 'The risks of underwater birth', *American Journal of Obstetrics and Gynecology* 190:1211-1215. Sarah Nguyen, Carl Kuschel, Rita Teele and Claire Spooner (2002) 'Water Birth – A Near-Drowning Experience' *Pediatrics* 110:411-413 (Note: Chest X-ray showed 'bilateral interstitial and alveolar edema with bilateral pleural effusions.'). Lee Carpenter and Phil Weston (2011), 'Neonatal respiratory consequences from water birth', *Journal of Paediatrics and Child Health*, 48:419-423. P. Lorenz, P. Waibel, A. Malzacher (2019), 'Aspiration pneumonia in a newborn following water birth', Case of the month. Swiss Society of Neonatology. https://ww.neonet.ch/application/files/COTM_2019_03. L. Franzin, C. Scolfaro, D, Cabodi, M. Valera, P.A. Tovo (2001) 'Legionella pneumophila in a newborn after water birth: a new mode of transmission', *Clinical Infectious Diseases* 33:e103-104. https://doi.org/10.1086/323023. Manu Kaushik, Brittany Bober, Leonard Eisenfeld and Naveed Hussain (2015) 'Case Report of Haemophilus Parainfluenzae Sepsis in a Newborn Infant Following Water Birth and a Review of Literature', AJP Reports 5:e188-e192. *American Journal of Perinatology* https://www.ncbi.nlm.nih.gov/pmc/articles/PMC4603865 (Note: Initial chest X-ray showed 'a right lower lobe infiltrate, compatible with pneumonia.')

95   Georgina O'Halloran 'Birthing pool baby died from breathing water', *Irish Times*, 10 June 2010. Taguhito Nagai, Hisanori Sobajima, (2003) 'Neonatal Sudden Death Due to Legionella Pneumonia Associated with Water Birth in a Domestic Spa Bath', *Journal of Clinical Microbiology* 41:2227-2229. Roger W. Byard and Jane M. Zuccollo (2010) 'Forensic Issues in Water Birth Fatalities', *The American Journal of Forensic Medicine and Pathology* 31:258-260 (Note: The baby was infected by Pseudomonas aeruginosa during water birth.)

CHAPTER 26: CHILDREN DROWNING ON SCHOOL TRIPS

1   'Deaths of Stoke Poges Middle School pupils at Land's End.' *The Times*, 16 July 1985. Julian Fulbrook (2005) *Outdoor Activities, Negligence and the Law.* Ashgate Publishing, Aldershot. p. 57-58 (Drownings at Land's End.)

2   Martin Wainwright 'Drowned schoolgirl a non-swimmer, inquest told', *The Guardian* 7, May 2003. 'Student died in school pool accident', BBC News, 5 October 1999. Ben Quinn 'Schoolboy footballer dies in river accident.' *The Guardian*, 17 November 2008

3   Adam Sage 'Divers search lake for school trip girl', *The Times*, 4 July 2001

4   Grant Woodward 'I told the teacher she was still in the water', *Yorkshire Evening Post*, 8 May 2003

5   'Boy, 8, drowned at Scout camp', BBC News, 18 February 2000. Jaya Narain 'Girl Drowns at Her Swimming Class with Four Helpers Close By', *The Daily Mail*, 15 November 2001

6   Nicola Woolcock 'No one realised that drowning girl was missing', *The Daily Telegraph*, 21 March 2001. 'Death river did not look dangerous, teacher tells inquest', *Northampton Chronicle*, 28 August 2007. 'Lake death schoolgirl, 11, disappeared as her teachers looked after injured boy', *The Daily Mail*, 15 January 2008. Richard Savill, 'Coroner calls for safety improvements at Ten Tors inquest.' *The Daily Telegraph*, 26 October 2010. Tim Baker, 'Schoolgirl who drowned at Drayton Park theme park was on ride without teacher, inquest told', *The Evening Standard*, 4 November 2019

7   'Mystery of school trip boy found drowned in a pool', *The Daily Mail*, 29 June 1996. 'Inquest into waterfall death; Neath: Hearing to open after boy, 16, dies at beauty spot', *Western Mail*, Cardiff, 1 August 2002. Ian Herbert 'School outing ends in tragedy as boy dies in raging torrent', *The Independent*, 28 May 2002

8   Richard Savill, 'River death teacher had no safety training', *The Daily Telegraph*, 22 August 2002. Helen Carter 'Safety watchdog blames teacher for boy's drowning on school trip', *The Guardian*, 10 March 2005. Linus Gregoriadis 'Girl's drowning blamed on teacher', *The Guardian*, 8 June 2001

9   'Student died trying to save teacher', BBC News, 21 August 2002

10   Martin Wainwright, 'Tragedy highlights perils of river walks', *The Guardian*, 5 March 2002

11   Steven Morris 'Schoolgirl who drowned was forced to cross swollen stream, inquest hears', *The Guardian*, 7 December 2009

12   Lara Keay 'Rugby coach "was taking photos instead of supervising" when talented player, 17, drowned after friend pushed him into lake on school trip to Canada', *Daily Mail* 8 January 2018

13   'Girl, 11, died on water ride after joking with friends about who could get the wettest, inquest hears', *Daily Telegraph*, 4 November 2019

14   Richard Alleyne, 'Seaside tragedy of girl dance star', *The Daily Mail*, 1 August 1997

15   Andrew Norfolk, 'Teacher "ignored" girl's warning on drowning friend', *The Times*, 8 May 2003

16   Claire Jones, 'Events surrounding the Lyme Bay tragedy where four Plymouth students died shocked the city and the country', *Plymouth Herald*, 22 March 20-13. Julian Fulbrook (2005) *Outdoor Activities, Negligence and the Law*, Ashgate Publishing, Aldershot. p. 24-26 (Drownings in Lyme Bay.)

17   'Report of the investigation of the capsize of a school boat on Fountain Lake, Portsmouth with the loss of one life on 16 September 1999', Marine Accident Investigation Branch, Southampton

18   Helen Johnstone, 'Pupil drowned unnoticed in survival class', *The Times*, 11 September 1998

19   Nick Britten, 'Schoolgirl describes last moments of her friend lost in flooded cave', *The Daily Telegraph*, 18 November 2005

20   'Inquest hears of earlier accident on school trip', *The Daily Mail* 19 February 2002

21   David Sanderson '"Not enough supervision" on school outing when girl drowned', *The Times*, 17 January 2008

22   Kevin Bocquet, 'Teachers blamed for school trip deaths', BBC News, 8 March 2002

23   Elizabeth Judge, 'School criticised over boat trip girl's death', *The Times*, 21 March 2001

24   Jaya Narain, 'Teacher is jailed over boy who drowned during a school trip', *The Daily Mail* 24 September 2003

25   Nigel Bunyan 'We're not going to die, Mummy, are we?' *The Daily Telegraph*, 24 September 2003

26   Phil Revell 'Lest we forget', *The Guardian*, 17 October 2000. Julian Fulbrook (2005) *Outdoor Activities, Negligence and the Law*, Ashgate Publishing, Aldershot. p.66 (Drowning at Shell Island, North Wales.). 'School "was warned before pupil drowned"', *The Yorkshire Post*, 6 May 2003. Julian Fulbrook (2005), *Outdoor Activities, Negligence and the Law*, Ashgate Publishing, Aldershot. p.50-51 (Drowning at Le Touquet, France.)

27   Simon Midgley 'Boss is jailed over canoe deaths.' *The Independent*, 9 December 1994

28   Nigel Bunyan "Teacher who let boy jump to death is jailed', *The Daily Telegraph*, 24 September 2003

29   Nicole Martin in Boulogne 'British teacher convicted over girl's drowning', *The Daily Telegraph*, 6 April 2001

30   'Manslaughter teacher wins appeal', BBC News, 16 April 2002

31   Michael Collins (1998) *Health and Safety of Pupils on Educational Visits*. Department for Education and Environment, London. p.69-52

32   Andrew Brookes (2003), 'Outdoor education fatalities in Australia 1960-2002. Part 2. Contributing circumstances: supervision, first aid, and rescue.' *Australian Journal of Outdoor Education* 7:34-42

33   Jon Henley in Paris and Steven Morris 'Last moments of girl who died on school trip', *The Guardian*, 7 July 2001

34   Christian Gysin and Susie Boniface 'Death trip teachers could be charged', *Daily Mail*, 6 July 2001

35   'Teachers advised to boycott trips.' BBC News, 26 July 2001

36   'School trips without risk? Where's the fun in that, asks explorer.' *Worcester News*, 3 April 2009

37   *The Lion and Albert* by Marriott Edgar, in *Stanley Holloway Monologues* (1979) Elm Tree Books, London.

38   Carol Sarler 'Building character? Forget it', *The Times*, 18 November 2005

39   Julian Fulbrook *Outdoor Activities, Negligence and the Law*. p.24-26 and 30

40   Simon Midgley 'Boss is jailed over canoe deaths', *The Independent*, 9 December 1994

41   Orders of the Day. *Activity Centres (Young Persons' Safety) Bill*. Order for Second Reading read. House of Commons Hansard Debates for 27 January 1995

42   Jan Bradford (2000) 'From Lyme Bay to Licensing', http://www.aals.org.uk/lymebay01.html

43   'Adventure activities licensing', Health and Safety Executive. http://www.hse.gov.uk/aala/

44   Jan Bradford (2000) 'From Lyme Bay to Licensing', http://www.aals.org.uk/lymebay01.html

45   Rowland Woolven Pete Allinson and Peter Higgins (2007) 'Perception and Reception: The Introduction of Licensing of Adventure Activities in Great Britain', *Journal of Experiential Education* 30:1-20

46   David Ashton 'The Annual Report of the Adventure Activities Licensing Service for the year 1 April 2008 to 31 March 2009', http://www.hse.gov.uk/aala/annual-report-0809.pdf

47   Phil Revell 'A taste of risk without any recklessness', *The Times Educational Supplement* 7 February 1997

48   Julian Fulbrook (2005) *Outdoor Activities, Negligence and the Law*, Ashgate Publishing, Aldershot. p.63-65 (Drownings at Stainforth Beck.)

49 Marcus Bailie 2003) 'Lessons learned from Stainforth Beck.' Adventure Activity Licensing Authority 20 August 2003, http://www.outdoor-learning.org/news/stainforthbeck.htm

50 Nigel Bunyan 'Police inquiry into why girls were allowed to river walk', *The Daily Telegraph* 12 October 2000

51 Martin Wainwright 'Boy tells of peril of flood river', *The Guardian*, 20 February 2002

52 Kevin Bicquet 'Teachers blamed for school trip deaths', BBC News, 8 March 2002

53 Paul Stokes 'Drowning fear of girl who died on river walk', *The Daily Telegraph*, 22 February 2002

54 Helen Carter 'Teacher's rescue of third pupil in beck tragedy', *The Guardian*, 28 February 2002

55 'Teacher saw girl washed away', BBC News, 27 February 2002

56 'Weeping teacher tells of nightmare over drowned girl', *The Yorkshire Post*, 26 February 2002

57 'Schoolgirl's body found in river.' BBC News, 30 October 2000 'River Ribble drownings raise questions of safety', *The Westmorland Gazette* 13 October 2000 (Note: Thirty members of the local Cave Rescue Organisation took part in the search. One of their duty controllers, an outdoor instructor with an AALA Licence said he believed that if the group had been with a qualified adviser, the river walk would have been cancelled.)

58 Marcus Baillie 'Lessons learned from Stainford Beck?' Adventure Activity Licensing Authority 20 August 2003. http://www.outdoor-learning.org/news/stainforthbeck.htm

59 Kevin Bicquet 'Teachers blamed for school trip deaths', BBC News, 8 March 2002. Phil Revell 'Trips that end in tragedy', *The Guardian*, 11 March 2002

60 Julian Fulbrook (2005) *Outdoor Activities, Negligence and the Law*. Ashgate Publishing, Aldershot. p.64-65 (Drownings at Stainforth Beck.)

61 Sydney Young, 'Adventure hols leader is swept to death saving boy from river.' *The Daily Mirror*, 26 June 1998

62 Stephen Garsed (2005) 'Glenridding Beck investigation full report.' Health and Safety Executive. www.hse.gov.uk/aala/glenridding-beck-investigation-full-report.pdf

63 Julian Fulbrook (2005), *Outdoor Activities, Negligence and the Law*. Ashgate Publishing, Aldershot. p.216 (Drowning at Glenridding Beck.)

64 Stephen Garsed (2005), 'Glenridding Beck investigation full report', Health and Safety Executive. www.hse.gov.uk/aala/glenrlidding-beck-investigation-full-report.pdf

65 Nigel Bunyan, 'We're not going to die, Mummy, are we?' *The Daily Telegraph* 24 September 2003.

66 Nigel Bunyan 'Teacher who let boy jump to death in pool is jailed', *The Daily Telegraph*, 24 September 2003

67 Stephen Garsed (2005) *'Glenridding Beck investigation full report.'* Health and Safety Executive. www.hse.gov.uk/aala/glenridding-beck-investigation-full-report.pdf

68 Marcus Bailie (1996) 'Risk Assessments, Safety Statements, and all that Guff.' *Far Out – Practical and Informative Adventure Education* 1:6-7, quoted by Julian Fulbrook in *Outdoor Activities, Negligence and the Law*, p.213 (Drowning at Glenridding Beck.)

69 Andrew Norfolk, 'Boy on school trip drowned in cave notorious for flooding', *The Times*, 6 November 2007

70 'Nidderdale', http://ukcaving.com/wiki/index.php/Nidderdale

71   'Manchester Hole', http://ukcaving.com/wiki/index.php/Manchester_Hole

72   'Council "failed boy, 14, who drowned on school trip to Dales cave"', *The Yorkshire Post*, 14 April 2010

73   'Re: Manchester Hole death in the news' http://ukcaving.com/board/index.php?topic=10030.0

74   'Nidderdale', http://ukcaving.com/wiki/index.php/Nidderdale

75   'Our fight for survival in flooded cave – teacher', *The Yorkshire Post*, 8 November 2007

76   'The chilling words of an instructor to 11 children on tragic caving trip', *The Press* (York), 18 November 2005

77   Andrew Norfolk 'He took a deep breath and went into the water. He never came out', *The Times*, 18 November 2005

78   'Cave rescuer found body of schoolboy, inquest told', *The Yorkshire Evening Post*, 5 November 2007

79   'Council denies safety breaches over boy's death in flooded cave', *The Yorkshire Post*, 27 February 2009

80   'Council cleared over schoolboy's potholing death', *The Guardian*, 25 May 2010

81   *Education outside the Classroom* (2005). Second Report of Session 2004-2005: Report, Together with Formal Minutes, Oral and Written Evidence. The Stationery Office, London.

82   *Memorandum submitted by the Adventure Activities Licensing Authority to the House of Commons Education and Skills Committee: Written Evidence: Ev 140 Education outside the Classroom* (2005). (Note: In addition to the 2 adults and 22 children who died of drowning, 1 teacher and 12 children died when their mini-bus ran into the back of a stationary vehicle on the hard shoulder of the M40, 1 teacher died on a French ski slope, 1 teacher died of natural causes, 1 child was murdered, 1 child died from acute asthma, 1 child died after a drinking session, 2 children died from sudden cardiac arrhymias, 3 children died in falls, 4 children died in road traffic accidents, and 6 children died in skiing or sledging accidents.)

83   *Memorandum submitted by the Adventure Activities Licensing Authority to the House of Commons Education and Skills Committee: Written Evidence: Ev 137. Education outside the Classroom* (2005).

84   Sir Chris Bonington, quoted by Paul Revell in 'Lest we forget', *The Guardian*, 17 October 2000

85   'School trips without risk? Where's the fun in that, asks explorer', *The Worcester News*, 3 April 2009

86   Julian Fulbrook (2011) *Outdoor Activities, Negligence and the Law.* p.65 (Drowning at Gullett Quarry in 2001.) Sam Greenhill and Chris Brooke 'Fatal short cut of pupil drowned on trip to the alps; Police quiz teachers after 17-year-old is swept away by river', *The Daily Mail*, 17 July 2003

87   Tim Gill 'The end of zero risk in childhood', *The Guardian*, 3 July 2011

88   Andrew Brookes (2002) 'Outdoor education fatalities in Australia 1960-2002 Part 1. Summary of incidents and introduction to fatality analysis.' *Australian Journal of Outdoor Education* 7:20-35

89   'The Birth of Outward Bound', http://www.outwardbound.net/about/history/ob-birth.html

90   J.W. Hogan (1977) 'The Need for Adventure Versus Safety.' Chapter in *Children, the Environment and Accidents*, R.H. Jackson (ed.) Pittman Medical Publishing, Tunbridge, Kent, p.114

91   J.W. Hogan (1977) 'The Need for Adventure Versus Safety.' in *Children, the Environment and Accidents*. p.118

92   Rob Hogan (2002) 'The Crux Of Risk Management In Outdoor Programs – Minimising The Possibility Of Death And Disabling Injury', *Australian Journal of Outdoor Education* 6:71-79

93   Stephen Garsed (2005) '*Glenridding Beck investigation full report*.' Health and Safety Executive. www.hse.gov.uk/aala/glenridding-beck-investigation-full-report.pdf

94   '*Glenridding Beck – Conclusions*.' http://products.ihs.com/Ohsis-SEO/829430.html

95   *Education outside the Classroom* (2005). Second Report of Session 2004-2005: Report, Together with Formal Minutes, Oral and Written Evidence. The Stationery Office, London. p.3. *Transforming Education Outside the Classroom* (2010) Sixth Report of Session 2009-10: Report, together with formal minutes, oral and written evidence', The Stationary Office, London, p.15

96   'Teacher cleared over girl's death.' BBC News, 14 October 2004

97   Nigel Bunyan 'Teacher who let boy jump to death in pool is jailed.' *The Daily Telegraph*, 24 September 2003

98   'Council denies safety breaches over boy's death in flooded cave', *Yorkshire Post*, 27 February 2009

99   'North Yorkshire County Council acquitted over death of Tadcaster schoolboy', *The Press*, (York) 26 May 2010

100   'North Yorkshire County Council acquitted over death of Tadcaster schoolboy', *The Press*, (York) 26 May 2010

101   'Group Safety at Water Margins' (2003), Department for Education and Skills and the Central Council for Physical Education, London

102   'Council for Learning Outside the Classroom: History.' https://www.lotc.org.uk/about/history. 'LOtC Quality Badge.' https://www.lotc.org.uk/lotc-accreditations/lotc-quality-badge

103   'Training for leadership of outdoor learning and educational visits.' http://www. outdooreducationaladvisers.co.uk/training

104   'About OEAP: National Guidance for the management of outdoor learning, off-site visits and learning outside the classroom.' https://oeapng.info/guidance-documents

105   About OEAP: National Guidance for the management of outdoor learning, off-site visits and learning outside the classroom.' https://oeapng.info/guidance-documents '7i Group Safety at Water Margins'. (Note: Recently, the Council for Learning Outside the Classroom and the Outdoors Educational Advisers' Panel announced their intention to develop a much closer partnership, strengthening and extending the support they offer to schools. 'CLOtC and OEAP embark on a closer partnership to strengthen support for schools.' https://www.lotc. org.uk/clotc-and-oeap-embark-on-a-closer-partnership.

106   *Planning and Leading Visits and Adventurous Activities – Guidance for schools and colleges teaching children and young people from 5 to 18 years*, RoSPA, https://www. rospa.com/docs/ school-college-safety/school-visits-guide

107   Dave Grant 'From Lyme Bay to Licensing to De-Regulation? The current state of safety within the outdoor activities sector.' Rural Policy Centre. Scotland's Rural College March

2013. 'Schools and adventurous activity providers – have your say on the Adventure Activities Licensing review.' https://www.lotc.org.uk/schools-and-adventurous-activity-providers

108 'Health and safety on educational visits.' November 2018, https://gov.uk/government/publications/health-and-safety-on-educational-visits

109 'InfoLog – AALA – HSE.' http://www.hse.gov.uk/aala/articles/index.html

110 'The Investigation of Accidents to General Aviation Aircraft.' (2010) Air Accidents Investigation Branch, Department of Transport.

111 Caroline Davies 'Schools urged to give pupils swimming tests after boy drowns on trip', *The Guardian*, 10 January 2018

112 Regulation 28 Report to Prevent Future Deaths sent to The Rt. Hon Damian Hinds, Secretary of State for Education.

113 Polly E. Bijur (1995) 'What's in a name? Comments on the use of the terms "accident" and "injury"', *Injury Prevention* 1:9-11. Ronald M. Davis and Barry Pless (2001) 'BMJ bans "accidents": Accidents are not predictable', Editorial in *British Medical Journal* 322:1320-132. (Note: This editorial provoked a massive correspondence for and against the decision to ban the use of the word 'accident'.)

114 Phil Revell 'How can we stop these tragedies?' *The Times Literary Supplement*, 13 July 2001 (He was quoting Val Sumner, spokesman for the Royal Life Saving Society.)

115 Ambrose Bierce The Devil's Dictionary, in *The Collected Writings of Ambrose Bierce*. The Citadel Press, New York. 2nd Edn. 1963. p.192

CHAPTER 27: THE GLOBAL DEATH TOLL OF DROWNING

1 '2004 Indian Ocean earthquake and tsunami', Wikipedia. https://en.wikipedia.org/wiki/2004_Indian_Ocean_and tsunami

2 *Global Report on Drowning: Preventing a Leading Killer.* (2014) Edited by David Meddings, Adnan Hyder, Joan Ozanne-Smith and Aminur Rahman. World Health Organisation. http://www.who.int/violence_injury_prevention/publications/drowning

3 Adam Taylor 'Why the number of refugees drowning in the Mediterranean keeps rising', *The Washington Post*, 3 June 2016

4 Mark Whittaker 'The Silent Epidemic', BBC World Service, 12 February 2013. http://www.bbc.co.uk/programmes/p0143r60

5 *'Children Drowning: Drowning Children.'* (2009) Unicef Bangladesh. swimsafe.org/wp-content/uploads/2009/09/Drowning-in-Bangladesh.pdf

6 M. Kapil Ahmed, Mizanur Rahman and Jeroen van Ginneken (1999) 'Epidemiology of child deaths due to drowning in Matlab, Bangladesh.' *International Journal of Epidemiology* 28:306-311. Aminur Rahman, A. K. M. Fazlur Rahman, Shumona Shafinaz and Michael Linnan (2005) 'Bangladesh Health and Injury Survey: Report on Children.' http://unicef.org/bangladesh/Bangladesh_Health_and_Injury_Survey. A. Hyder, S. Arifeen, N. Begum, S. Fishman, S. Wali and A. Baqui (2003) 'Death from drowning: defining a new challenge

for child survival in Bangladesh.' *International Journal of Injury Control and Safety Promotion* 10:205-210

7   Aminur Rahman, A. K. M. Fazlur Rahman, Shumona Shafinaz and Michael Linnan (2005) 'Bangladesh Health and Injury Survey: Report on Children.' http://unicef.org/bangladesh/ Bangladesh_Health_and_Injury_Survey

8   A. Rahman, S. Mashreky, S. Chowdhury, M. Glashuddin, I. Uhaa, S. Shafinaz, M. Hossain, M. Linnan and F. Rahman (2009) 'Analysis of the childhood fatal drowning situation in Bangladesh: exploring prevention measures for low-income countries.' *Injury Prevention* 15:75-79

9   'Ganges Delta.' https://en.wikipedia.org/wiki/Ganges_Delta/pdf

10 J. Mayaux, M. Ali, J.Chakraborty and A. de Francisco (1997) 'Flood control embankments contribute to the improvement of health status of children in Bangladesh', *Bulletin of the World Health Organisation* 75:533-539. '1998 Bangladesh floods', Wikipedia https://en.wikipedia. org/wiki/1998_Bangladesh_floods.pdf

11 Peter Foster 'Bangladesh cyclone death toll hits 15,000', *The Daily Telegraph*, 18 November 2007

12 Michael Linnan, Cuong Viet Pham, Linh Cu Le, Phuong Nhan Le and Anh Vu Le (2003). 'Report to UNICEF on the Vietnam Multi-center Injury Survey.' Hanoi School of PublicHealth, Vietnam. http://www.swimsafe.org/wpcontent/uploads/2009/09/ VietnamUNICEFfinalVSMI. Favid Penson and John Pearn 'Going to school – Drowning on the way.' World Conference on Drowning Prevention, Da Nang, Vietnam 2011

13 Michael Linnan, Cuong Viet Pham, Linh Cu Le, Phuong Nhan Le and Anh Vu Le (2003) '. Report to UNICEF on the Vietnam Multi-center Injury Survey.' Hanoi School of Public Health, Vietnam. http://www.swimsafe.org/wpcontent/uploads/2009/09/ VietnamUNICEFfinalVSMI

14 Orapin Laosee, Ratana Somrontong, Tracie Reinten-Reynolds, Michael Linnan and Chitri Sitti-amorn. 'Child drowning in Thailand.' World Conference on Drowning Prevention, DaNang, Vietnam 2011.

15 Wijaya Godakumbura 'Drowning in a developing country in Asia. Its nature and the support that such countries need.' World Conference on Drowning Prevention, Da Nang Vietnam 2011.

16 James Vaughan (2014) 'Global drowning: The silent epidemic in need of serious attention.' Royal National Lifeboat Institution. http://www.rnli.org/aboutus/international/documents/ rnliinternational-thinkpiece.pdf

17 Tim Weiner and Lydia Polgreen 'Grief as Haitians and Dominicans Tally Flood Toll', *New York Times*, 28 May 2004

18 David Szpilman 'Drowning deaths in Brazil: Can we trust our database of death certificates concerning place and circumstance?' World Conference on Drowning Prevention, Da Nang Vietnam 2011

19 D. Sewduth 'The data related to the cases of drowning in South Africa and the implications of this', World Conference on Drowning Prevention, Oporto, Portugal 2007

20 Job Kania 'Drowning Mortality in Kenya', World Conference on Drowning Prevention, Potsdam, Germany 2013

21   'Lake Victoria disaster: many dead after pleasure boat sinks', *The Guardian* 25 November 2018. 'Uganda boat capsize: Dozens feared dead in Lake Albert', BBC News, 20 May 2019. Hudson Apunyo 'Eight drown in Lake Kyoga.' *New Vision*, Uganda 30 January 2019. Jerome Dralega 'Desk research into communities at risk of drowning – Case of Uganda in East Africa.'. World Conference on Drowning Prevention, Da Nang, Vietnam 2011

22   Sia E. Msuya, Emma L. Msuya, Mary V. Mosha, Jackie Urio and Robert K. Stallman (2011). 'Establishing a surveillance system to record drowning incidence in a low income country: A top down-bottom up approach.' World Conference on Drowning Prevention, Da Nang, Vietnam 2011

23   'Centre for Injury Prevention and Research, Bangladesh', International Life Saving Federation. http://www.ilfs.org/about/members/ciprb

24   A.K.M. Fazlur Rahman, Aminur Rahman, Saidur Rahman Mashreky and Michael Linnan 'Evaluation of PRECISE: a comprehensive Child Injury Prevention Program in Bangladesh. The First Three Years: 2006-2008.' https://www.unicef.org/evaluation/files/injury-prevention-programme-evaluation.pdf

25   'International Drowning Research Centre- Bangladesh (IDRC-B)', http://www.ciprb.org/centres/international-drowning-research

26   A. Rahman et al (2009) 'Analysis of the childhood fatal drowning situation in Bangladesh: exploring prevention measures for low-income countries.' *Injury Prevention* 15:75-79

27   'Saving of Lives from Drowning (SoLiD), Bangladesh', John's Hopkins International Injury Research Unit. Baltimore, MD. http://www.jhph.edu/research/centers-and-institutes. Mike Ives 'Troubled waters.' *Johns Hopkins Magazine*, 1 March 2012, http://hub.jhu.edu/magazine/2012/spring/troubled-waters

28   Michael Linnan et al (2012) 'Child Drowning: Evidence for a newly recognised cause of child mortality in low and middle income countries in Asia.' Working Paper 2012-07, Special Series on Child Injury No. 2. Unicef Office of Research, Florence. 'SwimSafe: Preventing child drowning in Asia through teaching survival swimming skills. Interventions Part 1: Creches.' http://swimsafe.org/2013/05/interventions-part-1

29   Saidur Rahman Mashreky, Kamran ul Baset, Fazlur Rahman and Aminur Rahman, 'The social autopsy – A tool for community awareness after a drowning event', World Conference on Drowning Prevention, Da Nang, Vietnam 2011

30   Danyel Walker, Aminur Rahman, Jahanjir Hossain and Saidur Rahman Mashreky 'Behavioural change communications for drowning prevention in low literacy environments', World Conference on Drowning Prevention, Da Nang, Vietnam 2011

31   'SwimSafe: Preventing child drowning in Asia through teaching survival swimming skills.' http://swimsafe.org/swimsafe-projects/bangladesh.'Preventing drowning: an implementation guide', David Meddings (ed.) (2017), World Health Organisation, Geneva. p.95-96

32   'SwimSafe: Preventing child drowning in Asia through teaching survival swimming skills.' http://swimsafe.org/swimsafe-projects/bangladesh

33   David Bergman 'Bangladesh: Where swimming lessons are a matter of life and death', *The Independent*, 23 March 2012

34   Raffat Binte Rashid 'In Bangladesh, children learn how to swim – and how to survive.' Unicef: Bangladesh. http://www.unicef.org/infobycountry/bangladesh_70629.html

35 Michael Linnan et al (2012) 'Child Drowning: Evidence for a newly recognised cause of child mortality in Low and Middle income countries in Asia.' Working Paper 2012-07, Special Series on Child Injury No. 2. Unicef Office of Research, Florence.

36 Jahangir Hossain, Tom Mecrow, Aminur Rahman, Nahida Nusrat, Justin Scarr and Michael Linnan. 'Moral Hazard and SwimSafe – The early results are in and it does not increase risk-taking', World Conference on Drowning Prevention, Da Nang, Vietnam 2011. T. S. Mecrow, M. Linnan, A. Rahman, J. Scarr, S. R. Mashreky, A. Talab and A. K. Rahman (2015), 'Does teaching children to swim increase exposure to water or risk-taking when in water? Emerging evidence from Bangladesh.' *Injury Prevention*, 21:185-188. Abantee Harun, Shumona Shafinaz, Fazlur Rahman, Aminur Rahman and Michael Linnan, 'The process of learning natural swimming.' World Conference on Drowning Prevention, Da Nang, Vietnam 2011

37 Tom Stefan Mecrow, Aminur Rahman, Michael Linnan, Justin Scarr, Saidur Rahman Mashreky, Abu Talab and A. K. M. Fazlur Rahman (2013) 'Children reporting rescuing other children in rural Bangladesh: a descriptive study.' *Injury Prevention*, 21:51-55

38 Saad Hammadi in Sirajganj 'Teaching children swimming and CPR to save lives in Bangladesh', *The Guardian* 28 July 2016

39 Tom Mecrow, Aminur Rahman, Nahida Nusrat, Fazlur Rahman and Justin Scarr 'Barriers to CPR training in a rural LMIC setting', World Conference on Drowning Prevention, Da Nang, Vietnam 2011

40 Nahida Nusrat, Tom Mecrow, Aminur Rahman, Fazlur Rahman Justin Scarr and Michael Linnan, 'Large-scale community training in CPR as a basis for a community response system in an LMIC', World Conference on Drowning Prevention, Da Nang, Vietnam 2011

41 *Global Report on Drowning: Preventing a Leading Killer.* (2014) Edited by David Meddings, Adnan Hyder, Joan Ozanne-Smith and Aminur Rahman. World Health Organisation. p.14 http://www.who.int/violence_injury_prevention/publications/drowning

42 Karen McVeigh in Barisal '"Superstition" prevents action against child drowning in Bangladesh', *The Guardian*, 30 December 2016

43 N. N. Borse, A. A. Hyder, P. K. Streatfield, S. E. Arifeen and D. Bishar (2011) 'Childhood drowning and traditional rescue measures: case study from Matlab Bangladesh', *Archives of Disease in Childhood* 96:675-680. Karen McVeigh in Barisal '"Superstition" prevents action against child drowning in Bangladesh', *The Guardian*, 30 December 2016

44 Tom Mecrow, Aminur Rahman, Nahida Nusrat, Fazlur Rahman and Justin Scarr 'Barriers to CPR training in a rural LMIC setting.' World Conference on Drowning Prevention, Da Nang, Vietnam 2011. T. S. Mecrow, A. Rahman, S. R. Mashreky, F. Rahman, N. Nhusrat, J. Scarr and M. Linnan (2015) 'Willingness to administer mouth-to-mouth ventilation in a first response program in rural Bangladesh.' BMC International Health and Human Rights 15:19

45 Laura Fennimore 'RNLI lifeguards give lifesaving training to help tackle drowning in Bangladesh.' 'Royal National Lifeboat Institute: International Development.' 28 February 2012 http://rnli.org/aboutus/International/Pages/international-development.aspx. Jahangir Hossain, Tom Mecrow, Steve Wills, Aminur Rahman and Fazlur Rahman 'Effectiveness of

SeaSafe lifeguarding in Cox's Bazar beach, Bangladesh.' World Conference on Drowning Prevention, Penang, Malasia 2015

46   Philly Byrde 'Bangladesh lifesavers join RNLI flood rescue training in Scotland.' https://rnli. org/news-and-media/2016/september/21/bangladesh-lifesavers

47   William Kremer 'Pete Peterson: The ex-POW teaching Vietnam to swim', BBC World Service, 2 March 2013

48   'SwimSafe: Preventing child drowning in Asia through teaching survival swimming skills. Da Nang, Vietnam'. http://swimsafe.org/swimsafe-projects/danang. Ross Cox, Amy Peden and Tamara Rubin 'The use of portable pools to increase swim learning: The SwimSafe DaNang Experience.' World Conference on Drowning Prevention, Da Nang, Vietnam 2011

49   'SwimSafe. Preventing child drowning in Thailand.' http://swimsafe.org/SwimSafe-Thailand-information-brochure.pdf. 'SwimSafe programme, Bangladesh, Thailand and Viet Nam.' *Preventing drowning: an implementation guide.* Edit. David Meddings. (2017). World Health Organization, Geneva. p.39-40. Andrew Stenning 'Life lessons: taking on Thailand's shocking drowning statistics.' *The Guardian,* 9 October 2013

50   Soewarta Kosen, Michael Linnan, Tracy Reinten-Reynolds, Ibu Ingan and Ibu Endang 'Drowning mortality in the Asian Tsunami and correlates of survival.' World Conference on Drowning Prevention, Da Nang, Vietnam 2011

51   Michael Linnan, Tracie Reinten-Reynolds, Soewarta Kosen, Ibu Ingan and Ibu Endang 'Protective effect of swimming in a tsunami disaster and its protection of caregivers and children.' World Conference on Drowning Prevention, Da Nang, Vietnam 2011

52   '2004 Indian Ocean earthquake and tsunami.' Wikipedia. https://en.wikipedia.org/wiki/2004_Indian_Ocean_and tsunami

53   'The tsunami's impact on women.' Oxfam Briefing Note. www.preventionweb.net/files/1502_bn050326tsunamiwomen.pdf

54   Tom Parker 'Tsunami prompts women's swimming lessons.' BBC News, 6 May 2005

55   Christine Fonfe and Michael Fonfe 'Breaking Cultural Barriers to Asian Women Swimming.' World Conference on Drowning Prevention, Potsdam, Germany 2013

56   Christina Fonfe and Michael Fonfe 'The Sri Lanka Women's Swimming Project.' World Conference on Drowning Prevention, Da Nang, Vietnam 2011

57   Christina Fonfe and Michael Fonfe 'The Sri Lanka Women's Swimming Project.' World Conference on Drowning Prevention, Da Nang, Vietnam 2011

58   Laura Davies 'Ten years of the Sri Lanka Women's Swimming Project.' http://blogs.fco.gov. uk/lauradavies/2014/12/18/ten.

59   Pete Grant 'Swimming instructor rewarded by Queen after causing a splash in Sri Lanka.' *Bucks Free Press,* 18 June 2015

CHAPTER 28: A DROWNING ACCIDENT EXPLAINED

1   Frank Golden and Michael Tipton *Essentials of Sea Survival.* p.71-75

2   A. C. Guyton and J. E. Hall *Textbook of Medical Physiology.* p.494-495 (Note: This normal

recruitment of additional capillaries ensures that pulmonary blood pressure remains low.)

3   Mike Stroud (2004) *Survival of the Fittest*. Yellow Jersey Press, London. p. 42-43

4   Lauralee Sherwood *Fundamentals of Physiology*. p.156-158

5   Lauralee Sherwood *Fundamentals of Physiology*. p.353-355. M. J. Parkes (2005) 'Breath-holding and its breakpoint.' *Experimental Physiology* 91:1-15 (Note: The normal partial pressure of carbon dioxide in arterial blood is 40 mm Hg. A rise to 55 mm Hg triggers an overwhelming urge to breathe.)

6   Emilio Agostoni (1963) 'Diaphragm activity during breath holding: factors related to its onset.' *Journal of Applied Physiology* 18:30-36

7   Frank Golden and Michael Tipton *Essentials of Sea Survival*. p.83-84

8   Lauralee Sherwood *Fundamentals of Physiology*. p.356

9   'Larynx.' Wikipedia. https://en.wikipedia.org/wiki/Larynx

10   A. C. Guyton and J. E. Hall *Textbook of Medical Physiology*. p.804-805

11   Lauralee Sherwood. *Fundamentals of Physiology*. p.428-429

12   A. C. Guyton and J. E. Hall *Textbook of Medical Physiology*. p.805

13   Frank Golden and Michael Tipton *Essentials of Sea Survival*. p.84

14   Lauralee Sherwood *Fundamentals of Physiology*. p.428 (Note: On rare occasions, laryngospasm persists after a swimmer regains the surface. In 2007, a skilled white-water rafter described his narrow escape after his raft overturned in an Oregon river, plunging him underwater. Before he could regain the surface, he had to push the raft out of his way Desperate to take a breath, he found himself unable to pull air into his lungs. Instead of grabbing the raft, he grabbed his throat, but no matter how hard he tried, he could neither breathe nor cough. He knew that he would die if he blacked out. Supported by his life jacket, he managed to float downstream until the current carried him towards the bank. As he climbed out of the river, the laryngospasm suddenly relaxed, and he was able to breathe again. Walt Bammann (2007) 'Dry Drowning', http://www.nrs.com/safety_tips/dry_drowning.asp)

15   F. G. Banting, G. E. All, J. M. Janes, B. Leibel and D. W. Lougheed (1938) 'Physiological Studies in Experimental Drowning (A).' *Canadian Medical Association Journal* 39:226-22 (Note: A larger volume of water sometimes brought about such a sustained laryngospasm that the animal died of asphyxiation. It used to be thought that 10-15% of the victims of fatal drowning suffered such prolonged laryngospasm that they died from asphyxiation without inhaling water [so-called 'dry drowning']. More recent research indicates that almost all victims of fatal drowning inhale water into their lungs.). P. Lunetta, J.H. Modell and A. Sajantila (2004) 'What is the incidence and significance of "dry-lungs" in bodies found in water?' *The American Journal of Forensic Medicine and Pathology* 25:291-301. Angela D. Levy et al (2007) 'Virtual Autopsy: Two- and Three-dimensional Multidetector CT Findings in Drowning with Autopsy Comparison' *Radiology* 243:862-868). (Using CT scans, American pathologists confirmed that all drowning victims had water in their mastoid and nasal sinuses, and in their lungs.)

16   A. C. Guyton and J. E. Hall *Textbook of Medical Physiology*. p.516

17   A. C. Guyton and J. E. Hall *Textbook of Medical Physiology*. p.516

18   'Oxygen saturation (medicine).' https://en.wikipedia.org/wiki/Oxygen_saturation

19   Roger H. Fuller (1963) 'The Clinical Pathology of Human Near-Drowning.' *Proceedings of the Royal Society of Medicine* 56:33-38

20   David A. Neidhart and Robert M. Greendyke (1967) 'The significance of diatom demonstration in the diagnosis of death by drowning.' *The American Journal of Clinical Pathology* 48:377-382. M. S. Pollanen (1997) 'The diagnostic value of the diatom test for drowning. 11. Validity: analysis of diatoms in bone marrow and drowning medium.' *Journal of Forensic Sciences* 42:286-290 (Note: Diatoms provide post-mortem forensic evidence of drowning when identical species are found in the water at the drowning site and in the victim's lungs, bone marrow, liver or kidneys at post-mortem.)

21   M. Noguchi, Y. Kimula and T. Ogata (1985) 'Muddy lung.' *American Journal of Clinical Pathology* 83:240-244

22   Donnie P. Donagan, Joseph E. Cox, Michael C. Chang and Edward F. Haponik (1997) 'Sand Aspiration with Near-Drowning', *American Journal of Respiratory and Critical Care Medicine*, 156:292-29

23   Frank Golden and Michael Tipton *Essentials of Sea Survival.* p.84 (Note: The reaction is particularly fierce if the water is contaminated by oil, agricultural chemicals or sewage. John Pearn (1985) 'The management of near drowning.' *The British Medical Journal* 291: 1447-1452)

24   H.J.H. Colebatch and D.F.J. Halmagyi (1961) 'Lung mechanics and resuscitation after fluid aspiration.' *Journal of Applied Physiology* 16:684-696

25   C. Cot (1931) quoted by J. M. Modell, M. Bellefleur and J. H. Davis (1999) 'Drowning Without Aspiration: Is This an Appropriate Diagnosis?' *Journal of Forensic Sciences* 44:1119-1123 (Note: In animal experiments, French scientists added methylene blue dye to the water and saw blue discolouration spread throughout the lungs, while X-rays of animals drowned in bismuth milk showed shadowing throughout the lungs.)

26   J. H. Modell and J. H. Davies (1969) 'Electrolyte changes in human drowning victims.' *Anesthesiology* 30:414-420

27   A. Joseph Layon and Jerome H. Modell (2009) 'Drowning: Update 2009.' *Anesthesiology* 110:1390-1401

28   Charles E. Rath (1953) 'Drowning Haemoglobinuria.' *Blood* 8:1099-1104. Michael Tsokos, Glenda Cains and Roger W. Byard (2008) 'Hemolytic Staining of the Intima of the Aortic Root in Freshwater Drowning.' *The American Journal of Forensic Medicine and Pathology* 29:128-130 (Note: In 1962, a Norwegian boy was rescued from an ice-covered river 'apparently dead' after a submersion lasting 22 minutes. So much free haemoglobin was found in his blood plasma that he was not expected to survive the extensive destruction of his red blood cells. Once in hospital, he was given a 3-hour-long exchange transfusion of 3,000 ml of freshly drawn blood – and he survived. Tone Dahl Kvittingen and Arne Naess (1963) 'Recovery from drowning in fresh water', *British Medical Journal* 1(5341):1315-1317.)

29   T.A. Ports and T. F. Deuel (1977) 'Intravascular coagulation in fresh-water submersion: report of three cases.' *Annals of Internal Medicine* 87:60-61

30   M.Schwameis et al (2015) 'Asphyxia by Drowning Induces Massive Bleeding due to Hyperfibrinolytic Disseminated Intravascular Coagulation.' *Critical Care Medicine* 43:2394-2402

31   Frank Golden and Michael Tipton *Essentials of Sea Survival.* p. 80 and 86 (Note: This spreading deterioration of lung function explains why inhalation of small amounts of seawater can be fatal even without submersion. Survivors of boating accidents sometimes drown by inhaling wave splash even while their life jackets are holding them at the surface.)

32   H. G. Swann and N. R. Spafford (1951) 'Body salt and water changes during fresh and sea water drowning', *Texas Reports on Biology and Medicine* 9:356-392

33   J. H. Modell and J. H. Davies (1969). 'Electrolyte changes in human drowning victims.' *Anesthesiology* 30:414-420. (Note: The great majority of drowning victims inhale less than 1500ml of water, often very much less.) A. Joseph Layon and Jerome H. Modell (2009) 'Drowning: Update 2009.' *Anesthesiology* 110:1390-1401

34   Lauralee Sherwood. *Fundamentals of Physiology.* p.335-336

35   Sylvia Wrobel (2004) 'Bubbles, Babies and Biology: The Story of Surfactant', *The FASAB Journal* Vol 18 No 3 1624e (Federation of American Societies for Experimental Biology) http://www.fasebj.org/content/18/13/1624e.full)

36   Frank Golden and Michael Tipton, *Essentials of Sea Survival.* p.82

37   A. C. Guyton and J. E. Hall, *Textbook of Medical Physiology*, p.479-481

38   Philippe Lunetta and Jerome H. Modell (2007) 'Macroscopical, Microscopical, and Laboratory Findings in Drowning Victims.' Chapter in *Forensic Pathology Reviews*, Vol 3, Michael Tsokos (ed.) Humana Press, Totowa, NJ. p.. 3-77

39   Paul Fornes, Gilbert Pepin, Didier Heudes and Dominique Lecomte (1998) 'Diagnosis of Drowning by Combined Computer-assisted Histomorphometry of Lungs with Blood Strontium Determination' *Journal of Forensic Sciences* 43:772-776. (Note: Pathologists sometimes describe this forced enlargement of alveoli as 'emphysema aquosum'.)

40   Philippe Lunetta and Jerome H. Modell 'Macroscopical, Microscopical, and Laboratory Findings in Drowning Victims.' Chapter in *Forensic Pathology Reviews*, Vol 3, M. Tsokos (ed.) Humana Press, Totowa, NJ. p.3-77 (Note: Sub-pleural haemorrhages [Paltauf's spots] are a common finding at post-mortem examination in victims of drowning. Many drowning victims have heavy, water-laden lungs which overfill their chests and push against their ribs. However, water never totally fills the lungs.)

41   Frank Golden and Michael Tipton, *Essentials of Sea Survival*, p.83. (Note: Oedema fluid contains plasma proteins with the same osmotic pull as blood plasma, and once oedema fluid collects in the alveoli, it tends to remain there.) D. S. Cohen, M. A. Matthay, M. G. Cogan and J. F. Murray (1992 'Pulmonary edema associated with salt water near-drowning: new insights.' *The American Review of Respiratory Disease* 146:794-796. Debra Perina (2003) 'Noncardiogenic pulmonary edema', *Emergency Clinics of North America* 21:385-393

42   Frank Golden and Michael Tipton, *Essentials of Sea Survival*, p.83 and 86. (Note: Oedema fluid collects in the alveoli whether a drowning victim inhales fresh water or seawater, but in seawater drowning, the strong osmotic pull of seawater draws additional fluid out of the bloodstream into the alveoli, resulting in severe water-logging of the lungs.)

43   'Ventilation/perfusion ratio', https://en.wikipedia.org/wiki/Ventilation/perfusion_ratio

44   John Pearn (1985), 'The management of near drowning', *British Medical Journal* 291:1447-1452

45   C. Pancaro, E. Diaz, P. Lindholm and M. Ferrigno (2007) 'Cerebral Oxygenation and Neurological Problems during Prolonged Breath-holds', Abstract of paper presented at the Undersea and Hyperbaric Medical Society Scientific Meeting, Hawaii

46   Lauralee Sherwood, *Fundamentals of Physiology*, p.353-356

47   P.T. Marshall and G.M. Hughes (1980) *Physiology of mammals and other vertebrates*. 2nd Edn. Cambridge University Press, Cambridge. p.73-74. (Note: carbon dioxide makes up 4% of expired air, but only 0.04% of atmospheric air.)

48   A. C. Guyton and J. E. Hall, *Textbook of Medical Physiology*. p.390-391

49   A. C. Guyton and J. E. Hall, *Textbook of Medical Physiology*. p.399

50   Stephen M. Roth 'Why does lactic acid build up in muscles? And why does it cause soreness?' *Scientific American* 23 January 2006. Kerry Brandis 'Lactic acidosis.' http://anaesthesiamcq.com/AcidBaseBook/ab8_1.php

51   Lauralee Sherwood, *Fundamentals of Physiology*, p.407-413 (Note: A pH as low as 6.33 has been recorded after a non-fatal drowning accident. H. Opdahl (1997) 'Survival put to the acid test: extreme arterial blood acidosis (pH 6.33) after near-drowning.' *Critical Care Medicine* 25:1431-1436.)

52   'Pulmonary gas pressures.' https://en.wikipedia.org/wiki/_Pulmonary_gas_pressures

53   A. C. Guyton and J. E. Hall, *Textbook of Medical Physiology*. p.493

54   D. F. J. Halmagyi and H. J. H. Colebatch (1961) 'Ventilation and circulation after fluid aspiration', *Journal of Applied Physiology* 16:135-140

55   Frank Golden and Michael Tipton, *Essentials of Sea Survival*. p.87

56   John B. West (1990) *Respiratory Physiology – the essentials*, 4th Edn. Williams and Wilkins, Baltimore, MD. p.35

57   Lauralee Sherwood, *Fundamentals of Physiology*, p.206-207

58   Frank Golden and Michael Tipton, *Essentials of Sea Survival*, p.87

59   'Tricuspid valve', Wikipedia https://en.wikipedia/wiki/Tricuspid_valve

60   Norbert F. Voelkel et al (2006), 'Right Ventricular Function and Failure', *Circulation* 114:1883-1891

61   Frank Golden and Michael Tipton, *Essentials of Sea Survival*, p.87

62   Lauralee Sherwood, *Fundamentals of Physiology*. p.229

63   'Sinoatrial node', Wikipedia, http://en.wikipedia.org/wiki/Sinoatrial_node

64   A. C. Guyton and J. E. Hall, *Textbook of Medical Physiology*. p.252

65   S.W. Davies and J.A. Wedzicha (1993) 'Hypoxia and the heart', *British Heart Journal* 69:3-5

66   Lauralee Sherwood, *Fundamentals of Physiology*, p.211-214

67   A. C. Guyton and J. E. Hall, *Textbook of Medical Physiology*. p. 154

68   Frank Golden and Michael Tipton *Essentials of Sea Survival*, p.87

69   Steven White (2010) 'Hypoxic brain injury.' Fact sheet. Headway – the brain injury association. https://www.headway.org.uk/media/2804/hypoxic-brain-injury.pdf

70   E.H. Koo, J.L. Boxerman and M.A. Murphy (2011) 'Cortical blindness following a near-drowning incident.' *Journal of Neuro-Ophthalmology* 31:347-349

PART 3: COPING WITH A DROWNING ACCIDENT

CHAPTER 29: RESCUE

1   Amy Teimann (2013), 'Would you recognize a drowning child if you saw one right in front of you?' http://doingrightbyourkids.com/2013/05/20/would-you (Note: This includes a brief clip of a rescue on Orchard Beach, New York.) Mario Vittone and Francesco A. Pia, 'It Doesn't Look Like They're Drowning', *On Scene, the Journal of US Coast Guard Search and Rescue*, Fall 2006

2   David S. Smith and Sara J. Smith (1994) *Water Rescue: Basic Skills for Emergency Responders.* Mosby-year Book Lifeline, St. Louis, MO. p.46

3   'Rescue methods – RoSPA' The Royal Society for the Prevention of Accidents. https://www.rospa.com/en/Leisure-Safety/Water/Advice/Children. 'How to rescue someone from drowning.' The Royal National Lifeboat Institute. 24 March 2017. https://rnli.org/magazine/magazine-featured-list

4   Mary Donahue 'How to rescue a drowning victim using a reaching assist or a shepherd's crook.' https://marydonahue.org/how-to-rescue-a-drowning-victim

5   Ian Key 'Sister is saved by 99p net', *The Daily Mirror*, 11 August 2011

6   John. H. Pearn and Richard. C. Franklin (2009) '"Flinging the squaler" Lifeline Rescues for Drowning Prevention', *International Journal of Aquatic Research and Education*, 3:315-321

7   *Games from the Scouting for Boys handbook 1908* p.25, http://younginfo.net/Scouting/GamesFromScoutingForBoys.html

8   'River death parents welcome lifebelt changes', BBC News, 20 March 2019

9   Frank Pia (1974), 'Observations on the drowning of nonswimmers', *Journal of Physical Education*, July–August 1974, p.164-167

10   Frank Golden and Michael Tipton, *Essentials of Sea Survival*, p.68-70

11   Sara Dixon, 'Father drowns saving son washed out to sea', *The Daily Express*, 10 April 2012

12   David S. Smith and Sara J. Smith, *Water Rescue: Basic Skills for Emergency Responders*, p.109

13   Tony Miles, Martin Ford and Peter Gathercole *The Practical Fishing Encyclopedia.* p. 206-208

14   Mary Donahue 'How to rescue a drowning victim using a reaching assist or a shepherd's crook.' https://marydonahue.org/how-to-rescue-a-drowning-victim

15   'Failure to recover a man overboard', Health and Safety Executive May 1999, http://www.hse.gov.uk/aala/articles/man-overboard.htm

16   Dennis K. Graver *Aquatic Rescue and Safety.* p.180.

17   'Ten-year-old boy rescues couple from capsized boat', *The Daily Telegraph*, 2 August 2012

18   Adebola Lamuye, 'Ex-Army man in Thames plunge rescue', *The Evening Standard*, 25 February 2019

19   Andrew Ffrench, 'Chef in river rescue', *Oxford Mail*, 10 April 2001

20   'Woking pensioner saves man from drowning in canal', http://www.getsurrey.co.uk 5 November 2009

21   'Hero leaps into canal to save drowning toddler… despite the fact that he could not swim', *The Daily Mail* 15 July 2010

22   'Boy, 7, saved by holiday heroes.' *The Daily Record*, Glasgow 14 August 2000

23   Emma Ray 'Honeymoon man saves tot from pool', *Coventry Evening Telegraph*, 21 August 2001

24   Chris Brooke, '"Superdad" leaps from balcony into pool to rescue drowning son… while lifeguards "stand and watch"', *The Daily Mail*, 22 June 2010

25   Steven Morris 'Baby blown into freezing sea saved by 'amazing' response from rescuers', *The Guardian*, 28 January 2013

26   Barney Davies, '"Ironman" Hero Saves Boy, 4, from Thames; Youngster pulled out with "seconds to spare" after he slips through railings', *Evening Standard*, 14 December 2016

27   John L. Hunsucker and Scott J. Davison (2013) 'Time Required For a Drowning Victim to Reach Bottom', *Journal of Search and Rescue* 1:19-28

28   Paul Cheston, 'Friend of drowning boy: I tried to grab his hair but he slipped away', *Evening Standard*, 20 December 2011

29   Martin Barwood, Victoria Bates, Geoffrey Long and Michael Tipton (2011) '"Float First": trapped air between clothing layers significantly improves buoyancy on water immersion in adults, adolescents and children', *International Journal of Aquatic Research and Education* 5:147-163

30   M. J. Barwood, V. Bates, G. Long and M. J. Tipton (2010) '"Float First": an assessment of the buoyancy provided by seasonal clothing assemblies before and after swimming', Department of Sport and Exercise Science, University of Portsmouth.

31   Tom Wilkinson 'Rowing coach dies after falling off his bike and into the river', *The Daily Mirror* 2 March 2015

32   Frank Golden and Michael Tipton, *Essentials of Sea Survival*, p.59-64. Jeffrey Pollinger 'Cold water immersion: the shocking reality', *Coast Guard News*, 27 June 2008

33   Louise Aiken 'Mum plunges into North-east loch to save drowning son', *Evening Express*, Aberdeen 12 August 2015

34   Martin Barwood, Victoria Bates, Geoffrey Long and Michael Tipton 'Safety behaviour to avoid drowning – Should we 'Float First' on accidental immersion?' World Conference on Drowning Prevention, Da Nang, Vietnam 2011. Heather Bowes, Claire E. Eglin, Michael Tipton and Martin Barwood (2016), 'Swim performance and thermoregulatory effects of wearing clothing in a simulated cold-water survival situation', *European Journal of Applied Physiology* 116:759-767

35   Austswim, *Teaching Swimming and Water Safety*, p.165-167

36   David S. Smith and Sara J. Smith, *Water Rescue: Basic Skills for Emergency Responders*. p.100

37   Fred Lanoue (1963) *Drownproofing: A New Technique for Water Safety*. Prentice-Hall Inc., NJ. 'Drownproofing', Wikipedia https://en.wikipedia.org/wiki/Drownproofing

38   Stephen Templin, 'SEAL Training 9: Drownproofing', http://www.stephentemplin.com/blog/seal-training-9-drownproofing

39   Gert Froese and Alan C. Burton (1957), 'Heat Losses From the Human Head', *Journal of Applied Physiology*, 10:235-241. Lorenz E. Wittmers and Margaret V. Savage (2001), 'Cold Water Immersion', Chapter in *Medical Aspects of Harsh Environments*, Vol 1, Edit. Kent B. Pandolf and R.E. Burr Publishers, p.532

40   'Philip Guston – Clothes Inflation Drill', https://www.pinterest.com/pin/15347429977390...

41   L.G. Pugh (1978), 'Isafjordur trawler disaster: medical aspects', *British Medical Journal*, 1(6595):826-829. C.J. Brooks (2003), *Survival in Cold Waters: Staying Alive*. A Report Prepared for Marine Safety Directorate, Transport Canada. https://apps.dtic.mil/dtic/fulltext

42   Frank Golden and Michael Tipton, *Essentials of Sea Survival*, p.108-112

43   Frank Golden and Michael Tipton, *Essentials of Sea Survival*, p.96. Jess Staufenberg 'Mother drowns holding toddler son above water until he is rescued from lake in Utah', *The Independent*, 27 August 2016. (Note: Three adults and two children were on a moving houseboat when the little boy fell overboard. No one was wearing a life jacket.)

44   Frank Golden and Michael Tipton, *Essentials of Sea Survival*, p.114-116

45   R. L. Sawyer and P. Barss (1998), 'Stay with the boat or swim for the shore? A comparison of drowning victim and survivor responses to immersion following a capsize or swamping', Fourth World Conference on Injury Prevention and Control, Amsterdam, Netherlands. W.R. Keatinge (1969), *Survival in Cold Water*, p. 8-9 and 94

46   M. B. Ducharme and D. S. Lounsbury (2007) 'Self-rescue swimming in cold water: the latest advice', *Applied Physiology, Nutrition and Metabolism* 32:799-807

47   'Gudlaugur Fridþórsson: Icelandic Seaman Who Survived After 6 Hours in Ice-Cold Water.' *The Viking Rune*. March 2009, https://www.vikingrune.com/2009/03/true-viking

48   W. R. Keating, S. R. Coleshaw, C. E. Millard and J. Axelsson (1986), 'Exceptional case of survival in cold water', *British Medical Journal* 292:171-172

49   'Gudlaugur Fridþórsson: Icelandic Seaman Who Survived After 6 Hours in Ice-Cold Water', *The Viking Rune*, March 2009, https://www.vikingrune.com/2009/03/true-viking

50   'Cruise ship fall: Woman rescued after 10 hours in sea off Croatia', BBC News, 20 August 2018

51   Ian Sample 'How did Kay Longstaff survive 10 hours at sea after falling overboard?' *The Guardian*, 20 August 2018

52   The National Aquatics Safety Company (2011) Lifeguard Textbook, p.27-29 nascoaquatics. com/wp-content/uploads/.../lifeguard-textbook-2012.pdf

53   Frank Pia (1974) 'Observations on the drowning of nonswimmers', *Journal of Physical Education*, July-August 164-167

54   H. Turner, H. Vogelsong and R. Wendling (2003), 'Is in-service training for lifeguards necessary?' *Parks and Recreation* 38:43-45. David C. Schwebel, Heather N. Jones, Erika Holder and Francesca Marciani (2011) 'The influence of Simulated Drowning Audits on Lifeguard Surveillance and Swimmer Risk-Taking at Public Pools', *International Journal of Aquatic Education and Research*, 5:210-218

55   Peter Fenner, Stephen Leahy, Andrew Bukh and Peter Dawes (1999), 'Prevention of drowning: visual scanning and attention span in Lifeguards', *The Journal of Occupational Health and Safety – Australia and New Zealand* 15:61-66. David V. C. Schwebel, Sydneia Lyndsay and Jennifer Simpson (2007), 'Brief Report: A Brief Intervention to Improve Lifeguard Surveillance at a Public Swimming Pool', *Journal of Pediatric Psychology* 32:862-868

56   Frank Pia (1984), 'The RID Factor as a cause of drowning', *Parks and Recreation* 19:52-57. Kat Lay, 'Boy drowned "while lifeguard was chatting"', *The Times*, 20 April 2013

57   Tom Griffiths, 'From the Bottom Up', *Aquatics International*, October 2006

58   L. Patterson, 'Factors affecting lifeguard recognition of the submerged victim: implications for lifeguard training, lifeguarding systems and aquatic facility design', Paper given at International Life Saving Federation, World Conference on Drowning Prevention 2007. Porto, Portugal

59   Tom Griffiths 'From the Bottom Up', *Aquatics International*, October 2006. Jerome H. Modell(2010) 'Prevention of Needless Deaths from Drowning', *Southern Medical Journal* 103:650-653

60   'Pool lifeguard refused to believe boy was dying', *The Daily Telegraph*, 30 August 2001

61   'Council fined after pool death', BBC News, 28 January 2004

62   Nigel Bunyan 'Underwater cameras save pool girl from drowning', *The Daily Telegraph*, 1 September 2005

63   Jenny Page, Victoria Bates, Geoff Long, Peter Dawes and Mike Tipton (2011) 'Beach Lifeguards: Visual Search Patterns, Detection Rates and the Influence of Experience', *Ophthalmic and Physiological Optics* 31:216-224 M. Tipton, T. Reilly, A. Rees, G. Spray and F. Golden (2008) 'Swimming performance in surf: the influence of experience', *International Journal of Sports Medicine* 29:895-898

64   Mark Duell 'Dramatic moment RNLI lifeguards on a JET SKI rescue three teenagers who had drifted out to sea on makeshift raft', *The Daily Mail*, 1 September 2016

65   'RNLI Facts – Loud and Clear 2012-2013', The Lifeboat Fund. www.thelifeboatfund.org.uk

66   Emma Palmer, 'Boy, 15, hurts neck diving from jetty into 3ft-deep water', *Southend Standard*, 22 July 2009

67   L.A. Robles (2006) 'Cervical spine injuries in ocean bathers: wave-related accidents', *Neurosurgery* 58:920-923

68   Peter Wernicki, Peter Fenner and David Szpilman 'Immobilisation and Extraction of Spinal Injuries', Chapter in *Handbook on Drowning*, p.291-297

69   'Defensive Block with Torpedo Buoy – RLLS', Royal Life Saving Society video on YouTube

70   'Rescue buoy', wikipedia, http://en.wikipedia.org/wiki/Rescue_buoy

71   'Surf Life Saving Wales rescue techniques', Royal National Lifeboat Institution video on YouTube

72   'The birth of surf lifesaving', National Museum of Australia. https://nma.gov.au/exhibitions/between-the-flags/birth-of surf-lifesaving

73   'United States Lifesaving Association (USLA) History', https://www.usla.org/page/HISTORY/United-States-Lifesaving-Association

74   P. J. Fenner, S. L. Harrison, J. A. Williamson and B. D Williamson (1995), 'Success of Surf Lifesaving Resuscitations in Queensland 1973-1992', *Medical Journal of Australia* 163:580-583

75   'Swim between the flags', http://www.watersafety.nsw.gov.au/beach-safety/... Jeff Wilks, Monica de Nardi and Robert Wodarski (2007) 'Close is not Close Enough: Drowning and Rescues Outside Flagged Beach Patrol Areas in Australia', *Tourism in Marine Environments* 1:57-62

76   B. Brighton, S. Sherker, R. Brander, M. Thompson and A. Bradstreet (2013) 'Rip current related drowning deaths and rescues in Australia 2004-2011', *Natural Hazards and Earth Systems Sciences* 13: 1069-1075

77   'American Lifeguard Rescue and Drowning Statistics for Beaches. 2015 National Lifesaving Statistics.' United States Lifesaving Association, http://arc.usla.org/Statistics/current.asp

78   '5 Swimmers Are Drowned In a Riptide', *The New York Times*, 1 June 1994

79   C. M. Branche and S. Stewart (Editors) (2001) *Lifeguard Effectiveness: A Report of the Working Group.* Centers for Disease Control and Prevention. National Center for Injury Prevention and Control, Atlanta, GA.

80   'Phuket beach drowning deaths double in wake of lifeguard contract failure', *The Phuket News*, 29 September 2018. 'Missing backpacker drowned', BBC News, 1 November 2019

81   'Royal National Lifeboat Institution', Wikipedia, https://en.wikipedia.org/wiki/Royal_National_Lifeboat_Institution

82   Jessica Frank-Keyes 'Grandparents and young children rescued from rising tide on Norfolk beach which has seen "several fatalities"', *Eastern Daily Press*, Norfolk 2 August 2019. 'Mile out to sea… man drifting in a toy dinghy' *Metro*, 2 August 2017. 'Mud-stuck sailor rescued from capsized dinghy in Poole', BBC News, 27 April 2014. 'Stromness lifeboat called to oil rig in Scapa Flow', *The Orcadian*, Orkneys, 25 November 2019

83   Chris Millar, 'Five are plucked to safety after eight days on life raft', *Evening Standard*, 15 September 2004. Larisa Brown, 'Prince William flies RAF helicopter to save girl, 16, who got into trouble while helping sister who was swept away on a riptide', *The Daily Mail*, 17 August 2012. Elizabeth McCormack, Chris A. Turner and Michael J. Tipton (2009), 'The prediction of survival time in water', University of Portsmouth and US Coastguard Research and Development Center, New London, CT.

84   Frank Golden and Michael Tipton *Essentials of Sea Survival*, p. 258-264

85   'Swift Water Rescue Exercise At Newcastle Emlyn.' http://mawwfire.goc.uk/press_media_eng/News_details.asp?id=720

86   Carla St-Germain and Andrea Zaferes, 'Ice Rescue', Section in *Handbook on Drowning*, Bierens, p.250

87   Vic Calland (2000) *Safety at Scene*, Mosby, Edinburgh, p.164-166

88   David S. Smith and Sara J. Smith, *Water Rescue: Basic Skills for Emergency Responders*, p.200

89   'RoSPA ice warning as freezing conditions continue', 7 January 2009, Press Release by the Royal Society for the Prevention of Accidents

90   Neil Millard, 'Matt the cocker spaniel saved from a watery grave by 17 firefighters and a very long ladder', *The Daily Mail*, 31 December 2009

91   'RoSPA ice warning as freezing conditions continue', 7 January 2009, Press Release by The Royal Society for the Prevention of Accidents

92   A. H. Idris et al (2003), 'Recommended Guidelines for Uniform Reporting of Data From Drowning', *Circulation* 108:2565-2574

93   'Norwich teenager recalls final moments of brave sister who drowned in lake', *Eastern Daily Press*, 15 August 2015

94   Dennis K. Graver, *Aquatic Rescue and Safety*, p.144-145

95   David S. Smith and Sara J. Smith, *Water Rescue: Basic Skills for Emergency Responders*, p.149

96   Dilvin Yasa, 'What drowning really looks like', http://www.essentialkids.com.au/health/health-wellbeing/what-drowning-really-looks-like-20150112-12mcp8 (Reprinted on Samuel

Morris Foundation website 13 January 2015)

97   David S. Smith and Sara J. Smith, *Water Rescue: Basic Skills for Emergency Responders*, p.149

98   Frank Pia (1974) 'Observations on the drowning of nonswimmers', *Journal of Physical Education*, July-August 164-167

99   Alison M. Dahl and Doris I. Miller (1978) 'Body Contact Swimming Rescues – What are the Risks?' *American Journal of Public Health* 68:150-152

100   Dennis K. Graver, *Aquatic Rescue and Safety*, p.128

101   Frank Pia (1974), 'Observations on the drowning of nonswimmers', *Journal of Physical Education* July-August 164-167

102   Frank Golden and Michael Tipton, *Essentials of Sea Survival*, p.59-75

103   Frank Golden and Michael Tipton, *Essentials of Sea Survival*, p.66

104   David S. Smith and Sara J. Smith, *Water Rescue: Basic Skills for Emergency Responders*, p.4

105   Christian Gysin, 'Mother dies trying to save her sons from riptide: Beach tragedy... 24 hours after lifeguard duty ends for the season', *The Daily Mail*, 5 September 2012

106   Lucy Bannerman, 'The hell started. I went under. I thought, I can't do this' *The Times*, 8 September 2012

107   Ilan Kelman (2003), '*U.K. Drownings*', The Martin Centre, Cambridge

108   Graham Keely, 'Father who couldn't swim dies after trying to save his son from rough sea' *The Times*, 3 September 2012

109   Christian Gysin and Jaya Narain, 'Two British doctors swept to their death by waves as they tried to rescue their children from the water at holiday beach off Tenerife', *Daily Mail*, 7 April 2014

110   A. Turgut and T. Turgut (2012), 'A study of rescuer drowning and multiple drowning incidents,' *Journal of Safety Research*, 43:129-132

111   Richard Franklin and John Pearn (2010), 'The rescuer who drowns', *Injury Prevention* 16:A81

112   Richard C. Franklin and John H. Pearn (2010), 'Drowning for love: the aquatic victim-instead-of-rescuer syndrome: drowning fatalities involving those attempting to rescue a child', *Journal of Paediatrics and Child Health* 47:44-47

113   Kevin Moran and Teresa Stanley (2013) 'Readiness to Rescue: Bystander Perceptions of Their Capacity to Respond in a Drowning Emergency', *International Journal of Aquatic Research and Education*, 7:290-300. Caleb Starrenburg 'Would-be rescuers losing their lives', 5 January 2014. http://www.stuff.co.nz/national/9574440/Would-be-rescuers-losing-their-lives

114   'Helicopter rescues dog owner in Powys river', BBC News, 11 September 2010

115   'Man dies trying to rescue dog from icy River Lune', BBC News, 27 November 2010

116   'Police salute fallen heroes', *The Blackpool Gazette*, 4 January 2008. 'Blackpool statue honours drowned officers and 999 services', BBC News, 27 June 2013

117   Joanna Bale 'Best friends drown trying to save dog', *The Times*, 30 May 2005

118   'Girl watches father drown trying to save pet dogs', *The Daily Mail*, 10 July 2007

119   Auslan Cranb, 'Baby girl orphaned after parents die trying to save pet dogs', *The Daily Telegraph*, 11 May 2009

120   Patrick Mulchrone, 'The Quiet Hero; Shy firefighter dies in pond rescue bid', *The Daily Mirror*, 7 September 1999

121   'Fire service in the clear over death', *Manchester Evening News*, 22 October 2004

122   Andy Gilchrist, General Secretary of the Fire Brigades Union 'Chronicle of a death foretold', *Firefighter*, December 2004

123   David Lister, 'Fireman hauled over the coals for saving a woman's life', *The Times*, 26 March 2007

124   Marion Scott, 'Hero fireman Tam escapes the sack', *The Sunday Mail* (Glasgow), 13 May 2007

125   'River rescue firefighter honoured', BBC News, 10 August 2007

126   Julian Fulbrook, *Outdoor Activities, Negligence, and the Law*, p.100

127   'Boys drowned "while lifeguard was distracted"', *Daily Mail*, 26 April 2006

128   Rosie Cowan 'Met officer summons in pool deaths', *The Guardian*, 24 March 2005. Neeta Dutta, 'Officer cleared over pool deaths', *Hendon and Finchley Times*, 15 June 2006.

129   'Met Police fined over pool deaths', BBC News, 13 July 2007

130   'Schoolboy dies after pond rescue', BBC News, 4 May 2007

131   Paul Britton, 'Hero, 10, died saving sister', *Manchester Evening News* 15 September 2007

132   'Boy drowned as police support officers "stood by"', *The Guardian*, 21 September 2007

133   Paul Britton, 'Hero, 10, died saving sister', *Manchester Evening News*, 15 September 2007

134   Paul Britton, 'Hero, 10, died saving sister', *Manchester Evening News*, 15 September 2007

135   'Police defend drowning death case', BBC News, 21 September 2007

136   'Police defend drowning death case', BBC News, 21 September 2007

137   John Connelly (2007) 'Swimming rescues by Irish police officers.' International Life Saving Federation, World Conference on Drowning Prevention 2007, Porto, Portugal.

138   'Man drowned feeding swans on lake after epileptic seizure', BBC News, 22 February 2012

139   Euan Stretch '999 crews: We can't rescue drowning man from 3ft pond... it's too DEEP', *The Daily Mirror*, 22 February 2012

140   'Firemen refused to go in 3ft-deep lake as man floated face down', *The Daily Telegraph*, 22 February 2012

141   'Firemen refused to go in 3ft-deep lake as man floated face down', *The Daily Telegraph*, 22 February 2012

142   Ben Quinn, 'Man drowned in shallow lake after firefighters "not allowed" to rescue him', *The Guardian*, 22 February 2012

143   'Heroic Acts by Police Officers and Firefighters', Crown Prosecution Service: Legal Guidance, http://cps.gov.uk/legal/H_toK/heroic_acts_by_police_officers...

CHAPTER 30: FIRST AID AND RESUSCITATION AT THE SCENE

1   'First aid – Recovery position', NHS Choices, http://www.nhs.uk

2   Frank Golden and Michael Tipton, *Essentials of Sea Survival*. p.224-225 (Note: Thin plastic 'space blankets' do little to warm a drowning victim on their own, even with the shiny side on the inside, because a rescued swimmer's skin is extremely cold. They are best used as an outer layer over insulating layers of dry cloth.)

3   Frank Golden and Michael Tipton, *Essentials of Sea Survival*, p.84-86

4   Charles E. Putman, Anna M. Tummillo, Daniel A. Myerson and Paul J. Myerson (1975) 'Drowning: another plunge.' *American Journal of Roentgenology,* 125:543-548

5   Frank Golden and Michael Tipton, *Essentials of Sea Survival,* p.82-83 and 86

6   Kerry Mcqueeney, 'Man hauls himself out of water after plunging off boat… only to DROWN on dry land hours later', *Daily Mail,* 16 May 2012

7   John H. Pearn (1980), 'Secondary drowning in children', *British Medical Journal* 281:1103-1105

8   Jacque Wilson, 'You can drown after you leave the pool', CNN 10 June 2014. https://edition.cnn.com/2014/06/09/health/secondary-drowning

9   'ABC of Basic Life Support' http://en.wikipedia.org/wiki/ABC_(medicine). Demetrios N. Kyriacou, Edgardo L. Arcinue, Corinne Peek and Jess F. Kraus (1994) 'Effect of Immediate Resuscitation on Children with Submersion Injury' *Pediatrics* 94:137-142

10   N. Manolios and I. Mackie (1988) 'Drowning and near-drowning on Australian beaches patrolled by life-savers: a 10-year study, 1973-1983', *Medical Journal of Australia* 148:165-167, 170-171

11   David Szpilman and Marcio Soares (2004) 'In-water resuscitation – is it worthwhile?' *Resuscitation* 63:25-31

12   'Statements on Positioning a Patient on a Sloping Beach', International Life Saving Federation, http://www.ilsf.org

13   'Airway management', Wikipedia, http://en.wikipedia.org/wiki/Airway_management

14   Robert A. Berg et al (2010) 'Part 5: Adult Basic Life Support. 2010 American Heart Association Guidelines for Cardiopulmonary Resuscitation and Emergency Cardiovascular Care', *Circulation* 122:S685-S705

15   Dennis K.Graver *Aquatic Rescue and Safety,* p.162

16   Marc D. Berg et al (2010) 'Part 13: Pediatric Basic Life Support. 2010 American Heart Association Guidelines for Cardiopulmonary Resuscitation and Emergency Cardiovascular Care Science', *Circulation,* 122:S862-S875

17   Frank Golden and Michael Tipton, *Essentials of Sea Survival,* p.269

18   Tom P. Aufderheide and Keith G. Lurie (2004), 'Death by hyperventilation: A common and life-threatening problem during cardiopulmonary resuscitation', *Critical Care Medicine,* 32: S345- S351

19   Bernard I. Lewis (1953) 'The hyperventilation syndrome', *Annals of Internal Medicine* 38:918-927

20   Volker Wenzel, Ahamed H. Idris, Michael J. Banner, Ronnie S. Fuerst and Kelly Tucker (1994), 'The Composition of Gas Given by Mouth-to-Mouth Ventilation During CPR', *Chest* 106:1806-1810

21   Dennis K. Graver, *Aquatic Rescue and Safety,* p.193

22   Bart Jan Meursing, 'The History of Resuscitation', Chapter in *Handbook on Drowning,* p. 14-21. Robert Smirke 'A man recuperating at the receiving house of the Royal Humane Society, after resuscitation', (1774) Watercolour in the collection of the British Library, https://www.bl.uk/collection-items/a-man-recuperating-after-rescue

23   *The Swimmers and Skaters Guide. The Derby Chap Books: 1820-1840.* Thomas Richardson, Derby. p.17

24   Jonas A. Cooper, Joel D. Cooper and Joshua M. Cooper (2006) 'Cardiopulmonary Resuscitation: History, Current Practice, and Future Direction', *Circulation*, 114:2839-2849

25   Arthur C. Guyton and John E. Hall, *Textbook of Medical Physiology*, p.503. Robert H. Woods (1906) 'On Artificial Respiration', *Transactions of the Royal Academy of Medicine in Ireland* 24:136-141

26   Gene L. Colice (1994) 'Historical Perspective on the Development of Mechanical Ventilation', Chapter in *Principles and Practice of Mechanical Ventilation*, Martin J. Tobin (ed.) McGraw-Hill Inc. New York, p.16

27   Julius H. Comroe (1979) "'... In Comes the Good Air." Part 11. Mouth-to-Mouth Method', *American Review of Respiratory Disease* 119:1025-1031

28   Mickey S. Eisenberg, Peter Baskett, and Douglas Chamberlain (2007), 'A history of cardiopulmonary resuscitation', Chapter in *Cardiac Arrest: The Science and Practice of Resuscitation Medicine*, Norman A. Paradis et al. (ed.) Cambridge University Press, Cambridge. Julius H. Comroe (1979) "'... In Comes the Good Air" Part 1. Rise and Fall of the Schafer Method', *American Review of Respiratory Disease*, 119:803-809. Thomas F. Baskett (2007), 'The Holger Nielsen method of artificial respiration', *Resuscitation*, 74:403-405. G.H.Gibbens (1942) 'Artificial Respiration at Sea', *British Medical Journal* 2(4277):751-752 (Note: Dr Frank Eve published his 'rocking method' in 1932. The method was widely used by the British Navy throughout the Second World War.)

29   James O. Elam (1977), 'Rediscovery of Expired Air Methods for Emergency Ventilation', Chapter in *Advances in Cardiological Resuscitation*, Peter Safar (ed.) Springer Verlag, New York, p.263-265

30   James O Elam, Elwyn S. Brown and John D. Elder (1954), 'Artificial Respiration by Mouth-to-Mask Method – A Study of the Respiratory Gas Exchange of Paralysed Patients Ventilated by Operator's Expired Air', *New England Journal of Medicine*, 250:749-754

31   Peter Safar, Lourdes A. Escarraga and Francis Chang (1959), 'Upper airway obstruction in the unconscious patient' *Journal of Applied Physiology* 14:760-764

32   Peter Safar, Lourdes A. Escarraga and James O. Elam (1958) 'A Comparison of the mouth-to-mouth and mouth-to-airway methods of artificial respiration with the chest-pressure arm-lift methods', *New England Journal of Medicine* 258:671-677

33   *The Journal of the American Medical Association*, 17 May 1958 – quoted by Mickey S. Eisenberg, Peter Baskett and Douglas Chamberlain (2007) 'Cardiac Arrest: A history of cardiopulmonary resuscitation', Section in *The Science and Practice of Resuscitation Medicine*, Norman A. Paradis et al (ed.) Cambridge University Press, Cambridge, p.10

34   E. E. T. Taylor "Kiss of Life", Letter, *British Medical Journal*, 7 August 1965, p.366

35   A. P. McCormack, S. K. Damon and M. S. Eisenberg (1989) 'Disagreeable physical characteristics affecting bystander CPR', *Annals of Emergency Medicine*, 18:283-285

36   N. Manolios and I. Mackie (1988) 'Drowning and near-drowning on Australian beaches patrolled by life-savers: a 10-year study, 1973-1983', *Medical Journal of Australia* 148:165-167, 170-171. Frank Golden and Michael Tipton, *Essentials of Sea Survival*, p.289

37   Dennis K. Graver, *Aquatic Rescue and Safety*, p.84

38   David Roberts, 'I saved my own son's life', *Medical Monitor*, 18 February 1998

39   Mark Henning, David Bogod and Alan Aitkenhead (1996) *Essential Anaesthetics for medical students*, Arnold, Hodder Headline Group, London, p.9

40   Dennis K. Graver, *Aquatic Rescue and Safety*, p.193

41   Mark Henning, David Bogod and Alan Aitkinhead (1996) *Essential Anaesthesia for Medical Students*, Arnold, Hodder Headline Group, London, p.9-11

42   B. A. Sellick (1961), 'Cricoid pressure to control regurgitation of stomach contents during induction of anaesthesia', *The Lancet* 2:404-40

43   Frank Golden and Michael Tipton, *Essentials of Sea Survival*, p.269-270

44   W. R. Keatinge and M. G. Hayward (1981), 'Sudden Death in Cold Water and Ventricular Arrhythmia', *Journal of Forensic Scientists* 26:459-461. Frank Golden and Michael Tipton *Essentials of Sea Survival*, p.269

45   Robert A. Berg et al (2010), 'Part 5: Adult Basic Life Support. 2010 American Heart Association Guidelines for Cardiopulmonary Resuscitation and Emergency Cardiovascular Care', *Circulation* 122:S685-S705. 'Bee Gees song 'Stayin' Alive' helps doctors perform CPR', *The Daily Telegraph*, 17 October 2008

46   Tom P. Aufderheide et al (2005), 'Incomplete chest wall decompression: a clinical evaluation of CPR performance by EMS personnel and assessment of alternative manual chest compression-decompression techniques', *Resuscitation* 64:353-362

47   Lauralee Sherwood, *Fundamentals of Physiology*, p.205-207

48   Robert A. Berg et al (2010), 'Part 5: Adult Basic Life Support. 2010 American Heart Association Guidelines for Cardiopulmonary Resuscitation and Emergency Cardiovascular Care', *Circulation* 122:S685-S705

49   Tom P. Aufderheide et al (2004), 'Hyperventilation-Induced Hypotension During Cardiopulmonary Resuscitation.' *Circulation* 109:1960-1965

50   L. A. Geddes, M. K. Boland, P. R. Taleyarkhan and J. Ritter (2007) 'Chest compression force of trained and untrained CPR rescuers', *Cardiovascular Engineering*, 7:47-50

51   'CPR: More Rib Fractures, But Better Survival Rates', *Science Daily*, 1 June 2007

52   Robert A. Berg et al (2010) 'Part 5: Adult Basic Life Support. 2010 American Heart Association Guidelines for Cardiopulmonary Resuscitation and Emergency Cardiovascular Care', *Circulation* 122:S685-S705

53   Lauralee Sherwood, *Fundamentals of Physiology*, p.22

54   D. J. Fodden, A. C. Crosby and K. S. Channer (1996), 'Doppler measurement of cardiac output during cardiopulmonary resuscitation', *Journal of Accident and Emergency Medicine* 13:379-382

55   Kevin Moran and Teresa Stanley (2011), 'Toddler parents training, understanding, and perceptions of CPR', *Resuscitation* 82:572-576

56   Sebastian Berger, 'Swimmer is brought back from dead after heart stops for hour', *The Daily Telegraph*, 18 January 2008

57   W. B. Fleming, J. T. Hueston, J. L. Stubbe and J. D. Villiers (1960), 'Two episodes of cardiac arrest in one week: Full recovery after cardiac massage', *British Medical Journal* 1(5167):157-160

58   W. B. Kouwenhoven, James R. Jude and G. Guy Knickerbocker (1960), 'Closed-chest cardiac massage.' *Journal of the American Medical Association* 173:1064-1067

59   Peter Safar, Torrey C. Brown, Warren J. Holtey and Robert J. Wilder (1961), 'Ventilation and Circulation with Closed-Chest Cardiac Massage in Man', *Journal of the American Medical Association* 176:574-576

60   Nina Tjonsland and Peter Baskett (2002), 'The Resuscitation Greats: Åsmund Laerdal', *Resuscitation* 53:115-119. Bjørn Lind (2007) 'The birth of the resuscitation mannequin, Resusci Anne, and the teaching of mouth-to-mouth resuscitation', *Acta Anaesthesiologica Skandinavica* 51:1051-1053

61   Jeremy Grange, 'Resusci Anne and L'Inconnue: The Mona Lisa of the Seine', *BBC News Magazine*, 16 October 2013

62   Nina Tjomsland, Tore Laerdal and Peter Baskett (2005), 'Resuscitation Great:. Bjorn Lind – the ground-breaking nurturer', *Resuscitation* 65:133-138

63   'Laerdal Resusci-Junior with Water Rescue Kit.' http://www.watersafety.com. Gavin D. Perkins (2005) 'In-water resuscitation: a pilot evaluation', *Resuscitation* 65:321-324

64   Ian Jones, Richard Whitfield, Michael Colquhoun, Douglas Chamberlain, Norman Vetter and Robert Newcombe (2007)'At what age can schoolchildren provide effective chest compressions? An observational study from the Headstart UK schools training programme', *British Medical Journal* 334: 1201

65   Wayne Crenshaw '7-year-old Warner Robins girl credited with saving boy from drowning', *Macon Telegraph*, Georgia 31 May 2015

66   John Pearn (1992), 'The urgency of immersions' *Archives of Disease in Childhood* 67:257-261

67   Jeannette Marchant, Nicholas G. Change, Lawrence T. Lam, Fiona E. Fahey, S. V. Soundappan, Danny T. Cass and Gary J. Browne (2008), 'Bystander basic life support: an important link in the chain of survival for children suffering a drowning or near-drowning episode', *The Medical Journal of Australia* 188:484-485

68   John M. Field et al. (2010), 'Part 12: Cardiac Arrest in Special Situations. Part 12.11: Drowning. 2010, American Heart Association Guidelines for Cardiopulmonary Resuscitation and Emergency Cardiovascular Care Science', *Circulation*, 122 (18 suppl 3)

69   Judith Fisher (1986), 'Recognising a cardiac arrest and providing basic life support' *British Medical Journal* 292:1002-1004

70   Michael Sayre et al (2008)'Hands-Only (Compression-Only) Cardiopulmonary Resuscitation: A Call to Action for Bystander Response to Adults Who Experience Out-of-Hospital Sudden Cardiac Arrest: A Science Advisory for the Public From the American Heart Association Emergency Cardiovascular Care Committee', *Circulation*, 117:2162-2167

71   Bentley J. Bobrow et al (2010) 'Chest Compression-Only CPR by Lay Rescuers and Survival From Out-of-Hospital Cardiac arrest', *Journal of the American Medical Association*, 304:1447-1454. Michael Hüpfl, Harald F. Selig and Peter Nagele (2010)'Chest-compression-only versus standard cardiopulmonary resuscitation: a meta-analysis', *The Lancet* 376:1552-1557

72   Rebecca Smith, 'Skip the "kiss" when giving the kiss of life doctors recommend', *The Daily Telegraph* 15 October 2010

73   Carlos Federico Arend (2000), 'Transmission of Infectious Diseases through Mouth-to-Mouth Ventilation: Evidence-Based or Emotion-based Medicine?' *Archives of Brazilian Cardiology (Arquivos Brasileiros de Cardiologia)* 74: 86-97

74 Kenneth Heilman and Carl Muschenheim (1965) 'Primary cutaneous tuberculosis resulting from mouth-to-mouth respiration', *The New England Journal of Medicine* 273: 1035-1036

75 Frank Pia (2006), 'Management of Physical and Psychological Responses During Administration of CPR to a Drowned Person', Section in *Handbook on Drowning*, p.302-305. Tina Robins, 'Mum saves tot from drowning in water trough', *Wiltshire Gazette and Herald*, 14 June 2017

76 Lech Mintowt-Czyz, 'Boy who came back from the dead', *Evening Standard* 14 July 2004. Bill Billiter, 'Phone Advice Saves Baby Who Fell in Pool', *Los Angeles Times*, 24 April

77 'History and science of defibrillation', http://www.resuscitationcentral.com/defibrillation/history-science/

78 Steve Beerman and Bo Løfgren (2006), 'Automated External Defibrillators in the Aquatic Environment', Chapter in *Handbook on Drowning*, p.331-335

79 C. G. Alex, 'Olympic swimmer thanks Newquay Waterworld staff who saved her from drowning', *Cornish Guardian*, 29 April 2015

80 David J. Steedman (1994), *Environmental Medical Emergencies*. Oxford University Press, Oxford, p.11-14. P. Nordberg, T. Ivert, M. Dalén, S. Forsberg and A, Hedman (2014) 'Surviving two hours of ventricular fibrillation in accidental hypothermia', *Prehospital Emergency Care* 18:446-449. G. G. Giesbrecht and J. S. Hayward (2006) 'Problems and complications with cold-water rescue', *Wilderness & Environmental Medicine* 17:26-30

81 C. S. Beck, W. H. Pritchard and H. S. Feil (1947) 'Ventricular fibrillation of long duration abolished by electric shock', *Journal of the American Medical Association* 135:985-986

82 Claude S. Beck, Elden C. Weckesser and Frank M. Barry (1956), 'Fatal heart attack and successful defibrillation: New concepts in coronary heart disease', *Journal of the American Medical Association* 161:434-436

83 Mark E. Silverman (2009), 'Restoring Life: The Story of Human Defibrillation and Modern CPR', Chapter in *The Textbook of Emergency Cardiovascular Care and CPR*, Edit. John M. Field Walters Kluwer/Lippincott Williams and Wilkins, p.197

84 Dave Beaudouin (2002) , 'W. B. Kouwenhoven: Reviving the body electric', *Johns Hopkins Engineer* Fall 2002: 27-32

85 W. B. Kouwenhoven, W. R. Milner, G. G. Knickerbocker and W. R. Chesnut (1957), 'Closed chest defibrillation of the heart', *Surgery* 42:550-561. Ramsey Flynn 'A Dying Dog, a Slow Elevator, and 50 years of CPR', *Hopkins Medicine Magazine* 18 February 2011. Brian Dawson, Emerson A. Moffat, William J. Glover and H. J. C. Swan (1962) 'Closed-Chest Resuscitation in a Cardiac Catheterisation Laboratory', *Circulation* 25:976-984

86 M. F. ORourke, E. Donaldson and J.S. Geddes (1997), 'An airline cardiac arrest program', *Circulation* 96:2849-2853

87 Terence D. Valenzuela et al (2000) 'Outcomes of Rapid Defibrillation by Security Officers after Cardiac Arrest in Casinos', *New England Journal of Medicine* 343:1206-1209

88 Steve Beerman and Bo Løfgren (2006) 'Automated External Defibrillators in the Aquatic Environment', Section in *Handbook on Drowning*, p.331-335

89 Frank Golden and Michael Tipton, *Essentials of Sea Survival*, p.272

90 Frank Golden and Michael Tipton, *Essentials of Sea Survival*, p.273

91   Frank Golden and Michael Tipton, *Essentials of Sea Survival*, p.253-25. G. G. Giesbrecht and G. K. Bristow (1992), 'A second postcooling afterdrop: more evidence for a convective mechanism', *Journal of Applied Physiology* 73:1253-1258

92   Frank Golden and Michael Tipton, *Essentials of Sea Survival*. p. 273-274

93   B. A. Gooden (1992) 'Why some people do not drown: Hypothermia versus the diving response', *The Medical Journal of Australia* 157:629-632. Brett A. Gooden (1994) 'Mechanism of the human diving response', *Integrative Physiological and Behavioural Science* 29:6-16

94   F. St C. Golden, M. J. Tipton and R. C. Scott (1997) 'Immersion, near-drowning and drowning.' *British Journal of Anaesthesia* 79:214-225

95   J. H. Modell, A. H. Idris, J. A. Pineda and J. H. Silverstein (2004) 'Survival after prolonged submersion in freshwater in Florida', *Chest*, 125:1948-1451

96   A. M. Kemp and J. R. Sibert (1991), 'Outcome in children who nearly drown: a British Isles study', *British Medical Journal* 302:931-933

97   Lech Mintowt-Czyz, 'Back from the dead: Joe drowned... then came to life after seven-hour resuscitation', *The Evening Standard*, 14 July 2004

98   W. R. Keatinge, 'Hypothermia: dead or alive?' *British Medical Journal* 302:3-4

99   J. P. Wyatt, G. S. Tomlinson and A. Busuttil (1999), 'Resuscitation of drowning victims in south-east Scotland', *Resuscitation* 41:101-104

100   David Steedman, Timothy Rainer and Ciro Campanella (1997), 'Cardiopulmonary resuscitation following profound immersion hypothermia', *Journal of Accident and Emergency Medicine* 14:170-172

101   Jonas Hilmo, Torvind Naesheim and Mads Gilbert (2014), '"Nobody is dead until warm and dead": Prolonged resuscitation is warranted in arrested hypothermic victims also in remote areas – A retrospective study from northern Norway', *Resuscitation*, 85:1204-1211

102   'Bag valve mask', Wikipedia. http://en.wikipedia.org/wiki/Bag_valve_mask

103   'Tracheal tube', Wikipedia https://en.wikipedia.org/wiki/Tracheal_tube

104   A. D. Simcock (1986), 'Treatment of near drowning – a review of 130 cases', *Anaesthesia* 41:643-648

CHAPTER 31: HOSPITAL TREATMENT AND INTENSIVE CARE

1   J. H. Modell, S. A. Graves and E. J. Kuck (1980), 'Near-drowning: Correlation of level of consciousness and survival.' *Canadian Anaesthetists' Society Journal* 27:211-215

2   John Pearn (1992), 'The urgency of immersions', *Archives of Disease in Childhood* 67:257-261

3   J. Pearn, J. Nixon and I. Wilkey (1976) 'Freshwater drowning and near-drowning accidents involving children: a five-year total population study', *Medical Journal of Australia* 2(25-26):942-946

4   A. M. Kemp and J. R. Sibert (1991) 'Outcome in children who nearly drown: a British Isles study', *British Medical Journal* 302:931-933

5   Demetrios N. Kyriacou, Edgardo L. Arcinue, Corinne Peek and Jess F. Kraus (1994), 'Effect of Immediate Resuscitation on Children with Submersion Injury', *Pediatrics* 94:137-142

6   Martin M. Monti, Steven Laureys and Adrian M. Owen (2010), 'The vegetative state', *British Medical Journal* 341:292-296

7   Joseph A. Layon and Jerome H. Modell (2009) 'Drowning: Update 2009', *Anesthesiology* 110:1390-1401

8   'Pulse Oximetry', Wikipedia http://en.wikipedia/wiki/Pulse_oximetry

9   'Oxygen Saturation (medicine)', https://en.wikipedia.org/wiki/Oxygen_saturation

10   'Bag valve mask', Wikipedia. http://en.wikipedia.org/wiki/Bag_valve_mask

11   Frank Golden and Michael Tipton, *Essentials of Sea Survival*, p.83-86

12   Joseph A. Layon and Jerome H. Modell (2009) 'Drowning: Update 2009', *Anesthesiology* 110:1390-1401

13   'Bag valve mask', Wikipedia. http://en.wikipedia.org/wiki/Bag_valve_mask

14   Martin J. Tobin and Patrick J. Fahey 'Management of the patient who is "fighting the ventilator"', Chapter in *Principles and Practice of Mechanical Ventilation*. Edited by Martin J. Tobin (1994) McGraw-Hill, New York. p.1149-1162

15   N. Manolios and I. Mackie (1988) 'Drowning and near-drowning on Australian beaches patrolled by life-savers: a 10-year study, 1973-1983', *Medical Journal of Australia* 148:165-167, 170-171

16   F. St. C. Golden, M.J. Tipton and R.C. Scott (1997), 'Immersion, near-drowning and drowning', *British Journal of Anaesthesia* 79:214-225

17   R. S. Watson, P. Cummings, L. Quan, S. Bratton and N. S. Weiss (2001) 'Cervical spine injuries among submersion victims', *The Journal of Trauma* 51:658-662

18   C. J. Hinds and D. Watson (1999), 'Circulatory support', Section in *ABC of Intensive Care*, Edit. Mervyn Singer and Ian Grant. BMJ Publishing, London. p.16-19

19   A.D. Simcock (1986), 'Treatment of near drowning – a review of 130 cases', *Anaesthesia* 41:643-648

20   *ABC of Intensive Care*. Edit. Graham R. Nimmo and Mervyn Singer (2011) BMJ Books, London.

21   Charles E. Putman, Anna M. Tummillo, Daniel A. Myerson and Paul J. Myerson (1975) 'Drowning: another plunge', *American Journal of Roentgenology* 125:543-548

22   Frank Golden and Michael Tipton, *Essentials of Sea Survival*, p.82-83

23   Acute respiratory distress syndrome.' Wikipedia. https://en.wikipedia.org/wiki/Acute_respiratory_distress_syndrome. Norman L. Fine, Daniel A. Myerson, Paul J. Myerson and Joseph J. Pagliaro (1974) 'Near-Drowning Presenting as the Adult Respiratory Distress Syndrome', *Chest* 65:347-349. Davide Chiumello, E. Carlesso and Luciano Gattinoni (2006) 'Management of ARDS', Chapter in *Handbook on Drowning*, p.410-416

24   A. D. Simcock (1986) 'Treatment of near drowning – a review of 130 cases', *Anaesthesia* 41:643-648

25   F. St. C. Golden, M.J. Tipton and R.C. Scott (1997) 'Immersion, near-drowning and drowning', *British Journal of Anaesthesia* 79:214-225

26   Andrea Rossi and Marco V. Ranieri 'Positive end-expiratory pressure', Chapter in *Principles and Practice of Mechanical Ventilation* (1994), Edited by Martin J. Tobin. McGraw-Hill, New York, p.259-303

27   A. D. Simcock (1986) 'Treatment of near drowning – a review of 130 cases', *Anaesthesia* 41:643-648

28   H. Onarheim and V. Vik (2004) 'Porcine surfactant (Curosurf) for acute respiratory failure after near-drowning in 12 year old.' *Acta Anaesthesiologica Scandinavica* 48:778-781a. N. Fettah, D. Dilli, S. Beken, A. Zenciroglu and N. Okumus (2014), 'Surfactant for acute respiratory distress syndrome caused by near drowning in a newborn', *Pediatric Emergency Care* 30:180-181

29   'Cardiopulmonary bypass', Wikipedia. https://en.wikipedia.org/wiki/Cardiopulmonary_bypass. K.Visser and R. Schepp (2013), 'The treatment of accidental hypothermia and drowning with the aid of extracorporeal circulation/ECMO', *Journal of Cardiothoracic Surgery* 8(Suppl 1):083

30   Nick Britten, 'Hale and Hearty: the boy who died for seven hours', *The Daily Telegraph*, 15 July 2004

31   S. Steen, Q. Liao, L. Pierre, A. Paslevicius and T. Sjöberg (2002), 'Evaluation of LUCAS, a new device for automatic mechanical compression and active decompression resuscitation', *Resuscitation* 55:285-299. Stig Steen, Tryave Sjöberg, Paul Olsson and Marie Young (2005) 'Treatment of out-of-hospital cardiac arrest with LUCAS, a new device for automatic mechanical compression and active decompression resuscitation', *Resuscitation*, 67:25-30. Hans Friberg and Malin Rundgren (2009), 'Submersion, accidental hypothermia and cardiac arrest, mechanical chest compressions as a bridge to final treatment: a case report', *Scandinavian Journal of Trauma, Resuscitation and Emergency Medicine*, https://doi.org/10.1186/1757-7241-17-7

32   Frank Golden and Michael Tipton, *Essentials of Sea Survival*, p.89-93

33   Ramesh K. Batra and Jonathan J. Paddle (2009), 'Therapeutic hypothermia in drowning induced hypoxic brain injury: a case report', *Cases Journal* 2:9103, http://www.casesjournal.com/content/2/1/9103. David S. Warner and Johannes Knape (2006) 'Brain Resuscitation in the Drowning Victim', Chapter in *Handbook on Drowning*, p.437, 439 and 447-449

34   Wilfred G. Bigelow (1984), *Cold Hearts: The Story of Hypothermia and the Pacemaker in Heart Surgery*, McClelland and Stewart, Toronto

35   Duncan Gray (1987) 'Survival after burial in an avalanche', *British Medical Journal*, 294:611-612. Ulrich Althaus, Peter Aeberhard, Peter Schüpbach, Bernhard H. Nachbur and Wilfried Mühlemann (1982) 'Management of profound accidental hypothermia with cardiopulmonary arrest', *Annals of Surgery*, 195:492-495

36   Frank Golden and Michael Tipton, *Essentials of Sea Survival*, p.265

37   Frank Golden and Michael Tipton, *Essentials of Sea Survival*, p.107

38   Ortrud Vargas Hein, Andeas Triltsch, Christoph von Buch, Wolfgang J. Kox and Claudia Spies (2004). 'Mild hypothermia after near drowning in twin toddlers', *Critical Care*, 8:R353-R357

39   Linda Quan and Dennis Kinder (1992), 'Pediatric Submersions: Prehospital Predictors of Outcome', *Pediatrics*, 90:909-913. Saleh M. Al-Mofadda, Ali Nassar, Abdullah Al-Turki and Abdullah A. Al-Sallounm (2001) 'Pediatric near drowning: The experience of King Khalid University Hospital', *Annals of Saudi Medicine*, 21:300-303

40   James P.Orlowski (1988) 'Drowning, Near-Drowning, and Ice-Water Drowning', Editorial,

Journal of the American Medical Association 260:390-391. 'Kiss of life for a cold corpse', Editorial (1990) *The Lancet* 335:1435

41 Robert G. Bolte, Philip G. Black, Robert S. Bowers, J. Kent Thorne and Howard M. Corneli (1988) 'The use of extracorporeal warming in a child submerged for 66 minutes', *Journal of the American Medical Association* 260:377-379

42 S. K. Hughes, D. E. Nilsson, R. G. Bolte, R. O. Hoffman, J. D. Lewine and E. D. Bigler (2002) 'Neurodevelopmental outcome for extended cold water drowning: a longitudinal case study', *Journal of the International Neuropsychological Society* 8:588-595

43 M. Gilbert, R. Busund, A. Skagseth, P. A. Nilsen and J. P. Sølbe (2000), 'Resuscitation from accidental hypothermia of 13.7 degrees C with circulatory arrest', *The Lancet* 355:375-376

44 'Frozen Woman: A "Walking Miracle"', CBS News, 11 February 2009

45 P. B. Crone and G. H. Tee (1974), 'Staphylococci in swimming pool water', *The Journal of Hygiene (London)*, 73:213-220

46 Clifford Dobell (1960) *Antony van Leeuwenhoek and his 'Little Animals'*, Dover Books, New York

47 Ryan Watts, Rhys Blakely, George Greenwood and Dylan Lewis, 'No river safe for bathing', *The Times*, 3 August 2019

48 'Marine organism', https://en.wikipedia.org/wiki/Marine-microorganism. Jay M. Fleisher, David Kay, Roland L. Salmon, Frank Jones Mark D. Wyer and Alan F. Godfree (1996) 'Marine Waters Contaminated with Domestic Sewage: Nonenteric Illnesses Associated with Bather Exposure in the United Kingdom', *American Journal of Public Health* 86:1228-1234

49 A. D. Simcock (1986) 'Treatment of near drowning – a review of 130 cases', *Anaesthesia* 41:643-648

50 Roger H. Fuller (1963) 'The Clinical Pathology of Human Drowning', *Proceedings of the Royal Society of Medicine* 56:33-38

51 C. Langton Hewer (1962), 'Drowning', Letter, *The Lancet*, 279:636

52 M. Noguchi, Y. Kimula and T. Ogata (1985) 'Muddy Lung', *American Journal of Clinical Pathology* 83:240-244

53 Donnie P. Dunagan, Joseph E. Cox, Michael C. Chang and Edward F. Haponik (1997) 'Sand Aspiration with Near-drowning: Radiographic and Bronchoscopic Findings', *American Journal of Respiratory and Critical Care Medicine* 156:292-295. N. Kapur, A Slater, J.McEniery, M.L. Greer, I.B. Masters and A.B. Chang (2009) 'Therapeutic Bronchoscopy in a child with sand aspiration and respiratory failure from near-drowning – case report and literature review', *Pediatric Pulmonology* 44:1043-1047. H. Mangge, B. Plecko, H. M. Grubbauer et al (1993) 'Late-onset miliary pneumonitis after near drowning', *Pediatric Pulmonology* 15:122-124

54 Peter T. Ender and Matthew J. Dolan (1997) 'Pneumonia Associated with Near-Drowning', *Clinical Infectious Diseases* 25:896-907. A. Robert, P. E. Danin, H. Quintard et al. (2017), 'Seawater drowning-associated pneumonia: a 10-year descriptive cohort in intensive care unit', *Annals of Intensive Care* 7:45. doi:10.1186/s13613-017-0267-4

55 M. Van Berkel, J. J. Bierens et al (1996), 'Pulmonary oedema, pneumonia and mortality in submersion victims; a retrospective study in 125 patients', *Intensive Care Medicine* 22:101-107

56 Peter T. Ender and Matthew J. Dolan (1997), 'Pneumonia Associated with Near-Drowning',

Clinical Infectious Diseases 25:896-907. D. F. Viera, H. K. F. Van Saene and D.R. Miranda (1984) 'Invasive pulmonary aspergillosis after near-drowning', *Intensive Care Medicine* 10:202-203

57   A. J. Peabody (1980) 'Diatoms and Drowning – A Review', *Medicine, Science and the Law* 20:254-261

58   Piet Leroy, Annick Smismans and Tatjana Seute (2006) 'Invasive Pulmonary and Central Nervous System Aspergillosis After Near-Drowning of a Child: Case Report and Review of the Literature', *Pediatrics* 118:509-513

59   Christian Garzoni, Stephane Emonet, Laurence Legout, Rilliet Benedict, Pierre Hoffmeyer, Louis Bernard and Jorge Garbino (2005) 'Atypical Infections in Tsunami Survivors', *Emerging Infectious Diseases* 11:1591-1593

60   David S. Warner and Johannes Knape (2006), 'Brain Resuscitation in the Drowning Victim', Chapter in *Handbook on Drowning*. p. 447. A. C. J. M. De Pont, C. P. C. De Jager, W. M. Van den Bergh and M.J. Schultz (2011) 'Recovery from near drowning and postanoxic status epilepticus with controlled hypothermia', *The Netherlands Journal of Medicine* 69:196-197

61   T. A. Ports and T. F. Deuel (1977), 'Intravascular coagulation in fresh-water submersion: report of three cases', *Annals of Internal Medicine* 87:60-61. Michael Schwameis et al (2015) 'Asphyxia by Drowning Induces Massive Bleeding Due To Hyperfibrinolytic Disseminated Intravascular Coagulation', *Critical Care Medicine* 43:2394-2402. M. Levi, C. H. Toh, J. Thachil and H. G. Watson (2009) 'Guidelines for the diagnosis and management of disseminated intravascular coagulation', *British Journal of Haematology* 145:24-33

62   S. Timothy Spicer, David Quinn, Nyein N. Nyi, Brian J. Nankivell, James M. Hayes and Elliott Savdie (1999) 'Acute Renal Impairment after Immersion and Near-Drowning', *Journal of the American Society of Nephrology* 10:382-386

63   Magnus W. Prull, Gülseren Sürmeci, Irini M. Breker and Marc van Bracht (2012) 'Case Report: Resuscitation of a 55-year old physician,' *Applied Cardiopulmonary Pathophysiology* 16:121-126

64   Linda Quan and Dennis Kinder (1992) 'Pediatric Submersions: Prehospital Predictors of Outcome', *Pediatrics* 90:909-913

65   Bradley Peterson (1977) 'Morbidity of Childhood Near-Drowning', *Pediatrics* 59:364-370

66   'Reperfusion injury', https://en.wikipedia.org/wiki/Reperfusion_injury

67   Robin S. Howard, Paul A. Holmes and Michalis A. Koutroumanidis (2011) 'Hypoxic-ischaemic brain injury', *Practical Neurology* 11:4-18

68   'Glasgow Coma Scale', Wikipedia, http://en.wikipedia/wiki/Glasgow_Coma_Scale

69   Robin S. Howard, Paul A. Holmes and Michalis A. Koutroumanidis (2011) 'Hypoxic-ischaemic brain injury', *Practical Neurology* 11:4-18

70   J. H. Waugh, M. J. O'Callaghan and W. R. Pitt (1994) 'Prognostic factors and long-term outcomes for children who have nearly drowned', *Medical Journal of Australia* 161:598-599

71   Matthew J. Biggart and Desmond J. Boh (1990), 'Effect of hypothermia and cardiac arrest on outcome of near-drowning accidents in children', *The Journal of Pediatrics* 117:179-183

72   Harald Siebke, Tore Rød, Harald Breivik and Bjorn Lind (1975) 'Survival after 40 minutes submersion without cerebral sequelae', *The Lancet* 305:1275-1277. John Pearn (1977)

'Neurological and psychometric studies in children surviving freshwater immersion accidents', *The Lancet* 309:7-9. C. Eich, A. Brauer, D. Kettler (2005) 'Recovery of a hypothermic drowned child after resuscitation with cardiopulmonary bypass followed by prolonged extracorporeal membrane oxygenation', *Resuscitation* 67:145-148

73   Robert G. Bolte, Philip G. Black, Robert S. Bowers, J. Kent Thorne and Howard M. Corneli (1988), 'The use of extracorporeal warming in a child submerged for 66 minutes', *Journal of the American Medical Association* 260:377-379

74   Klaus Brinkbäumer, 'Ein perfecter Unfall', *Der Speigel*, 17 February 2007

75   'Skier revived from clinical death', BBC News, 28 January 2000

CHAPTER 32: THE PROBLEM OF WATER IN THE LUNG

1   R. K. Haugen (1963) 'The Café Coronary: Sudden Deaths in Restaurants', *Journal of the American Medical Association* 186:142-143

2   Henry Heimlich (2012) 'Historical Essay: The Heimlich Maneuver', *American Bronchoesophological Association Spring Newsletter* 2012

3   Henry J. Heimlich (1975) 'Pop goes the cafe coronary' *Emergency Medicine* 6:154-155

4   Henry Heimlich (2012) 'Historical Essay: The Heimlich Maneuver', *American Bronchoesophological Association Spring Newsletter* 2012

5   Henry J. Heimlich (1975) 'A Life-Saving Maneuver to Prevent Food-Choking', *Journal of the American Medical Association* 234:398-401

6   Henry Heimlich (2012) 'Historical Essay: The Heimlich Maneuver', *American Bronchoesophological Association Spring Newsletter* 2012

7   Henry J. Heimlich (1975) 'A Life-Saving Maneuver to Prevent Food-Choking', *Journal of the American Medical Association* 234:398-401

8   Henry J. Heimlich (1981) 'Subdiaphragmatic Pressure to Expel Water From the Lungs of Drowning Persons', *Annals of Emergency Medicine* 10:476-480

9   Tod Spivak 'Fighting for Air: Drowning and the Heimlich Maneuver', *Houston Press*, 10 October 2007

10   James O Elam, Elwyn S. Brown and John D. Elder (1954) 'Artificial Respiration by Mouth-to-Mask Method – A Study of the Respiratory Gas Exchange of Paralysed Patients Ventilated by Operator's Expired Air.' *New England Journal of Medicine* 250:749-754. Archer S. Gordon, Charles W. Frye, Lloyd Gittelson, Max S. Sadove and Edward J. Beattie Jr (1958), 'Mouth-to-mouth versus artificial respiration for children and adults', *Journal of the American Medical Association* 167:320-328. James O. Elam, David G. Greene, Elwyn S. Brown and John H. Clements (1958) 'Oxygen and carbon dioxide exchange and energy cost of expired air resuscitation', *Journal of the American Medical Association* 167:328-334. David G. Greene and James O.Elam (1960) 'Expired air resuscitation during cardiac emergencies', *Journal of the American Medical Association* 173:375. James O. Elam, Arne Ruben, David G. Greene and Theodore J. Bittner (1961) 'Mouth-to-nose Resuscitation During Convulsive Seizures', *Journal of the American Medical Association* 176:565-569

11   W. B. Kouwenhoven, James R. Jude and G. Guy Knickerbocker (1960), 'Closed-chest Cardiac Massage', *Journal of the American Medical Association* 173:1064-1067. James R. Jude, W. B. Kouwenhoven and G. Guy Knickerbocker (1961), 'A New Approach to Cardiac Resuscitation', *Annals of Surgery* 154:311-317

12   Peter Safar, Torrey C. Brown, Warren J. Holtey and Robert J. Wilderl (1961), 'Ventilation and Circulation with Closed-Chest Cardiac Massage in Man', *Journal of the American Medical Association* 176:574-576 (Note: Soon after this data was collected, the Dean of Johns Hopkins Medical School prohibited further use of medical student volunteers in CPR research.)

13   Peter Safar, Lourdes A. Escarraga and James O. Elam (1958) 'A comparison of the mouth-to-mouth and mouth-to-airway methods of artificial respiration with the chest-pressure arm-lift methods', *New England Journal of Medicine* 258:651-677

14   *'New Pulse of Life'* [motion picture], Pyramid Films. Revised edition of the 1962 production. Dennis K. Graver *Aquatic Rescue and Safety*. p.180-181

15   Terry L. Vanden Hoek (Chair) et al (2010) *'2010 American Heart Association Guidelines for Cardiopulmonary Resuscitation and Emergency Cardiovascular Care. Part 12: Cardiac Arrest in Special Situations. Part 12.11: Drowning'*, *Resuscitation* 122:S829-861

16   J Dennis K. Graver, *Aquatic Rescue and Safety*. p.180-181

17   A.S. Simcock (1986) 'Treatment of near drowning – a review of 130 cases', *Anaesthesia* 41:643-648 (Note: Only part of this liquid is inhaled water. The rest is oedema fluid formed after water damage to the lungs.)

18   J. Z. B. S. Werner, P. Safar, N. G. Bircher, W. Stezoski, M. Scanlon and R. D. Stewart (1982), 'No Improvement in Pulmonary Status By Gravity Drainage or Abdominal Thrusts After Sea Water Near Drowning in Dogs', *Anesthesiology* 57:A81

19   Modell and F. Moya (1966) 'Effects of volume of aspirated fluid during chlorinated fresh water drowning', *Anesthesiology* 27:662-67

20   Ron Word, 'UF allows research on drowning dogs', *St Petersburg Evening Independent*, 13 February 1986. Ron Word 'Committee approves request to drown dogs for research.' *Associated Press* 12 February 1986

21   Henry Heimlich, 'Drowning experiments cruel', *Palm Beach Post*, 13 March 1986

22   Thomas Francis, 'Heimlich's Maneuver', *Cleveland Scene*, 11 August 2004

23   'Researchers may not conduct controversial drowning study on dogs', *Associated Press,* 20 February 1986

24   Arne Ruben and Henning Ruben (1962) 'Artificial respiration: Flow of water from the lung and the stomach', *The Lancet* 1:780-781 (Note: The anaesthetists worked on the early trials of mouth-to-mouth resuscitation. Henning Ruben was the inventor of two important pieces of resuscitation equipment: the first ambulant suction pump and the first self-inflation resuscitator, the Air Mask Bag Unit or AMBU. Henning Rubin (1959) 'Self-contained resuscitation equipment', *Canadian Medical Association Journal* 80:44-45. Henning M. Rubin, James O Elam, Arne Ruben and David Greene (1961) 'Investigation of upper airway problems in resuscitation.' *Anesthesiology* 22:271-279.)

25   Thomas F. Baskett (2007) 'The Holger Nielsen method of artificial respiration', *Resuscitation* 74:403-405 (Note: The Holger Nielsen Method was widely used in Scandinavia for decades

and became the method of choice in America and Britain during the 1950s.) Frank C. Eve (1932), 'Actuation of the inert Diaphragm by a gravity method', *Lancet* 220:995-997. (Note: The casualty was laid face down on a stretcher, which was rocked head down then head up, forcing the abdominal organs to slide back and forth like a piston, so moving the diaphragm up and down.) G. H. Gibbens (1942) 'Artificial respiration at sea', *British Medical Journal* 2(4277):751-752 (Note: Eve's Rocking Method was adopted by the Royal Navy during World War 11.)

26   Mark Heining, David Bogod and Alan Aitkenhead (1996), *Essential Anaesthesia for Medical Students*. Arnold, London.

27   Joseph P. Ornato (1986) 'The Resuscitation of Near-Drowning Victims', *Journal of the American Medical Association* 256:75-77. L. Quan (1993) 'Drowning issues in resuscitation.' *Annals of Emergency Medicine* 22:366-369. Peter Rosen, Michael Stoto and Jim Harley (1995) 'The use of the Heimlich maneuver in near drowning: Institute of Medicine Report', *The Journal of Emergency Medicine* 13:397-405. A. Joseph Layon and Jerome H. Modell (2009), 'Drowning: Update 2009', *Anesthesiology* 110:1390-1401. David Szpilman and Anthony Handley (2014), 'Positioning of the Drowning Victim', Chapter in *Drowning: Prevention, Rescue, Treatment*. p.629. Robyn Meyer, Andreas Theodorous and Robert Berg (2014) 'Paediatric Considerations in Drowning', Chapter in *Drowning: Prevention, Rescue, Treatment*. p.643

28   'Discussion on Artificial Respiration', (1960) *Proceedings of the Royal Society of Medicine* 53:311-315

29   James O. Elam, Arne M. Ruben and David G. Greene (1960) 'Resuscitation of Drowning Victims', *Journal of the American Medical Association* 174:13-16

30   N. Manolios and I. Mackie (1988) 'Drowning and near drowning on Australian beaches patrolled by life-savers: a 10 year study', *Medical Journal of Australia* 148:165-171

31   Henry J. Heimlich (1981) 'Subdiaphragmatic Pressure to Expel Water From the Lungs of Drowning Persons', *Annals of Emergency Medicine* 10:476-480

32   James P. Orlowski (1987) 'Vomiting as a Complication of the Heimlich Maneuver', *Journal of the American Medical Association* 258:512-513

33   H. J. Heimlich and E. A. Patrick (1988) 'Using the Heimlich maneuver to save near-drowning victims', *Postgraduate Medicine* 84:71-73

34   Henry H. Heimlich (1981) 'Subdiaphragmatic Pressure to Expel Water From the Lungs of Drowning Persons', *Annals of Emergency Medicine* 10:476-480

35   Terry L. Vanden Hoek (Chair) et al (2010) '*2010 American Heart Association Guidelines for Cardiopulmonary Resuscitation and Emergency Cardiovascular Care. Part 12: Cardiac Arrest in Special Situations. Part 12.11: Drowning.*' *Resuscitation* 122:S829-861

36   Joseph Redding, C. Carl Voigt, and Peter Safar (1960) 'Drowning treated with intermittent positive pressure breathing', *Journal of Applied Physiology* 15:849-854

37   J. H. Modell and F. Moya (1966) 'Effects of volume of aspirated fluid during chlorinated fresh water drowning', *Anesthesiology* 27:662-672

38   Phillipe Lunetta and Jerome H. Modell (2005) 'Macroscopical, Microscopical, and Laboratory Findings in *Drowning Victims: A Comprehensive Review*', *Forensic Pathology Reviews* Vol. 3. Edited by Michael Tsokos (Hamburg) Humana Press, Totowa, NJ. p.3-77

39   David Szpilman, Joost J.L.M. Bierens, Anthony J. Handley and James P. Orlowski(2012) 'Drowning', *New England Journal of Medicine* 366:2102-2110

40   John Pearn (1985), 'The management of near drowning', *British Medical Journal* 291:1447-1452

41   Lauralee Sherwood, *Fundamentals of Physiology. A Human Perspective*. p. 338-339

42   Linda Quan (1993), 'Drowning issues in resuscitation', *Annals of Emergency Medicine* 22:366-369

43   http://www.cdc.gov/healthywater/swimming/pools/mahc/committees

44   Alison Rice 'Dr John Hunsucker', *Aquatics International*, February 2013

45   John L. Hunsucker, 'The Heimlich Maneuver and Drowning', *Splash*, World Waterpark Association, May-June 1996

46   Lifeguard Textbook (1999, revised 2011), National Aquatics Safety Company, http://nascoaquatics.com/wp-content/uploads/2012/02/lifeguard-textbook-2012.pdf

47   J. L. Hunsucker and S. J. Davison (2011) 'Analysis of rescue and drowning history from a lifeguarded waterpark environment', *International Journal of Injury Control and Safety Promotion* 18:277-284

48   Tod Spivak, 'Fighting for Air: Drowning and the Heimlich Maneuver', *Houston Press*, 10 October 2007

49   Lifeguard Textbook (1999, revised 2011), National Aquatics Safety Company, http://nascoaquatics.com/wp-content/uploads/2012/02/lifeguard-textbook-2012.pdf

50   J. L. Hunsucker, 'Opinion: To Heimlich First or Not?' *Illinois Parks & Recreation*, July/August 1998

51   Tod Spivak 'Fighting for Air: Drowning and the Heimlich Maneuver', *Houston Press*, 10 October 2007

52   Pamela Mills-Sen 'Water Rescue Sequence. The Controversial Role of the Heimlich Maneuver', United States Lifesaving Association 2000 https://www.usla.org/resmgr/lifeguard-library/heimlich_article_mills_sen (Note: This article was originally intended for publication in *Funworld*, the official magazine of the International Association of Amusement Parks and Attractions (IAAPA). It was published as a special supplement, '*Funworld Special Report*', which was mailed to IAAPA members on 30 March 2000.)

53   J. L. Hunsucker and S. J. Davison (2011), 'Analysis of rescue and drowning history from a lifeguarded waterpark environment', *International Journal of Injury Control and Safety Promotion* 18:277-284

54   'An Open Letter To Our Clients, The Public and The Press', NASCO, November 2009, http://nascoaquatics.com/?page_id-1273

55   '2010 American Heart Association Guidelines for Cardiopulmonary Resuscitation and Emergency Cardiovascular Care Science' *Circulation*, 122 Issue 18 Supplement 3: S639-S946

56   Robert A. Berg et al (2010) 'Part 5: Adult Basic Life Support. 2010 American Heart Association Guideline for Cardiopulmonary Resuscitation and Emergency Cardiovascular Care', *Circulation* 122:S685-S705

57   Marc B. Berg et al (2010) 'Part 13: Pediatric Basic Life Support. 2010 American Heart Association Guidelines for Cardiopulmonary resuscitation and Emergency Cardiovascular Care', *Circulation* 122:S862-875

58   Terry L. Vanden Hoek et al (2010) 'Part 12: Cardiac Arrest in Special Situations.' 2010 American Heart Association Guidelines for Cardiopulmonary Resuscitation and Emergency Cardiovascular Care', *Circulation* 122: S829-S861

59   Robert A. Berg et al (2010) 'Part 5: Adult Basic Life Support. 2010 American Heart Association Guidelines for Cardiopulmonary Resuscitation and Emergency Cardiovascular Care', *Circulation* 122:S685-S705. Terry l. Vanden Hoek et al (2010) 'Part 12: Cardiac Arrest in Special Situations. Part 12.11: Drowning. 2010 American Heart Association Guidelines for Cardiopulmonary Resuscitation and Emergency Cardiovascular Care', *Circulation* 122: S829-S861

60   J. H. Modell (2010) 'Prevention of needless deaths from drowning.' *The Southern Medical Journal* 103:650-653

61   P. F. Mahoney, L. Williams and J. I. Andrews (1993) 'Successful resuscitation from sea water drowning', *Archives of Emergency Medicine* 10:120-122

62   D. Redmond and T. S. Mallikarjun (1984), 'Resuscitation from drowning', *Archives of Emergency Medicine* 1:113-115

63   'Proficiency Testing requirements for the 2010-2011 season', *Surf Life Saving Australia*, June 2010

64   A. D. Simcock (1986), 'Treatment of near drowning – a review of 130 cases.' *Anaesthesia* 41:643-6748

65   J. L. Hunsucker 'Opinion: To Heimlich First or Not', *Illinois Parks and Recreation*, July/August 1998

66   Don Harris, 'Lifeguard Discussion List', 13 August 2004. http://listserv.icors.org/scripts/wa-ICORS.exe?A2=ind0408&L=lifeguard

67   H.J. Heimlich (1981) 'Subdiaphragmatic Pressure to Expel Water From the Lungs of Drowning Persons', *Annals of Emergency Medicine* 10:476-480

68   David S. Warner, Joost J. L. M. Bierens, Stephen B. Beerman and Laurence M. Katz (2009), 'Drowning: A Cry for Help', *Anesthesiology* 110:1211-1213

69   Joost Bierens (2011) 'Lack of evidence blocks development of drowning resuscitation guidelines', World Conference on Drowning Prevention, Da Nang, Vietnam

70   '2005 American Heart Association Guidelines for Cardiopulmonary Resuscitation and Emergency Cardiovascular Care. Part 1: Introduction', *Circulation* (2005) 112:IV-1-IV-5

71   John L. Hunsucker, 'NASCO Position Paper. The Use of the Sub-Diaphragmatic Thrust in Drowning', National Aquatic Safety Company, February 2011

CHAPTER 33: INQUEST VERDICTS AND DROWNING STATISTICS

1   A guide to Coroners and Inquests', Ministry of Justice. January 2010

2   'The role of the Procurator Fiscal in the investigation of deaths' Crown Office and Procurator Fiscal Service, Edinburgh.

3   'Inquests in England and Wales', Wikipedia. http://www.en.wikipedia.org/wiki/Inquests_in_England

4   'The role of the Procurator Fiscal in the investigation of deaths', Crown Office and Procurator

Fiscal Service, Edinburgh

5   'Inquests in England and Wales', Wikipedia, http://www.en.wikipedia.org/wiki/Inquests_in_ England..

6   *The Inquest Handbook: a Guide for bereaved families, friends and advisers*, Section 4.3 Conclusions (2016) https://www.inquest.org.uk (Note: 'Death by misadventure' is similar to 'Accidental Death', but it implies that an intended legal action went wrong.)

7   Wills Robinson, 'Boy, 2, drowned in family pool as his parents got ready for their wedding day' *The Daily Mail*, 9 January 2014

8   Mia de Graaf ,'Is this Britain's first ice bucket drowning death? Teenager drowns in quarry after tombstoning as part of internet craze', *The Daily Mail*, 25 August 2014

9   'Liverpool teacher drowned in "tragic" kayaking accident', *Liverpool Echo* 23 November 2012

10   Karen Britton, 'Tragedy of grandfather who drowned on boat trip in Australia,' *The Macclesfield Express*, 21 February 2014

11   'Veteran sailor died in the Bristol Channel because he wore the wrong shoes', *Bristol Post*, 9 May 2015

12   'Coroner's Inquests – A Guide for Learners', Health Education England. www.oxforddeanery. nhs.uk/pdf/

13   Nick Lavigueur 'Body of missing Huddersfield diver… recovered from Cumbria lake', *Huddersfield Examiner*, 31 January 2013

14   Neil Atkinson 'Huddersfield man's fatal heart attack in Lake District dive', *Huddersfield Examiner*, 27 February 2014

15   'Windsurfer died of heart trouble while out in the Solent', *The News*, Portsmouth 8 May 2015

16   'Coroners: Legal Guidance: Crown Prosecution Service', http://www.cps.gov.uk/legal/a_ to_c/coroners/#a09

17   Emma Lidiard, 'Rocks in rucksack of Ullswater ferry jumper – inquest', *The Westmorland Gazette*, 3 November 2010

18   *The Inquest Handbook: a Guide for bereaved families, friends and advisers*, Section 4.3 Conclusions. (2016) https://www.inquest.org.uk

19   Zaiba Malik, 'Watery grave', *The Guardian*, 15 December 2004

20   'Drowned 70-year-old had fallen victim to fraudsters', *Bristol Post*, 9 September 2011

21   Martin Shaw, 'Mystery remains over death of Milnsbridge teenager drowned in Huddersfield canal', *Huddersfield Examiner*, 25 April 2013

22   'Coroners: Legal Guidance: Crown Prosecution Service', http://www.cps.gov.uk/legal/a_to_c/ coroners/#a09. *The Inquest Handbook: a guide for bereaved families, friends and advisers*. Section 4.3 Conclusions. (2016). https://www.inquest.org.uk

23   'Kent man drowned in car while birdwatching', *Kent Online*, 10 June 2010

24   'Sick Penrith woman "dropped" in "ice cold" sea', BBC News, 21 April 2011

25   'Preventing Future Deaths Reports', Coroners: Legal Guidance: Crown Prosecution Service. http://www.cps.gov.uk/legal/a_to_c/coroners/#a28.   https://www.judiciary.gov.uk/wp-content/uploads/2014/.../Richardson-2013-0261.pdf

26   Chief Coroner (2013) 'Law Sheet No. 1: Unlawful Killing.' https://www.judiciary.gov.uk/wp-content/uploads/JCO/Documents/coroners

27   Patrick Barkham, '"Drunk bully" guilty of killing schoolmate', *The Guardian*, 10 March 2005

28   David Brown 'Drunk teenagers threw student to drown in icy river', *The Times*, 5 September 2008

29   Rachel McDermott, 'Nine-month-old baby drowned in bath when father went to get towel and nappy', *The Daily Mirror*, 26 February 2015

30   James Nixon and John Pearn (1977), 'Non-accidental immersion in bath-water: another aspect of child abuse', *British Medical Journal* i:271-272. A. M. Kemp, A. M. Mott and J. R. Sibert (1994) 'Accidents and child abuse in bathtub submersions', *Archives of Disease in Childhood* 70:435-43

31   Vicky Smith, 'Gross and criminal neglect: Mum left her baby to die in bath while she boozed… then tried to blame it on her older son', *The Daily Mirror*, 31 July 2012

32   Chief Coroner (2013) 'Law Sheet No. 1: Unlawful Killing.' https://www.judiciart.gov.uk.wp-content/uploads/JCO/Documents/coroners

33   Jan Dressler et al (2011) 'Neonatal Freshwater Drowning After Birth in the Bathroom', *American Journal of Forensic Medicine and Pathology* 32:119-123

34   Jay Dix, Michael Graham and Randy Hanzlich (2000), *Asphyxia and Drowning: An Atlas*, CRC Press, Boca Raton, Florida. Michel H. A. Piette and Els A. De Letter (2006), 'Drowning: Still a difficult autopsy diagnosis', *Forensic Science International* 163-1-9

35   Derrick J. Pounder (1992) 'Bodies from Water', Department of Forensic Medicine, University of Dundee

36   Vincent J. DiMaio and Dominick DiMaio (2001) 'Death by Drowning'. Chapter 15 in *Forensic Pathology*. 2nd Edn. CRC Press, Boca Raton, Florida. p.399-407

37   P. Lunetta, A. Pentilla and A Sajantilla (2002) 'Circumstances and macropathologic findings in 1590 consecutive cases of bodies found in water', *American Journal of Forensic Medicine and Pathology* 23:371-376

38   John Pearn (2009), 'Drowning and near drowning. Part 1 of 5.' Philippine Life Saving. http://www.philippinelifesaving.org/articles/drowning-near-drowning

39   Frank Golden and Michael Tipton, *Essentials of Sea Survival*, p.61 and 89

40   W. Keatinge and M. Hayward (1981) 'Sudden Death in Cold Water and Ventricular Arrhythmia', *Journal of Forensic Sciences* 26:459-461

41   Sonia Sharma, 'Man suffered heart attack in river fall', *News Chronicle*, Newcastle 23 February 2007

42   John Pearn (2009) 'Drowning and near drowning. Part 5 of 5', Philippine Life Saving. http://www.philippinelifesaving.org/articles/drowning-near-drowning

43   W. Lawler (1992), 'Bodies recovered from the water: a personal approach and consideration of the difficulties', *Journal of Clinical Pathology* 45:654-659

44   Andrea Zaferes and Walt Hendrick (2006) 'Homicidal Drowning', Chapter in *Handbook on Drowning*. p. 633-635

45   'Brides in the Bath Murders', The Metropolitan Police, London, http://content.met.police.uk/Article/Brides-in-the-Bath. Jane Robins (2010), *The Magnificent Spilsbury and the Case of the Brides in the Bath*, John Murray, London.

46   Michel H. A. Piette and Els A. De Letter (2006), 'Drowning: Still a difficult autopsy diagnosis', *Forensic Science International* 163:1-9

47   'Family sea deaths were accidents', BBC News, 22 August 2005

48   'Harold Holt drowned, coroner finds', *Sydney Morning Herald*, 2 September 2005

49   '1974: "Drowned" Stonehouse found alive', BBC Home. On This Day: 24 December 1974. https://news.bbc.co.uk/onthisday/hi/dates/stories/...

50   J. W. Passmore, J. O. Smith and A. Clapperton (2007), 'True burden of drowning: compiling data to meet the new definition', *International Journal of Injury Control and Safety Promotion* 14:1-3

51   John D. Langley and David J. Chalmers (1999), 'Coding the circumstances of injury: ICD-10 a step forward or backwards?' *Injury Prevention* 5:247-253. Clare Griffiths and Cleo Rooney (2003) 'The effect of the introduction of ICD-10 on trends in mortality from injury and poisoning in England and Wales', *Health Statistics Quarterly* 19:10-21

52   Gordon S. Smith and John D. Langley (1998) 'Drowning surveillance: How well do E codes identify submersion fatalities?' *Injury Prevention* 4:135-137

53   John D. Langley and Gordon Smith (1996), '"Hidden drownings": a New Zealand study', in *Proceedings of the International Collaborative Effort on Injury Statistics*, Vol 2. National Center for Health Statistics, Hyattsville, MD. https:www.cdc.gov/nchs/data/ice/ice95v2/c05.pdf

54   David Gunnell, Keith Hawton and Nav Kapur (2011) 'Coroners' verdicts and suicide statistics in England and Wales', Editorial, *British Medical Journal* 343.d6030

55   Anita Brock and Clare Griffiths (2003), 'Trends in suicide by method in England and Wales, 1979 to 2001', *Health Statistics Quarterly*, 20:7-18

56   K. R. Linsley, Kurt Shapiro and T.P Kelly (2001), 'Open verdict v. suicide – importance to research', *The British Journal of Psychiatry*, 178:465-468

57   Chris Hill and Lois Cook (2011) 'Narrative verdicts and their impact on mortality statistics in England and Wales', *Health Statistics Quarterly* 49:81-100. David Gunnell, Keith Hawton and Nav Kapur (2011) 'Coroners' verdicts and suicide statistics in England and Wales', Editorial, *British Medical Journal* 343.d6030

58   Chris Hill and Lois Cook (2011), 'Narrative verdicts and their impact on mortality statistics in England and Wales', *Health Statistics Quarterly* 49:81-100

59   R.. Carroll, K. Hawton, N. Kapur, O. Bennewith and D. Gunnell (2012), 'Impact of the growing use of narrative verdicts by coroners on geographical variations in suicide: analysis of coroners' inquest data', *Journal of Public Health* 34: 447-453

60   'Elderly twins' sea suicide pact', BBC News, 9 December 2004

61   Simon de Bruxelles, 'Mother texted boyfriend then drowned in her car', *The Times*, 19 January 2006

62   David Sapsted, 'Row led to family's river deaths', *The Daily Telegraph*, 22 July 2006

63   'Cliff death plunge: open verdict', *Bridlington Free Press* 13 September 2001

64   Philippe Lunetta, Tsung-Hsueh Lu and Gordon S. Smith (2011), 'Standard World Health Organisation (WHO) data on drowning: A cautionary note concerning undetermined drowning', World Conference on Drowning Prevention, Da Nang, Vietnam 2011

65   Gordon S. Smith et al (1996), 'International Comparisons of Injury mortality databases: evaluation of their usefulness for drowning prevention and surveillance.' http://www.cdc.gov/nchs/data/ice/ice95v2/c06.pdf

66   Clare Griffiths, Oliver Wright and Cleo Rooney (2006) 'Trends in injury and poisoning mortality using the ICE on injury statistics matrix, England and Wales 1979-2004', *Health Statistics Quarterly* 32:5-18

67   'Drowning Statistics – RoSPA. Accidental drowning deaths:1983-2013', http://www.rospa. com/leisure-safety/statistics/drowning

68   'Drowning Statistics – RoSPA' http://www.rospa.com/leisure-safety/statistics/drowning

69   'Drowning Statistics – RoSPA' Accidental drowning deaths: for the years 1983 to 2013', http://www.rospa.com/leisure-safety/statistics/drowning. John Connolly (2011), 'Suicide by drowning is the 21st century's rescue challenge: A review', World Conference on Drowning Prevention, Da Nang, Vietnam 2011

70   'RoSPA – Number of UK drownings at lowest since records began' 11 June 2015. http:// www.wired-gov.net/wg/news.nsf/articles/RoSPA

71   'One more drowning is still one too many', 12 June 2015, http://www.rlss.org.uk/news/

72   Laura Connor, 'Devastated mum of drowned student M… R… warns of dangers of water in heatwave', *Daily Mirror*, 18 June 2015. Dan Bean, 'Hard-hitting video serves as river warning', *The Press*, York, 17 June 2015 (Note: The video 'Beneath the Surface – The Families' Stories' includes interviews with relatives of drowning victims. RLSS UK Films. The video can be viewed on YouTube.) 'Dorset's emergency services back Drowning Prevention Week', *Dorset Echo*, 25 June 2015

73   'Drowning Statistics – R0SPA.' http://www.nationalwatersafety.org.uk/waid/info/waid_ fatalincidentreport_2014.xls

74   www.nationalwatersafety.org.uk/waid/info/waid_fatalincidentreport_2009.xls
        www.nationalwatersafety.org.uk/waid/info/waid_fatalincidentreport_2010.xls
        www.nationalwatersafety.org.uk/waid/info/waid_fatalincidentreport_2011.xls
        www.nationalwatersafety.org.uk/waid/info/waid_fatalincidentreport_2012.xls
        www.nationalwatersafety.org.uk/waid/info/waid_fatalincidentreport_2013.xls
        www.nationalwatersafety.org.uk/waid/info/waid_fatalincidentreport_2014.xls
        www.nationalwatersafety.org.uk/waid/info/waid_fatalincidentreport_2015.xls
        www.nationalwatersafety.org.uk/waid/info/waid_fatalincidentreport_2016.xls
        www.nationalwatersafety.org.uk/waid/info/waid_fatalincidentreport_2017.xls
        www.nationalwatersafety.org.uk/waid/info/waid_fatalincidentreport_2018.xls

CHAPTER 34: THE TRAGIC AFTERMATH OF A DROWNING ACCIDENT

1   Nancy J. Rigg 'Coping with a Sudden and Traumatic Drowning Death' 10 August 2010. https://drowningsupportnetwork.wordpress.com

2   'What to do after someone dies', A UK government leaflet available on the internet. https:// www.gov.uk/after-a-death/print. The Bereavement Advice Centre. https://bereavementadvice. org.uk. Childhood Bereavement Network http://www.childhoodbereavementnetwork.org.uk

3   J. Nixon and J. Pearn (1977) 'Emotional sequelae of parents and sibs following the drowning or near-drowning of a child.' *Australian and New Zealand Journal of Psychiatry* 11:265-268

4   Debra Applebaum and G. Leonard Burns (2010) 'Unexpected Chidhood Death: Post-traumatic Stress Disorder in Surviving Siblings and Parents', *Journal of Clinical Child Psychology* 20:114-120

5   R. Krell and L. Rabkin (1979), 'The effects of sibling death on the surviving child: a family perspective.' *Family Process* 18:471-477

6   Jackie Ellis, Chris Dowrick and Mari Lloyd-Williams (2013) 'The long-term impact of early parental death: lessons from a narrative study', *Journal of the Royal Society of Medicine* 106:57-67

7   R. C. Franklin and J. H. Pearn (2011), 'Drowning for love: The aquatic victim-instead-of-rescuer syndrome: drowning fatalities involving those attempting to rescue a child', *Journal of Paediatrics and Child Health* 47:44-47

8   'Myths about organ donation', https://www.organdoation.nhs.uk/helping-you-to-decide-about-organ-donation. C.J. McNamee, D.L. Modry. D. Lien and A. Alan Conlan (2003) 'Drowned donor lung for bilateral lung transplantation.' *The Journal of Thoracic and Cardiovascular Surgery* 126:910-912

9   Emily Andrews, 'Mother saved after receiving two kidneys from 16-month-old toddler who drowned in pond', *The Daily Mail* 10 February 2009

10   'Saved by her dead son… now we're friends', *The Daily Mirror*, 25 June 2009

11   Mark Foster, 'Canoe victim's gift of life', *The Northern Echo*, 26 November 2012

12   'The Chagall Windows, All Saints Church, Tudeley', http://www.tudeley.org/chagallwindows.htm

13   'Mother of drowned student advises those on night out about river dangers', ITV News, 28 January 2016

14   Jaymi McCann, 'Mother's crusade to teach babies how to survive in water', *The Daily Express*, 9 April 2017

15   'Mum launches campaign in memory of swimming star son', *This is Gloucestershire*, 14 May 2010, http://www.gloucestercitizen.co.uk/Mum-launches-campaign. The video 'Shallow Water Blackout – This video could save your life!' can be viewed on YouTube.

16   'Schoolboy drowned in waterfall', BBC News, 8 December 2006

17   Eryl Crump, 'How mum of teenager who drowned in a Conwy river is helping to keep children safe in the water', *Daily Post*, North Wales 13 July 2016

18   'FMU scholarship established in memory of drowning victim', ,http://www.wmbfnews.com/story/11531307/fmu-scholarship

19   Veasey Conway 'Additional funding for 'Chase' Lee memorial scholarship extends son's positive influence', *Morning News*, Florence, SC 23 March 2015

20   Charles Courtenay-Clack, 'A quiet ponder in Postman's Park', *Time Out*, 21 October 2010

21   Kaya Burgess, 'Name of man who died saving boy is added to "everyday heroes" memorial', *The Times*, 12 June 2009

22   Kate Ashley-Griffiths, '"Non-fatal drowning" victims hidden casualties in pool accidents', *Herald Sun*, Melbourne, 12 December 2016

23   'Effects of hypoxic/anoxic brain injury', Headway – the brain injury association, Nottingham. https://www.headway.org.uk

24   Reid A. Abrams and Scott Mubarak (1991) 'Musculoskeletal Consequences of Near-Drowning in Children', *Journal of Pediatric Orthopaedics* 11:168-175

25   'Disorders of consciousness', www.nhs.uk/conditions/vegetative-state/Pages/Introduction. aspx

26   Jo-ann Morris, 'A mother's heartbreak: Water safety.' The Samuel Morris Foundation. 6 March 2010. Lydia Slater 'The little boy who came back from the dead: His parents thought they had lost him. But an hour after Jago's body was plucked from a pond, medics found a pulse.' *The Mail on Sunday*, 30 October 2012

27   Sanjay V. Desai, Tyler J. Law, Dale M. Needham (2011) 'Long-term complications of critical care', *Critical Care Medicine* 39:371-379

28   Dan Gordon 'Living with Tragedy of Near-Drowning', *Los Angeles Times*, 15 June 1997. William M. Feinberg and Peggy C. Ferry (1984) 'A Fate Worse Than Death: The Persistent Vegetative State', *The American Journal of Diseases of Children* 138:128-130

29   The Jago Worrall Foundation, http://thejagoworrallfoundation.org. The Samuel Morris Foundation, http://samuelmorrisfoundation.org.au. The Drowning Prevention Foundation, http://drowningpreventionfoundation.com

30   Dan Gordon 'Living with Tragedy of Near-Drowning', *Los Angeles Times*, 15 June 1997

31   Richard W. Swanson (1993), 'Psychological Issues in CPR', *Annals of Emergency Medicine* 22:350-353

32   David Schottke and Garry L. Briese (1997), *First Responder: Your First Response in Emergency Care*, 2nd Edn. Jones and Bartlett Publishers, Sudbury, Mass. p. 276-277

33   David S. Smith and Sara J. Smith, *Water Rescue: Basic Skills for Emergency Responders*. p.299

34   Steve Walker 'A river of tears for children lost', *The Daily Telegraph*, 26 November 2009

35   J.K. Kleboom et al (2015) 'Outcome after resuscitation beyond 30 minutes in drowned children with cardiac arrest and hypothermia: Dutch nationwide retrospective cohort study', *British Medical Journal* 350:h418

36   Ian Maconochie and Charles D. Deakin (2015), 'Resuscitating drowned children', Editorial *British Medical Journal* 350:h535

PART 4: THE IMPORTANCE OF PREVENTION

1   'Franklin's Philadelphia. In Case of Fire', https://www.ushistory.org/franklin/philadelphia/ fire.htm

2   John Wilson and Wim Rogmans 'Prevention of Drowning', Section in *Handbook of Drowning*. p.86

3   'List of countries by traffic-related death rate', Wikipedia, https://en.wikipedia.org/wiki/ List_of_countries_by_traffic-related_death_rate. Gordon Smith (2006) 'The Global Burden of Drowning.' in *Handbook of Drowning*, Edit. Joost J.L.M. Bierens, Springer Verlag, Berlin. p.57

4   Rebecca Mitchell, Ann Williamson and Jake Oliver 'Calculating estimates of drowning morbidity and mortality adjusted for exposure to risk', World Conference on Drowning

Prevention 2011. Da Nang, Vietnam.

5   'Yellow fever', Wikipedia. https://en.wikipedia.org/wiki/Yellow_fever

6   Frederick P. Rivara (2009), 'Prevention of Drowning: The Time Is Now', *Archives of Pediatric and Adolescent Medicine* 163:277-278

CHAPTER 35: A DOZEN WAYS OF PREVENTING DROWNING ACCIDENTS

1   Ruth A. Brenner (2002) 'Childhood drowning is a global concern: Prevention needs a multifaceted approach', *British Medical Journal* 324:1049-10

2   Andrej Michalsen (2006), 'Risk Assessment and Perception', Section in *Handbook of Drowning*, p.93-97

3   'Baby bath-seat risk warning', BBC News, 2 February 2015

4   R. Byard, C. de Konong, B. Blackbourne, J. Nadeau and H. F. Krous (2001), 'Shared bathing and drowning in infants and young children', *Journal of Paediatrics and Child Health* 37:542-544

5   Tom Griffiths, 'Killer Parties' *Aquatics International Magazine*, 1 March 2008

6   'Child Holiday Swimming Pool Safety', Royal Society for the Prevention of Accidents (RoSPA), http://www.rospa.com/leisure-safety/water/advice

7   N. Manolios and I. Mackie (1988), 'Drowning and near-drowning on Australian beaches patrolled by life-savers: a 10-year study 1973-1983', *Medical Journal of Australia* 148:165-7, 170-1. Kevin Moran (2010), 'Watching parents, Watching Kids: Water Safety Supervision of young children at the Beach', *International Journal of Injury Control and Safety Promotion* 16:215-221

8   Austswim (2001), *Teaching Infant and Preschool Aquatics*, p.117-119

9   'Two five-year-old girls swept out to sea on blow-up swan after RNLI warning to parents over "deadly" inflatables', *Daily Telegraph*, 2 June 2019

10   Diane C. Thompson and Fred Rivara (2000), 'Pool fencing for preventing drowning of children', Cochrane Database of Systematic Reviews, https:// www.ncbi.nim.nih.gov/pubmed/1079674

11   K. J. Fisher and K. P. Balanda (1997), 'Caregiver factors and pool fencing: an exploratory analysis' *Injury Prevention*, 3:257-261

12   C. Blum and J. Shield (2000), 'Toddler drowning in swimming pools', *Injury Prevention* 6:288-290

13   N. Milliner, J. Pearn and R. Guard (1980) 'Will fenced pools save lives? A 10-year study from Mulgrave Shire, Queensland', *Medical Journal of Australia* 2:510-511

14   Luke Morrison, David J. Chalmers, John D. Langley, Jonathan C. Alsop and Catriona McBean (1999), 'Achieving compliance with pool fencing legislation in New Zealand: a survey of regulatory authorities', *Injury Prevention* 5:114-118

15   A. A. Ellis and R. B. Trent (1977), 'Swimming pool drownings and near-drowning among California preschoolers', *Public Health Reports* 223:73-77

16   Don Scott Brown (2003) 'Children are Drowning Without a Sound... Is the Fire Service Listening ?' Orange County Fire Authority. http://www.usfa.fema.gov/pdf/efop/efo36275.

pdf. G. J. Wintemute (1990) 'Swimming Pool Owners' Opinions of Strategies for Prevention of Drowning', *Pediatrics* 85: 63-69. G. J. Wintemute and M. A. Wright (1991) 'The attitude-practice gap revisited: risk reduction beliefs and behaviors among owners of residential swimming pools', *Pediatrics* 88:1168-1171. H. Morganstern and R.T. Bingham (2000), 'Effects of pool-fencing ordinances and other factors on childhood drowning in Los Angeles County 1990-1995', *The American Journal of Public Health* 90:595-601

17   G. J. Wintemute, J. F. Kraus, S. P. Teret and M. Wright (1987), 'Drowning in childhood and adolescence: a population-based study', *American Journal of Public Health* 77:830-832

18   W. Barry, T.M. Little and J.R. Sibert (1982) 'Childhood drownings in private swimming pools: an avoidable cause of death', *British Medical Journal* 285:542-543. Jennifer Smith 'Coroner calls for new law to fence off garden swimming pools after "adventurous" boy found drowned on his third birthday', *The Daily Mail*, 15 April 2015

19   'The design and safety of swimming pool covers', *International Journal for Consumer and Product Safety* 1:163-174

20   L. Quan and P. Cummings (2003) 'Characteristics of drowning by different age groups', *Injury Prevention* 9:163-168

21   Committee on Sports Medicine and Fitness and Committee on Injury and Poison Prevention (2000) 'Swimming Programs for Infants and Toddlers', *Pediatrics* 105:868-870

22   Ruth A. Brenner et al (2009), 'Association Between Swimming Lessons and Drowning in Childhood: A Case-Control Study', *Archives of Pediatrics and Adolescent Medicine* 163:203-210

23   Jeffrey Weiss (2010), 'Policy Statement – Prevention of Drowning', Committee on Injury, Violence and Poison Prevention, *Pediatrics* 126:178-185. 'Taking Children Swimming', RoSPA. http://rospa.com/leisuresafety/adviceandinformation/watersafety

24   Virginia Hunt Newman, *Teaching an infant to swim*, p.21

25   Virginia Hunt Newman, *Teaching an infant to swim*, p.38-47. Austswim (2001) *Teaching Swimming and Water Safety*, p.35-37

26   Frank Golden and Michael Tipton, *Essentials of Sea Survival*, p.52-53

27   Virginia Hunt Newman, *Teaching an infant to swim*, p.52-53. Austswim, *Teaching Infant and Preschool Aquatics*, p.94-108

28   Gerald N. Goldberg, Elmer S. Lightner, Wayne Morgan and Sid Kemberling (1982) 'Infantile Water Intoxication After a Swimming Lesson', *Pediatrics* 70:599-600

29   Harvey Barnett, 'A behavioral approach to pediatric drowning prevention', Presentation at World Conference on Drowning Prevention 2011, Da Nang, Vietnam. Michael J. Silverman, 'Defeating Drowning With the "Flip and Float"', ABC News 22 June 2006. Sue Pryor 'Swimming to a safe place.' Presentation at World Conference on Drowning Prevention 2011, Da Nang, Vietnam.

30   'Learning to swim early saved my baby from drowning', *The Liverpool Echo*, 12 May 2008

31   K. Moran and T. Stanley (2006), 'Toddler drowning prevention: teaching parents about water safety in conjunction with their child's in-water lessons', *International Journal of Injury Control and Safety Promotion* 13:254-256. Jennifer D. Blitvich, Kevin Moran, Lauren A, Petrass, G. Keith McElroy and Teresa Stanley (2012), 'Swim Instructor Beliefs About Toddler and Pre-

School Swimming and Water Safety Education', *International Journal of Aquatic Research and Education* 6:110-121. B. A. Morrongiello, M. Sandomierski, D. C. Schwebel and B. Hagel (2013) 'Are parents just treading water? The impact of participation in swim lessons on parents' judgments of childrens' drowning risk, swimming ability, and supervision needs', *Accident Analysis and Prevention* 50:1169-1175

32   B. A. Blanksby, H. E. Parker, S. Bradley and V. Ong (1995), 'Children's readiness for learning front crawl swimming', *Australian Journal of Science and Medicine in Sport* 27:34-37. H. E. Parker and B. A. Blanksby (1997), 'Starting age and aquatic skill learning in young children: mastery of prerequisite water confidence and basic aquatic locomotion skills', *Australian Journal of Science and Medicine in Sport* 29:83-87. Stephen J. Langendorfer (2013) 'Which Stroke First?' Editorial, *International Journal of Aquatic Research and Education* 7:286-289. Robert Kieg Stallman (2014) 'Which Stroke First? No Stroke First!' *International Journal of Aquatic Research and Education* 8:5-8

33   R. K. Stallman, M. Junge and T. Blixt (2008), 'The teaching of swimming based on a model derived from the causes of drowning', *International Journal of Aquatic Research and Education* 2:372-382

34   Austswim, *Teaching Swimming and Water Safety*, p.165-169

35   Kevin Moran (2014) 'Can You Swim in Clothes? An Exploratory Investigation of the Effect of Clothing on Water Competency', *International Journal of Aquatic Research and Education* 8:338-350. Kevin Moran (2015) 'Can You Swim in Clothes? Reflections on the Perception and Reality of the Effect of Clothing on Water Competency', *International Journal of Aquatic Research and Education* 9:116-135

36   John Connolly, 'Drowning: The First Time Problem', *International Journal of Aquatic Research and Education* 8:66-72

37   Frank Golden and Michael Tipton, *Essentials of Sea Survival*, p.59-64. J. M. Graham and W. R. Keating (1978) 'Deaths in cold water', *British Medical Journal*

38   M. J. Barwood, V. Bates, G. Long and M. J. Tipton (2010), '"Float First" An assessment of the buoyancy provided by seasonal clothing assemblies before and after swimming', University of Portsmouth. http://www.rospa.com/occupationalsafety/info/bnfl/float-first.pdf, 'Radical "Float First" call to prevent drowning deaths', RoSPA 29 April 2010. Sally Guyoncourt, 'Float before you swim if you fall into cold water, says the RNLI', https://inews.co.uk/news/uk/float-first-swim-fall-cold-water

39   'Teaching survival skills', Royal Life Saving W. A., https://royallifesavingwa.com.au. 'Save your child's life with survival swimming lessons', 26 July 2012, http://www.swmontanamagazine.com/save-your-childs-life-with-survival-swimming-lessons

40   P. L. Kjendlie, T. Pedersen, T. Thorensen, T. Setlo, K. Moran and R. K. Stallman (2013), 'Can you swim in waves? Children's swimming, floating, and entry skills in calm and simulated unsteady water conditions', *International Journal of Aquatic Research and Education* 7:301-31

41   Peter Whipp (2002), 'Teaching swimming in secondary schools: Is there a case for differentiation?' Australian Association for Research in Education. Presentation at AARE 2002 Conference, Brisbane. http://www.aare.edu.au/02pap/whi02118.html

42   Ofsted (2000) '*Swimming in Key Stage 2: An inspection report on standards and provision*',

Office for Standards in Education, London. Claire Phipps 'Sink or swim?' *The Guardian*, 14 November 2000

43   'School Swimming', *Hansard*, 18 December 2000. http://hansard.parliament.uk/commons/2000-12-18/debates/49687

44   Ofsted Report (2007), *Reaching the Key Stage 2 standard in swimming*. Office for Standards in Education, London

45   'Inclusion of Swimmers with a Disability', Amateur Swimming Association (now Swim England), https://www.swimming.org/library/documents/477/download

46   'Save School Swimming. Save Lives', The 2012 School Swimming Census and Manifesto. Amateur Swimming Association. http://swimmingtrust.com/wp.../ASA_School_Swimming_full_manifesto.pdf

47   'Concern over end of swimming scheme', BBC News Scotland, 22 June 2015

48   'Water Safety Policy in Scotland – A Guide', RoSPA 4 April 2014. Judith Duffy 'Call for 'water safety minister' in wake of drowning deaths', The *Herald*, Glasgow 28 August 2016. Water Safety Scotland (2018), 'Scotland's Drowning Prevention Strategy 2018-2026. https://www.watersafetyscotland.org.uk (Note: This document was developed and written by members of the Water Safety Scotland Strategy Subgroup. Its members – Carlene McAvoy, Kenny MacDermid, Michael Avril, Gillian Barclay and Elizabeth Lumsden, and representatives of RoSPA, RLSS, RNLI and Scottish families – brought together Scotland's drowning prevention initiatives and safety programmes.)

49   Ofsted Report (2007) *Reaching the Key Stage 2 standard in swimming*. Office for Standards in Education, London. (Note: The choice of the words 'pupils from some black ethnic minorities' is discussed in: Amani Saeed, Elliott Rae, Rob Neil, Vivienne Connell-Hall and Frank Munro 'To BAME or not to BAME: the problem with racial terminology in the civil service.' Civil *Service World* 4 October 2019.) Andy Akinwolere 'Why do so few black children learn to swim?' The *Daily Telegraph*, 12 August 2015

50   'Alice Dearing: Black Swimming Association launched to get BAME people swimming and safe in the water', BBC Sport, 2 March 2020

51   C. Irwin, R. Irwin, N. Martin and S. Ross (2010), 'Constraints Impacting Minority Swimming Participation: Phase 11', University of Memphis. http://www.usaswimming.org/-Rainbow/Documents/8ff56da3-ef9c=47ab-a83e-57b7

52   R. A. Brenner, A. C Trimble, G. S. Smith, E. P. Kessler and M. D. Overpeck (2001) 'Where children drown, United States, 1995', *Pediatrics* 108:85-89. Gitanjali Saluja, Ruth A. Brenner, Ann C. Trumble, Gordon S. Smith, Tom Schroeder and Christopher Cox (2006) 'Swimming Pool Drownings Among US Residents Aged 5-24 years: Understanding Racial/Ethnic Disparities', *American Journal of Public Health* 96:728-733. Julie Gilchrist and Erin M. Parker (2014), 'Racial/Ethnic Disparities in Fatal Unintentional Drowning Among Persons Aged <29 Years – United States, 1999-2010', Centers for Disease Control and Prevention MMWE, 16 May 2014

53   Ron Claiborne and Enjoli Francis, '6 Teens Drown While Wading in Louisiana's Red River', ABC News,1 August 2010. Finlo Rohrer, 'Why don't black Americans swim?' BBC News, Washington, 3 September 2010

54   Lan Trinh, 'Swim lessons help minority children break cycle', CNN 10 May 2012

55   Janet Smith (2005), *Liquid Assets. The lidos and open air swimming pools of Britain*, English Heritage, Swindon.p.12 (Note: Peerless Pool was one of the first public swimming pools built since the Romans left Britain.)

56   Ian Gordon and Simon Inglis (2009) *Great Lengths: The historic indoor swimming pools of Britain*. English Heritage, Swindon. John Bourn (2007) *The budget for the London 2012 Olympic and Paralympic Games*. The Stationary Office London. Swim England, 'National Lottery thanked for funding schemes to help get nation swimming', 8 November 2019, https://www.swimming.org/swimengland/national-lottery-thanked

57   Tristan Harris 'Historic Grade 11 listed Moseley Road Baths in £74,100 lottery win', South Birmingham Radio, 19 August 2018. 'Historic Pools of Britain.' https://historicpools.org.uk (Note: Historic Pools of Britain was established in 2015 to give representation to all indoor and outdoor pools across the country. Members celebrate, champion and campaign for historic pools.)

58   Dan Bloom, 'Swimming in urban decay: Eerie images of Britain's forgotten pools left derelict after the Victorian golden age of public bathing', *Daily Mail*, 25 October 2013. Dan Bloom, 'When the water dries up: Haunting images of Edwardian swimming baths that are now lying empty and abandoned', *Daily Mail*, 12 February 2015. Lucy Lynch, 'Nostalgia:Coventry swimming pools we have loved and lost', *Coventry Telegraph*, 18 September 2015

59   Jon Slater, 'Too busy and broke to spend hours in the pool. Trips to distant leisure centres can take a huge bite out of a packed timetable', *Times Educational Supplement*, 1 August 2003

60   Rehema Figueiredo, 'Tessa Jowell leads campaign to stop gym company closing swimming pools', *Evening Standard*, 2 November 2015. Miranda Bryant 'Olympic champions join battle to save London pools from closure', *Evening Standard*, 5 January 2016

61   Sport England (2013) 'Swimming Pools: Design Guidance Note.' https://www.sportsengland.org/media/4187/swimming-pools-dgn-2013.pdf

62   Jenny Scott 'Rebecca Adlington: Why are we closing swimming pools?' BBC News, 17 February 2014

63   'Swimming top participation sport in England.' Amateur Swimming Association 30 January 2015. Simon Shibli (2007) 'A guide to swimming participation statistics in England to assist ASA regions in the preparation of their strategic plans for 2009-2013.' https://www.google.com/search?client=fitefox-b-d&q=a+guide+to+swimming+participation

64   'National Pool Lifeguard Qualification (NPLQ).' Royal Life Saving Society. https://www.rlss.org.uk/national-pool-lifeguard-qualifications

65   Keith Bibb 'Dogged Misconception', *Aquatics International*, 28 March 2014

66   W. A. Harrell (1999), 'Lifeguards' vigilance: effects of child-adult ratio and lifeguard positioning on scanning by lifeguards', *Psychological Review* 84:193-197

67   Joshua Brener and Michael Oostman, 'Lifeguards Watch, But They Don't Always See!' *World Waterpark Magazine*, May 2002. Tom Griffiths, 'From the Bottom Up', *Aquatics International*, October 2006

68   J. H. Modell (2010), 'Prevention of needless death from drowning', *Southern Medical Journal*, 103:650-653

69   Mary Donahue, 'How to rescue a drowning victim using a reaching assist or a shepherd's crook', https://marydonahue.org/how-to-rescue-a-drowning-person-using

70   'Rescue buoy', Wikipedia, https://en.wikipedia.org/wiki/Rescue_buoy

71   'Removing a casualty from the water', Royal Life Saving Western Australia. https://royallifesavingwa.au/your-safety/out-and-about/removing-a-casualy-from-the-water

72   A. Kemp and J. R. Sibert (1992) 'Drowning and near drowning in children in the United Kingdom: lessons in prevention', *British Medical Journal* 304:1143-1146

73   P. J. Fenner, S. L. Harrison, J. A. Williamson and B. D. Williamson (1995), 'Success of Surf Lifesaving Resuscitations in Queensland 1973-1992', *Medical Journal of Australia* 163:580-583. Roy Ballantyne, Neil Carr and Karen Hughes (2005), 'Between The Flags: An Assessment of Domestic and International University Students' Knowledge of Beach Safety in Australia', *Tourism Management* 26:4-26

74   Tom Brooks-Pollock, 'Britain's beaches "becoming more dangerous"', *Daily Telegraph*, 12 September 2014. Claire Lomas, 'Family of five rescued from rip current by lifeguards in Devon', *Daily Telegraph*, 4 November 2015. Kathryn Riddell, 'Heroic lifeguards rescue 18-year-old swimmer in rip tide off Sunderland coast', *Evening Chronicle*, Newcastle-upon Tyne, 29 June 2018. Veronica Rocha, '408 people rescued from strong rip currents at L.A County beaches', *Los Angeles Times* 29 July 2015. Chris Brewster, Richard E, Gould and Robert W. Brander (2019) 'Estimations of rip current rescues and drowning in the United States', *Natural Hazards and Earth System Sciences* 19:389-397

75   Tomas Leclerc, Juan Canabal and Heather Leclerc (2008) 'The Issue of In-Water Rescue Breathing: A Review of the Literature.' *International Journal of Aquatic Research and Education* 2:36-46. David Szpilman and Márcio Soares (2004), 'In-water resuscitation – is it worthwhile?' *Resuscitation* 63:25-31. M. Tipton, T. Reilly, A. Rees, G. Spray and F. Golden (2008) 'Swimming performance in surf: the influence of experience', *International Journal of Sports Medicine* 29:895-898

76   'National Vocational Beach Lifeguard Qualification (NVBLQ). Royal Life Saving Society. https://www.rlss.org.uk/national-vocational-beach-lifeguard-qualification

77   Ben Pike 'Bondi Beach dangers fuelled by swimmer ignorance, lifesavers say', The *Sunday Telegraph*, Sydney 5 July 2017

78   Diane Taylor, 'Verdict of misadventure given on seven drownings at Camber Sands', The *Guardian*, 30 June 2017

79   Chloe Chapman, 'Camber Sands: First picture of drowned Brazilian teenager released as friends appeal to send his body home', *Evening Standard*, 25 July 2016. 'Exclusive: Hero died trying to save teenager at Camber Sands', ITV News 30 June 2017

80   Nadia Khomami, 'Five men dead after being pulled from the sea at Camber Sands', The *Guardian*, 25 August 2016

81   Diane Taylor, 'Camber Sands beach to get lifeguards after seven deaths last year', The *Guardian*, 16 May 2017

82   K. N. Asher, F. P. Rivara, D. Felix, L. Vance and R. Dunne (1995) 'Water safety training as a potential means of reducing risk of young children's drowning', *Injury Prevention* 1:228-233. Peter Barss (1995) 'Cautionary notes on teaching water safety skills', *Injury Prevention* 1:218-219

83   Virginia Hunt Newman, *Teaching an infant to swim*.

84   Austswim, *Teaching Infant and Preschool Aquatics*, p.14

85   K. Moran and T. Stanley (2006), 'Parental perceptions of toddler water safety, swimming ability and swimming lessons', *International Journal of Safety Control and Safety Promotion* 13:139-143

86   J. R. Blitvich, K. Moran, L. A. Petrass, G. K. McElroy and T. Stanley (2012), 'Swim Instructor Beliefs About Toddler and Preschool Swimming and Water Safety Education', *International Journal of Aquatic Research and Education* 6:110-121. B. A. Morrongiello, M. Sandomierski, D. C. Schwebel and B. Hagel (2013), 'Are parents just treading water? The impact of participating in swim lessons on parents' judgments of children's drowning risk, swimming ability, and supervision needs', *Accident Analysis and Prevention* 50:1169-1175

87   K. Moran and T. Stanley (2006), 'Toddler drowning prevention: teaching parents about water safety in conjunction with their child's in-water lessons', *International Journal of Safety Control and Safety Promotion* 13:254-256

88   Howard Mintz and Julie Patel, '4-year-old drowns in wave pool Great America', *San Jose Mercury News*, 7 December 2007. 'Wave pool Safety Act: SB 107 – Bill Analysis', 17 June 2008, ftp://www.ihc.ca.gov/pub/07-08/bill/sen/sb_0101

89   Tom Griffiths, 'The 'Note and Float' drowning prevention program', World Congress on Drowning Prevention 2011, Da Nang, Vietnam.

90   Christopher Biswick and Melissa Matucci Lindberg 'Vested Interests.' *Aquatics International*, May 2011

91   Kevin Moran (2009) 'Parents, Pals, or Pedagogues? How Youth Learn About Water Safety', *International Journal of Aquatic Research and Education* 3:121-134. Kevin Moran (2010) 'Risk of Drowning: The "Iceberg Phenomenon" Re-visited' *International Journal of Aquatic Research and Education* 4:115-126

92   R. E. G. Sloan and W. R. Keatinge (1973) 'Cooling rates of young people swimming in cold water', *Journal of Applied Physiology* 35:371-375. National Water Safety Forum, 'Cold water shock dangers need including in swimming and classroom lessons', 18 July 2017, https://www.nationalwatersafety.org.uk/news/posts/2017/july/cold-warer-shock. Frank Golden and Michael Tipton, *Essentials of Sea Survival*, p.54-77

93   Cheshire Fire and Rescue Service, 'Stay safe by staying off frozen water', https://www.cheshirefire.gov.uk/public-safety/outdoor-safety/ice-safety

94   Ofsted Report *Swimming in Key Stage 2: An inspection report on standards and provision*. http://www.sportdevelopment.info/index.php/brow

95   R. Stallman, M. Junge and T. Blixt (2007) 'The teaching of swimming based on a model derived from the causes of drowning', *International Journal of Aquatic Research and Education* 2:372-382

96   'Body of boy, nine, found in lake.' BBC News, 13 June 2007

97   Sian Powell 'Lake death prompts call over safety; Children lack water skills, says accident watchdog', *Coventry Evening Telegraph*, 15 June 2007

98   Royal Life Saving Society, 'Runners and walkers at risk of drowning', https://www.org.uk/News/runners-and-walkers-at-risk-of-drowning

99   'Slips cause majority of coastal water deaths, RNLI warns', BBC News, 30 May 2019. 'RNLI "float" advice after 18 deaths off Welsh coast', BBC News, 30 May 2019

100   J. Howland, R. Hingson, T.W. Mangione, N. Bell and S. Bak (1996), 'Why are most drowning victims men ?' *American Journal of Public Health* 86:93-96

101   Kevin Moran, Robert Keig Stallman, Per-Ludvik Kjendlie, Dagmar Dahl, Jennifer Blitvich, Lauren A. Petrass, G. Keith McElroy, Toshiaki Goya, Keisuke Teramoto, Atsunori Matsui and Shuji Shimongata (2012), 'Can You Swim? An Exploration of Measuring Real and Perceived Water Competency', *International Journal of Aquatic Research and Education*. 6:122-135

102   Narbeh Minassian, 'Cheshunt teenager R… L… "told friend he was struggling to swim in lake" shortly before drowning', *Hertfordshire Mercury*, 19 June 2018. 'Top young cricketer drowns after friends urged him into river even though he couldn't swim', ITV News, 5 December 2016

103   Ian Clarke, 'Less than 24 hours after a double drowning at Bawsey Pits, sunseekers return to the water', *Eastern Daily Press*, 17 July 2013

104   Press Release 'Quarry Operators Issue Whitsun Holiday Warning Following Tragic Drowning', Mineral Products Association, 21 May 2014

105   Rod Mills, 'Teen swept away in river after 'shortcut' home from festival', *Daily Express*, 30 July 2013

106   'Teenage canoeist's body recovered at Grandtully rapids', BBC News, 9 April 2010

107   'Teenager paralysed from swimming pool dive "fights back"', ITV News, 5 September 2014

108   'Teenage boy killed in 30ft "tombstone" cliff leap', *Daily Mail*, 15 May 2009

109   D. Morgan, J. Ozanne-Smith and T. Triggs (2009), 'Direct observation measurement of drowning risk exposure for surf beach bathers', The *Journal of Science and Medicine in Sport* 12:457-462

110   'Teenager's death propels jet-skis into safety row', The *Independent*, 20 August 1997.

111   Tom Matthews, 'Wallington County Grammar schoolboy 'drowned after being pushed into a lake by a friend', *Croydon Advertiser*, 8 January 2018

112   Andrew Harris, Senior Coroner, London Inner South jurisdiction 'Regulation 28 Report to Prevent Future Deaths.' 18 January 2018

113   Torill Hindmarsh and Mats Melbye 'Good swimmers drown more often than non-swimmers: How open water swimming could feature in beginning swimming.' World Conference on Drowning Prevention 2011, Da Nang, Vietnam.

114   'Swimmer drowned during training', BBC News, 7 February 2007
Albert B. Craig (1961) 'Underwater Swimming and Loss of Consciousness', *Journal of the American Medical Society* 176:255-258

115   Kevin Moran (2006), 'Re-thinking Drowning Risk: The Role of Water Safety Knowledge, Attitudes and Behaviours in the Aquatic Recreation of New Zealand Youth', PhD Thesis at Massey University. Laurence Steinberg (2007) 'Risk Taking in Adolescence: New Perspective From Brain and Behavioral Science.' *Current Directions in Psychological Science* 16:55-59 Colin Barras 'It's not teenagers' fault, it's their brains', *Daily Telegraph*, 25 September 2007

116   L. Quan and P. Cummings (2003), 'Characteristic of drowning by different age groups', *Injury Prevention* 9:163-168

117   'Non-swimmer boasted he was going to jump into river', *Walesonline*, 16 April 2005. Paul Cheston 'Friend of drowning boy: I tried to grab his hair but he slipped away', *Evening Standard*, 20 December 2011. Taylor Geall, 'Boy who drowned in River Thames while playing game of

"dare" with his pals "couldn't swim"', *Daily Mirror*, 1 September 2016

118 'Drinking and drowning' (1979) Editorial. *British Medical Journal* 1(6156):70-71. T. Driscoll, J. Harrison and M. Steenkamp (2004) 'Review of the role of alcohol in drowning associated with recreational aquatic activity.' *Injury Prevention* 10:107-113

119 'Durham river deaths: Safety work follows student drownings.' BBC News, 24 June 2015

120 Sean O'Neill, 'Threat of drinking fines after student drownings', *The Times*, 13 February 2013

121 'Legacy leaves a safer generation of children', *Royal Life Saving Society News*, 23 July 2015

122 Mark Foster 'Don't drink and drown campaign follows river deaths', The *Northern Echo*, 3 October 2014. Dan Bean, 'Inquest into soldier's drowning "highlights need for greater river awareness"', The *Press*, York 6 March 2015

123 'Don't Drink and Drown', http://www.rlss.org.uk/dont-drink-and-drown

124 'Don't Drink and Drown Campaign', https://shropshirefire.gov.uk/news/dont-drink-and-drown

125 Gordon S. Smith et al (2001), 'Drinking and Recreational Boating Fatalities: A Population-Based Case-Control Study', *Journal of the American Medical Association* 286:2974-2980

126 Jonathan Howland, Gordon S. Smith, Thomas Mangione and Ralph Hingson (1993) 'Missing the Boat on Drinking and Boating', *Journal of the American Medical Association* 270:91-92. J. Howland, T. W. Mangione and S. Minsky (1996) 'Perceptions of risks of drinking and boating among Massachusetts boaters.' *Public Health Reports* 111:372-377

127 Kevin Moran (2006), 'Re-thinking Drowning Risk: The Role of Water Safety Knowledge, Attitudes and Behaviours in the Aquatic Recreation of New Zealand Youth', PhD Thesis, Massey University, New Zealand, Published (2009) by VDM Verlag, Saarbrucken

128 Kevin Moran (2006), 'Re-thinking Drowning Risk: The Role of Water Safety Knowledge, Attitudes and Behaviours in the Aquatic Recreation of New Zealand Youth.' PhD Thesis, Massey University, New Zealand. Published (2009) by VDM Verlag, Saarbrucken

129 P. Cummings, B. A, Mueller and L. Quan (2011), 'Association between wearing a personal floatation device and death by drowning among recreational boaters: a matched cohort analysis of United States Coast Guard data', *Injury Prevention* 17:156-159

130 C. S. Jones (1999), 'Drowning among personal watercraft passengers: the ability of personal flotation devices to preserve life on Arkansas waterways, 1994-1997', *Journal of the Arkansas Medical Society* 96:97-98

131 E. Cassell and S. Newstead (2015), 'Did compulsory wear regulations increase personal flotation device (PFD) use by boaters in small power recreational vessels? A before–after observational study conducted in Victoria, Australia', *Injury Prevention* 21:15-22

132 Lyndal Bugeja, Erin Cassell, Lisa R. Brodie and Simon J. Walter (2014), 'Effectiveness of the 2005 compulsory personal floatation (PDF) wearing regulations in reducing drowning deaths among recreational boaters in Victoria, Australia', *Injury Prevention* 20:387-392. Jacqui Wise 'Life jackets should be compulsory for all recreational boaters, say researchers', *British Medical Journal* 28 June 2014

133 'Kawarau River deaths spark coroner's call for lifejackets', *Mountain Scene*, Queenstown, New Zealand, 18 August 2011

134 Royal National Lifeboat Institution '*CHOOSE IT. WEAR IT. The RNLI guide to lifejackets*

and buoyancy aids, https://rnli.org/media/rnli/downloads/rnli-guide-to-lifejackets-and-buoyancy-aids. H. Lunt, D. White, G. Long and M. Tipton (2014), 'Wearing a crotch strap on a correctly fitted lifejacket improves lifejacket performance', *Ergonomics* 57:1256-1264

135   Jonathan Brown, 'Loch Gairloch canoe tragedy claims four as little girl dies', *The Independent*, 28 August 2012. 'Father in canoe tragedy: I thought they were proper life jackets', *Daily Telegraph*, 29 August 2012

136   Kevin Moran (2008) 'Rock-based Fishers' Perceptions and Practice of Water Safety', *International Journal of Aquatic Research and Education* Vol. 2:No. 2, Article 5. DOI: 10.25035/ijare.02.02.05

137   Kevin Moran (2011) 'Rock-Based Fisher Safety Promotion: Five Years On', *Aquatic Research and Education* Vol. 5:No. 2, Article 4. DOI: 10.25035/ijare.05.02.04. Kevin Moran (2017) 'Rock-Based Fisher Safety Promotion: A Decade On', *Aquatic Research and Education* Vol. 10:No. 2, Article 1. DOI: 10.25035/ijare.10.02.01

138   *'Trawling – Which lifejacket for you?'* Royal National Lifeboat Institution 2005

139   Devin Lucas, Jennifer M. Lucas, Philip Somervell and Theodore D. Teske (2011) 'Worker satisfaction with personal flotation devices (PFDs) in the fishing industry: Evaluations in actual use', *Applied Ergonomics* 43:747-752

140   Fishing Industry Safety Group (2013), 'Personal Flotation Devices. Overview of Project Group', http://www.fisg.org.uk/Projects/PFD.php

141   'Scottish fisherman who nearly drowned backs Personal Flotation Device sea safety campaign', Scottish Fisherman's Federation, 21 May 2015, https://www.sff.co.k/scottish-fisherman-who-nearly-drowned

142   Marine Accident Investigation Branch Report, 'Man overboard from creel fishing vessel Annie T with loss of 1 life.' 3 November 2016. https://gov.uk/maib-reports/man-overboard-from-creel-fishing-vessel (Note: This report was communicated to all fishing vessels: 'SAFETY FLYER TO FISHING VESSELS AND SMALL CRAFT. Annie T (CY 1), fatal man overboard accident, 4 October 2015.' https://www.gov.uk/government/publications/safety-lessons-lifejackets-can-help-survival-chances)

143   Marine Accident Investigation Branch (2016), *'Lifejackets: a review'*, https://assets.publishing.service.gov.uk

144   Rita Campbell, 'Nine deaths at sea this year lead to calls for compulsory lifejackets law', *The Press and Journal*, Aberdeen 3 November 2016

145   Tim Oliver, 'Life Jackets Law Shock', *Fishing News*, 28 January 2019

146   G. S. Bell, A. Gaitatzis, C. L. Bell, A. L. Johnson and J. W. Sander (2008), 'Drowning in people with epilepsy: How great is the risk?' *Neurology* 71:578-582. 'Warning of epilepsy drowning risk', BBC News, 25 August 2008

147   L. D. Budnick and D.A. Ross (1985), 'Bath-tub drownings in the United States, 1979-1981', *American Journal of Public Health* 75:630-633

148   K. D. Fitch and A. R. Morton (1971), 'Specificity of Exercise in Exercise-induced Asthma', *British Medical Journal* 4(5787):577-581

149   'Exercise-induced asthma – Symptoms and causes – Mayo Clinic', https://www.mayoclinic.org/diseases-conditions/exercise-induced-asthma

150   M. J. Ackerman, D. J. Tester and C. J. Porter (1999), 'Swimming, a gene-specific arrhythmogenic trigger for inherited long QT syndrome', *Mayo Clinic Proceedings* 74:1088-1094

151   'Long QT syndrome – Diagnosis and treatment', Mayo Clinic, https://www.mayoclinic.org/dideases-conditions/long-qt-syndrome

152   Domenico Corrado, Cristina Basso, Maurizio Schiavon and Gaetano Thiene (1998), 'Screening for Hypertrophic Cardiomyopathy in Young Athletes', *New England Journal of Medicine* 339:364-369

153   John M. Lippmann and John H. Pearn (2012), 'Snorkelling-related deaths in Australia, 1994-2006', *Medical Journal of Australia*, 197: 230-232. Editorial 'The life aquatic: Could an ageing cohort of sub-aqua enthusiasts be a worry for insurers?' *International Travel and Health Insurance Journal* 3 April 2017

154   A. C. Queiroga and A, Peden (2013), 'Drowning Deaths in Older People: A 10 year analysis of drowning in people aged 50 years and over', Royal Life Saving – Australia, Sydney

155   'Driving in heavy rain and floods. Stay safe on the road when it's wet, wet, wet', The Automobile Association, https://www.theaa.com/driving-advice/seasonal/driving-through-flood-water

156   Andrew Brookes, 'What is aquaplaning and how to avoid it', Royal Automobile Club, 2 November 2017. https://wwwrac.co.uk/drive/advice/winter-driving

157   Press Release 'Pensioners more likely to take flood risks than make a U-turn', GOV.UK. https://www.gov.uk/news/pensioners-more-likely-to-take-flood-risks. Tom Eden, 'Scots couple rescued from car swept away by Storm Dennis flood waters', The *Daily Record*, 16 February 2020

158   J. S. Becker, H. L. Taylor, B. J. Doody and K.C.Wright (2015), 'A Review of People's Behaviour in and around Floodwater', *Weather, Climate, and Society* 7:321-332

159   J. S. Becker, H. L. Taylor, B. J. Doody and K.C.Wright (2015), 'A Review of People's Behaviour in and around Floodwater', *Weather, Climate, and Society* 7:321-332 Andy Whelan and Samuel Underwood 'Mother and daughter die in floods', The *Argus*, Brighton 10 October 2008

160   'Don't underestimate the power of floodwater', Royal Life Saving Society 22 October 2014,. https://www.rlss.org.uk/news/dont-underestimate-the-power-of-floodwater

161   Elizabeth Walker 'Cars and water… avoiding a tragedy', *Staying Alive*, RoSPA February 2008

162   G. G. Giesbrecht and G. K. McDonald (2010), 'My car is sinking: automobile submersion, lessons in vehicle escape', *Aviation, Space, and Environmental Medicine* 81:779-84 Jaap Molenaar and John Stoop, 'Submerged Vehicle Rescue', Chapter in *Handbook on Drowning*, p.239-241

163   Gerren K. McDonald, Cheryl A Moser and Gordon Giesbrecht (2019), 'Public knowledge, attitudes and practices of vehicle submersion incidents: a pilot study', *Injury Epidemiology* 6:21

164   Dana Hunter, 'Instant Peril: Flash Floods (and How to Survive Them)', *Scientific American*, 28 December 2016

165   Amanda Perthen and Tom Worden, 'British pensioners drowned in 5ft flash flood as they sat at Spanish market stall', *Daily Mail*, 23 October 2011

166   Sharon T. Ashley and Walker S. Ashley (2008), 'Flood Fatalities in the United States', *Journal of Applied Meteorology and Climatology* 47:805-818

167   Caleb Starrenberg, 'Would-be rescuers losing their lives', http://www.stuff.co.nz/ipad-editors-picks/9574440

168   Kevin Moran and Teresa Stanley (2013), 'Readiness to Rescue: Bystander Perceptions of Their Capacity to Respond in a Drowning Emergency', *International Journal of Aquatic Research and Education* 7:290-300

169   Mary Donahue, 'How to rescue a drowning victim using a reaching assist or a shepherd's crook', https://marydonahue.org/how-to-rescue-a-drowning-victim-using-a-reaching-assist-or... 'Be someone's Lifeline: Know how to use a throw bag', Royal National Lifeboat Institution, https://rnli.org/magazine/magazine-featured-list/2017/june/be-someones-lifeline

170   'Rookie Lifeguard', The Royal Life Saving Society, https://www.rlss.org.uk/awards-activities/rookie-lifeguard

171   'Survive and Save', The Royal Life Saving Society, https://www.rlss.org.uk/survive-and-save

172   'Schools can now easily teach children vital lifesaving skills with new course for teachers', The Royal Life Saving Society, 21 January 2013, http://www.rlss.org/news/schools-can-now-easily-teach

173   Jackie Grant, 'Little Leisel, 8, saves pal from drowning just days after attending water safety course', *Daily Record*, Glasgow, 24 October 2014

174   'Scottish schoolgirl recognised for bravery by Royal Life Saving Society UK', http://www.rlss.org.uk/news/scottish-schoolgirl-recognised

175   'National Pool Lifeguard Qualification (NPLQ)', The Royal Life Saving Society, https://www.rlss.org.uk/national-pool-lifegyard-qualification

176   Emily Payne, 'Student saves parents and girlfriend from drowning after boat capsizes on Majorca holiday... weeks after training as lifeguard', *Daily Mail*, 13 November 2014

177   Jake Wallis Simons, 'Joining the RNLI: saving lives at sea', *The Daily Telegraph*, 12 April 2013

178   'Youth Education – Helping You Teach Children Water Safety', Royal National Lifeboat Institute, https://rnli.org/youth-education

179   Adele Robinson, 'Fire Service Teams Trained For Water Rescue', Sky News, 28 December 2013

180   Bjorn Lind (1961), 'Teaching mouth-to-mouth resuscitation in primary schools', *Acta Anaesthesiologica Scandinavica* 9:63-81

181   Ian Jones et al (2007), 'At what age can schoolchildren provide effective chest compressions? An observational study from the Headstart UK schools training programme,' *British Medical Journal*, 334:1201-1203

182   Wayne Crenshaw '7-year-old Warner Robins girl credited with saving boy from drowning', *The Telegraph*, Macon, Georgia, 31 May 2015

183   'Trio save two-year-old boy found unconscious in park pool', *Jersey Evening Post*, 2 July 2015

184   Kevin Moran and Teresa Stanley (2011) 'Toddler parents training, understanding, and perceptions of CPR', *Resuscitation* 82:572-576. (Note: This review reported that many parents were prepared to stop CPR after only five minutes.)

185   John Pearn (1992), 'The urgency of immersions.' *Archives of Disease in Childhood* 67:256-258

186   J. Marchant, N. G. Cheng, L. T. Lam, F. E. Fahy, S. V. Soundappan, D. T. Cass and G. J. Browne (2008), 'Bystander basic life support: an important link in the chain of survival for children suffering a drowning or near-drowning episode', *Medical Journal of Australia* 188:484-485

187 Lech Mintowt-Czyz, 'The boy who came back from the dead', *The Evening Standard*, 14 July 2004

188 John Pearn (1992), 'The urgency of immersions', *Archives of Disease in Childhood*, 67:257-261

189 John Pearn and James Nixon (1977), 'Swimming pool immersion accidents: an analysis from the Brisbane Drowning Study', *The Medical Journal of Australia* i:432-437. (Note: This paper was reprinted twenty years later as part of a series of Injury Classics. Injury Prevention (1997) 3:307-309.)

190 'Report Relating to Riverside Safety Adjacent to Kingston Bridge and the Gazebo Public House', Undertaken by Quadriga Health and Safety Ltd, Reading for Royal Kingston. December 2005. Andrea Corrie (2014) *Into the Mourning Light*, Perfect Publishers Ltd, Cambridge.

191 'Pub partners will promote water safety as Nicholson's chain adopts RNLI campaign', https://rnli.org/news-and-media/2017/may/31/respect-the-water

192 James Scott, 'Family of Hitchin toddler who died following swimming pool accident help create awareness film', *The Comet*, Hertfordshire, 16 August 2014

193 Paul Cassell, 'Gran's desperate bid to save drowning toddler', *Berkshire Live*, 27 February 2013

194 'Parent's warning over water risks for small children', ITV News, 6 August 2014. '*The Danger Age*' Produced by Jenni Thomas for The Angus Lawson Memorial Trust. Video on YouTube

195 'Divers find bodies in park lake', BBC News, 23 June 2005

196 James Rush and Chris Pleasance, 'Redditch boy R… F…, 15, drowned at busy beauty spot', *Daily Mail*, 17 March 2014

197 'Families of tragic teenagers speak out to back drowning charity', https://www.rlss.org.uk/News/families-of-trgic-teenagers. 'RLSS' UK's Mike Dunn on ITV Central', Video on YouTube, 24 June 2014

198 Drowning Prevention Week RLSS, '*Filling Up*', Video on YouTube, 29 May 2014. Drowning Prevention Week RLSS. '*Behind the Scenes of Filling Up*', Video on YouTube, 25 June 2014. Charity Digital Editorial 'Artem creates hard-hitting short film supporting Drowning Prevention Week for the Royal Life Saving Society', 18 June 2014. https://charitydigital.org.ul/topics/topics/artem-creates-hard-hitting-short-film

199 'Hard-hitting film launched for Drowning Prevention Week 2015', Royal Life Saving Society, 17 June 2015, https://www.rlss.org.uk/news/hard-hitting-film

200 Drowning Prevention Week RLSS, '*Beneath the Surface: the Families' Stories*', Video on YouTube, 17 June 2015. Rob Parsons, 'Video: Mother of tragic Megan keeps memory alive with drowning awareness film', *The Yorkshire Post*, 17 June 2015

201 Nick Irving, 'Tearful mum watches schoolgirl's lifeless body being dragged from water after freak speedboat accident', *Daily Mirror*, 31 May 2016

202 Debate on 'Emily's Code: Pleasure Vessel Safety', *Hansard*, Volume 622, 1 March 2017 'Family launches Emily's Code to save lives', Royal Yachting Association, 4 March 2017

203 'Chesterfield boy drowned in quarry, inquest hears', BBC News, 29 January 2010

204 Anthony Bond, 'Derbyshire's Blue Lagoon dyed black to deter people taking a dip in water that is nearly as toxic as bleach', The *Daily Mail*, 12 June 2013

205 'Tombstoners Leap 60 Feet Into Sea', Pirate FM News, 31 August 2016

206 Carl Stroud, 'Tombstoning tragedy: Man, 39, dies after 65ft "tombstone" plunge at notorious spot known as Dead Man's Cove', The *Sun*, 14 October 2016

207 RoSPA 'Tombstoning – 'Don't jump into the unknown', https://www.rospa.com/Leisure-Safety/Water/Advice/tombstoning

208 'Body of missing Melksham worker found in river', *Wiltshire Gazette & Herald*, 28 December 2009. 'Godmanchester man's river death still a mystery', The *Hunts Post*, Huntingdon 8 June 2010, https://www.huntspost.co.uk/lifestyle/Godmanchester-man-s-river-death. 'Mum calls for safer rivers after son drowned on way home from night out in Bath', *Wales on Line*, 14 August 2011 https://walesonline.co.uk/news/wales-news/mum-calls-safer-rivers. 'Council pledge over River Avon safety fencing in Bath', 22 April 2013, *Somerset Live*, Bath, http://stories.rssing.com/chan-5528425/all_p109.htm

209 Royal Society for the Prevention of Accidents 'Water Safety Review, River Avon, Bath', June 2011 https://www.whatdotheyknow.com/request/143581/responses

210 'River Safety. August 2016 Update', https://www.bathnes.gov.uk/services/environment/river-safety

211 Gareth Davies, 'Brighton man died in river', The *Argus*, Brighton, 22 November 2014. 'Student, 21, drowned in city centre river just after posting Facebook message to his girlfriend about his best year ever', *Daily Mail*, 27 January 2014. 'Man Who Drowned In River Avon Named By Police', The *Bath Echo*, 14 April 2014. Anthony Bond, 'Teenager missing after night out close to river that claimed the lives of five people', *Daily Mirror*, 15 September 2014. Paddy Dinham, 'Girlfriend of rugby player found in the river pays tribute to him', *Daily Mail*, 27 November 2016 (Note: Article lists 8 other victims, including a 60-year-old man who drowned in June 2016). Liam Trim, 'Tributes to 'selfless and dependable Bath Spa student found dead in River Avon', *Somerset Live*, 31 October 2018. Samuel J. Turnpenny, Kesney E. Boon, Marcus Parish, Joel M.T. Peachey and Julia S.Schinle (2018), 'Reducing accidental drownings in the City of Bath: A study of student river safety awareness', University of Bath School of Management, https://www.researchgate.net/publication/333450419_Reducing_accidental_drowning

212 'Drowned student's father says River Avon is 'Bath's serial killer', BBC News, 14 March 2017

213 '"Very cruel accident" claimed East Lancs teen's life', *Lancashire Telegraph*, 24 September 2014

214 '*Dying for a Dip*' water safety campaign, Lancashire Fire and Rescue Service, 4 June 2015 http://www3.lancashire.gov.uk/corporate/enewsviewer/index.asp?i

215 '*James Drowned, I nearly did*', Video produced by Lancashire Fire and Rescue Service, as part of their '*Dying for a Dip*' water safety campaign, Loaded on YouTube 20 April 2015. Keith Perry 'Mother watched her son drown in Lancashire reservoir', *Daily Telegraph*, 24 June 2014. 'Inspirational Pendle parents awarded MBEs in New Year's Honours list', *Burnley Express*, 28 December 2018. Canal & River Trust, 'Lancashire Fire & Rescue Service launch safety campaign', 21 June 2019 (Note: The victim's mother saw the accident, too far away to realise that it was her son who was in peril. Since his death, she has worked to educate young people about the dangers of open water. She was awarded an MBE in the New Year Honours List 2018. In June 2019, she unveiled two throwlines beside the reservoir, then firefighters gave a rescue demonstration and taught schoolchildren how to use them.)

216 Prime Minister's Questions, *Hansard*, 25 June 2014 , https://hansard.parliamenr.uk/Commons/2014-06-25/debates/1406

217   'All-Party Parliamentary Group on Water Safety and Drowning Prevention', https://powerbase.info/index.php/all-Party_Parliamentary_Group

218   Robert Goodwill 'The National Drowning Prevention Strategy', Department of Transport, London, February 2016. 'A Future without drowning: The UK Drowning Prevention Strategy 2016-2026', National Water Safety Forum. June 2016

219   'Recommendations to ensure all children leave primary school able to swim', March 2017, Swim-England-Curriculum-Swimming-and-Water-Safety-Review-Group-Report-2017.pdf

220   'All Party Parliamentary Group on Swimming launches', 7 September 2017, https://alotusrises.com/2017/09/07/all-party-parliamentary-group

221   Swim England, 'Learn to Swim: A guide to the Learn to Swim Programme'. Swimming Teachers Association, 'STA Launches New Swimming Academy Programme', April 2016. Dave Chandler, 'Let's Be Honest About School Swimming', STA.co.uk, 10 December 2018

222   Richard Sutcliffe, Matthew Terry and Kathryn Crowther 'Swimming Skills Literature Review', RNLI October 2016

223   'Learn to Swim Safe', Swim England and the RNLI 2018

224   'Water Safety Toolkit', Local Government Association 2017

225   'A practical implementation guide to setting up a Water Action Group and designing a local Water Safety Plan', RLSS June 2018

226   Poppy Middlemiss, 'Suicide prevention: Bridge signs', City of London Planning and Transportation Committee. https://democracy.cityoflondon.gov.uk/documents/s66299/Suicide-prevention

227   'Suicide Prevention Action Plan', City of London. January 2016, www.democracy.cityoflondon.gov.uk/documents/s74528/Appendix1_City

228   National Water Safety Forum, 'A future without drowning: The UK Drowning Prevention Strategy 2016-2026, Biennial Review 2016/17', 2017

229   Jessica Elgot, 'Teenage vlogger rebuked by police and RNLI after Tower Bridge stunt', The Guardian, 24 July 2015

230   National Water Safety Forum 'A future without drowning: The UK Drowning Prevention Strategy, 2016-2026, Biennial Review 2016/17', 2017

231   Nicole Klynman and Will Skinner, 'The Bridge Pilot', National Suicide Prevention Alliance, 21 September 2017. https://www.nspa.org.uk/wp-content/uploads/2017/10/City-of-London City of London Planning and Transportation Committee, 'Suicide Prevention: Bridge Signs', 5 July 2016

232   Poppy Middlemiss, 'Suicide prevention: Bridge signs', City of London Planning and Transportation Committee, https://democracy.cityoflondon.gov.uk/documents/s66299/Suicide-prevention

233   Poppy Middlemiss 'Suicide prevention: Bridge signs', City of London Planning and Transportation Committee. https://democracy.cityoflondon.gov.uk/documents/s66299/Suicide-prevention

234   Ruth Bloomfield and Russell Lynch, 'Volunteers aim to intercept suicide jumpers on Waterloo Bridge' Evening Standard, 6 July 2016

235   Lifeboats News Release, 'Duke of Cambridge visits Tower RNLI to discuss male suicide issue',

12 May 2016, https://rnli.org/news-and-media/2016/may/12/duke-of-cambridge-visits

236   'Tidal Thames Water Safety Forum Drowning Prevention Strategy', May 2019

237   Robert Jobson and Bonnie Christian, 'Prince William travels down River Thames as he launches new campaign to prevent people drowning in rivers', *Evening Standard*, 21 May 2019

238   National Water Safety Forum 'A future without drowning: The UK Drowning Prevention Strategy 2016-2026. Biennial Review 2016/17', 2017

239   Monica Perezi, 'Running or walking? Don't slip up', 27 April 2016, https://chieffireofficers.wordpress.com/2016/04/…

240   Chief Fire Officers Association, '*Be Water Aware:* CFOA Drowning Prevention and Water Safety Week', 25 April-1 May 2016, http://www.cfoa.org.uk/21142 (Note: CFOA's Drowning Prevention and Water Safety Week is repeated every year.) 'RNLI says "fight your instincts, not the water" to help stay alive', *Lifeboats News Release*, 25 May 2017. Mark Blunden, 'Lifeboat crews warn Londoners: Cooling off in the Thames will only get you in deep water', *Evening Standard*, 14 August 2018

241   Royal Life Saving Society, 'About *Don't Drink and Drown*', https:// www.rlss.org.uk/about-dont-drink-and-drown

242   National Water Safety Forum, 'A future without drowning: The UK Drowning Prevention Strategy 2016-2026, Biennial Review 2016/17', 2017

243   RLSS, 'National Water Safety Management Programme', https://www.rlss.org.uk/national-water-safety-programme-programme

244   David Walker (2019), *Safety at Inland Waters*, RoSPA, https://www.rospa.com/leisre-safety/water/inland

245   'Coast & beaches to visit in the UK', https://www.nationaltrust.org.uk/coast-and-beaches

246   National Water Safety Forum, 'A future without drowning: The UK Drowning Prevention Strategy 2016-2026, Biennial Review 2016/17', 2017. 'Safety campaign brings RNLI lifeguards to Three Cliffs Bay', 21 March 2016. https://www.nationaltrust.org.u/news/safety-campaign-brings-rnli. 'Sharp rise in rescues at Quantock beach due to "increasingly dangerous" conditions', 24 October, 2019https://www.cornwalllive.com/news/cornwall-news/sharp-risel

247   Jane Kerr, Mehul Kotecha, Caroline Turley and Matt Barnard, '*Personal narratives of serious incidents at sea and on the coast*', NatCen Social Research, London. August 2018

248   Maritime and Coastguard Agency, '*Managing Beach Safety*', August 2019, https://www.gov.uk/government/publications/managing-beach-safety

249   'About us – Maritime and Coastguard Agency – GOV.UK.', https://www.gov.uk/government/organisations/maritime-and-coastguard-agency. 'Maritime and Coastguard Agency.' https://en.wikipedia.org/wiki/Maritime_and_Coastguard_Agency. 'Lone Fisherman rescued after boat sank off Anvil Point', http://hmcoastguard.blogspot.com/2020/09/lone-fisherman-rescued

250   'Seafarer working and living rights', https://www.gov.uk/seafarer-working-and-living-rights. 'Health and safety on ships', https://www.gov.uk/health-and-safety-on-ships. 'Maritime safety: weather and navigation', https://www.gov.uk/maritime-safety-weathe-and-navigation. 'Emergency and life-saving equipment on ships,' https://www.gov.uk/emergency-and-lifesaving-equipment-on-ships

251   BBC 'Water: Working On or Near', https://www.bbc.co.uk/safety/recources/aztopics/working-on-water. BBC 'Boats: Working on', https://www.bbc.co.uk/safety/resources/aztopics/working-on-boats. BBC 'Diving', https://www.bbc.co/uk/safety/recources/aztopics/diving BBC 'Adventure Activities & High Risk Sports', https://www.bbc.co.uk/safety/resources/aztopics/adventure-activities-and-high-risk-sport

252   Water Safety Scotland (2018) *Scotland's Drowning Prevention Strategy 2018-2026*, https://www.watersafetyscotland.org.uk/strategy. Water Safety Scotland (2020) *Two-Year Review: Scotland's Drowning Prevention Strategy 2018-2026*. https://www.watersafetyscotland.org.uk/media/1649/water-safety-scotland

253   Michael Wright, Eshani Ghosh and Fawzia Ibrahim, 'Assessing Inland Accidental Drowning Risk', RoSPA, 3 September 2013, https://rospa.com/leisure-safety/inland-water-risk-assessment

254   'British Canoeing', https://en.wikipedia.org/wiki/British_Canoeing

255   'About the RYA – Royal Yachting Association', https://www.rya.uk/about-us/Pages/hub.asp

256   'About the Angling Trust', https://www.anglingtrust.net/page.asp?section

257   'British Sub-Aqua Club', https://en.wikipedia.org/wiki/British_sub-aqua_club

258   'UK Canyon Guides', https://www.canyonguides.org/contact/about

259   'British Caving Association', https://en.wikipedia.org/wiki/British_Caving_Association

260   'British Rowing', https://en.wikipedia.org/wiki/British_Rowing

261   'Swim England', https://en.wikipedia.org/wiki/Swim_England. 'Scottish Swimming', ushttps://www.scottishswimming.com/about-us.asp. 'Swim Wales', https://britishswimming.org/about-us/swim-wales. 'Swim Ulster', http://www.swimulster.net/About-Us.asp

262   'British Triathlon Federation', https://en.wikipedia.org/wiki/British_Triathlon_Federation

263   'SH2OUT', Royal Life Saving Society https://www.rlss.org.uk/pages/category/sh2out. Simon Griffiths, 'View from the Water – Why some swimmers are wary of SH2OUT', *Outdoor Swimmer* 3 June 2019 https://outdoorswimmer.com/blogs/why-some-swimmers-are-wary

264   Baroness Tanni Grey-Thompson (2017) *Duty of Care in Sport*, April 2017. https://www.gov.uk/government/publications/duty-of-care-in-sport-review (Note: In 2005, Australian researchers reported 150 serious injuries and forty-eight deaths during sporting activities in the state of Victoria between July 2001 and June 2003. Motor sports, power boating and horse riding had the highest rates of serious injury, but more than two-thirds of the deaths were due to drowning. B. J. Gabbe, C. F. Finch, P. A. Cameron and O. D. Williamson (2005), 'Incidence of serious injury and death during sport and recreation activities in Victoria, Australia', *British Journal of Sports Medicine* 39:573-577)

265   Jan Bradford 'From Lyme Bay to Licensing', Adventure Activities Licensing Authority, April 2000. Dave Grant 'From Lyme Bay to Licensing to De-Regulation? The current state of safety within the outdoor activities sector', Rural Policy Centre, Scotland's Rural College (SRUC). March 2013. 'Schools and adventurous activity providers – have your say on the Adventure Activities Licensing review.' https://www.lotc.org.uk/schools-and-adventurous-activity-providers

266   David Walker 'Coasteering safely', *Staying Alive*, RoSPA, August 2011. Adventure Activities Licensing Authority 'Coasteering', AALA Note: 6.20 (rev 2). March 2015 (Note: The National

Coasteering Charter is a representative body for coasteering providers rather than a national governing body.)

267   The Council for Learning Outside the Classroom, 'History', https://www.lotc.org.uk/about/history. The Council for Learning Outside the Classroom, 'LOtC Quality Badge', https://lotcqualitybadge.org.uk

268   Outdoors Education Advisers' Panel, 'Training for leadership of outdoor learning and educational visits', http://www.outdooreducationaladvisers.co.uk/training

269   Outdoors Education Advisers' Panel, 'National guidance for the management of outdoor learning, off-site visits and learning outside the classroom', https://oeapng.info/guidance-document (Note: The initial version of 'Group Safety at Water Margins' was published by the Department of Education and Skills and the Central Council for Physical Education in 2003.)

270   Outdoors Education Advisers' Panel, 'Underpinning Legal Framework', January 2019 https://oeapng.info/downloads/legal-framework-and-employer-systemsl

271   Department of Education, 'Health and safety on educational visits', November 2018 https://www.gov.uk/government/publications/health-and-safety-on-educational-visits. 'CLOtC and OEAP embark on a closer partnership to strengthen support for schools', 1 November 2019. https://www.lotc.org.uk/clotc-and-oeap-embark-on-a-closer -partnership

272   'Rugby star James Haskell backs RNLI campaign in Brighton', The Argus, Brighton, 16 August 2013

273   Ross Mcleod (2018), 'RNLI: 'Respect the Water' Campaign – Putting Prevention First', International Journal of Aquatic Research and Education: Vol. 11: No. 2, Article 11.

274   Kim Horton, 'Top RNLI tips on how to survive if you fall into deep water', Gloucestershire Live, 3 June 2017. 'Float to live: How to survive cold water shock.' ITV News 8 August 2019. https://www.itv.com/news/border/2019-08-08/float-to-live-how-to-survive-cold-water-shock 'River Thames fatalities: RNLI says 'fight your instincts not the water',. https://rnli.com/news-and-media/2017/may/26/river-thames-fatalities. 'Cornwall intern Lucy shares her favourite campaign. Float to live, a lifesaving campaign',. https://stuffadvertising.com/cornwall-intern-lucy-shares-her-favourite-campaign

275   'The RNLI urges people to know how to FLOAT as campaign hits Northern Ireland', Lifeboats News Release, 30 May 2019, https://rnli.org/news-and-media/2019/may/30/the-rnli-urges-people

276   'RNLI's 'Float to Live' advice saves man's life on the Thames', Lifeboats News Release, 11 July 2011https://rnli.com/news-and-media/2019/july/11/rnli-float-to-live-advice

277   'Boy saved by using 'Float to Live' advice seen on 'Saving Lives at Sea' Lifeboats News Release, 1 August 2020. https://rnli.org/news-and-media/2020/august/01/boy-saved-by-using…

278   www.nationalwatersafety.org.uk/waid/info/waid_fatalincidentreport_2016.xls. 'Accidental drowning deaths continue to fall across the UK: 2019 WAID report', National Water Safety Forum, 10 June 2020, https://nationalwatersafety.org.uk/news/pots/2020/june/accidental-drowning-deaths (Note: In the UK, annual reports of 'Accidental drowning deaths' include only those drowning fatalities given an Inquest verdict of 'Accidental Death' or 'Death by Natural Causes'. This excludes drowning deaths given an 'Open Verdict' or a 'Narrative Verdict' at Inquest, and those where suicide or homicide is suspected or confirmed – as explained in

Chapter 3, p.312-313.)

279   '255 people lost their lives in accidental drowning in the UK in 2017', https://nationalwatersafety. org.uk/news/posts/2018/may/255-people

280   '2018 UK Water Related Fatalities Published', *National Water Safety Forum*, 1 May 2019. https://nationalwatersafety.org.uk/news/posts/2019/april/2018-uk-water-related-fatalities-published

281   'Respect the Water campaign targets accidental drowning along the coast', *Lifeboats News Release*, 8 June 2016 https://rnli.org/new-and-media/2016/june/08/respect-the-water

INDEX